Microelectronics Interconnection and Packaging

Edited by
Jerry Lyman, Editor
Packaging & Production,
Electronics

Microelectronics Interconnection and Packaging

Edited by
Jerry Lyman, Editor
Packaging & Production,
Electronics

Electronics
MAGAZINE BOOKS

 McGraw-Hill Publications Co.
1221 Avenue of the Americas
New York, New York 10020

ELECTRONICS BOOK SERIES

Also published by *Electronics*
- **Microprocessors**
- **Large scale integration**
- **Basics of data communications**
- **Applying microprocessors**
- **Circuits for electronics engineers**
- **Design techniques for electronics engineers**
- **Memory design: microcomputers to mainframes**
- **New product trends in electronics, number 1**
- **Personal computing: hardware and software basics**

Library of Congress Cataloging in Publication Data

Main entry under title:
Microelectronics interconnection and packaging.
              (Electronics magazine books)
Articles from Electronics magazine.
Includes index.
1. Integrated circuits. 2. Microelectronic
packaging. I. Lyman, Jerry, 1926-
II. Electronics.
            TK7874.M483   621.381'73   79-21990
McGraw-Hill Book Company ISBN 0-07-019184-0
McGraw-Hill Publications Company ISBN 0-07-606600-2

Copyright © 1980 by McGraw-Hill, Inc. All rights
reserved. Printed in the United States of America. No
part of this publication may be reproduced, stored in
a retrieval system, or transmitted, in any form or by
any means, electronic, mechanical, photocopying,
recording, or otherwise, without the prior written
permission of the publisher.

Contents

Part 1: Lithography and Processing for Integrated Circuits   1

Electron-beam lithography draws a finer line   3
   T.H.P. Chang, M. Hatzakis, A.D. Wilson, and A.N. Broers

Scanning electron-beam system turns out IC wafers fast   13
   E.V. Weber and H.S. Yourke

X-ray lithography gains ground   19
   Jerry Lyman

X-ray lithography breaks the VLSI cost barrier   21
   Gregory P. Hughes and Robert C. Fink

E-beam projector takes giant steps toward practicality   29

Lithography chases the incredible shrinking line   30
   Jerry Lyman

Demands of LSI are turning chip makers towards automation, production innovations   42
   Jerry Lyman

Plasma processes set to etch finer lines with less undercutting   54
   C.J. Mogab and W.R. Harshbarger

Part 2: Thick- and Thin-film Hybrids   59

Hybrid circuit technology keeps rolling along   61
   Jerry Lyman

Why the design nod goes to resistors made as thin-film monolithic networks   75
   Donald B. Bruck and Allen L. Pollens

Thin-film layers shrink rf inductors to chip size   81

Part 3: Printed-Circuit-Board Technology   83

New methods and materials stir up printed wiring   85
   Jerry Lyman

New substrate causes a stir   94
   Jerry Lyman

Porcelain-on-steel boards can launch a thousand chips   95
   Murray Spector

Mass molding a boon to fast logic   99
   Jerry Lyman

Fine-line printed circuits catch on   101
   Jerry Lyman

Designer must plan early for flat cable   103
   James A. Henderson

Flat cable aids transfer of data   108
   Joseph B. Marshall

Flexible circuitry consolidates hardware for interconnections   114
   Peter Maheux

Flexible circuits bend to designer's will   116
   Jerry Lyman

Philips says yes to resistless etchless printed-circuit boards   125

Handy breadboard systems speed the development of prototypes   126
   Jerry Lyman

Assembling a complex breadboard can be as easy as 1,2,3   134
   Bernard J. Carey and Harry Grossman

Getting rid of hook: the hidden pc-board capacitance   140
   Wallace Doeling, William Mark, Thomas Tadewald, and Paul Reichenbacher

Part 4: Automatic Wiring Technology   147

Techniques of automatic wiring multiply   148
   Jerry Lyman

High-speed wire-and-solder technique tests connections as it makes them   153
   Bob Whitehead

Impulse-bonded wiring is economical alternative to multilayer boards   160
   F.G. Schulz, D.C. French, B.E. Criscenzo, and P.T. Klotz

Planar stitch welding yields higher board density at lower cost   167
   Rene Lemaire and Ray Calvin

Part 5: IC Packages and Connectors   171

Chip carriers are making inroads   173
   Jerry Lyman

Film carriers star in high-volume IC production   175
   Jerry Lyman

Beam tape plus automated handling cuts IC manufacturing costs   183
   Doug Devitt and Jim George

Growing pin count is forcing LSI package changes   187
   Jerry Lyman

Packaging technology responds to the demand for higher densities   198
   Jerry Lyman

64-pin QUIP keeps microprocessor chips cool and accessible   207
   William Lattin, Terry Mathiasen, and Steven Grovender

Zero-insertion-force connector ousts conventional backplane   211
   James Taylor

Conductive elastomers make small, flexible contacts   217
   Leonard S. Buchoff

Part 6: Environmental Factors Affecting Interconnections and Packages   221

The thermal demands of electronic design   223
   Stephen E. Grossman

Plastic power ICs need skillful thermal design   227
   Sumner B. Marshall and Raymond F. Dewey

Analysis can take the heat off power semiconductors   230
   Forest B. Golden

Forced air cooling in high-density systems   237
   Gordon M. Taylor

Liquid cooling safeguards high-power semiconductors   240
   John A. Gardner Jr.

Network analog maps heat flow   246
   Carl J. Feldmanis

Air through hollow cards cools high-power LSI   252
   Lou Laermer

Heat exchangers cool hot plug-in pc boards   258
   Benjamin Shelpuk

Thermal characteristics of ICs gain in importance   265
   Robert Bolvin

Heat pipes cool gear in restricted spaces   269
   Alan J. Streb

Measuring thermal resistance is the key to a cool
semiconductor 273
   Bernard S. Siegel
Countering the effects of vibration on
board-and-chassis systems 279
   David S. Steinberg
What happens to semiconductors in a nuclear
environment? 282
   David K. Meyers

Part 7: Computer Aided Design **285**
Computer-engineer partnerships produce precise
layouts fast 286
   Richard Larson
Symbolic layout system speeds mask design for ICs 292
   Robert P. Larsen

**Part 8: Automatic Testing of Interconnections
and Packages** **297**
Automated circuit testers lead the way out of
continuity maze 299
   Jerry Lyman
In-circuit testing pins down defects in pc boards early 308
   Fredrick A. Schwedner and Stephen E. Grossman
In-circuit tester using signature analysis adds digital
LSI to its range 313
   Craig Pynn

Index **319**

Preface

Today, as in the past, one of the primary problems in electronics, whether at the chip, hybrid or circuit-board level, is to get from point A to point B with the shortest, narrowest, lowest cost and most easily manufacturable interconnection. This has led to increased circuit density at the IC-chip, thick- and thin-film-hybrid and printed-circuit levels. In addition, the same pressure for highly packed interconnections has led to the creation of a rival to the pc board—automatic wiring.

A secondary and related problem is interfacing between differing interconnection techniques. For instance, integrated circuit chips are regularly called on to interface with connections on hybrids and printed circuit boards. In this case, these micron-sized IC interconnects must be successfully mated with either screened, evaporated, or plated connectors with widths measured in mils.

This book, consisting of up-to-date articles from the pages of *Electronic* magazine, covers the two aspects of microelectronics—interconnections and packaging—in 8 parts. Part 1 starts with a description of the lithography and processing needed to create LSI and VLSI chips with circuit details in the low micron range.

Part 2 moves up to the next level of interconnection and packaging—thick- and thin-film technology—describing the processes and materials needed to fabricate these dense packaging mediums.

Moving on to the next higher level of interconnection, Part 3 describes the broad field of printed-circuit technology. This treatment covers substrates, plating methods, flexible circuits, and even breadboards. A direct competitor of the pcb is automatic wiring, typified by Wire-Wrap and Multiwire. This subject is covered in Part 4.

Part 5 addresses itself to the chip-to-board and chip-to-hybrid interface. Subjects covered include film carriers, chip carriers, connectors, and comparisons of packaging methods.

Environmental factors that affect circuit interconnects and packages are considered in Part 6. These include thermal, vibration, and radiation effects. Part 7 covers the computer-aided design of IC masks and pc-board artwork. Finally, Part 8 discusses automatic testing of high-density interconnections for both unloaded and loaded circuit boards.

—Jerry Lyman

Part 1
Lithography and Processing for Integrated Circuits

Electron-beam lithography draws a finer line

Before LSI circuits can became any more complex, line widths must drop below 1 µm—which means switching from optical to electron-beam fabrication

by T.H.P. Chang, M. Hatzakis, A.D. Wilson, and A.N. Broers, *IBM Corp., Thomas J. Watson Research Center, Yorktown Heights, N.Y.*

☐ The width of the narrowest line in a large-scale-integrated circuit sets one very obvious limit to the circuit's density and complexity, and this limit is now under pressure from advances in the rest of LSI technology. Conventional optical lithographic techniques for producing circuit patterns on chips can manage about 2-micrometer-wide lines at best, however, and any success in the submicrometer region will require another form of pattern making, such as electron-beam lithography.

To bring electron-beam lithography to the point where it is useable for volume production of extremely dense LSI circuits, progress had to be made in several different areas. High-resolution electron resists needed developing, and improvements needed to be made in the performance of the systems used for forming and controlling the beam, for moving the workstage carrying the wafer, for generating circuit patterns by computer, and for registering the layers of device and metalization circuit patterns precisely

The result is the kind of system shown in Fig. 1—one that either makes masks or exposes a circuit pattern directly on a wafer with a higher degree of accuracy and resolution than optical systems. The technology also lends itself to automation, which for direct wafer exposure could mean very rapid turnaround times. Figure 2 shows a surface-acoustic-wave transducer with a line width of 0.1 µm that has been made by direct exposure to an electron beam.

Optical printing

What makes it difficult to refine optical lithography further? Lithography of any kind in semiconductor manufacture is the art of defining on a semiconductor wafer the intricate patterns needed for the fabrication of microcircuits. These patterns are formed by first coating the wafer with a resist—a thin film of light-sensitive

1. An electron-beam system. Developed at IBM Research, the Vector Scan One system (VS1) uses electron-beam lithography for fabricating high-resolution experimental devices. The electron-beam column is shown at the right, the control panel and electronics at the left.

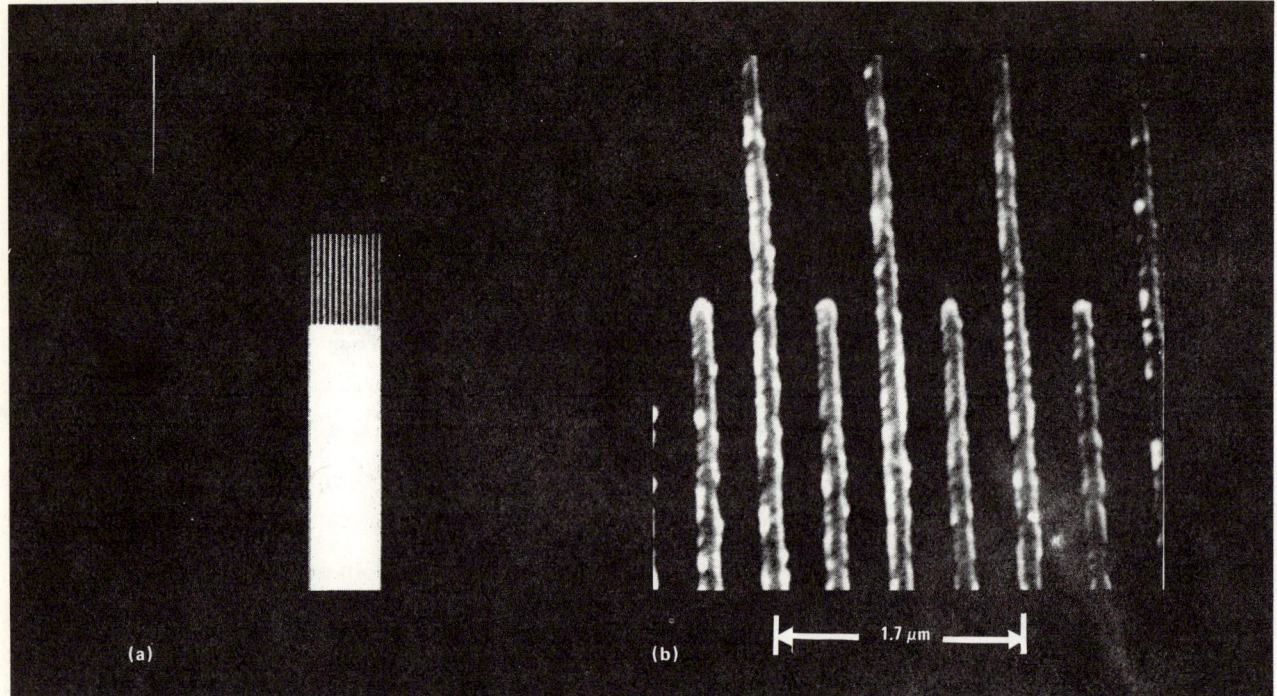

2. Fine lines. This 4.1-GHz acoustic surface-wave transducer (a) gets its submicrometer line widths from electron-beam lithography. Scanning electron micrograph (b) displays details of 0.1-μm metal lines made by this technique. (Source: Hughes Research Labs.)

organic material—and then by shining ultraviolet light onto the coated wafer through a patterned mask. Registration marks on the wafer align with special windows in the mask. Several patterning steps are needed for even the simplest device.

This straightforward and economical contact printing technique is well established. However, the damage that results from bringing the mask into close proximity with the wafer limits device yields, and the diffraction effects that occur between mask and wafer limit line widths to about 2 μm and also limit the accuracy of pattern-to-pattern alignment.

UV projection lithographic systems, in which the wafer is isolated from the mask, alleviate these drawbacks to some extent, only to introduce a new one of their own—for resolutions much below 2 μm, the limited depth of focus of their special optics requires extraordinarily flat wafers.

What electron beams can do

The electron-beam methods overcome these problems and have distinct advantages of their own, chief among which is their high resolution. Lines up to 20 times narrower than the optical limit can be readily generated. Diffraction effects are negligible, because the equivalent wavelength of electrons in the 10-to-25-kilovolt energy range is less than 1 angstrom (10^{-4} μm).

Ultimately, resolution of an electron optical system is limited by aberrations of the electron lenses and the deflection systems and, where a high beam current of more than 1 microampere has to be used, by electron-electron interactions. Resolution of the exposed image, on the other hand, is limited by scattering of electrons in the resist coating and the substrate. In general, line widths below 0.1 μm will require special wafer preparation. To date lines as narrow as 0.008 μm have been made at International Business Machines Corp.'s Research Center, Yorktown Heights, N.Y.

Equally important is the use of a computer to control the beam of electrons directly. This feature has important implications for both mask making and direct wafer exposure.

In addition, an electron beam has much larger depth of focus than optical systems, and it can readily be used to detect structures on the surface of a sample in the same manner as in the scanning electron microscope. The capability can also be exploited to control the accurate overlay of one pattern on another.

Three resist requirements

To realize the full potential of electron-beam lithography, it has been necessary to develop resists specifically suited to electron exposure. These "electron" resists, like the photo resists used in UV printing, are polymeric solutions that can be spin-coated on to the sample prior to pattern writing. After pattern exposure, the resist is developed in a solution that dissolves away either the unexposed portion (negative resist) or the exposed portion (positive resist).

A material's suitability for use as an electron resist, whether positive or negative, depends on its:
- Sensitivity—defined as the minimum electrical charge required for its complete development and customarily measured in electron charge deposited per unit area (coulombs per square centimeter). It must be achieved without significant thinning of the exposed area for negative resists or the unexposed area for positive resists.
- Resolution—indicated by the minimum line width that can be developed in a resist layer of a given thickness.

TABLE 1: CHARACTERISTICS OF ELECTRON RESISTS MATERIALS				
Material	Type	Typical sensitivity (C/cm^2)	Resolution (minimum line width reported) (μm)	Compatibility with semiconductor fabrication processes
KTRF-KMER-KPR (Kodak)	negative	5×10^{-6} **	1	good
Silicones	negative	10^{-5}	0.5	fair
Epoxidized polybutadiene	negative	5×10^{-8} **	?	fair
Shipley AZ-1350	positive	5×10^{-5} *	1	good
Poly (α-methyl-styrene)	positive	10^{-4} *	?	poor
Poly (methyl methacrylate)	positive	5×10^{-5} * 5×10^{-6} **	<1,000 Å	good
Poly (butene-1 sulfone)	positive	10^{-6}	0.5	good
Polydiallylorthophthalate	negative	10^{-6} **	2	good
Poly glycidylmethacrylate ethyl acrylate	negative	5×10^{-7} **	0.5	good

*No thickness loss after development   **Significant thickness loss

■ Compatibility with fabrication processes—including resistance to chemical etching, adhesion to the substrate, temperature stability of the resist image, resistance to ion-etching methods, and so on.

Many materials have been tried as electron resists, and some are listed in Table 1. In general, negative resists have the highest sensitivity (although at considerable thickness loss). Also, the developed negative-resist pattern has sloping walls that preclude its use with processes requiring steep or undercut walls.

With both resist types, ultimate resolution is determined by electron scattering both within the resist layer and back from the silicon substrate rather than by the resist material itself. In general, when high aspect ratios (height/width) are required, a positive resist in combination with lift-off or electroplating processes results in better resolution than a negative resist in conjunction with a subtractive etching process.

In the lift-off process, a vapor of the desired material to be put down on the silicon wafer is evaporated in vacuum on to the wafer through the openings in the resist film. Some is also deposited on top of the resist and is subsequently removed by soaking the workpiece in a solvent—a process that works satisfactorily only if the resist has an undercut profile.

With electroplating, a thin conductive layer is applied to the workpiece prior to the application of the resist. After resist exposure and development, the workpiece is inserted in a plating bath, and metal is plated onto the resist-free regions. Since in this case the metal line profiles will conform to the resist profile, the resist profiles should be as close to vertical as possible in order to obtain uniform plating.

For subtractive processes (e.g. the chemical etching used in producing photomasks) or dry etching methods (e.g. ion milling, reactive ion etching, and sputter etching), negative-resist profiles are often acceptable, and their higher sensitivity can be used to advantage. In these processes the developed resist pattern protects an underlying film of the desired material, but good edge definition is generally harder to obtain.

The resist field remains very active. Many laboratories are making concentrated efforts to improve the sensitivity of positive resists and the contrast and edge sharpness of negative resists.

Projection systems

Broadly speaking, electron-beam lithographic systems can be grouped into projection systems and scanning systems. The projection systems have potentially a lower cost and higher throughput than the scanning-electron-beam systems.

Two types of beam projection systems have been developed specifically for semiconductor device fabrication: the 1:1 and reduction projection systems.

Work on one-to-one projection systems has been under way for many years. It began at Westinghouse and has been more recently pursued by Mullard in England and Thompson CSF in France.

A schematic of Mullard's system is shown in Fig. 3a. This projection system employs a photocathode masked with a thin metal pattern. Photoelectrons from the cathode are accelerated onto the sample by a potential of about 20 kilovolts applied between cathode and sample. A uniform magnetic field focuses these photoelectrons onto the sample (anode) with unity magnification. Sample and wafer can, in principle, be as large as desired, and samples 5 centimeters in diameter have been successfully exposed.

In the system, image position is detected by collecting

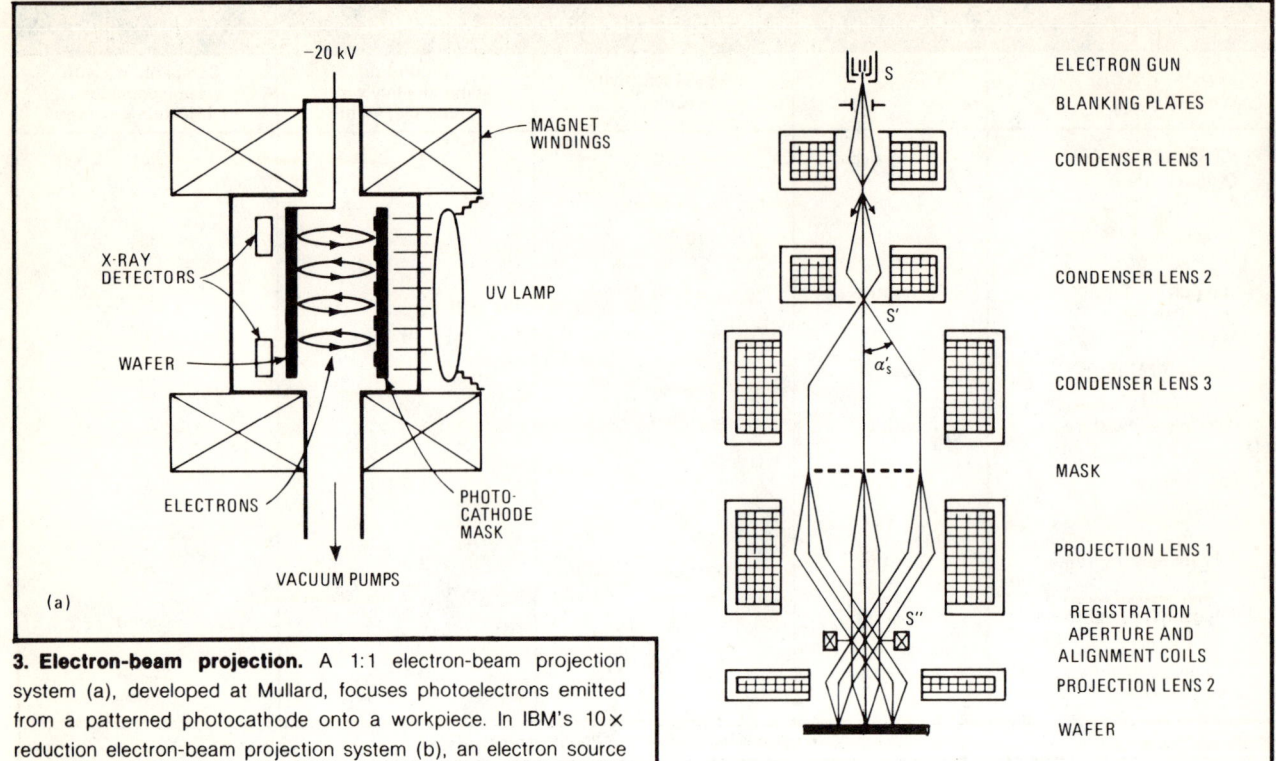

3. Electron-beam projection. A 1:1 electron-beam projection system (a), developed at Mullard, focuses photoelectrons emitted from a patterned photocathode onto a workpiece. In IBM's 10× reduction electron-beam projection system (b), an electron source illuminates a self-supported 10× mask. The electron image is then projected on to the workpiece demagnified 10 times.

characteristic X rays from marks on the wafer with the X-ray detectors shown. (During this process the photocathode is masked so that only the alignment marks are illuminated.) Magnetic deflection is then used to position the pattern with an accuracy of 0.1 μm. Image current density is about $10^{-5} A/cm^2$ (1-second exposure for 10^{-5} coulombs/cm^2 resist sensitivity) for cesium iodide photocathodes, which have the best lifetime and resistance to poisoning. The dominant aberration limiting resolution is chromatic aberration, and theoretical estimates of minimum line width vary from 0.5 μm to 1 μm. So far, the technique has been used to make operating semiconductor devices with lines 2 μm wide.

Because the wafer forms part of the imaging system, its flatness is critical if pattern distortion is to be avoided. An electrostatic chuck may be essential to hold the wafer for satisfactory performance. Another difficulty arises because scattered electrons from the sample are accelerated back onto the wafer into unwanted locations, giving rise to a background exposure that reduces the effective contrast in the image.

Figure 3b shows the reduction type of projection system developed at IBM Research. The basic concept for this system, which is the electron-optical analogy of reduction optical projection cameras, was first described by researchers at Tübingen University in Germany. The mask is a freely suspended metal foil. A special electron-optical system illuminates the mask and forms a sharp demagnified image of it on the wafer. The demagnification factor is 10×, and a field 3 millimeters in diameter and line widths of down to 0.25 μm can be produced.

For alignment, the system uses its scanning mode of operation. In this mode the illuminating beam is focused onto the mask rather than flooding it as in the case of image projection. The focused beam scans across the mask, and an image of this focused beam scans across the sample. Scattered electrons are collected from the sample to detect sample position, and correction is made by shifting the projected image with deflection coils placed between the two projection lenses.

Fabrication of the mask is the major obstacle to the successful application of reduction projection systems to device production. The possibility of replacing the foil mask altogether, by a photocathode/accelerating-structure combination, has been reported recently. If realized, it could open up some interesting new possibilities.

Scanning electron-beam systems

The most direct method for high-resolution pattern generation is scanning-electron-beam lithography. In this approach, the pattern is written with a small electron beam, which is generally controlled (deflected and turned on and off) by a computer.

Application of scanning electron-beam techniques to semiconductor processes emerged in the mid-1960s following the development of the scanning electron microscope at Cambridge University. By 1966 the high-resolution potential of electron-beam fabrication had been demonstrated both at universities like Tübingen, Cambridge, and Berlin and at companies like Westinghouse, IBM, Karl Zeiss, SRI, GE, AEI, and Mullard. Close behind them, electron-beam efforts were started at Texas Instruments, Hughes Research, Thomson CSF, Hitachi, Western Electric, and the University of California, Berkeley. In the late 1960s JEOL and the Cambridge Instrument Co. produced electron-beam equipment

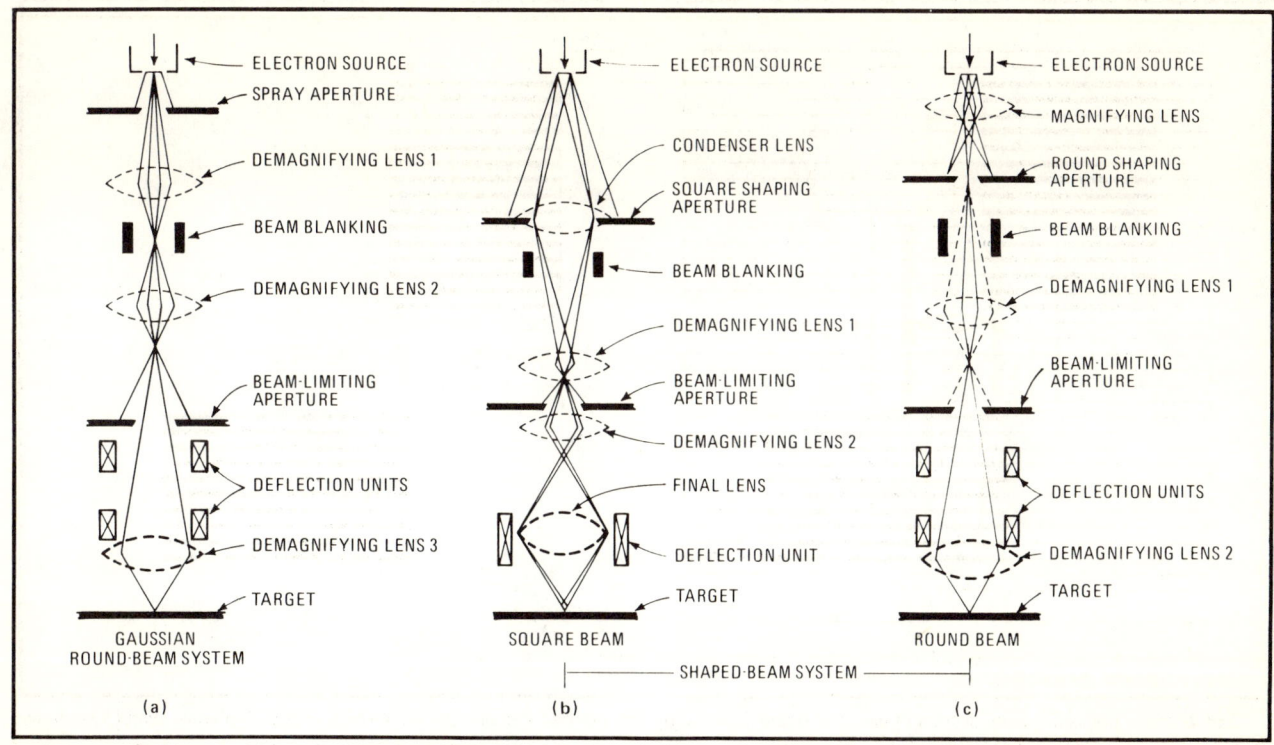

4. Beam forming. Electron beams for lithographic work on details of a mask or wafer can be Gaussian or formed into square or round shapes. In the Gaussian round-beam system (a), the probe-forming concept of the scanning electron microscope is used. The square (b) and round (c) beams instead use square or round apertures to modify the beam, which is then demagnified on to the workpiece.

capable of basic-process and experimental-device studies. By 1970 a good many other industrial companies had become involved in the technology. For example, in 1974 Bell Labs announced its scanning electron-beam mask maker. One of the latest developments is a system capable of direct device fabrication in an actual manufacturing environment—the EL1 from IBM's East Fishkill Laboratory.

The lithographic approach to be selected for a given operation will ultimately depend on the device requirements and cost considerations. There are still, however, several major unknowns that could significantly influence this selection. One is the in-plane wafer distortion associated with thermal processes. Should this distortion be excessive, it will not be possible to obtain adequate pattern overlay for high-resolution full-wafer printing systems.

Broadly speaking, electron-beam systems have two major components: the beam-forming and -deflection system, and the pattern-generation and -control systems.

Beam forming and deflection

Beam-forming systems use either the Gaussian round-beam approach or the shaped-beam (square or round) approaches of Fig. 4.

Gaussian systems use the conventional probe-forming concept of the scanning electron microscope (Fig. 4a). In general, two or more lenses focus the electron beam onto the surface of the workpiece by demagnifying the electron-gun source. High flexibility can be achieved since the size of the final beam can be readily varied by changing the focal length of the electron lenses. To ensure good line definition, the beam size is generally adjusted to about a quarter of the minimum pattern line width.

In the square-beam approach (Fig. 4b), an electron source illuminates a square aperture at the center of a condenser lens placed immediately after the gun. The condenser lens images the gun crossover (1:1) into the entrance pupil of a second condenser lens. This lens together with a third condenser lens demagnifies the square aperture to form the square beam. A fourth lens images the square beam (1:1) onto the target plane. The size of the square beam is generally equal to the minimum pattern line width. To achieve equivalent resolution, the edge slope of the intensity distribution of the square beam (defined as 10% to 90% points) is made equivalent to the beam size (50% intensity) of a Gaussian round beam.

In the case of the round shaped beam, a lens focuses a magnified image of the gun crossover onto a round aperture, and two condenser lenses demagnify the round aperture onto the plane of the target (Fig. 4c).

Round-beam systems (Gaussian or shaped) are generally simpler than square shaped-beam systems and have more flexibility. However, the square beam has more current in the spot (current is proportional to spot area for the same gun brightness) and therefore offers higher exposure speed in cases where speed is limited by beam current and/or beam stepping rate. Difficulties with the square-beam systems may arise when angle lines are required or when some lines have dimensions that are not integral multiples of the beam size. In the latter case, overexposure in the overlapping regions will occur.

Deflection systems for electron beams are generally

7

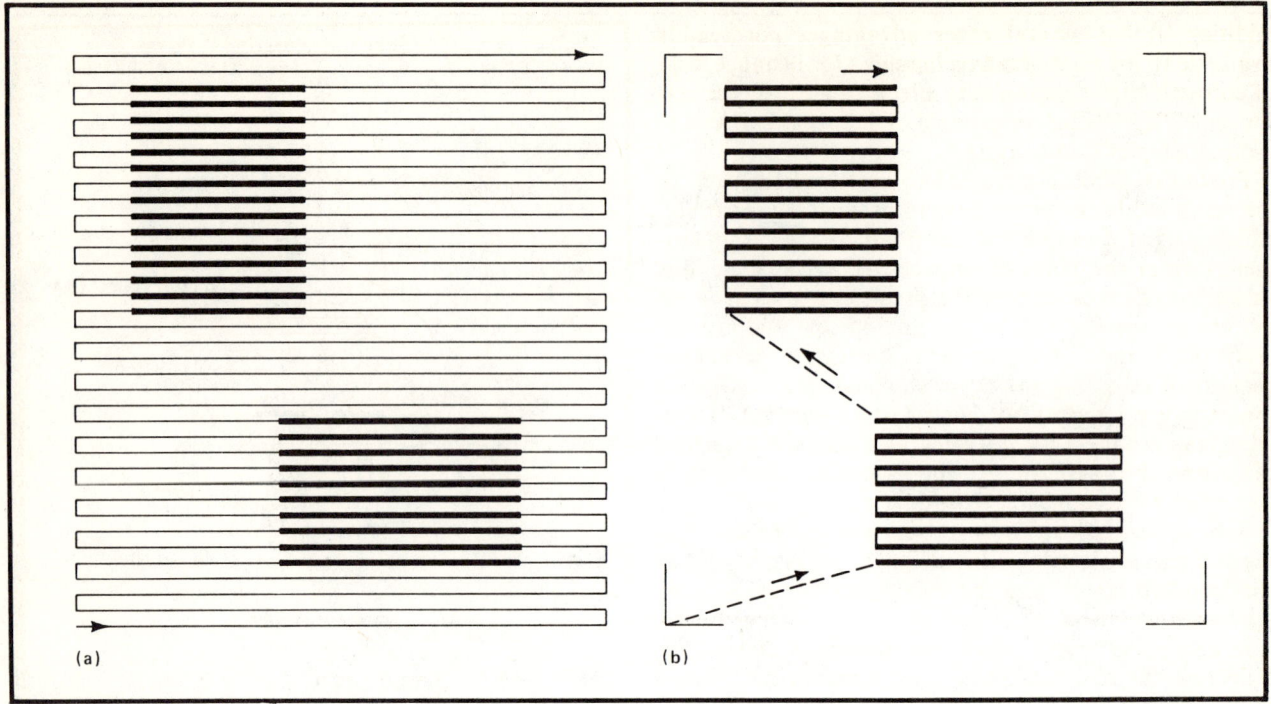

5. Chip scan. In the raster-scan system of electron-beam lithography (a), the beam scans the entire area of the chip while switching on and off according to the pattern data. In the vector scan system (b), the beam is scanned only over the parts where writing is required.

electromagnetic, though electrostatic systems and even a partly electromagnetic, partly electrostatic system exist.

Three electromagnetic approaches have been employed: a double-deflection pre-lens, single-deflection in-lens, and single-deflection post-lens. As the deflection field is inevitably coupled with the focusing field of the magnetic lens, the design of the deflection system must also take into consideration the effect of the lens field. Computer-aided-design programs have been developed to analyze this complex problem and to arrive at an optimization procedure. Distortion terms caused by field curvature and the isotropic astigmatism of the electron-lens system can be corrected dynamically, whereas other aberrations cannot.

Analysis of the three systems shows that without dynamic corrections an optimized pre-lens double-deflection system offers the best performance. With dynamic corrections, the performance of the pre-lens double-deflection and in-lens single-deflection systems rated nearly equal, and the post-lens single-deflection gave poorer results. For high beam currents of more than 1 microampere, when electron-electron interactions must also be considered, deflection *in* the lens is favored.

IBM's Vector Scan One (VS1) is a Gaussian round-beam system that uses a lanthanum hexaboride gun, three magnetic lenses, and a pre-lens double-deflection unit. It achieves exposures of 1-μm lines over a 4-by-4-mm field and 0.5-μm lines over a 3-by-3-mm field at a beam semiconvergent angle (α) of 1×10^{-2} radian without any dynamic corrections. The α value, together with the brightness of the LaB$_6$ gun of 1×10^6 A/cm^2 steradian, gives a current density of 300 A/cm^2. The measured distortion for a 2-by-2-mm field is 0.1 μm and the calculated distortion for a 4-by-4-mm field is 0.85 mm. Typical working distance of the lens is 5 cm.

A shaped-square-beam lithography system (EL1) designed by IBM's East Fishkill Laboratory employs a tungsten filament gun and a large-gap final lens with a single-deflection unit inside the lens. This unit, with dynamic focus corrections, can achieve a 5-by-5-mm field with a 2.5-μm square spot having an edge resolution of approximately 0.4-μm at a beam semiconvergent angle of 7×10^{-3} radians. Typical beam current is 3×10^{-6} A at a current density of 50A/cm^2.

A shaped-round-beam system designed by Texas Instruments (EBMII) features a lanthanum hexaboride gun, a pre-lens double-deflection unit, and a final lens with a focal length of 7.5 cm. With dynamic corrections, a field size of 6.35 by 6.35 mm has been achieved for 1 μm lines with a 3×10^{-3} radian beam semiconvergent angle.

Pattern generation and control

After the beam has been focused and shaped, it must be moved (scanned) over a wafer by a beam-writing technique.

The two basic beam-writing techniques are raster and vector (Fig. 5). In the raster technique (Fig. 5a), the beam is scanned over the entire chip area and is turned on and off according to the desired pattern. In the vector technique (Fig. 5b), the beam addresses only the pattern areas requiring exposure, and the usual approach is to decompose the pattern into a series of simple shapes such as rectangles and parallelograms.

Raster scan places less stringent requirements on the deflection system because the scanning is repetitious and distortions due to eddy currents and hysteresis can be readily compensated for. It can also handle both positive and negative images. Vector scan is more efficient but requires a higher-performance deflection system. In addition, it has several other advantages not readily available to the raster-scan technique—for instance, ease of correction for the proximity effects of electron scattering, and a significant compaction of data that can lead to a much simpler control system.

Proximity effects are created by scattered electrons in the resist and backscattered electrons from the substrate, which partially expose the resist up to several micrometers from the point of impact. As a result, serious variations of exposure over the pattern area occur when pattern geometries fall in the micrometer and submicrometer ranges. A vector-scan correction technique consists of adjusting the beam stepping rate and hence the exposure intensity for each pattern element. These adjustments can be readily integrated into the data that describes the pattern. A corresponding raster-scan correction method has yet to be developed.

Two main approaches have been used to expose and register patterns over the surface of the mask or wafer—the step-and-repeat and the continuously-moving-table approach.

In a step-and-repeat system, the pattern is exposed by deflecting the electron beam over a square field, typically a few millimeters on a side, with the workpiece stationary. After the exposure, the workpiece is moved to a new location, the beam is registered to the sample, and the pattern exposure repeats. This process continues until the whole wafer is covered. In each field location, the pattern can be generated using either the vector-scan or raster-scan technique.

The most widely used step-and-repeat systems employ vector scanning. The VS1 system is one example (Fig. 6a), and similar systems have been developed by Texas Instruments, Hughes Research, Mullard, Hitachi, Bell Northern, and several other establishments for their own research and development work. Several electron-microscope companies (Thomson CSF, Cambridge Scientific Instrument, Cambridge, Maryland, ETEC, Haywood, California, and others) have also produced commercial systems of this type

In VS1, the chip pattern is analysed in terms of rectangles, parallelograms, triangles, and others taken from a random-shape store of curved shapes or repetitive groups of shapes. Data is transferred from the computer to a high-speed pattern generator that drives dual-channel digital-to-analog converter units. These units consist of low-speed high-precision d-a converters, which address the beam to each pattern element, and high-speed low-accuracy d-a converters, which perform the fill-in of each element. In experimental tests, the d-a units together with a large-bandwidth deflection amplifier have been able to deflect the electron beam at a stepping rate of 10 megahertz (i.e. 100 nanoseconds for each beam location) and still give good exposures. An automatic registration system digitally enhances the signal-to-noise ratio of the registration signal.

The VS1 typically handles chip sizes 2,000 times the minimum pattern lines width, but it can also generate larger sizes by joining adjacent patterns. It has fabricated high-resolution devices with registration accuracy in the order of 0.1 µm for 1-µm line patterns.

One such device is the 8,192-bit FET RAM shown in

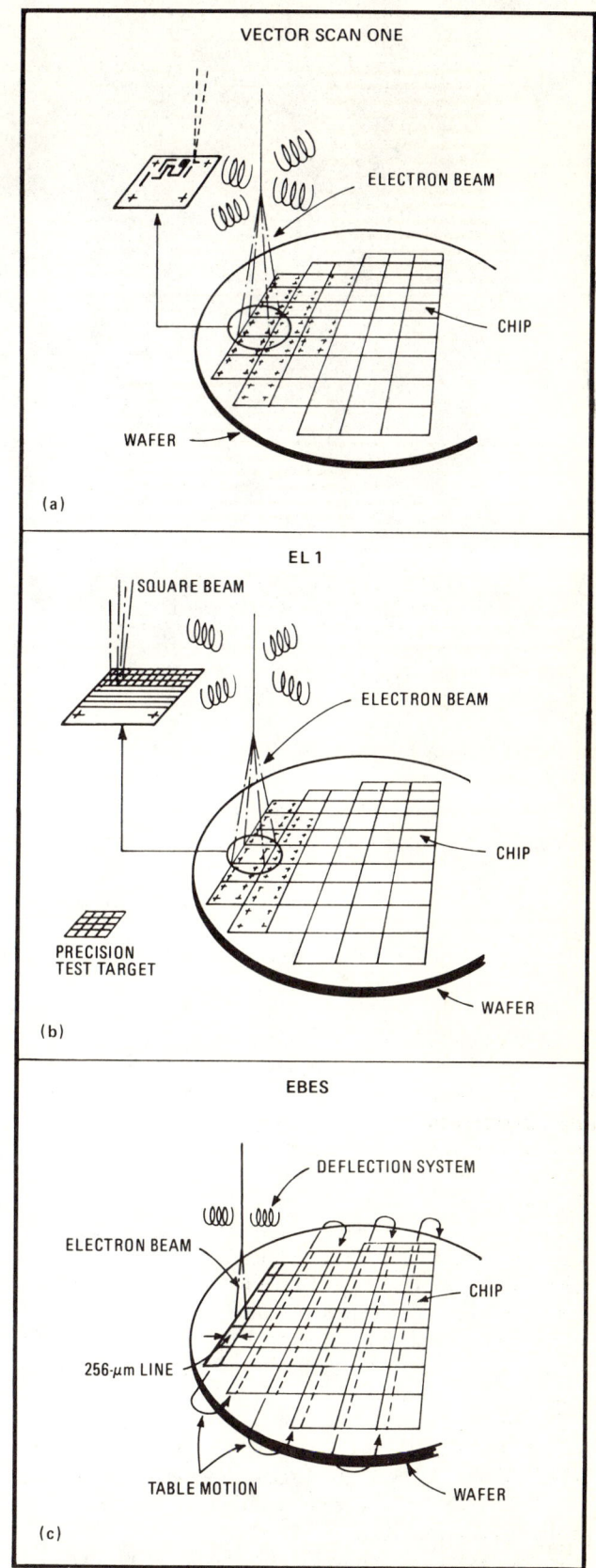

6. Pattern distribution. Two approaches are being applied to distribute patterns over wide surfaces. The step-and-repeat method of IBM's VS1 (a) and EL1 (b) exposes patterns over a square field. Then the workpiece is stepped to a new location. In Bell Telephone Labs' EBES (c), the table moves continuously while the beam scans a strip.

7. Random-access-memory chip. IBM's VS1 system produced this experimental 8,192-bit metal-oxide-semiconductor RAM. Metalization lines are 2 μm wide, and gate length is 1.25 μm. For a demonstration, the VS1 made the same unit with 0.6-μm line widths.

Fig. 7. The device has a channel length of approximately 1.25 μm and an access time of 90 ns. As a lithography exercise, the same 8-k chip has also been made with a line width of 0.6 μm, with all four pattern levels properly aligned and processed.

Pattern writing time of the VS1 system depends on pattern density. At a 10-MHz beam-stepping rate it takes 1 to 2 seconds to expose a typical LSI chip to a pattern occupying about 25% of its surface area.

An important extension of the step-and-repeat system is the incorporation of a precision table controlled by laser interferometry. Originally proposed by Thompson CSF, the idea is to detect errors in table position by laser interferometry and correct for them by deflecting the electron beam. A 1-megabit magnetic-bubble test chip, consisting of 2-μm bubbles contained within a 1-square-centimeter field, has been fabricated by the electron-beam system of Texas Instruments (EBMII), which uses this technique. The refinement allows fewer registration marks to be used per wafer and enables a composite pattern to be stitched together using the laser interferometer as the reference.

The square-beam EL1 system is also a step-and-repeat system. The high-throughput system is the first electron-beam system devoted to direct semiconductor device manufacturing—it has been operating for two years in a production environment.

In this system, shown in Fig. 6b, the pattern is exposed by the 2.5-by-2.5-μm shaped beam, the size of which equals the smallest dimension of the pattern. A stepped raster-scan technique is used so that distortions caused by eddy currents, thermal characteristics, etc., are repeated and can be readily corrected. Deflection distortion errors over a 5-mm field are measured by scanning the beam over a gold calibration grid and corrected by an automatic correction system to a less than 30-part-per-million deviation. An overlay accuracy better than 0.5 μm over a 5-mm field is achieved by the automatic registration system.

The EL1 system has achieved high-speed production of up to 22 2¼-inch wafers per hour of LSI circuits. Typical performance for producing each 5-mm chip is: registration time of 200 milliseconds, pattern write time of 0.96 s, and table move time of 250 ms.

In a continuously-moving-table system, the electron beam is raster-scanned in one direction, and the table moves continuously in the other direction, while a laser interferometer measures the table position and feeds information back to the control system of the electron beam.

This scheme is the basic principle of the electron-beam exposure system (EBES) developed at Bell Telephone Laboratories (Fig 6c). In EBES, the beam scanning width is fixed at 256 μm, and the pattern is formed by joining a number of these strips, which are typically 4 mm long. Pattern registration is performed once for the whole workpiece, but checks are made from time to time to correct for beam drift.

The system uses a 0.5-μm spot with a current density of 10 A/cm^2 based on reported gun brightness of 10^{-4} A/cm^2 steradian and beam semiconvergent angle of 1.3 × 10^{-2} radian. It has demonstrated a writing speed of

8. Electron-beam mask. High-resolution masks can be made by electron beam systems as in this enlarged portion of a mask generated by Bell Telephone Laboratories' EBES on a chromium surface. Minimum line width on the chromium mask is 0.5 μm.

2 cm²/min based on 0.5-μm addresses (digital X, Y coordinates or increments) and a 20-MHz stepping rate. A 0.25-μm address is also available at a quarter the writing speed. The table moves continuously at a speed of 2 cm/s with a laser interferometer that can measure increments of λ/24, where λ is the laser radiation wavelength. Normal time to complete a mask, for a 3-in. wafer is given as 40 minutes (0.5-μm addresses). A section of a typical electron-beam-generated mask is shown in Fig. 8. Commercial versions of this system are currently being offered by ETEC and Extrion, Gloucester, Mass.

In principle, the step-and-repeat and continuously-moving-table approaches can be applied equally well to both mask making and direct wafer exposure. At present, however, most device fabrication by electron-beam exposure has been with step-and-repeat systems—though this may be simply because there are more systems of this type available. In general, step-and-repeat systems, which do one complete chip at a time, are more suitable for high-resolution work since no pattern-joining error from table inaccuracy need be considered. Also, better registration accuracy has been achieved with this approach to date, because it is less sensitive to problems such as beam drift and wafer distortion.

The continously-moving-table approach places less stringent requirements on the performance of the electron optics and deflection system and can operate with a small deflection angle and at a relatively short working distance. In principle, this should allow higher beam-current densities. In addition, as the table is moving during the pattern-writing process, part of the table-movement time need not be considered an overhead as in the step-and-repeat system.

Factors affecting throughput

Throughput for scanning-electron-beam systems is determined by the speed of registration, table motion, and especially pattern writing speed. The same basic factors affect the pattern writing speed for the VS1, EL1 and EBES scanning systems.

Consider a typical 2,000-line chip, i.e. with side dimensions equal to 2,000 times the minimum pattern geometry, so that 2.5-μm lines create a 5-by-5-mm chip. The pattern writing time (T) for such a chip is given approximately by $T = N/f$, where N is the number of beam addresses in the chip and f is the beam stepping rate. The top graph of Fig. 9 shows writing time versus beam stepping rate for N values of 16×10^6, 4×10^6, and 64×10^6 representing the mode of operation for VS1 (25% coverage), EL1, and EBES respectively. (Beam address is taken as a quarter of the minimum pattern line width for VS1 and EBES and equal to it for EL1). Not unexpectedly, writing time shortens as the beam stepping rate increases.

In general, beam-stepping rate is limited both by the speed of the deflection system and by the combined effect of beam-current density and resist sensitivity.

Deflection speed is governed by the noise-bandwidth characteristics of the deflection system, digital-to-analog conversion rates, deflection unit impedance, etc. Typical working values reported to date for VS1, EL1, and EBES are 10 MHz, 5 MHz, and 20 MHz respectively.

The beam stepping rate is related to the beam-current density (j) and resist sensitivity (q) by this expression: $f = j/q$. Beam current density is a function of electron optical parameters such as gun brightness, lens and deflection-system aberrations, and electron-electron interaction effects. Published current densities for VS1, LS1 and EBES are 300 A/cm², 50 A/cm² and 10 A/cm² respectively.

Figure 9's two sets of curves show the effect on pattern writing time of deflection speed, beam-current density, and resist sensitivity. Writing time per chip for the three systems is listed in Table 2 (but note that these values in the table are based on simplified data without writing time overheads and do not represent the inherent limit of the three systems).

The writing speeds for the three systems could all be improved. For VS1, the main limitation at present is deflection speed, and improvement in the deflection drivers and d-a converters would be of direct benefit. In addition, the use of more than one beam size (small beam for outline and large beam for fill-in) would also improve speed. At the extreme, deflection system settling time and logic switching time in the pattern generator may present a limit.

For EL1, high speed can be achieved by improving resist sensitivity and deflection speed as neither is at its limit. However, a data-rate limitation may be encoun-

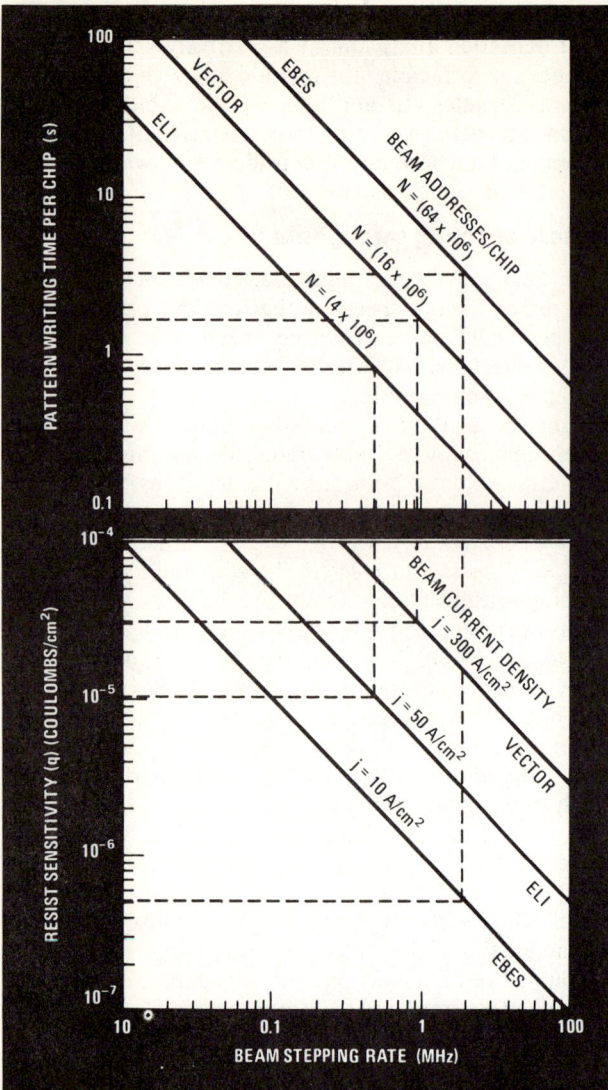

9. Beam stepping rate. The top graph indicates the effect of beam stepping rate on pattern writing time per chip for the VS1, EL1, and EBES. Effects of beam-current density and resist sensitivity on beam stepping rate are shown in the bottom graph, in which three vertical dotted lines indicate the operating stepping rates. Note that systems with low beam-current density require high resist sensitivity.

TABLE 2: WRITING TIME FOR TYPICAL ELECTRON-BEAM SYSTEMS		
System	Writing time (seconds/chip)	Resist sensitivity needed (coulombs/cm^2)
VSI	1.6	$\leqslant 3 \times 10^{-5}$
EL1	0.8	$\leqslant 1 \times 10^{-5}$
EBES	3.2	$\leqslant 5 \times 10^{-7}$

operation, though the optical printing system restricts resolution. Technical details for the mask-making operation have been resolved, and several systems have been developed for routine use. The main advantages of this approach are improved mask quality and faster turn-around time. Cost has been found to be competitive even when high-cost prototype electron-beam systems are used. Commercial systems for this application should be available in the near future and should widen its usage in the industry.

Direct wafer exposure can be divided into two categories: pattern line width in the optical range (down to 1 to 2 μm) and suboptical line width (1 μm and below).

In the optical range, electron-beam lithography has been used in a manufacturing environment, but it competes directly with present-day UV printing. Its main advantages are sharper pattern definition and greater overlay accuracy. In addition, it should improve yield by eliminating some of the defects caused by the use of masks and greatly reduce turnaround time when circuit customization is required.

For suboptical pattern line widths, the scanning electron-beam approach is at present the only proven fabricating technique. (The only other technologies that could compete in the future are electron-beam projection, X ray, and extensions of contact printing that use shorter-wavelength UV in combination with thin masks that can be conformed to the sample.) But although electron-beam lithography is ready for suboptical geometries, some of the associated processing technologies still require considerable development. Working circuits fabricated to date are chiefly special-purpose high-frequency devices, like microwave transistors, surface acoustic transducers, and high-density magnetic-bubble circuits. Only a few LSI components have been made, primarily for research and feasibility studies to verify the economics and technological benefits of small devices.

In general, the use of scanning-electron-beam lithography for direct wafer exposure still costs more than optical techniques. But the gap is narrowing rapidly, and with improvement in throughput and equipment cost, it should be possible to achieve cost-competitiveness in the future, including suboptical applications.

Lastly, electron-beam techniques can be used to produce masks for other high-resolution replication systems such as X ray, deep UV printing, and 1:1 photoelectron projection printing, which are still in the developmental stages. All these systems promise high-speed printing from a high-resolution mask produced by a scanning electron beam, along with the potential of low production cost. □

tered because in this instance each beam position must be individually specified in the data stream.

In EBES, resist sensitivity is already very high, and further speed increase would require some improvement in beam-current density. In principle, such improvement should be possible as discussed earlier. In addition, the number of beam addresses could be reduced by skipping scan lines where no pattern exists. However, table speed and data rate may pose limits.

Three areas of application

A scanning electron-beam system can be used for mask making for conventional optical printing systems, as well as for direct wafer exposure and mask making for advanced high-resolution replication systems.

The electron-beam approach to mask making requires no major changes in an existing optical lithographic

1. Electron-beam line. Electron-beam lithography systems of the EL-1 type on this semiconductor production line at IBM's East Fishkill, N.Y. facility have a throughput of 22 silicon wafer exposures per hour for 57-millimeter wafers with 2.5-micrometer lines and spaces.

Scanning electron-beam system turns out IC wafers fast

by E. V. Weber and H. S. Yourke, *IBM Corp., Hopewell Junction, N. Y.*

□ Sometimes the answer to achieving the ever smaller geometries needed in semiconductor pattern-making is 'think big.' So it is with the first scanning-electron-beam system that achieves the throughput necessary for commercial chip production. The EL-1 (Fig. 1) scans the wafer with a square beam that covers much more of the surface than the round beam of other scanning systems. Thus more of a pattern can be exposed in an equivalent time, and wafer throughput rises dramatically. This one-of-a-kind machine points the way to commercial wafer production by electron-beam lithography.

The principal advantage of electron-beam lithography over optical lithography systems is that it avoids the limitations of resolution and depth of focus imposed by the wavelength of light. Scanning systems, which rely on computer control rather than masks to form the pattern, also have a high degree of pattern flexibility. For example, etching different patterns on adjacent areas of a wafer is accomplished easily and quickly. Moreover, eliminating the mask used in contact and projection optical lithography systems saves time and reduces errors and defects.

Scanning systems also make it possible to control microscopic pattern distortions in real time. Finally, they excel in overlaying new patterns on previously etched levels of a chip. An example of a high-quality pattern of 2.5-micrometer dimensions that was produced by a scanning system is shown in Fig. 2.

Raising throughput

The principal obstacle to full deployment of scanning systems in integrated-circuit production has been the difficulty of achieving the necessary throughput for cost-effective use. The throughput for the two commercially available scanning machines is only about one exposure an hour. Of course, these machines are primarily intended for mask making, but their production rate gives an idea of the distance that is to be traveled to meet

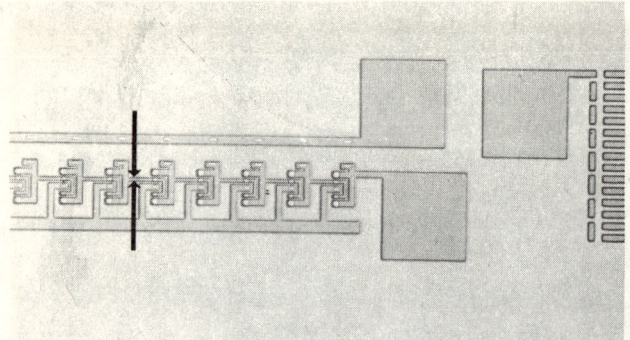

2. Small patterns. High-quality integrated-circuit chip was exposed by the electron-beam scan of the EL-1. Minimum conductor widths, such as in the line segment shown between arrows, are 2.5 μm. With minor modifications, this machine will be able to expose 1-μm lines.

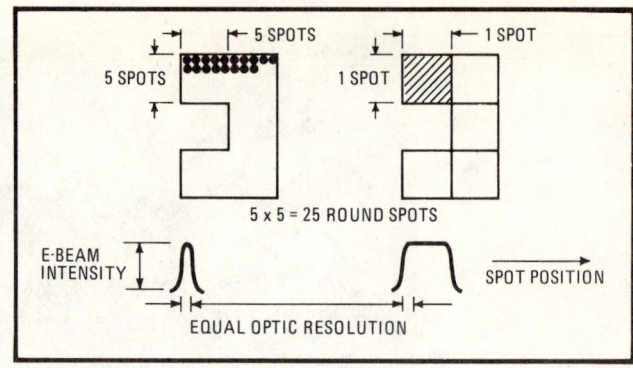

3. On the spot. Other electron-beam lithography systems scan the image of IC patterns with a small circular spot. In the EL-1, a square spot equivalent to many circular spots is scanned across the silicon surface of a wafer. This procedure drastically cuts exposure time.

IC makers' standards of 20 to 30 exposures an hour.

The alternative electron-beam solution to the throughput problem is a projection system. Masks are used in such systems to expose all image points on a wafer at one time. While projection-system development is still under way, the problems of mask fabrication and handling have slowed progress toward a practical machine.

The EL-1 may be seen as a compromise: it uses the principle of the bigger exposure area found in projection machines together with the maskless flexibility of scanning units. In most other scanning systems, the beam is pencil-shaped, which gives a limited, round coverage of the wafer area. The square beam of the EL-1 produces an image as big as the smallest pattern element to be constructed — much bigger than other scanning systems' beams. The result is a throughput of 22 wafer exposures an hour, based on 57-millimeter (2¼-inch) wafers with 2.5-micrometer geometry.

Putting it to work

First use of the EL-1, at IBM's East Fishkill facility in Hopewell Junction, N. Y., has been for the production of bipolar large-scale integrated circuits. The unit's ease of data transfer, quick turnaround, and ability to balance inventory by mixing different patterns on a wafer make electron-beam exposure economically attractive even for IC patterns with line widths thick enough to be exposed with light optics.

Rapid pattern exposure is achieved by scanning a small, square wafer area, called a field, with a square projected image, called the spot. The sides of the spot are as long as the smallest dimension of the pattern that is being exposed. At a given current density, a 2.5-μm square spot is equivalent to a 5-by-5 array of 0.5-μm round spots, but it requires only 1/25th the exposure time (Fig. 3).

The square spot scans the field in a stepped fashion, moving from one grid on the field to the next by magnetic and electric deflection. When the scan of the field is complete, the stepping table that holds the wafer moves it so that the adjacent unexposed field is within the deflection range of the beam.

The field-scan procedure (top of Fig. 4) produces an exposure with an edge gradient equal to the beam's edge slope (defined as the distance from the edge of the spot that the beam's intensity takes to rise from 10% to 90% of its full value). In practical implementations, the square-beam approach provides better than an order-of-magnitude advantage in writing speed over other scanning systems.

Because limiting a pattern to a design made on a grid equal to the minimum line width of that pattern would be unduly restrictive, an offset capability is included in the EL-1. Spots may be displaced from their nominal location in increments of ⅕ of a spot (bottom of Fig. 4).

The electron-optical column that focuses and shapes the beam is designed to provide the maximum beam current and field size at the required edge slope. For instance, placing the deflection coil within the projection lens minimizes chromatic and electron-electron-interaction effects that result from the large currents used. Dynamic correction of focus and astigmatism keeps edge slope to less than 1/10,000 the length of the field.

To obtain a workable lithographic tool, many engineering tradeoffs were necessary among such key performance parameters as throughput, edge slope, field size, current density, and overlay. Moreover, the applications in which these machines are used may influence these parameters. For example, field size when exposing chips smaller than 5 mm on a side is best adjusted to equal the chip size. On the other hand, larger chips are easily fabricated by the use of interstitial registration marks that make it possible to stitch adjacent fields together with optimum overlay and throughput.

A set of performance specifications for the EL-1's initial application at East Fishkill is listed in the table. As product requirements demand, the system may be adjusted to make exposures with dimensions that are in the 1-μm range.

On the spot

The square spot as big as the smallest pattern dimension is the chief design feature giving the system its rapid exposure rate. But to achieve this high throughput and to get good overlay (layer-to-layer registration) and large exposure field sizes, other features play an important role. Among them:

■ Advanced electron optics to obtain high beam current for quick large-field exposures with high resolution.

■ A combination of narrow- and wide-bandwidth deflec-

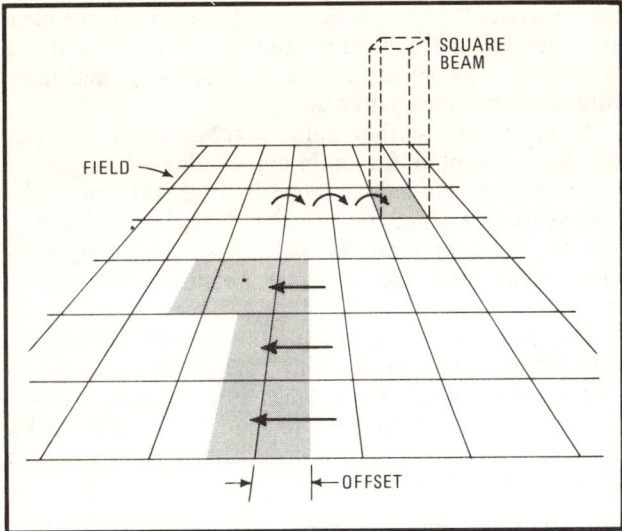

4. Checkerboard scan. For exposure purposes, a field is divided into a matrix of small squares. The shaped electron beam is stepped to each square in the matrix. It can be offset to expose lines centered on a grid of finer resolution than spot size.

TABLE: EL-I EXPOSURE SYSTEM SPECIFICATIONS	
Field size (maximum writing)	5 mm
Spot shape	square
Spot size (50% intensity)	2.5 μm
Edge slope (10 – 90%)	0.5 μm
Beam voltage	25 kV
Beam current (at 50 A/cm^2)	3 μA
Overlay (3σ)*	0.5 μm
Throughput for 57-mm wafer (76 chips)	22 wafer exposures per hour
Writing grid	2.5 μm
Writing grid offset capability	0.5-μm increments

*the 3σ error between the centerlines of two patterns on different layers designed to be coincident.

tion of the beam for a good signal-to-noise ratio.
■ A deflection cycle that repeats itself exactly, thereby ensuring that any errors are reproduced exactly—which aids error correction.
■ Automatic measurement of deflection errors and compensation for them in order to attain maximum pattern accuracy.
■ Four-mark registration with an associated deflection modification to give optimum overlay of patterns and to permit matching of boundaries on adjacent fields.
■ A three-step highly automated sequence for wafer handling and alignment, which boosts throughput.
■ Beam-deflection correction of errors in the position of the table that steps the wafer from field to field, thereby giving a high stepping speed and rapid settling.
■ Use of servomechanisms to maintain beam current, spot focus, and column alignment over long periods.

Combining deflections

Within a given exposure field, the electron beam is positioned chiefly by a large-range, narrow-bandwidth magnetic deflection. During writing, a bidirectional magnetic ramp (Fig. 5) deflects the beam in a bidirectional raster fashion. Superimposed on this ramp is a bucking sawtooth applied by a small-range, wide-bandwidth electrostatic deflection. This combination causes the beam to step. In addition, a small-range, moderate-bandwidth electrostatic deflection compensates for errors. Restricting these larger bandwidths to small ranges results in minimum deflection noise and minimum random pattern error.

The deflection cycle of the EL-1 is a three-part repetition. In standard operation, the cycle repeats even while a new wafer is being loaded. The object is a steady-state deflection, so that distortions due to eddy currents, thermal currents, etc., will repeat at all points.

During the registration part of the cycle (Fig. 6a), the computer scans the beam sequentially over the locations of four registration marks. During the writing part of the cycle (Fig. 6b), the beam scans sequentially over the entire field. Each possible location is addressed, with the computer blanking out the beam at points that do not require exposure. This approach ensures that pattern differences will not change the deflection history. In general, densities of LSI patterns are high enough so that, even though throughput is slightly less than for deflection that addresses only the points to be exposed, the decrease is negligible and is far outweighed by the increase in accuracy.

The move part of the cycle is the time that it takes the stepping table to move the wafer to the next exposure field. During the move, a special deflection (Fig. 6c) is always executed. Occasionally, it is used in conjunction with a focus fixture in the electron-optical column for automatic sensing and correction of focus.

Righting deflection errors

Repetitive deflection of the square beam gives the system its ability to define line widths and positions precisely. But repetition is not enough; the spots must be accurately positioned in order to obtain compatibility with different electron-beam or optical tools.

Accurate deflection is obtained by scanning a calibration grid on a special target and then compensating for sensed errors. The grid (Fig. 7) is an array formed by square openings in a layer of gold on a silicon substrate.

To measure deflection error, the grid is positioned on the stepping table, and the beam scans it. As the beam moves from the gold, which has a relatively high backscatter coefficient, to the silicon, which has a relatively low backscatter coefficient, the current changes in backscatter detectors. The time of this change is recorded.

A special program selects the times associated with selected grid points, edits and averages them to obtain centers, and then compares the results with a table of expected centers to obtain the deflection error. The table is based on calibrated locations of the grid marks on the target being used; thus the grid does not have to be perfect.

The error figure goes into programs that generate piecewise-linear corrections to the deflection. Usually, the error-measurement program applies previous corrections when directing the beam over the grid, and the new corrections are applied to the earlier ones. This iterative

15

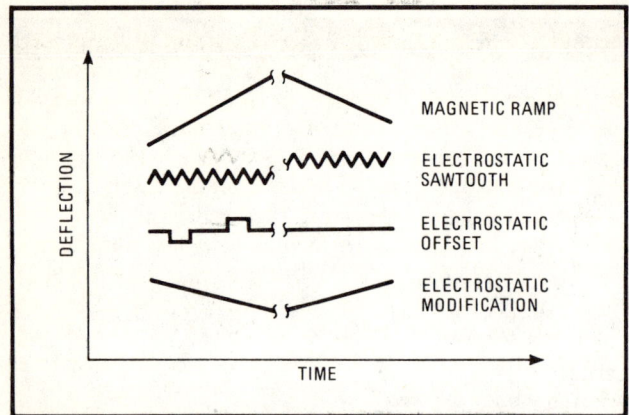

5. Combined scan. A combination of narrow-bandwidth magnetic deflection and wide-bandwidth electrostatic deflection precisely positions the electron beam. This combination scan reduces deflection error and noise to acceptable manufacturing levels.

procedure minimizes the accuracy required of the correction electronics, since it does not permit errors to accumulate. Typical errors before and after correction are illustrated in Fig. 8.

It also is necessary to coordinate the registration scan, when the beam checks four registration marks on the wafer, with the subsequent write scan. The calibration target comes into play here, too. After the correction process just described, which is a writing-scan correction, the program calculates the expected positions of several grid points designated as test registration marks—basing these calculations on a write scan of the grid. Then the electron beam operates in the registration-scan mode to locate these registration marks, and the difference between the observed and the calculated positions is sensed. This error information is stored and used during registration to adjust the sensed positions of the wafer registration marks so that they correspond with the writing scan.

These measurements and corrections of deflection errors bring the field being scanned on the wafer to within 30 parts per million of ideal. However, ideal deflection may not achieve optimum overlay of successive patterns required for the manufacture of semiconductor devices. Some causes of deviations from the ideal are: imperfect mechanical positioning of the wafer on the table, wafer distortions caused by processing, and inaccuracies in previous patterns.

As in any photolithographic process, accurate registration is essential for optimum overlay of successive patterns on a chip and to properly mesh adjacent exposure fields on a chip. With the EL-1, registration goes a step beyond simple mechanical adjustment to registration marks, and the scanning beam is adjusted to them.

The system achieves optimum overlay by locating the registration marks in the four corners of the previous pattern and adjusting magnification, rotation, translation, and shape of the field to match it to the marks. Typically, the marks are features formed as a byproduct of earlier processing.

Registration gives accuracy

The detection process is quite similar to the one used in deflection correction. It depends upon the fact that the energy distribution and quantity of back-scattered electrons are essentially constant when the beam scans a flat surface, but change when the beam crosses an edge formed by a change in material or by the topography of the wafer surface. Automatic gain-control circuitry compensates for wide signal variations between wafers of different types and between different processing levels on a given chip.

Each time the signal crosses a registration edge, special circuitry transmits the time to the control computer. To identify each of the marks, the computer edits the sequence of time samples and correlates it to a model. Then deflection modifications are generated to match the writing field to the registration marks.

The average error of magnification is used to obtain an approximation of height errors and to form a focus correction. Even though depth of focus is an order of magnitude greater than for light optics, this correction helps achieve accuracy.

The maximum size of the exposure field is limited, but chips can be made of more than one field—as large as a whole wafer, even—by stitching fields together. Marks

6. Three-step. A repeating three-part deflection cycle maintains distortion constant over all points in a field. First step (a) is a registration scan followed by a writing scan (b) over the entire field. Last is a move cycle (c) in which the beam steps to a new field.

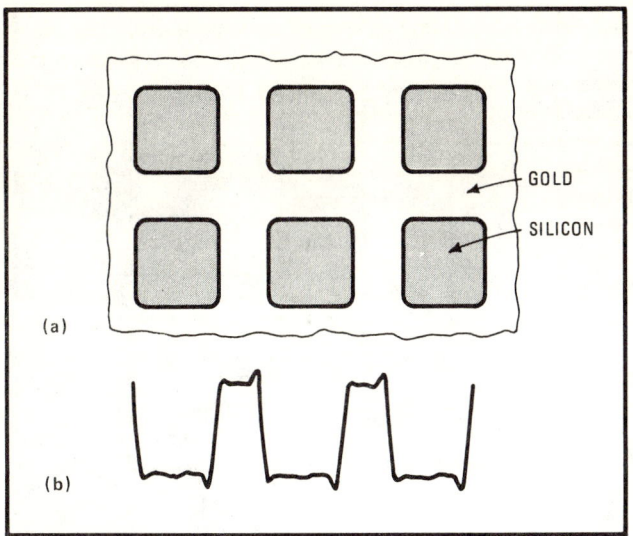

7. Calibration. A grid consisting of an array of square openings in a layer of gold on a silicon substrate (a) is the reference for the EL-1's circuitry for deflection error compensation. Signals (b) from the gold's high backscatter yield errors in grid position.

between adjacent fields are shared, thereby making their boundaries coalesce.

The system can write a new registration mark before an earlier one degrades to the point of unusability. Space is allocated for rewriting a mark three times, each at a new location. In some cases, rewriting can take place at a previously used location.

This four-mark registration is the final step in a three-part sequence designed to provide progressively finer adjustments of the wafer's alignment to the beam. First, as wafers enter the EL-1, a mechanical handler places them on carriers and subsequently positions them under the beam. The positioning locates the wafer relative to the beam to ±75 µm.

The second step is a registration process involving the entire wafer. It adjusts the magnification and rotation in the individual field to provide an alignment of wafer and beam within 2µm to 5 µm of each other. The necessary registration data is obtained by scanning two special marks on the wafer. Third, the four-mark registration procedure further positions and modifies the writing field to obtain an overlay of better than ±0.5 µm.

To provide rapid stepping of the table as it moves the wafer from one field to the next, a dc servo motor drive smoothly accelerates and decelerates to a stop without hunting for a precise position. Design emphasis was on rapid settling of the table to within ±7.5 µm of the desired position. Then a beam deflection in response to a position encorder achieves beam-to-table accuracy better than 2.5 µm. This two-prong approach makes it possible for the table to move 5 mm to another field and settle within 250 milliseconds.

Load and unload

Another design feature that speeds the production rate is a vacuum interlock combining the wafer transport with the valving mechanism that removes the air. The volume of air that must be removed is small enough so that the system can be rapidly pumped to an intermediate volume, then opened to the large main chamber, which is at vacuum, giving a transition from atmosphere to 10^{-6} torr in 4 s. This combination of transfer and interlock makes it possible to exchange a wafer on the stepping table with the next wafer that is at atmosphere in about 15 s.

Another feature that insures maintenance of high throughput is a set of servomechanisms that monitor and adjust alignment of the electron-optical column and parameters of the electron-beam gun during the period in which the table is stepping the wafer from one field to the next. These servomechanisms insure satisfactory exposure quality over long periods of time. Periodically, the edge definition of the beam is monitored by the scanning of a focus target.

The various components of EL-1 go together as indi-

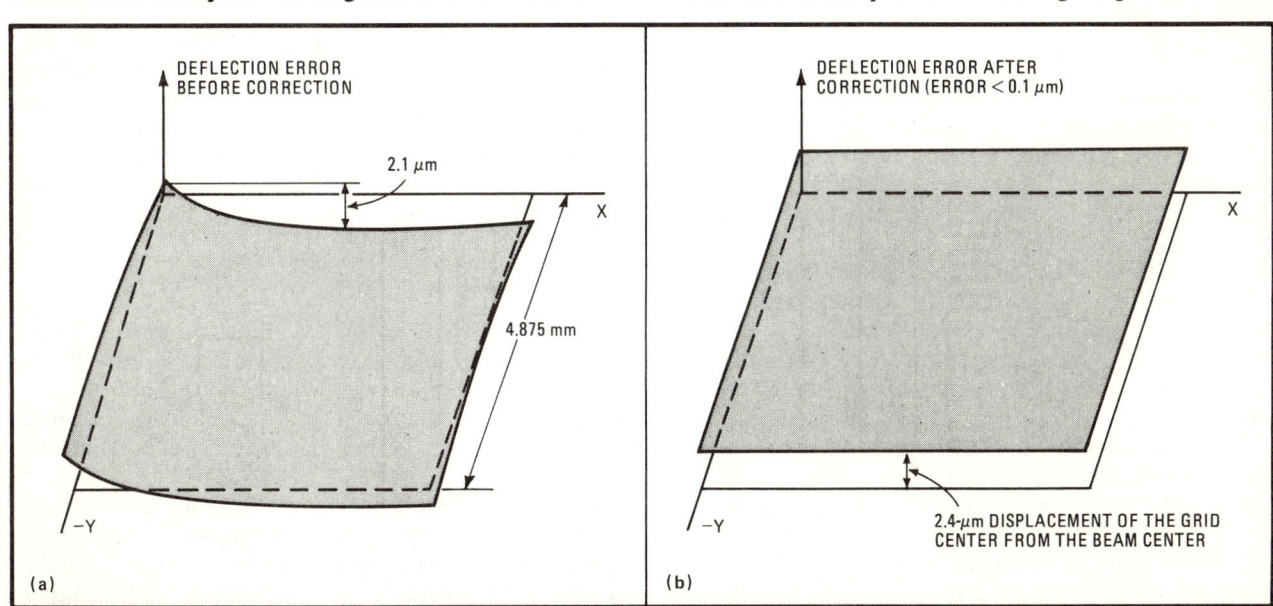

8. Deflection correction. Typical three-dimensional plots of X deflection error before and after correction (a and b, respectively) show how a scanning system can electronically correct itself. In this case, a 2.1-µm error is reduced to less than 0.1 µm.

17

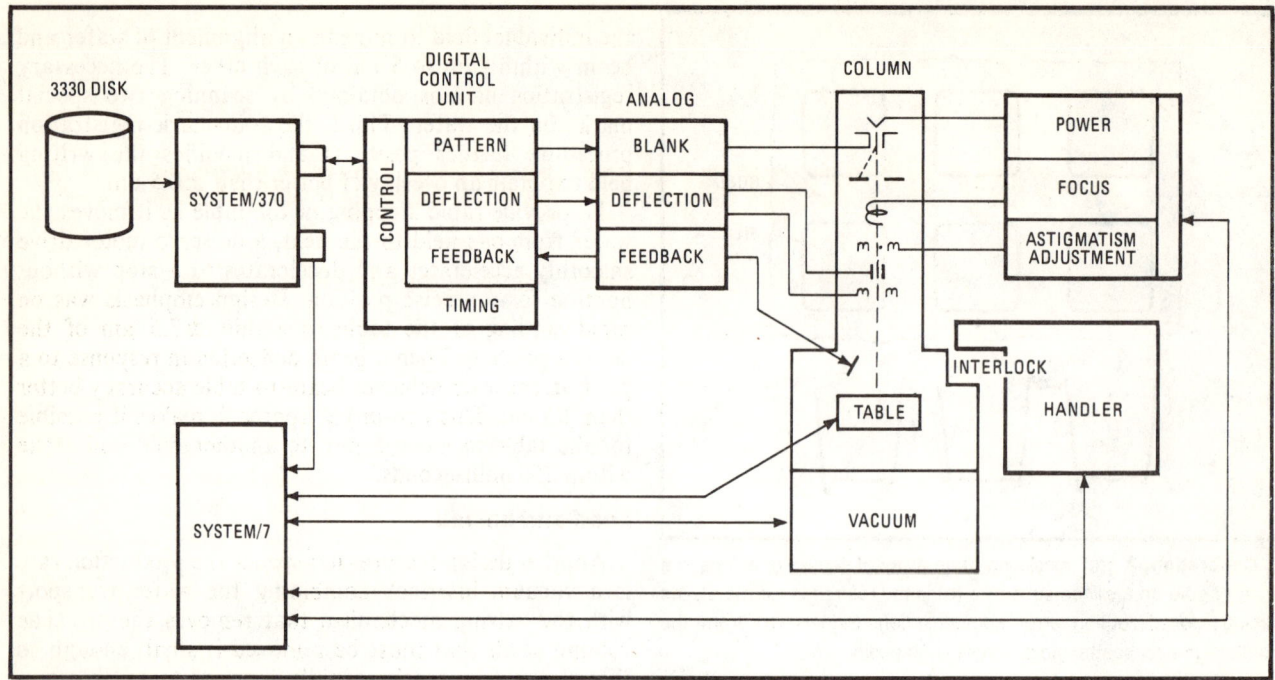

9. Automated electron-beam. The major components of the EL-1 are a vacuum beam-producing column with an automatic wafer handler, a digital control unit, analog correction circuitry, and an IBM System/370 plus magnetic-disk memory as the central controller.

cated in Fig. 9. A System/370 computer is the central control element. The principal information store is a 3330 disk storage, containing such key data as pattern descriptions, wafer maps describing pattern types and locations on the wafer, descriptions of deflection patterns to define field size, and the locations of registration marks.

As part of system preparation, a deflection path suitable for writing a pattern, is transferred from disk to the deflection memory in the digital control unit. Under control of the system clock in this unit's timing section, the deflection data causes a sequence of digital control signals to be transmitted to the deflection circuitry in the analog unit.

Beam drive

This circuitry in turn produces the appropriate drive for the beam to follow the paths of Fig. 6. Once the deflection has stabilized, the System/370 control program causes the table to move the calibration target under the beam, transmits a pattern to the pattern section of the digital control unit, and activates the feedback sensors. When the beam is in the vicinity of the selected grid marks, the pattern section unblanks it. Back-scatter signals are processed by hardware and software to form the set of deflection corrections, which are placed in the correction memory of the digital control unit.

When the deflection has been corrected, the system is ready to write on the wafers. The operator tells the control program for the System/370 which wafer map to use. The control program initiates the transfer of the first wafer to the stepping table via a System/7 computer, which interfaces and monitors the subsystems. The wafer registration mark specified in the wafer map is then moved under the beam.

After the three-step registration cycle is completed at the first field, the pattern defined by the wafer map is called out, the field is corrected on the basis of the registration information, and the pattern is written.

The wafer moves to the next field site, and the sequence of four-mark registration, write, and move repeats for each of the fields on the rest of the wafer. Patterns could differ at each exposure, depending on what is specified in the wafer map. When the last exposure is completed, the table may move to the focus target or the calibration target to collect data and determine whether focus or deflection has drifted enough to call for an update. Such updates to corrections are required only infrequently.

If the following batch of wafers requires a different field size, an appropriate deflection and the corrections previously acquired for that deflection are loaded. The corrections are checked, but the previous corrections usually are adequate, so writing proceeds without correction convergence. Numerous different deflection cycles are available to optimize field size and accommodate different registration marks.

Small dimensions

For the past three years, the EL-1 has been successfully exposing bipolar patterns on silicon wafers. High-quality images with minimum dimensions of 2.5 μm and layer-to-layer registration of well under 0.5 μm are being routinely achieved in the large-scale production of bipolar wafers.

In the future, this capability could easily be extended to pattern geometry with 1-μm detail. Addition of higher-speed data-conversion circuitry to raise throughput to even higher levels and a redesigned wafer staging area for handling larger wafers are other possible system improvements. □

X-ray lithography gains ground

With optical projection systems reaching their limits,
researchers are looking for a better way to make dense chips

by Jerry Lyman, Packaging & Production Editor

With ever denser LSI and VLSI chips in the offing, researchers for semiconductor firms, as well as Bell Laboratories, are taking a close look at X-ray lithography as a low-cost alternative to present systems.

This may seem slightly paradoxical, since before the technique can become viable commercially, one of the problems it must overcome is the high cost of present equipment. Also, the sensitivity of the resists used in X-ray lithography needs improving, and the X-ray power source must be boosted to shorten exposure time and raise output. And that still leaves the difficulty of aligning the successive circuit layers to be dealt with. Still, work is going on in all these areas.

At present, large-scale integrated circuits are being fabricated routinely in volume with 2- to 4-micrometer circuit details. These features are the end result of images exposed from electron-beam–generated precision masks employed in either an optical 1:1 projection or step-and-repeat projection lithography systems.

The IC industry would prefer to go to direct exposure by the electron-beam system itself, since the available optical projection lithography systems are operating close to their optical limits (1 to 2 µm). However, the only two electron-beam machine, at present commercially available, ETEC's Mebes and Varian Extrion's EBMG-20, can only expose one wafer per hour and are priced at about $1.5 million. Consequently, these machines are mainly being devoted to mask making.

Advantages. Many firms such as General Instrument, TI, IBM, Hughes, and Bell Labs are now looking at X-ray lithography as a low-cost, high-throughput alternative to this method. An X-ray system has certain inherent advantages over an electron-beam system. For one, there is an absence of the scattering and diffraction effects that limit the actual minimum resolution of electron-beam lithography to slightly less than 1 µm. Also, it is possible to expose patterns deep into a thick coating of resist. And the instrumentation is inherently simple. The X-ray lithography system developed at Bell Labs, Murray Hill, N. J., illustrates the simplicity of this method (see drawing). The system has relatively simple optics and needs none of the complex circuitry and equipment required for scanning electron-beam lithography.

A basic X-ray lithography setup must have three elements: a mask, an X-ray resist, and an X-ray source, seen in the small inset of the illustration. The mask and resist-coated wafer are separated by a small micrometer-sized gap, making it actually a proximity rather than a contact type of lithography.

Fragile. Mask fabrication is one of the touchiest areas. Instead of the rugged chrome-iron masks used in optical lithography, a fragile, ultrathin (2–10-µm) membrane of silicon, silicon nitride, silicon carbide, Mylar, or polyimide is used to form the mask substrate. The IC pattern is defined by scanning an electron

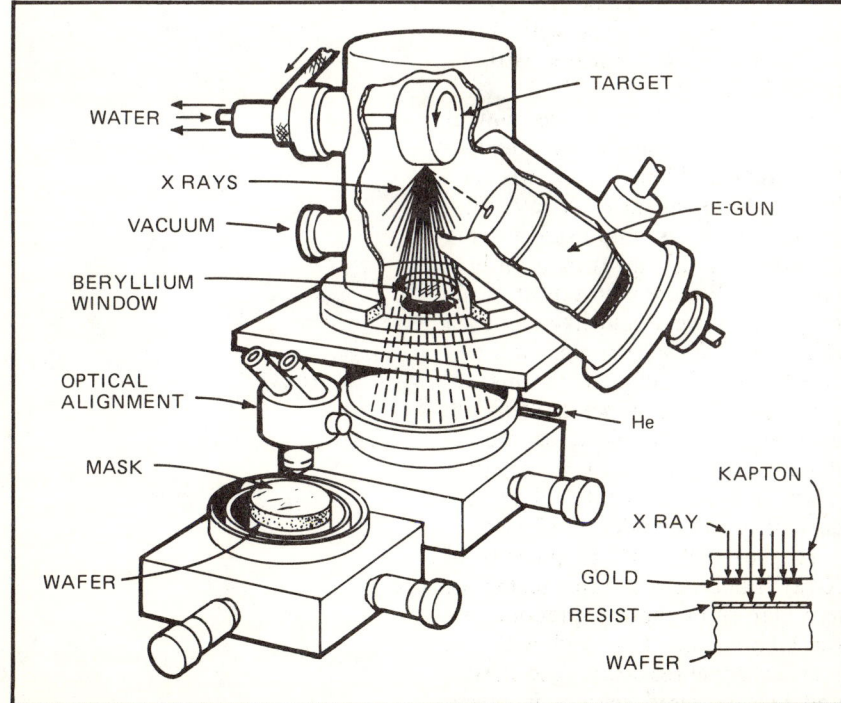

Simplicity. X-ray lithography system developed at Bell Labs has relatively simple optics and less complex circuitry than is required for scanning electron-beam lithography.

beam across a layer of gold that heavily absorbs X-ray radiation.

X-ray resists are liquid polymers that are sensitive to X-ray radiation and can be spun onto silicon wafers easily. Many of the same resists are used in both electron-beam and X-ray lithography. Polymethyl methacrylate is an example of a resist common to both processes. At present, most of the firms investigating X-ray lithography are favoring negative resists because of their speed advantage over positive types.

Most X-ray sources being used for lithography are in the 0.4-to-40-kilowatt range. The most heavily used source is an electron-beam–bombarded aluminum which has a wavelength of 8.3 angstroms. Work has been done at Bell Labs with a palladium source (4.37-angstrom wavelength) for even greater resolution. Since the exposure time of the resist-covered pattern is a function of both X-ray power and resist sensitivity, these two areas are being heavily researched. In fact, synchroton radiation and laser-generated plasma have both been suggested for extremely high-power sources of X-ray radiation.

Disadvantages. Perhaps X-ray lithography's most serious disadvantage is the difficulty in aligning successive layers of a wafer. First of all, no optical aligner exists that is accurate enough for submicrometer geometries. For instance, a 0.5-μm line would require an alignment of within 0.1 μm and only electron-beam systems at present have this accuracy. Secondly, the fragile masks require special handling. Finally, the stability of the mask itself can contribute to misalignment. Of course, single-layer devices (magnetic bubbles and microwave devices) require no alignment.

General Instrument Corp.'s Microelectronics division, Hicksville, N.Y., has been engaged in X-ray lithography for about two years, according to Gregory Hughes, its senior X-ray lithography engineer. GI is now into its second-generation 1-kw system combined with a modified Cobilt aligner that is being used to fabricate electrically alterable read-only memories with eight levels of masking by the X-ray lithography method.

Mask. GI has developed its own mask-making technique to eliminate the mask-breakage problem. Its masks start out on 60-mil-thick glass plates onto which a thin layer of liquid polyimide is spun. Gold is evaporated onto the polyimide, followed by a resist. The resist is then exposed by a scanning electron beam and developed. Next, ion-beam etching is used to remove the unwanted gold. The last step is to etch away the unnecessary glass and put the remaining material on a silicon ring.

Along with its development work on masks, GI is engaged in a joint effort with the Providence, R.I., division of Philip A. Hunt Chemical Corp. to develop a high-speed negative X-ray resist. The Long Island firm has furnished a low-powered X-ray system for Hunt Chemical to use in its evaluation.

With its 1-kw source and copolymer negative resists from Mead Co. of Rolla, Mo., GI has been able to make wafer exposures in about 35 minutes. Higher-powered sources at Bell Labs and Texas Instruments Inc., Dallas, have cut this time to a few minutes. GI has fabricated metal-oxide-semiconductor devices with 4- and 2-μm geometries, and at Bell Labs MOS devices with 1-μm details are being fabricated with X-ray lithography.

Hughes of General Instrument comments on the X-ray method: "Our real problem is that we need better processing after X-ray lithography, such as improved plasma etching of aluminum conductors.... We are now almost at the stage where we would like to buy a commercial X-ray system. Unfortunately, there is none on the market yet. I would like to see a $500,000 unit with a 10-to-16-kw source, 2-minute exposure with a negative resist, and an automatic alignment capability on the order of 0.1 μm."

Hughes and most proponents of X-ray lithography see this technique as a high-throughput production tool for high-volume LSI. And with the more sensitive resists, rugged masks, and high-powered sources in development, a reasonably priced commercial machine could go far. □

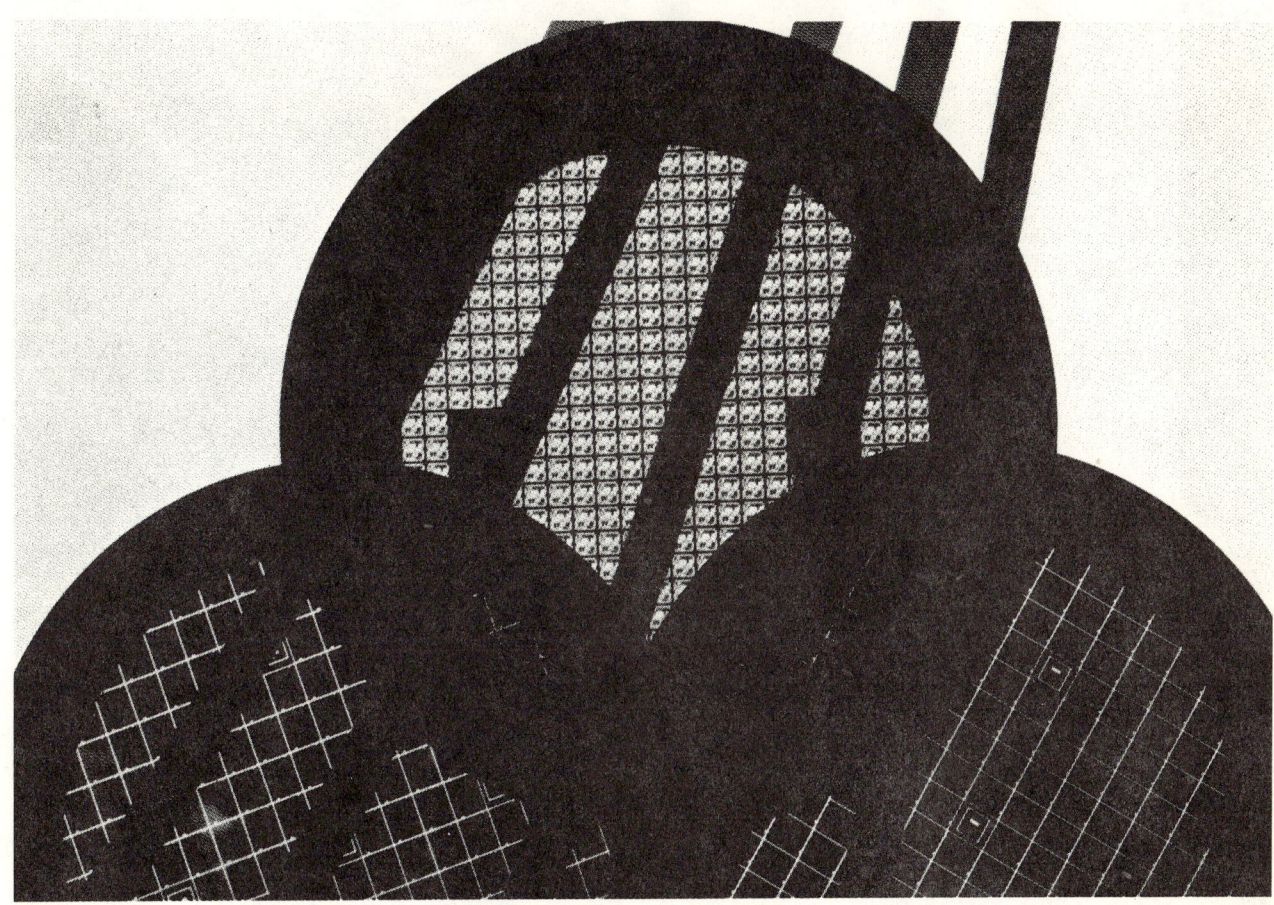

X-ray lithography breaks the VLSI cost barrier

Relatively simple X-ray systems promise mass production of very large-scale integrated circuits with 1-μm line widths

by Gregory P. Hughes and Robert C. Fink, *General Instrument Corp., Microelectronics Division, Hicksville, N. Y.*

☐ As the designers of integrated circuits struggle to cram more devices on to silicon chips, ever smaller geometries are called for. If present trends continue, lines 2 micrometers wide will be common after 1979 and 1-μm widths will be standard after 1982 (Fig. 1). To produce lines as fine as this, integrated-circuit makers will have to decide which lithographic process will be their best choice. And though the electron-beam and optical direct-step-on-wafer process are popular today, a third process in development, X-ray imaging, could be the most powerful technology yet (Fig. 2).

X-ray imaging has produced lines as narrow as 0.16 μm and can do so at a cost that is relatively low compared with that of competitive methods. Already shown to be economically and technically viable for single-mask devices like surface-acoustic wave devices and bubble memories, it is on the way to becoming equally suitable for producing complex multimask devices of all types. Moreover, it is the one lithographic technology that is easily downward-compatible in device geometry: a system using X rays today to produce devices with 1-μm geometries will work just as well tomorrow for 0.5-μm geometries.

Table 1 shows a Dataquest comparison of X-ray lithography with the other wafer-imaging techniques currently in use in either production or research: contact printing, projection lithography, and electron-beam lithography (see also "X-ray lithography's competi-

21

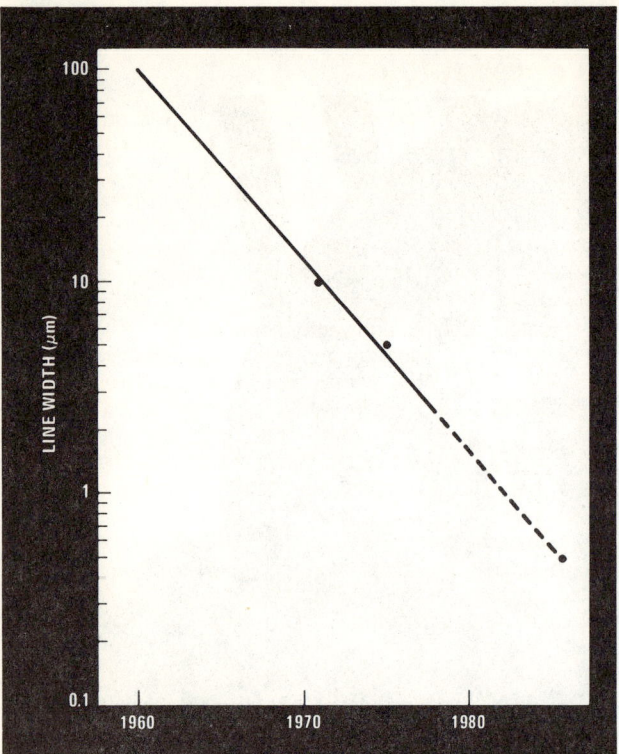

1. Narrowing lines. If minimum line widths for production large-scale integrated circuits continue to decrease, 1-micrometer-wide lines will be in production by 1982, followed by 0.16-μm-wide lines in 1986. Both widths are well within the capabilities of X-ray lithography.

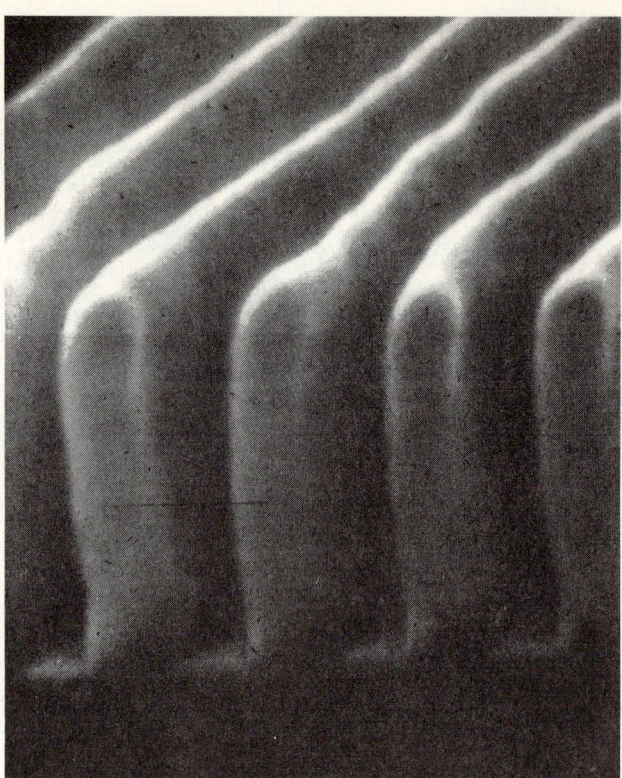

2. SEM. This scanning-electron micrograph shows a grating with 1,600-angstrom line width in PMMA photoresist that has been exposed to a copper X-ray source. (Courtesy, P. X. Flanders, Lincoln Laboratories, Massachusetts Institute of Technology.)

tion"). As it shows, the X-ray approach yields the narrowest lines at reasonable defect densities, throughput rates, and cost. The figure of merit listed calculates the cost-effectiveness or profitability of a process and is determined by the empirical relationship:

$$\frac{\text{throughput}}{(1 + 0.15 \text{ defects})^8 \times \text{cost} \times (\text{line width})^2}$$

In these terms, X-ray lithography is the most promising of all the choices available, one of the major reasons being the relative simplicity of the basic X-ray system.

Basic X-ray lithography

An X-ray lithography system resembles an ultraviolet-light one in that it consists of a radiation source, mask, and wafer coated with a photoresist (Fig. 3).

The radiation source is an X ray whose energy must be soft enough to be absorbed by and expose the photoresist yet great enough not to be absorbed by the X-ray window or the photomask substrate. The X-ray mask cannot be simply a glass plate because that would absorb too much of the X-ray flux; instead it must be a thin substrate that is relatively transparent to X rays, overlaid by a dark field area of a material that will absorb X rays. Gold is usually used for this absorber, the counterpart of the emulsion or chromium mask used in UV lithography. Lastly, the silicon wafer must be coated with a photoresist that will absorb enough of the X rays to be exposed in a short period of time.

Figure 4 shows a basic system. The radiation source is mounted in a vacuum chamber and passes its X rays first through a thin window (usually 25-μm-thick beryllium), then through a helium atmosphere at standard pressure, and finally through the mask, exposing the resist on the wafer. Typical geometries that have been proposed for 3-inch wafers place the X-ray source 20 to 50 centimeters away from the wafer and keep mask and wafer 3 to 25 μm apart to avoid the damage to both that could result from contact between them.

X-ray sources create a number of design problems unique to them. Since the wavelengths involved range from 2 to 50 angstroms, there are no convenient mirrors or lenses that can be used to collimate the rays. This means that the ideal X-ray source would be a point source far enough away from the target for the X rays to appear to arrive in a parallel beam.

Unfortunately, a series of compromises with distance and divergence must be considered since the intensity from a source is inversely proportional to the square of the distance from it. Moreover, the X-ray source is not a true point source. The size of the source and the divergence of the beam cause two types of distortion: penumbral and geometric, respectively.

Penumbral distortion can be determined from the geometry of Fig. 3. Because of the gap, s, between the wafer and the X-ray absorber (mask), the definition of the X-ray absorber becomes blurred by the X rays arriving from the finite-source geometry. The following relationship results:

$$\delta = s(d/D)$$

where δ is the minimum attainable line resolution, d is

X-ray lithography's competition

Conventional contact printing has long been the production mainstay of the integrated-circuit industry. This technique, which may be of the hard, soft, or proximity type, puts a photomask next to a wafer and an ultraviolet light source behind the mask. Resolution is excellent when the mask and wafer are placed in hard contact, but this causes many defects, such as abrading, spalling of resist films, plucking and transfer of resist film, scratching of masks, etc. A soft contact minimizes these defects but diminishes resolution.

The proximity mode of contact printing, in which there is a slight mask-to-wafer separation, further reduces defects but also diminishes resolution and causes feature distortion because the transmitted light is diffracted. The magnitude of this feature distortion depends on the actual spacing variations across a wafer.

To avoid these problems, projection wafer printing puts a large space between wafer and mask. It entered the production environment during 1973 in static systems that used refractive optics to flood-expose wafers. But the procedure's growth was limited by the impossibility of producing refractive lens systems with an effective spectral bandwidth for wafers larger than 2¼ inches. Scanning slit projection techniques with reflective optics yielded manufacturing systems capable of handling wafers up to 100 mm, or 4 in., across. These systems use the entire lamp output, and the optical system is inherently 1:1 because its design is telecentric. The masks used are the same as for conventional wafer printing.

Step-and-repeat projection printing is based on refractive optics. In this method, a small area is printed from a mask which is photocomposed at 2×, 4×, 5×, or 10×. The process is then repeated many times to cover a single wafer. With far-ultraviolet, resolution will be better.

Electron-beam technology is now generally accepted as the way to generate masks for fine-line lithography. Currently three approaches are being tried: flood exposure, raster scanning, and vector scanning.

Flood-exposure electron imaging has reached only the prototype stage. It uses ultraviolet light to illuminate a photocathode layer that is deposited as a patterned surface on an optical mask. The photocathode then generates a patterned emission of electrons. This technique becomes prohibitively difficult because of field nonuniformities.

Scanned electron-beam systems use raster- or vector-scanning schemes. In the raster-scanning schemes developed by Bell Labs [*Electronics*, May 12, 1977, p. 97], the scan lines are generated by periodically deflecting the beam over a line of limited length. The entire substrate is covered by moving it while the beam is scanning. In the vector-scanning approach, the electron beam is deflected on the substrate and modulated to draw a desired pattern [*Electronics*, Nov. 10, 1977, p. 96].

It is obvious that the demanding requirements of either raster- or vector-scanning electron-beam systems require an extremely large investment in the fabrication of precise, sophisticated equipment.

---

the source diameter, and D is the distance between the mask and the source. The effect of this distortion is to produce nonvertical walls in the photoresist and thus create an uncertainty for subsequent processing of the geometry being printed. In most X-ray lithography systems this is controlled to within 1,000 Å. Given a typical X-ray source 5 mm in size and a source-to-wafer distance of 20 cm, the gap between mask and wafer must be 4 µm or less for this control.

The geometric distortion in X-ray lithography arises from the fact that it is a projection system: the image does not lie directly on the wafer but is a projection of the mask on the wafer by the X-ray source. The degree of this distortion, Δ, is dependent on the location of the image on the wafer:

$$\Delta = s(\omega/D)$$

where D is the source-to-wafer distance and ω is the distance from any point of the wafer to the perpendicular projection of the X-ray source to the wafer, as shown in the left insert of Fig. 3.

This distortion looks like an error in the image compared to the mask. It is a serious problem with circuits that require superposition of one photo level on the next, but is much less serious with single-level high-resolution circuits like bubble memories and surface-acoustic-wave devices.

To be more precise, the problem is not the distortion but the variation in distortion from one photo level to the next. This variation arises from an inability to keep the gap spacing constant. If the variation in the gap is ds, the variation in the geometric distortion dΔ is:

$$d\Delta = Ds(\omega/D)$$

To hold dΔ to 1,000 Å over a 3-in. wafer when D is 20 cm requires a control of ds to less than 0.5 µm. This is difficult mechanically for the mask-and-wafer holder and also requires the flatness of a wafer to be controlled with great precision through the various steps of high-temperature wafer processing.

The main components of a basic X-ray system—X-ray sources, masks, photoresists, and mask-to-wafer alignment mechanisms—can be implemented in a variety of ways. To begin at the beginning, the choice of the X-ray source affects most of the rest of the X-ray system, and that in turn demands a consideration of the relative advantages of hard and soft X rays.

Sources

A wide range of X-ray wavelengths could in theory be used to improve upon the resolution of UV systems. They extend from very soft X rays, having approximately a 50-Å wavelength, to hard ones with approximately a 2-Å wavelength. Unlike sources in electron-beam or optical systems, none of these has any significant problem with diffraction. In fact, the diffractions in the exposures made with mask-to-wafer gaps as large as 1 millimeter barely differ from those made with a 1-µm gap.

Hard X rays have the advantage of not being easily absorbed in the vacuum window and X-ray mask substrate—and the disadvantage of not being easily absorbed in the photoresist, so that it takes a greater flux

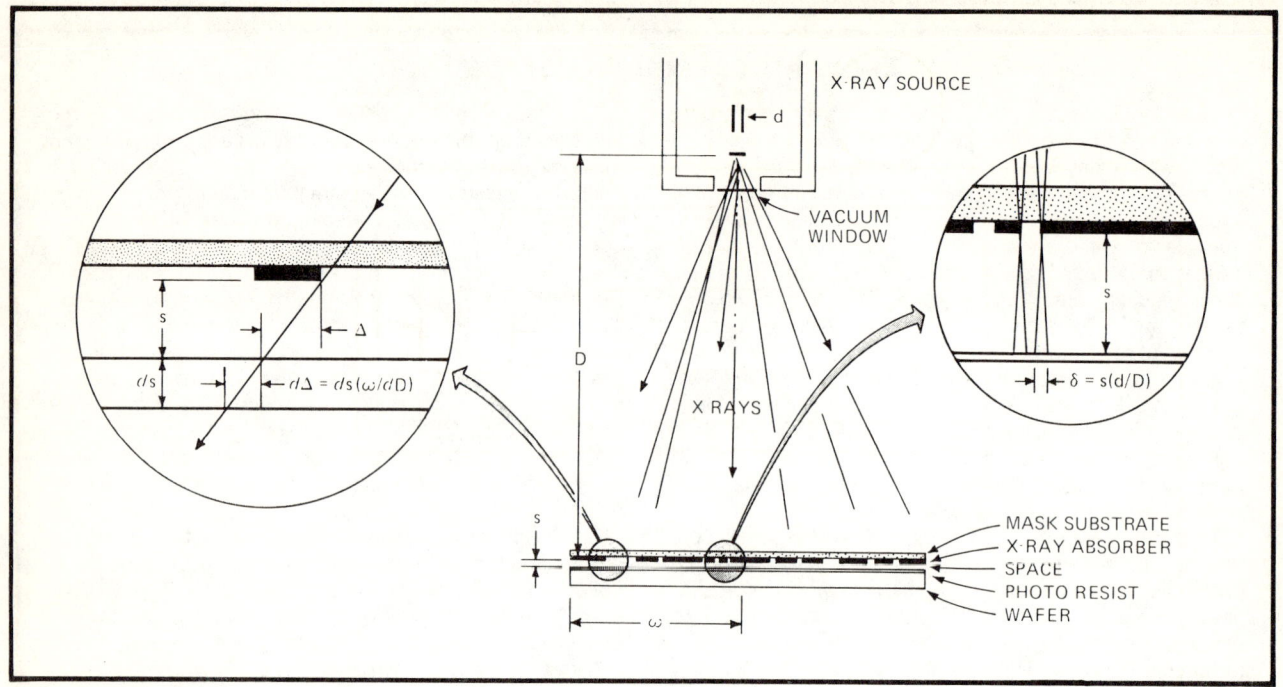

3. X-ray lithography. In X-ray lithography, the mask is kept out of contact with the wafer to avoid defects. But the mask-to-wafer gap gives rise to a geometric distortion that is variable from mask to mask (left, insert), as well as to penumbral distortion (right insert).

to cause a satisfactory exposure of the resist.

A further problem is created by the byproduct of the inelastic collision of X rays with matter: photoelectrons. These can have energies as great as the incident X rays but unlike them are not directional, so they can blur a mask pattern. This happens to the extent of the range of the photoelectron, which generally varies with the square of its energy and with the medium. Therefore, with hard X rays, photoelectron energy can significantly degrade pattern resolution—exactly the problem that causes the so-called proximity effect in electron-beam lithography.

Soft X rays have the opposite kind of problem, being much too easily absorbed by a vacuum window and X-ray masks. This cuts the flux output to the point that vacuum windows are not considered feasible—although those X rays that do get through to the photoresist will have a greater chance of getting absorbed in it and thus the exposures will require less incident flux. Even this creates a problem, however, since soft X rays tend to be absorbed in only the top layer of a photoresist.

Therefore most present-day systems are designed around a compromise, medium-energy X ray. Though systems have been designed around hard X rays from targets made of palladium ($\lambda = 4.3$ Å) and soft X rays from targets of carbon ($\lambda = 44$ Å), the aluminum X ray ($\lambda = 8.34$ Å) is generally favored today.

The maximum range of photoelectrons from the aluminum X ray is 0.1 μm, so that resolution is good. Typical systems with the aluminum source will absorb 70% of the original X-ray flux in the vacuum window and mask and allow 30% to be transmitted to the wafer. Most resists will then absorb about 5% of the X rays incident on them.

Once the X-ray wavelength is selected, a commercially feasible source with sufficient X-ray brightness must also be selected. A number of possibilities have been suggested and tested. These range from electron storage rings to the traditional electron-beam-generated source.

The electron-beam-generated X ray is created as an energetic electron knocks a tightly bound electron out of the target material it strikes. This vacancy is then filled by another bound electron falling into the vacancy and giving off an X ray whose energy is dependent on the difference between the two electrons' energy states. The energy of this characteristic X ray is also dependent on the material with which the electron is interacting.

A side effect is the so-called bremsstrahlung radiation coming from the electron-electron interaction when an incident electron is simply deflected, thus being accelerated and giving off electromagnetic radiation. Figure 5 shows a typical intensity versus energy curve of the characteristic peak superimposed upon the background bremsstrahlung when the incident electron has 20-kiloelectronvolt energy. Although the absolute intensity of the Al peak is two orders of magnitude greater than the bremsstrahlung, the bremsstrahlung still contributes about 10% of the power absorbed by the photoresist.

To obtain a useful source from electron-beam-generated X rays, the beam must be focused to a small spot to limit penumbra distortion. The flux of the beam must also be high to make exposure times short. The trouble with this procedure is that the system acts like an evaporator melting the aluminum target.

For the source not to evaporate, therefore, it must be water-cooled. But water-cooling it prevents it from handling more than 1.5 kilowatts of electron-beam power in a 6-mm-diameter spot. To handle more power, the source must also be continuously rotated out from under the hot spot to allow it to cool down. Typical systems with 10-cm-diameter rotating aluminum anodes can handle 15 kw. If a silicon-source X ray ($\lambda = 7$ Å) is used, though, the same rotating anode can handle 25 kw

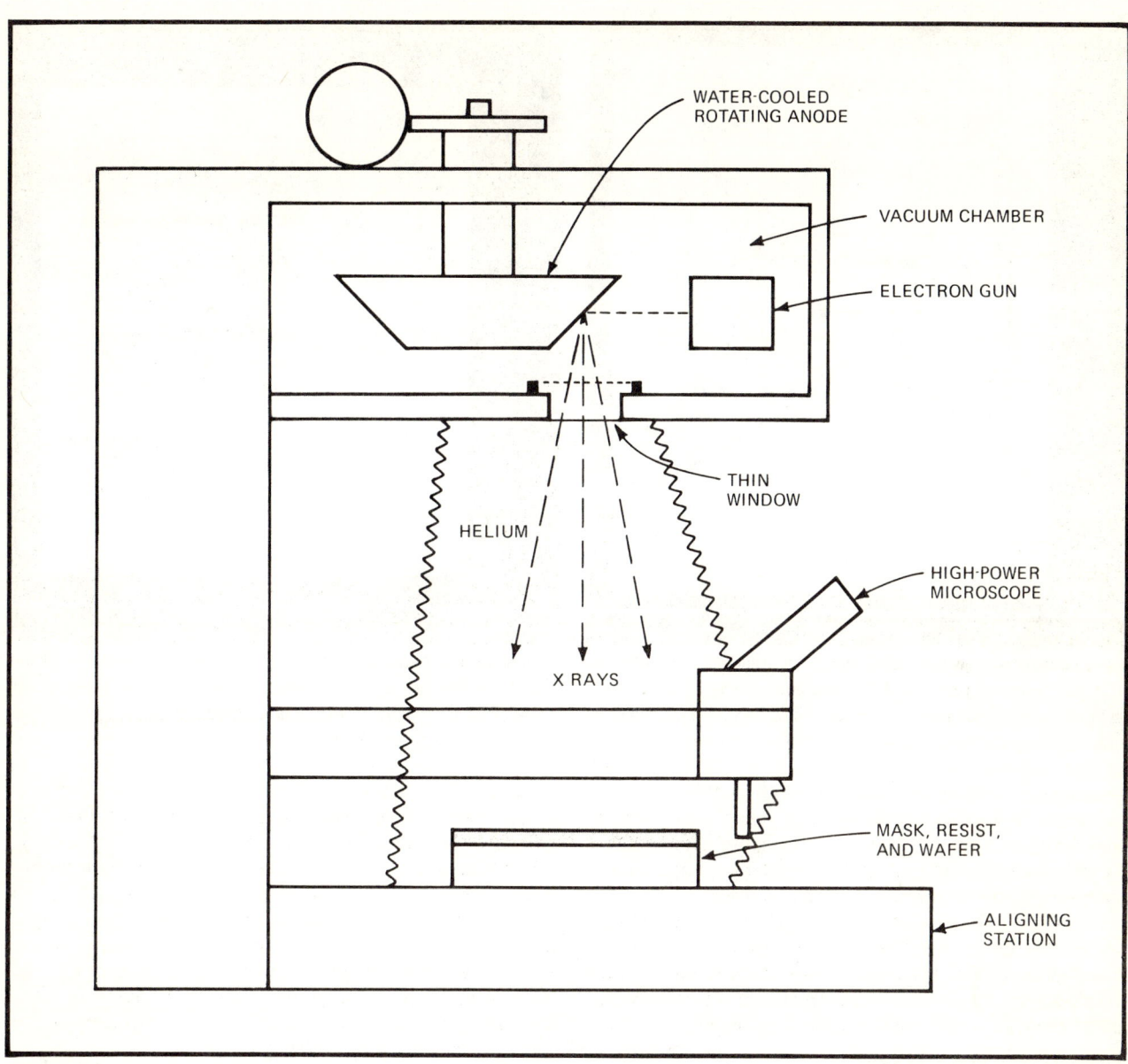

4. High power. Block diagram of a high-power X-ray lithography station shows the source mounted over the wafer-to-mask aligning mechanism. Optics for aligning are on a sliding mount so that they can be moved out of the way during exposures.

TABLE 1: COMPARISON OF PHOTOMASKING TECHNIQUES					
	Typical line width (μm)	Equipment cost ($\times$ 10^3)	Achievable defect level (per cm^2)	Throughput (wafers/hr)	Figure of merit ($\times 10^6$)
Contact	3	30	2.5	50	14
Projection	2	185	1	65	29
Ultraviolet projection	1	200	1	50	82
Step-and-repeat projection	1	350	1	20	19
Electron-beam	0.5	1,500	0.5	10	15
X-ray	0.3	300	1	20	218

SOURCE: DATAQUEST

TABLE 2: CHARACTERISTICS OF X-RAY RESIST MATERIALS			
Resist	Type	Sensitivity (mJ/cm^2)	Resolution (μm)
PMMA	positive	1,000	0.1
PBS	positive (electron-beam)	100	0.5
FPM	positive (electron-beam)	100	0.5
PGMA-EA	negative (electron-beam)	50	0.5
DCPA	negative	10*	0.5*
TI XR79	negative	1.5	0.5
Hunt experimental	negative	8	0.5

*when exposed to palladium radiation

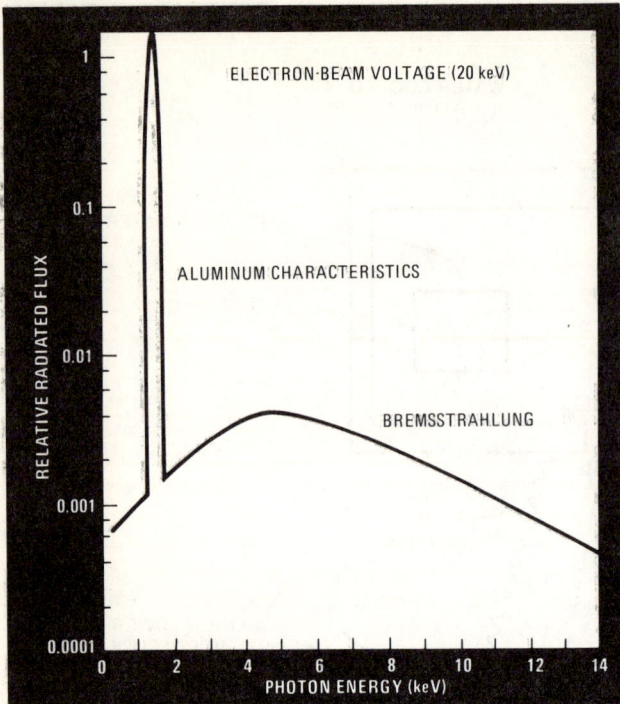
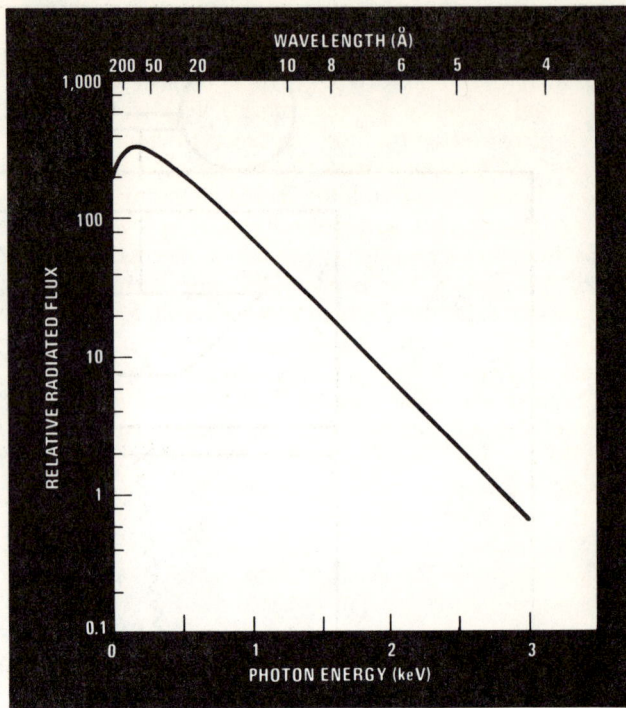

5. Electron-beam–generated X rays. An aluminum anode struck by a 20-keV electron gives off the characteristic radiation at 1.5 keV (8.34 Å) plus a background bremsstrahlung radiation. The bremsstrahlung contributes about 10% to the photoresist exposure.

6. Synchrotron radiation. The curve represents the unfiltered radiation from a 0.7-GeV electron storage ring with a 100-mA beam and 12-kilo-oersted magnetic field at a distance of 7 meters from the source. Note the peak intensity is in the very soft X-ray range.

because the melting point of silicon is higher.

Even with 15 kw of input electron-beam power, exposure time with some resists can be a problem. With a typical system, using a 15-kw rotating anode and a source-to-wafer distance of 20 cm, exposure will take 2 minutes with the fastest commercial photoresist PGMA-EA—Bell's electron-beam negative resist. It should be noted that there are other proprietary resists under development by Texas Instruments and Hunt/General Instrument that improve this time significantly.

The electron storage ring

Synchrotron radiation has often been suggested as the ideal X-ray source. The synchrotron continuously accelerates a pulse of very energetic electrons around a circular path, creating radiation of a spectrum that is fairly broad, depending on the beam's energy and radius of curvature. Figure 6 plots the intensity versus energy of the X-ray radiation typically given off by a synchrotron. The integrated energy output of a synchrotron is mostly at the soft end of the spectrum, but is several orders of magnitude greater than that of the most powerful rotating anode.

Synchrotron radiation also has the advantage of being very well collimated, since the electrons are traveling at a relativistic speed. When viewed from the laboratory frame of reference, the radiation from the accelerated electrons looks like a narrow cone emitted in the direction of the electrons' instantaneous velocity. In fact, the beam could be compared to a searchlight sweeping around a circle. However, though the extreme collimation of the beam means there are virtually no penumbral or geometric distortions, it does cause a problem when a large area like a wafer must be exposed. The beam can expose only 0.5-cm-high strips of wafer, and thus the exposure must be done in a scanning mode, strip by strip.

With such an intense source of radiation, one would expect the exposure times to be very short. However, the exposure time is highly dependent on the system design. Using no beryllium vacuum window and a 4-μm Mylar X-ray mask, the soft radiation is very effective. On the other hand, when a beryllium vacuum window is used, the soft radiation is not as effective and the exposure times are increased by a factor of five.

However, the cost of a synchrotron facility—$2 million after initial design—has by and large deterred industry from working with it as a production X-ray source rather than in its present research and development applications.

X-ray masks

Many different approaches have been taken in fabricating X-ray lithography masks. The primary problem is producing a thin substrate that is transparent to X rays. This generally requires a membrane that is 1 μm to 12 μm thick, for soft and hard X rays respectively. For aluminum X rays, a membrane thinner than 5 μm is typically desirable.

Two basic types of membranes have been tried—organic and inorganic films. Organic materials include Mylar, Kapton, Pyrolene, and polyimide. Inorganic types include silicon, silicon oxide, silicon metals, aluminum oxide, and silicon carbide. In general, inorganic membranes are very difficult to fabricate and very fragile (although SiC is reported to be very durable), whereas organic films are durable and easy to make but sometimes have an undesirably rough surface.

A primary question with both types of substrates is the dimensional stability of the masks with respect to temperature, stretching, and exposure to humidity. If superposition from one mask to another is to be within 0.1 μm, then the mask stability has to be better than 0.1 μm over 10 cm, or 1 part per million. (For the sake of comparison, low-expansion glass has a thermal expansion of 3.7 ppm/°C.) It is generally believed (although not proven) that the inorganic masks will be dimensionally stable enough, despite suffering from thermal expansion. Thick Kapton masks used with hard X rays have been shown to be stable to better than 1 ppm; and the thin polyimide masks, which are currently still under investigation, are at least as good as 10 ppm.

X-ray masks are similar to standard IC masks, except that the materials and orders are changed. In general, the thin film is either stretched across a frame, as is the case with many of the organic masks, or deposited on a flat substrate. At this point, either the substrate can be etched away to leave a blank mask substrate, or the X-ray absorber pattern can be created on the substrate, which is then etched away.

The X-ray absorber pattern can be created in a number of ways. Typically, it consists of two metal layers, first a thin layer of chromium for adhesion to the substrate and then a layer of gold. The thickness of the gold layer depends on the type of X rays and the contrast of the photoresist. For aluminum X rays, 5,000 Å of gold is typically used; for a softer X ray like copper, only 2,000 Å of gold is required, but for a hard X ray like palladium, 7,000 Å of gold will provide the same absorption. Patterns are created in the gold by one of four processes; ion-beam etching, electroplating, sputter etching, or liftoff.

X-ray resists

The field of X-ray resists is very new. When researchers started to work with X rays as an exposure source, they used a resist of polymethyl methacrylate (PMMA) that had been popular with electron-beam patterning. This was not a bad beginning. The basic mechanism for X-ray exposure is to absorb X rays by exciting electrons, which then have the energy of the incident X rays. Thus, a resist that is sensitive to electron radiation should be sensitive to X rays.

Since then, a number of good electron-beam resists have been developed for mask making, including Bell Laboratories' polybutene sulfone (PBS), which is a positive resist, and poly(glycidyl methacrylate)-co-ethylacrylate (PGMA-EA), which is a negative resist. These resists are relatively sensitive to X rays, but they are marginal from a direct wafer-processing point of view. Only recently have researchers begun to address the problem of producing a good X-ray resist.

The first successful X-ray resist, poly(2,3-dichloro-1-propylacrylate) was designed for harder X rays like palladium. The chlorine in DCPA increases its absorption of hard X rays. Most electron resists absorb 2% of the palladium X rays, but 1 μm of DCPA absorbs 12% and is 10 times more sensitive than PGMA-EA. Recently, two other X-ray resists have also been fabricated successfully

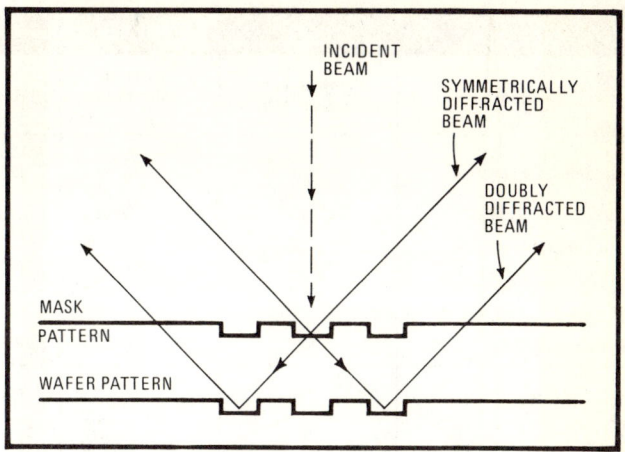

7. Interferometric alignment. In the diagram shown, the mask grating splits the incident beam into a transmitted beam incident upon the wafer and a diffracted beam that goes to the detector. The wafer grating then diffracts the incident beam to the detector, where it interferes with the beam diffracted from the mask.

for use specifically with the aluminum X ray.

Table 2 summarizes the characteristics of the various resists that have been used in X-ray lithography. Note that, with the new X-ray resists produced by Hunt/GI and Texas Instruments, exposure time for a 15-kw rotating anode system is well under 1 minute even when the distance from source to wafer is 30 cm. This will allow throughputs of better than 60 wafers per hour, a rate that is comparable to those of current UV contact and projection systems.

X-ray alignment

If X-ray lithography is to be used to manufacture a device requiring more than one precision mask, then precision alignment systems must be considered. Five approaches have basically been used in multiple-mask alignment systems. They are optical manual alignment, photoelectric reflection and transmission methods, the use of moiré patterns, and the interferometric technique.

The optical technique of aligning an image on the wafer with another on the mask works well, except when the magnification of the microscope system needs to be increased. This increase reduces the depth of field, making it hard to align a mask and wafer separated by a finite gap.

The photoelectric reflection technique is usually more accurate because it uses a photoelectric edge detector that has a greater depth of field and lets the system be automated. The detector and alignment system adjust the wafer so that the alignment marks on wafer and mask coincide. The limiting factor is the contrast of the image on the wafer. The photoelectric transmission method corrects this limitation by using a hole that has been etched into the wafer as the alignment mark. It has extreme contrast and has been demonstrated to be accurate to within 0.05 μm, or 500 Å.

The moiré pattern has often been suggested as an accurate method of alignment. To automate such a system, however, computer pattern recognition is needed. An outgrowth of this method is the simpler interferometric alignment system developed at Massa-

27

TABLE 3: EXISTING X-RAY SYSTEMS

Builder	Electron-beam power (kW)	Source-to-wafer drain (cm)	Resist	Throughput (chips/hr)	Devices reported produced	Source
Bell Laboratories	4.5	50	DCPA	15	charge-coupled devices	palladium
General Instrument Corp.	1.0	20	PGMA	2	MOS EE-PROM	aluminum
Hughes Research Laboratories	10	24	PGMA	5	C-MOS on sapphire	aluminum
IBM Corp.	2.5	10	PMMA	1	bubble devices	aluminum
Nippon Telegraph and Telephone	25	50	FPM	4		silicon (7.1 Å)
Sperry Univac	7	16	PBS	12	64-kilobit bubble circuit	aluminum
Texas Instruments Inc.	18	43	XR79	70	bubble memories	aluminum

chusetts Institute of Technology's Lincoln Laboratories. The basic idea here is that a monochromatic light beam can be diffracted into symmetric beams by a grating. If the mask and the wafer both have a grating, then two sets of diffracted beams occur and will interfere with each other. The interference is dependent on the phase relationship between the two patterns and causes an intensity difference in the symmetric beams except when the patterns are aligned with each other. This method has been demonstrated to be accurate to within 1% of the gratings' line width and has been used to superimpose one pattern on top of another to within better than 0.02 μm, or 200 Å (Fig. 7).

Existing X-ray systems

Although X-ray lithography alignment systems are not commerically available, many companies have built their own. All of these systems use the basic electron-beam–generated X-ray source but have different targets and power levels. Some are used for single-mask devices like bubble memories and thus do not require good alignment. The other systems have been used in a wide range of processes, from prototype fabrication of standard devices and to production of small test devices with line widths ranging from 1 to 2 μm.

A system at Bell Laboratories is at the pilot-line stage. With DCPA resist and only a 4.5-kw source, throughputs on the order of 15 per hour are reported. Bell Labs is producing a device of standard size at present to prove that X-ray lithography is at least competitive with conventional optical lithography.

A summary of the X-ray systems that have been described in the literature is given in Table 3. Most of these systems have high flux generated by rotating anodes.

The three main fine-line lithography processes that seem viable for very large-scale integration in the immediate future are electron-beam, direct-step-on-wafer, and X ray. Each of these processes is a radical change for today's typical wafer fabrication facility and thus will undergo a substantial learning curve.

The direct-step-on-wafer process will provide for small geometries of approximately 1 μm in pilot-line production. Because it allows for traditional geometries in the reticle, it will be particularly useful for manufacturers who do not have electron-beam masking equipment. The system is commercially available today and thus will allow development of processes that are compatible with 1-μm geometries.

Electron-beam lithography has already established a market in mask making. It is also starting to be used to write patterns directly on wafers—a process that will always maintain a market of its own, since it gives quick turnaround on prototype devices. The defect density of the method will probably always be the lowest of all three. This makes it well suited for optical devices where large areas of zero defects are required. It will also dominate the market in low-volume high-cost parts where mass production is not required.

X-ray lithography is now in the process of demonstrating itself as economically and technically viable not only for single-level mask devices but also for multimask devices of all types. Although X-ray lithography equipment is not commercially available today, it is nearly at a point of being so. Companies are already working on fabricating commercial equipment and commercial X-ray photoresists; interest has also been expressed in making commercial X-ray masks.

X-ray lithography will require costly masks made by electron-beam patterning, so that it will not have the fast turnaround time needed for prototype device development. But, once a working device pattern is generated, X-ray lithography will be able to replicate the pattern at low cost and high throughput. Thus, the X-ray approach produces the one kind of system that keeps the cost of lithography down, thus allowing for the high-volume, low-cost production of complex VLSI circuits with very small geometries. □

E-beam projector takes giant steps towards practicality

Well on its way in the long trek from concept to feasibility is an electron-beam image projector that promises to achieve the 1-micrometer line widths, 0.1-μm wafer alignments, and 0.2-μm mask alignments necessary for very large-scale integration. What's more, the still experimental technique will yield these line widths in production, says Peter Daniel, head of the development group at Philips' British Research Laboratory, Redhill, Surrey.

The Philips researchers have overcome several of the stumbling blocks that have prevented electron-beam projection from becoming as strong a candidate to replace optical lithography as are electron-beam pattern generation and X-ray lithography.

Achievements. What the development group has done is to perfect an X-ray method of aligning mask and wafer and to refine the magnetic focusing system to reduce image distortion radically. Also, a high-speed photocathode allows operation in a less-than-perfect vacuum.

Also at work on the technique is Electron Beam Microfabrication Corp., San Diego, Calif. President William Livesay says the company, which owns the rights to the initial electron-beam projector work of Westinghouse Electric Corp., has a U.S. Navy contract for a machine that will produce 0.5-μm line widths with 0.2-μm alignment accuracy at an output of 60 wafers per hour. Delivery may come in late 1979.

Philips has reached the point of turning out working circuits: simple shift registers with 1-to-2-μm gate windows and bubble-memory circuits.

In operation, a chrome-on-quartz mask 4 inches in diameter is coated with photoemissive caesium iodide and loaded into the imager vacuum chamber (see diagram). Illumination from an intense ultraviolet lamp causes electron emission from the unmetalized mask areas. The electrons are accelerated onto the slice by a 20-kilovolt electrical field and focused by a parallel magnetic field. Complete exposure can take as little as 30 seconds.

For alignment, current in the deflection coils is adjusted to shift the image, thus aligning mask and wafer grid markers. An ac signal, imposed on the deflection coil's current, causes the image to oscillate across the wafer. Because the wafer's grid marker is made of tantalum, this oscillation yields a modulated X-ray output signal proportional to the misalignment. Phase-sensitive photodetectors with the same grid patterns as mask and wafer pick up the signal and apply it to an analog servocorrection system.

Philips developed a focusing air-core magnet with a field that is uniform to within a few parts in 10^5 over the entire wafer surface. Such an accuracy keeps image distortion to less than 0.1 μm.

Also significant is the switch to caesium iodide, which is relatively insensitive to vacuum and deposition conditions, says Daniel. Thus it is possible to shorten the pump-down time since full vacuum is not necessary. Also, the high emitted electron energy gives higher resolutions.

Flatter wafers. Philips is attacking another important problem: wafer bowing, which can distort the image seriously in VLSI patterns. It has come up with an electrostatic chuck, which flattens the wafers.

Problems remaining to be solved include reduction of image contrast because of back-scattered electrons returning to the wafer surface and correction for proximity effects. However, the firm hopes for routine use in production within five years. □

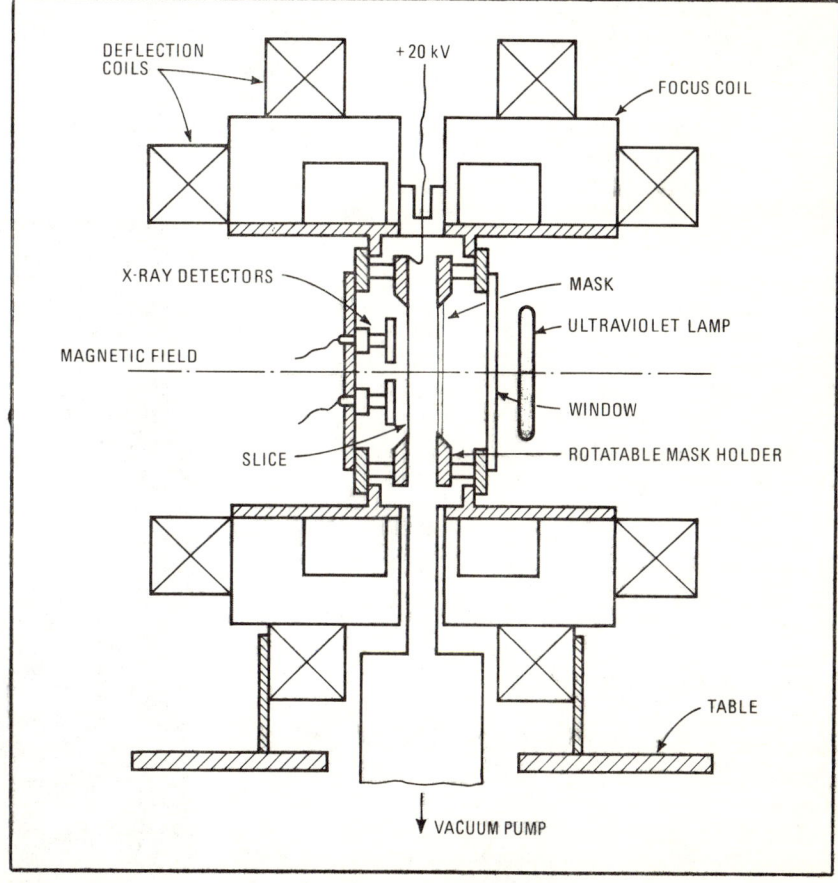

Finer lines. Electron-beam image projector is moving into the race for VLSI lithography equipment, now that Philips has solved such key problems as alignment and focusing.

Lithography chases the incredible shrinking line

With better optical, electron-beam, and X-ray systems on the way, mass production of chips with submicrometer line widths will soon be here

by Jerry Lyman, *Packaging & Production Editor*

☐ In laboratories and design centers in Silicon Valley, Japan Inc., and Europe, integrated-circuit designers are busy shrinking line widths and lengths on current LSI designs while creating incredibly finely detailed, dense semiconductor structures for the coming generation of very large-scale integrated circuits. The aim is to produce chips containing over 100,000 gates each—a shrinking of IC patterns that is already bringing increasing switching speeds and reducing chip power consumption. Already a few ICs that need the VLSI definition have appeared—mainly 64-K random-access memories made with metal-oxide-semiconductor technology—and more are in the development stage.

But if VLSI is ever to go into full-scale production, a whole group of major breakthroughs in materials, lithography, processing, and circuit design and testing must occur. This report will concentrate on the lithography aspect of VLSI manufacturing, since VLSI is not possible without the ability to make clean and accurate exposures on resist-covered silicon wafers of circuit patterns with minimum line widths of 2 micrometers or less.

A wide choice

Today the IC processing engineer has a wide range of lithographic methods to choose from. They include contact, proximity, 1:1 projection, step-and-repeat reduction projection, X-ray, and electron-beam systems. Table 1, which purposely omits proximity lithography, summarizes each technique's parameters.

At the present time, contact and proximity printing are still in heavy use, but they will undoubtedly be gradually phased out because of their poor yield. Projection printing, mainly 1:1, is currently the most popular. Scanning electron-beam lithography is at present used mainly to make wafer masks and reticles—in effect, chip masks—for projection printers; in the future it will be used for writing circuit patterns directly onto resist-covered wafers. X-ray lithography has potential as a high-resolution, high-throughput process but still has a number of many practical problems to be ironed out.

Contact and proximity printing are the oldest types of lithographic systems available. In contact printing, a glass mask bearing an emulsion or chromium-film pattern is first aligned with reference points on a resist-coated wafer, then pressed directly onto the wafer and exposed to ultraviolet light with a wavelength of 3,600 to 4,600 angstroms. In proximity, or soft contact, printing, a gap of several micrometers divides mask from wafer.

The resolution of contact printing is limited only by the wavelength of the light used, so that 1-μm-wide lines can easily be printed with this technique, making it suitable for LSI and VLSI chips. Proximity printing has a slightly poorer resolution that varies with the gap size. However, the defect density of chips made with both these techniques is extremely high due to the mask damage and wafer contamination caused by contact between wafer and mask.

In spite of these deficiencies, contact aligners are still

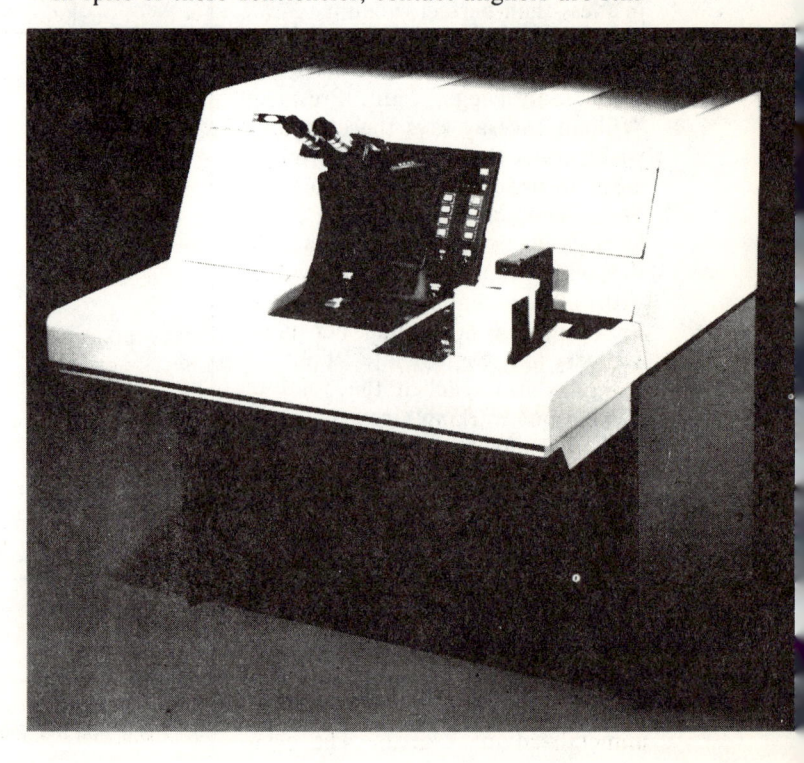

1. One-on-one. Most production projection lithography is done on Perkin-Elmer Micralign machines capable of resolving 3-μm-wide lines. With a deep-UV source, it is theoretically possible to achieve 1.5-μm resolution with this type of machine.

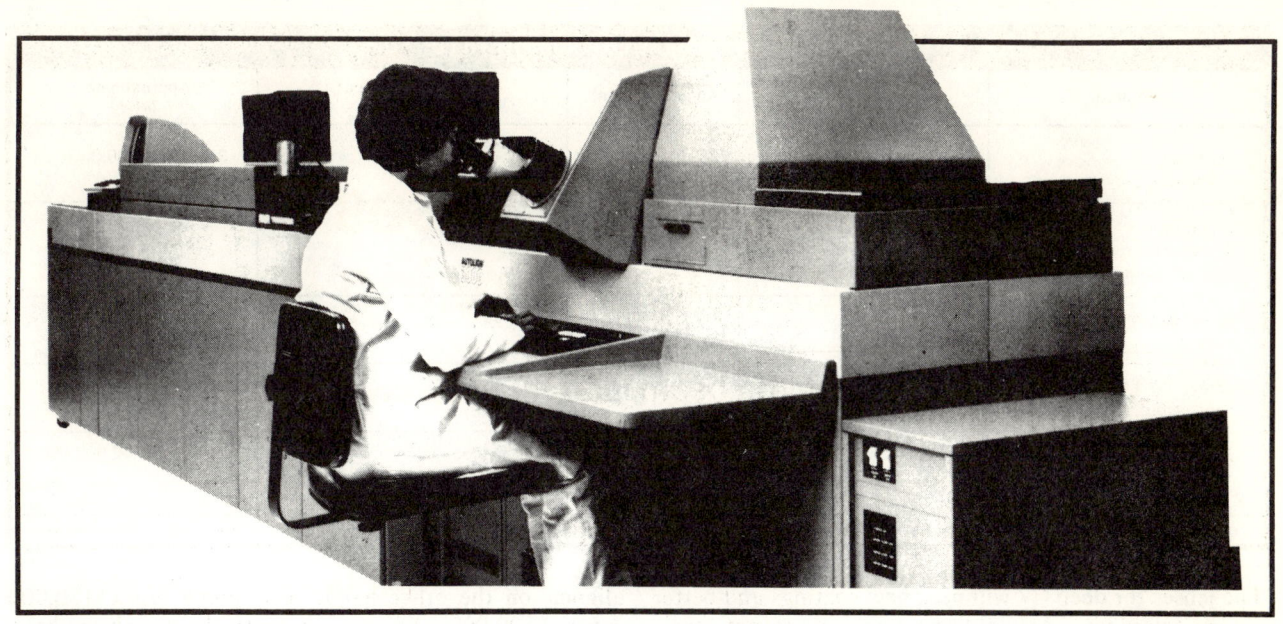

2. In line. A strong competitor to the Micralign is Cobilt's CA 3000, shown interfaced with an automatic in-line wafer-processing system. The CA 3000 features automatic mask-to-wafer alignment and can accommodate up to 92% of a 5-inch wafer.

in heavy demand, for their good resolution and low price ($50,000 to $100,000) make them suitable for custom LSI or research and development applications. For instance, Computervision Corp.'s Cobilt division in Santa Clara, Calif., has supplied many of its $42,000 CA 2020 contact aligners to companies engaged in magnetic-bubble work. Only one mask is normally required for this type of device, and its 1-μm details are exposed by the CA 2020 with an alignment accuracy of ±0.125 μm. The only other way of achieving this degree of accuracy would be by writing the bubble pattern directly on the chip with a scanning electron-beam machine or direct-step-on wafer projection system.

Canon Inc. in Japan has extended contact printing into the submicrometer range with a new deep-ultraviolet aligner, the PLA 520A. A combined contact and proximity aligner, it uses light in the 2,000-to-2,600-Å range, which has far less of a problem with diffraction effects than longer UV wavelengths. The PLA 520A can expose lines 0.5 μm wide in the contact mode or 2 μm wide in the proximity mode, with a gap of 20 μm between mask and wafer. This year's entire production of the $160,000 machine is already sold out.

So far as LSI and VLSI are concerned, however, contact/proximity lithography can never be a viable production method because of its high defect density—a problem neatly circumvented by projection systems.

Projecting an image

A sharp dividing line runs through the world of optical projection lithography. On one side of it are the well-established 1:1 reflective optical systems having resolutions in the 2- to 3-μm range, alignment accuracies of 0.5 to 1 μm, and throughputs of about 60 wafers per hour. On the other side are the reduction projection systems, or wafer steppers, with 1×, 5×, and 10× lenses. These machines have a better basic resolution (about 1.5 μm) and tighter alignment but lower throughput.

Perkin-Elmer Corp., Norwalk, Conn., delivered its first 1:1 optical projection printer, the Micralign, in 1973. Hundreds of these systems are now in use at almost every major IC manufacturer because they have much higher yields than contact aligners.

The Micralign has a complex reflective lens system that uses a mask to project an image the same size as the mask onto a wafer. Because it does not touch the wafer and therefore cannot damage it, the mask can be made of a hard material like chrome and thus achieve an extremely low defect density, thereby eliminating the chief drawback of the contact aligners. As for other specifications, the Perkin-Elmer machine exposes lines and spaces 2 μm wide with an alignment error of no more than ±1 μm—more than adequate for the 2- to 6-μm features of today's LSI.

A new 1:1 projector

The latest Perkin-Elmer aligner, the model 140 (Fig. 1), can expose 90% of the area of a wafer 100 millimeters, or 4 inches, in diameter. With an automatic wafer feeder, it can process about 60 wafers an hour—about half the throughput of a contact aligner.

Perkin-Elmer's main competition in this field is Cobilt's CA 3000 (Fig. 2), which first appeared in 1978. This system has a 3-μm average resolution and a ±0.5-μm alignment error, and it can expose 100% of a 4-inch wafer and 92% of a 5-inch wafer. Alignment of wafer with mask may be manual or automatic, and if automatic, will allow a throughput of up to 100 wafers per hour, says Ed Segal, Cobilt's marketing vice president.

Both the CA 3000 and the model 140 have a resolution of basically 3 μm. Perkin-Elmer's next goal is 2-μm resolution with an alignment error of ±0.25-μm, and a machine with these specifications will probably appear before the end of this year.

A deep-UV option—say a 2,400- to 3,000-Å source—could reduce the average resolution of this machine to 1.33 μm. However, before this UV dream machine is constructed, some major problems remain to be solved.

TABLE 1: SCANNING THE LITHOGRAPHY FIELD				
Method	Resolution (µm)	Alignment (µm)	Typical throughput (wafers/hr)	Approximate price
Scanning, 1:1 optical projection	2 (low volume) 2.5–3 (production)	± 1	60	$150,000–200,000
Scanning, 1:1 optical projection, deep ultraviolet	1–1.5 (not commercially available)	± 0.25	60	"about" $300,000
Direct-step-on-wafer, optical	1–2	± 0.25–0.5	10–60	$450,000–500,000
X-ray, contact/proximity	0.5–2	± 0.1–0.25	10–60	$150,000–200,000
Scanning electron beam, direct-write-on-wafer	0.2–1	± 0.05–0.1	1–2* (commercial machines) up to 22–30* (in-house types)	$1,000,000–2,000,000

*A function of resist, line width, wafer size, machine, etc.
Source: General Instrument Corp.

The lenses for deep UV will need new coatings and better optical tolerances. Available photoresists are not sensitive enough at these short wavelengths and require excessive exposure times. A light source has yet to be optimized for producing light at 2,400 Å. Finally, deep UV requires masks made not of glass but of fused silica, which is both fragile and expensive. These problems have deterred Cobilt from working on a deep-UV option for its projection aligner, though the company is developing a contact aligner with a deep-UV source.

The DSW alternative

For all their popularity, the 1:1 projection machines do have a serious limitation: if a wafer is distorted in processing, it will throw layer-to-layer registration of successive masks on the wafer out of specification, reducing the IC yield. In an effort to reduce this danger, a whole new class of projection aligners has sprung up in the world-wide centers of IC activity. They deal in smaller areas, aligning chip with mask, and aim at better than 3-µm resolution. Variously called reduction projection or direct-step-on-wafer (DSW) machines or wafer steppers, they all operate on the principle shown in Fig. 3—reduction of the size of the mask patterns.

A UV source is shone through the blown-up portion of an IC wafer pattern, commonly known as a reticle. The reticle's pattern is projected down through a reduction lens onto the surface of a resist-covered wafer. (The resist is a standard positive or negative UV type.) After exposure, the table is mechanically stepped to a new site for another exposure. This procedure is repeated until the reticle image is projected across the entire wafer surface. A high-resolution example of what is being done with a wafer stepper is shown in Fig. 4. This, of course, is the same step-and-repeat method that has been used in optical mask making for years.

The wafer stepper's main competition is the 1:1 projection machine, so a comparison of the two types is in order. The stepper boasts better resolution (1.5 µm versus 3 µm) and greater registration accuracy (it makes many alignment checks per wafer, not just one). It can handle much larger wafers, and reticles are much simpler to make than a 1:1 mask. The straight projection aligner, on the other hand, costs much less ($150,000 versus $500,000), has an inherently higher throughput (it makes one exposure instead of many), and is smaller.

Proponents of wafer stepping dispute the merits of high throughput. To quote Bill Tobey, marketing director of GCA/Burlington, Mass., division of GCA Corp.: "What is important is not wafer throughput per hour but good dice per month." Unfortunately, it is too soon to compare the two rivals on this score, although wafer stepper people expect a higher yield due to their system's potentially better layer-to-layer registration.

But no one disputes that the 1:1 systems are smaller. Most of the new reduction projection machines are quite large and a few approach an electron-beam lithography system in size and complexity. In any case, the potential advantages and sizable market for wafer steppers has brought forth many entries into this field.

The high steppers

A great many firms are convinced that the wafer stepper is the way to go for low micrometer resolution. In the U.S., GCA Corp. dominates with its 10× stepper, and Electromask and Optimetrix have entered similar $500,000 machines. All three offer minimum line widths of 1.5 µm and alignment accuracies of ±0.25 µm. Optimetrix and Ultratech also have 1× steppers with slightly higher resolution but lower prices. GCA Corp.'s Burlington, Mass., division already has orders for at least 50 of its $454,000 Mann 4800 DSW machines. At least 13 have been delivered. Prototype devices such as a bubble-memory with 1-µm details have already been exposed on a GCA stepper.

GCA's stepper has a 10× reduction lens (an optional 5× lens is also available) that can resolve usable line widths to 1.25 µm. Throughputs of 30 to 60 wafers per hour are possible with the 10× lens, and wafer sizes up to 5 inches in diameter can be accommodated.

The present system provides for one-time alignment of a wafer on the stage referenced to the system's optical axes. Mask-to-wafer alignment is through a television monitor with a push-button and joystick control. X-Y motion of the state is governed by a laser interferometer that controls the drive servos, the arrangement used in

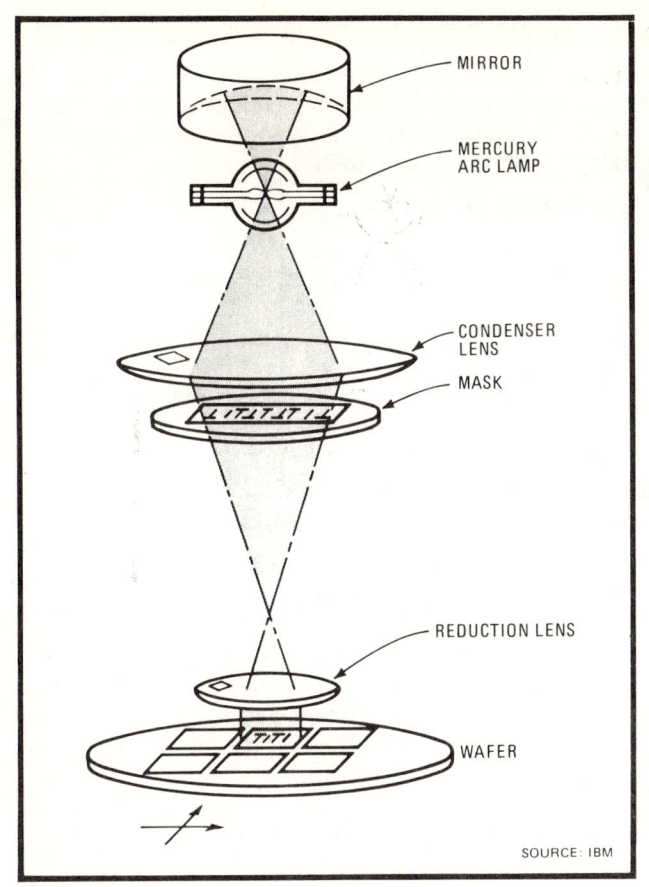

3. Stepdown. The step-and-repeat reduction projection system, shown in a simplified form, can expose 1-μm geometries. A reduced image of a 10×-2× mask is stepped across a resist-covered wafer. With this method it is possible to align at each exposure.

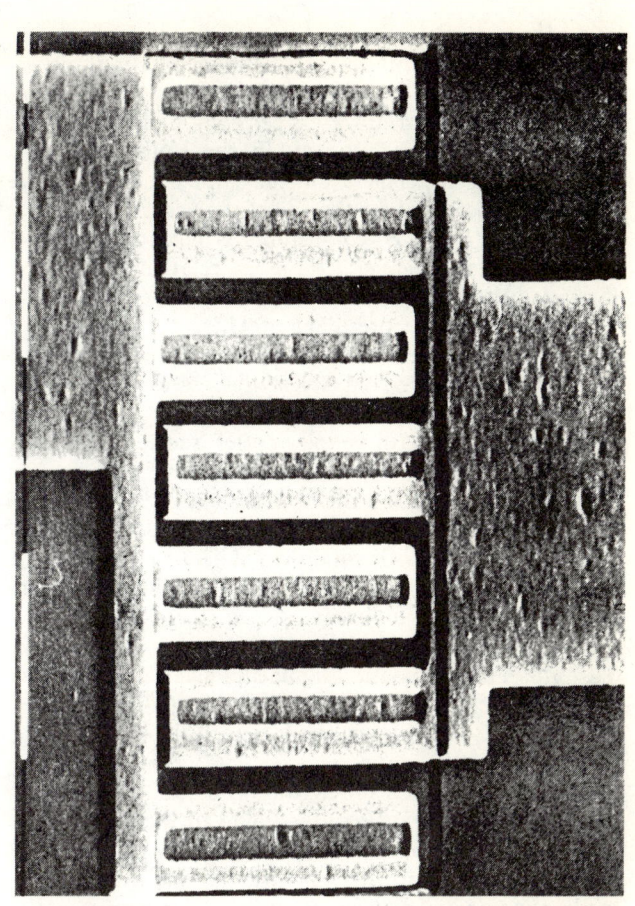

4. Chip shrink. This four-layer bipolar high-frequency transistor with 1-μm details is an example of the lithography possible with a direct-step-on-the-wafer projection. The transistor's wafer was exposed on a Philips Silicon Repeater SIRE-3, a system with 5× optics.

most scanning electron-beam lithography units. Alignment is accurate to within ±0.25 μm for 2 sigma values.

This off-axis alignment, shown in Fig. 5, contrasts with the through-the-lens alignment of other wafer steppers. In the future, GCA will have an automatic aligner option based on its off-the-axis technique. However, it is also evaluating the through-the-lens alignment method.

In the U. S., GCA's main competitors are three California companies—Electromask Inc. of Woodland Hills, Optimetrix Corp. of Mountain View, and Ultratech Corp. of Santa Clara. Electromask is the strongest, having already received orders for several of its $450,000 systems. It expects to ship its first 700 SLR system this month. The Electromask wafer stepper has a 10× lens system and a wafer stage controlled by a laser interferometer. Resolution is 1 to 1.25 μm depending on field size. Throughput is 20 to 45 3-inch wafers per hour, and 4- and 5-inch wafers can be handled.

The 700 SLR achieves alignment accuracy to ±0.25 μm with a through-the-lens manual alignment method, which aligns the reticle directly with the wafer. In contrast the off-axis method aligns the reticle with fixed targets in the top block of a closed-circuit TV camera, and the wafer with a separate set of targets and optics at the lower end of the camera (Fig. 5).

Frank Chase, Electromask's vice president of marketing, points out that by aligning through the lens there is only one uncontrollable variable—the lens distortion—whereas in the off-axis system there are many variables, such as the lens orthogonality and lens reduction ratio. This is why Electromask can claim a machine-to-machine compatibility of 1 μm. In addition, through-the-lens alignment permits every chip to be aligned if necessary to compensate for wafer distortion.

In the future, according to Chase, Electromask will add an automatic alignment capability to its stepper. Another possibility is a redesign of the lens and source to be suitable for a 3,650-Å UV line. This could push the 700 SLR to 0.5-μm resolution.

A 10× model

Optimetrix, which is a Cutler-Hammer affiliate, also has a 10× stepper capable of 1-μm resolution with an automatically aligned through-the-lens system. Alignment and layer-to-layer overlap are specified at less than ±0.2 μm. Throughput is 30 to 60 4-inch wafers per hour and a laser interferometer controls an air-bearing wafer stage. The first of the $400,000 to $500,000 machines will be delivered in the spring. Aside from excellent optical specifications, the new machine measures only 46 by 38 by 70 inches, which is much smaller than either the GCA or Electromask machines. Also, it supposedly can work out on the production floor and not just in a temperature-controlled environment.

A second machine in development at this firm is a 1:1 stepper having a resolution of 1.5 µm on a 38-millimeter-square field. Alignment accuracy is to ±1.5 µm and throughput is 60 to 100 wafers per hour. Delivery on this $300,000 to $400,000 unit is in the summer. A 7× system with a larger field but the same accuracy as the 10× system is slated for the end of the year.

Last of GCA's U. S. competitors is Ultratech's model 900 projection stepper. It is a relatively low-cost ($185,000), automatically aligned 1:1 stepper.

Martin Lee, vice president of Ultratech, says: "We started out to make a 10:1 stepper but found we needed too much hardware and space to create this type of system with the performance we wanted. Instead we went to a system with simpler optics and less temperature control. A smaller overall unit resulted."

This stepper uses a 3-inch-square 1× mask with four 10-by-10-mm images on it. This allows a user either to choose an image with minimum defect density out of four identical images or conceivably to print four different ICs from four different images. Resolution of the system is 2.5 µm and alignment accuracy is to within 0.25 µm. The automatic alignment system aligns every chip to the reticle. Throughput is about 40 4-inch wafers per hour. Resolution could be extended further, down to 1.5 µm, with a lens design.

The system is controlled by a HP 9825 calculator rather than the minicomputer found in the other U. S. steppers. Size of the aligner is 36 in. deep by 46 in. wide by 60 in. high. Ultratech at present has a feasibility model running and expects to accept orders in mid-1979.

Japanese and European manufacturers are not sitting idly by but have developed optical steppers to compete with their American counterparts. In Japan, Canon has two high-throughput 1:1 steppers, while Nikon has a 10:1 model that competes directly with GCA's 4810 DSW. In the Netherlands, Philips has an improved model of its 5:1 stepper, while Liechtenstein's Censor Inc. is developing a 10:1 stepper due out in 1980. All of these machines resolve features in the 1-to-2-µm range.

Overseas steppers

The two Japanese companies that have developed optical steppers did so in cooperation with the VLSI Cooperative Laboratory [*Electronics,* June 9, 1977, p. 104]. Both machines can operate in the ultraviolet only at 4,360 and 4,050 Å because their makers believe that it is not feasible to design well-corrected lenses for operation at deeper UV wavelengths. Canon Inc. has started to sell its unit, the EP211A, but Nippon Kogaku Ltd. (Nikon) is not in production yet.

The Canon aligner has unity ratio optics and projects a 30-mm-square image. It can expose a 4-inch wafer in seven shots. Minimum line width is 2 µm, alignment error is ±0.3 µm. Alignment is done on each step, exposure control is automatic, and a throughput of 46 4-inch wafers per hour is claimed.

Canon has also announced the FPA 112A. This model exposes a 22-mm-square image on each shot. Resolution of the narrow field lens is higher, permitting exposure of 1.5-µm lines. Other stepper characteristics are the same. Either unit costs 50 million yen, or about $250,000.

The Nikon aligner features 10:1 reduction optics for a resolution of 1 µm. It also can handle 4-in. wafers. Automatic focusing at each step maintains resolution despite wafer distortion. Image size is 10 mm square. The system is designed for either automatic alignment at the initial position or automatic alignment at each step;

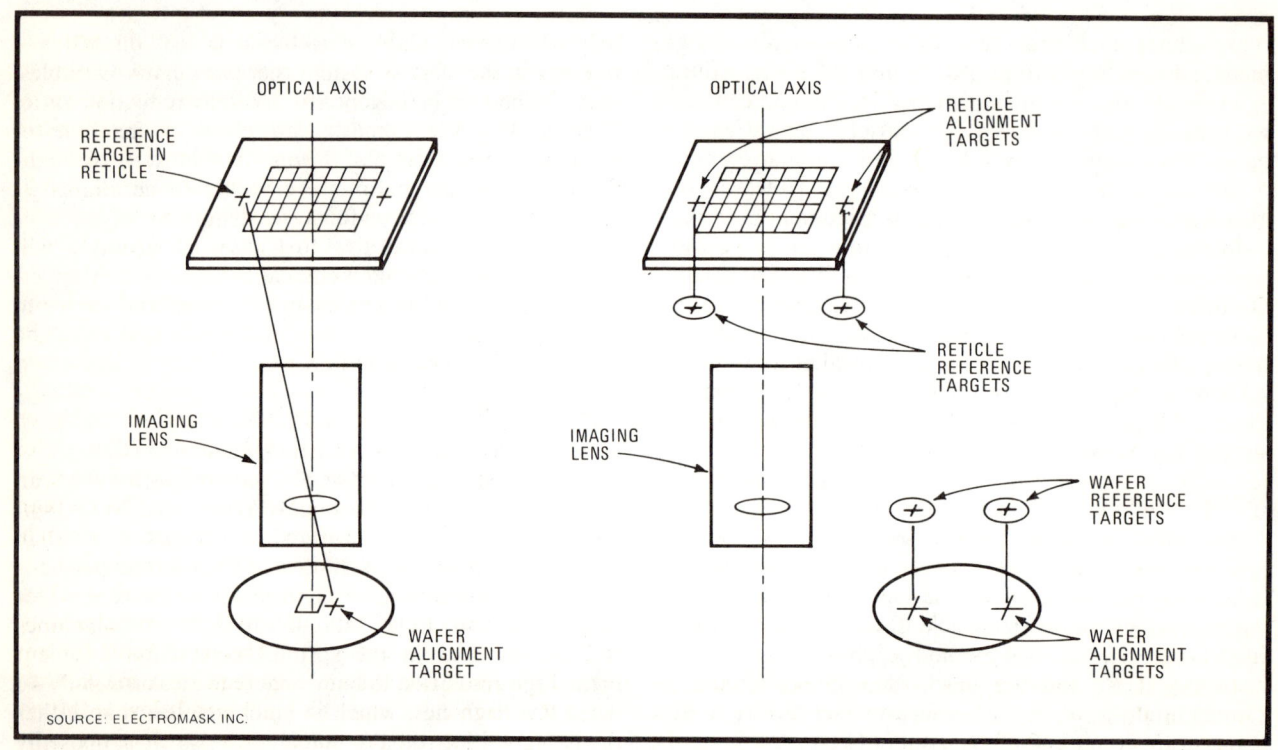

5. Alignment alternatives. Wafer steppers align wafer and reticle in one of two ways. Either the reticle is aligned with the wafer by eye directly through the lens (left) or alignment is referenced to off-axis targets. Most new steppers favor the first, easy-to-automate method.

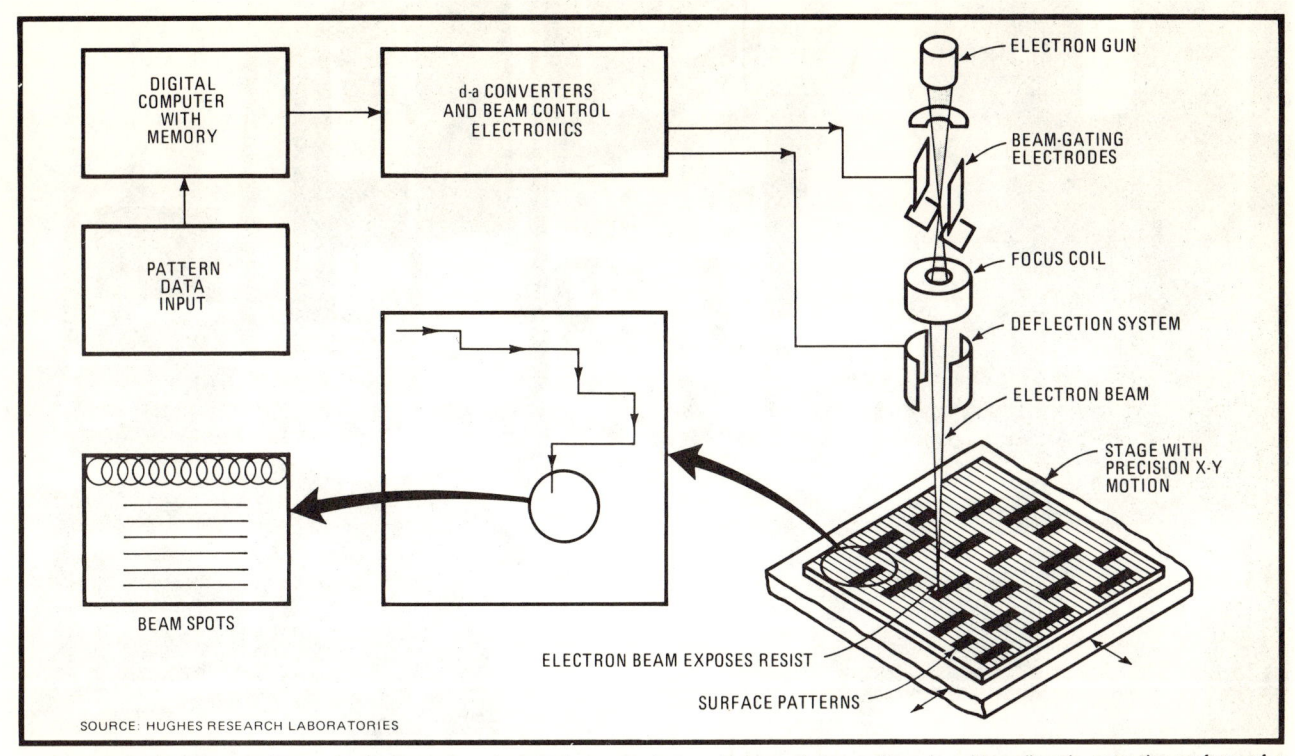

6. Scanning by electron beam. In the scanning electron-beam system, a high-resolution pattern is written directly onto the surface of a resist-covered wafer by a computer-controlled electron beam. This technique eliminates the delay and defects of optical masks.

either way, the time per step is less than 10 seconds.

Philips' unit is called SIRE 3—short for third-generation Silicon Repeater. It projects a 5:1 reduced image of a pattern directly onto a wafer, stepping the wafer after each exposure. The machine's automatic wafer-reticle alignment system ensures an overlap precision of ±0.25 µm. Minimum resolution attainable is 1.5 µm. Throughput is at least 50 4-inch-diameter wafers per hour. An unusual feature is the use of a helium-neon laser for the automatic alignment system; another is used for interferometric position sensing of the X and Y slider. Few steppers have a laser for autoalignment.

Delivery of the SIRE 3 is expected by the end of 1980. Estimated price is $500,000 to $600,000.

Censor's SRA-100 is a 10:1 reduction stepper featuring a resolution of 1 µm, autofocus, and autoalignment. Through-the-lens alignment at each step gives a registration accuracy of ±0.2 µm at 2-sigma probability. Throughput is 60 4-inch wafers per hour and unit cost is $450,000. The first SRA-100 will be assembled at Censor's Reno, Nev., facility in time to be demonstrated at Semicon West in 1980.

Going public?

Despite the large array of machines here and abroad, step-and-repeat imaging directly on the wafer is still mostly in the research and development phase at IC fabrication facilities. Often purchase of these systems is veiled in secrecy to mask out company plans.

In the U. S., however, the Motorola Semiconductor Group in Phoenix, Ariz., admits to purchasing a GCA direct-step system. Bill Howard, director of strategic planning for Motorola Semiconductor's IC division, says, "We consider step-and-repeat a strong contender for the mainstay of the early 1980s." But he emphasizes that the promise of direct-step equipment "precludes nothing else" and Motorola has an open mind. Direct-step registration down to 0.25 µm is possible, he says, but he expects other types of equipment to come along to go with direct steppers that will be more economical.

National Semiconductor Corp., Santa Clara, Calif., has been heavily into conventional projection printing with Perkin-Elmer Micralign systems. Now, however, this IC giant is evaluating two GCA steppers. It is also fairly certain that Fairchild Semiconductor, Mountain View, Calif., is evaluating direct stepping. Finally, at IBM Corp.'s Thomas J. Watson Research Center, Yorktown Heights, N. Y., scientists have built an experimental 5:1 direct stepper that has exposed 0.8-to-1.0-µm images on resist-covered wafers.

In Holland, Philips has built five versions of its SIRE 2 for in-house use. In France, the EFCIS (Société pour l'Etude et la Fabrication de Circuits Intégrés Spéciaux) program shared jointly by Thomson-CSF and the Commissariat à l'Energie Atomique (CEA) is banking on direct stepping for ICs with 2-to-3-µm design rules. CEA's research facility in Grenoble is centered around a GCA DSW 4800 machine with a 10× reduction. Next year, the machine will be modified to accept a French-designed 5× lens that should permit 1-µm resolution.

No matter which kind of projection system—1:1 or stepper—comes out on top, all masks and reticles must be made by electron beam. And the same holds for any other high-resolution system that requires a mask. Only when IC design rules move to 1 µm and below is the era of direct writing on the wafer with an electron beam likely to begin. In any case, scanning electron-beam lithography is vital to VLSI.

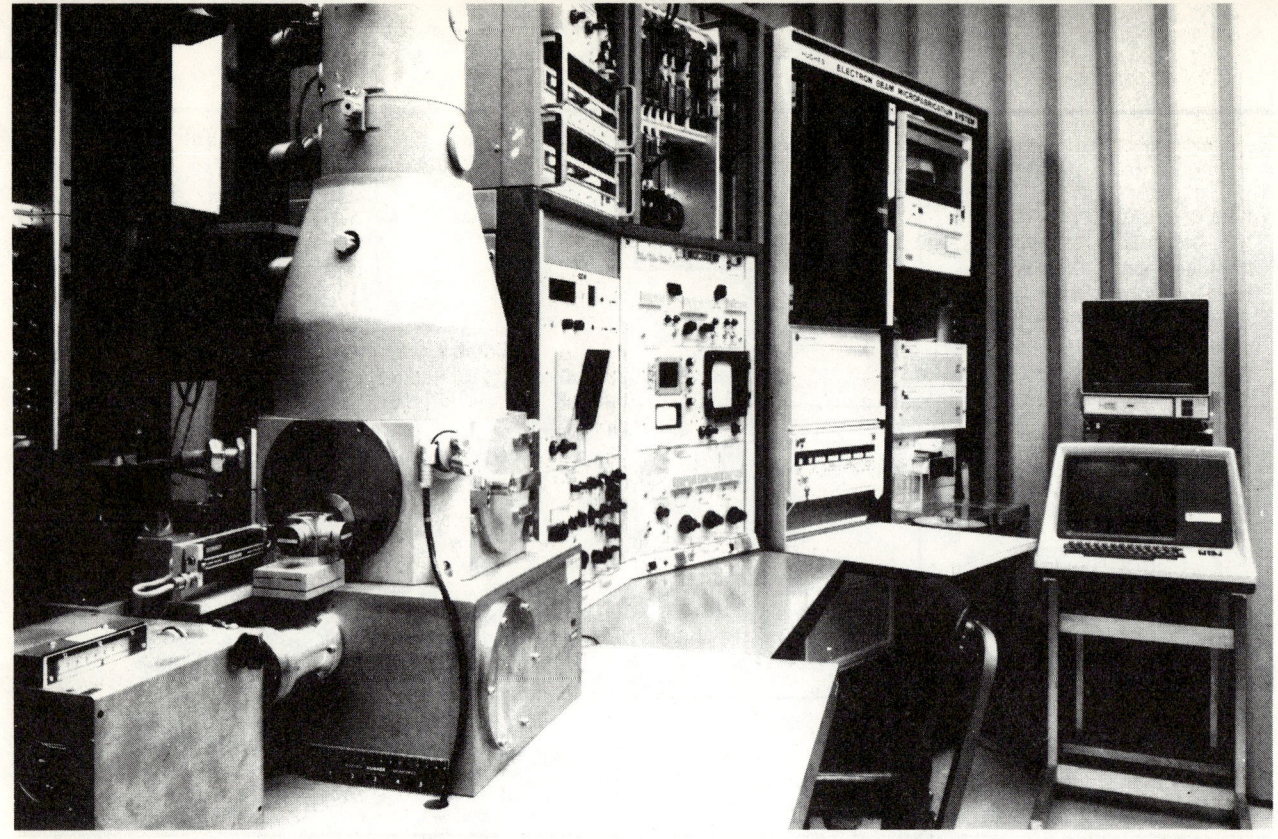

7. Home-brewed. Hughes Research Laboratories has built five of these vector-scanning electron-beam lithography machines. They are mainly used for writing circuit patterns in the submicrometer range directly on the wafer, for either discrete microwave devices or ICs.

Scanning electron-beam lithography is now a fact of life in major IC facilities worldwide. In this approach a computer-controlled electron beam scans an IC pattern with extremely high resolution and accuracy across a resist-covered silicon wafer or a resist-covered mask substrate (Fig. 6).

Electrons to the fore

Today, a wide range of commercial electron-beam machines exist, all in the $1 million to $2 million range, and more are in the works. Most machines sold are committed to making masks—sometimes around the clock—for use in projection and X-ray lithography, and already several IC firms in the U. S. are ordering their second unit for this purpose. The direct scanning of patterns on wafers, on the other hand, has been mainly confined, at least in the U. S., to firms that have developed their own electron-beam systems. Some of the leaders in this field include IBM, Texas Instruments, Bell Labs, and Hughes Research Labs, Malibu, Calif.

Finally, a different type of electron-beam lithography, one with both great potential and great technical problems, has surfaced at Philips' Redhill facility in England and Electron Beam Microfabrication Corp., San Diego, Calif.

The three greatest advantages of the scanning electron-beam system, in ascending order of importance, are its 0.2-to-1-μm resolution, its ability to align a pattern to within 0.05 μm, and its ability to correct for wafer distortion. Its greatest drawback is its throughput, which is invariably low. A complex function of the beam's characteristics, wafer and chip size, feature resolution, and resist, it ranges from 1 to 2 wafers per hour in commercial machines up to as high as 22 in IBM's EL-1 used in a production mode. Vector scanning, in which the beam scans selected areas of the chip, has a higher throughput than raster scanning, in which the beam moves across the entire surface, being switched on and off as the chip pattern requires.

Resolution is another problem. In electron-beam-fabricated structures, it is limited not by the electron optics, but by electron scattering in the resist. This is what sets the bottom limit of resolution to about 0.15 μm.

Finally, the initial cost of a scanning electron-beam machine is extremely high and there is a standing argument in the industry as to whether it can be cut. Undoubtedly, the improved scanners of the future will cost more than today's slower units.

Units for sale

Despite these disadvantages, electron-beam machines are selling at a rapid rate. Most sales in the U. S. have been made by three firms: ETEC Corp., Hayward, Calif., the Extrion division of Varian, Gloucester, Mass., and Cambridge Scientific Instruments Ltd., Melbourn, England, although GCA/Burlington and California's Electron Beam Microfabrication Corp. are developing high-throughput systems of the vector-scanning type to be available early 1980. The two commercial American machines are being used almost exclusively as mask makers. The British unit is a development tool capable of extremely fine submicrometer resolution when writing directly on the wafer, but the company has also sold electron-beam columns and scanning electron microscopes that have ended up or been converted into in-house electron-beam lithography systems.

The ETEC and Extrion machines are uprated licensed versions of Bell Laboratories' EBES [*Electronics*, May 12,

The electron-beam projector

One of the major obstacles preventing the scanning electron-beam-lithography direct-writing system from taking its place on IC fabrication lines is low throughput. A high-throughput alternative is the electron-beam projection system [*Electronics*, Nov. 23, 1978, p. 73], which two companies are in the process of developing. They are the Solid State Electronics division of Philips' Research Laboratories in Redhill, England, and Electron Beam Microfabrication Corp., San Diego, Calif.

This system is analogous to a 1:1 projection system, except that an electron-beam flood lamp replaces the ultraviolet-light source. To use it, a chrome-on-quartz mask is coated with photo-emissive caesium iodide and loaded into a vacuum chamber. Illumination from an intense ultraviolet lamp causes the unmetalized mask areas to emit electrons. These electrons are accelerated onto the wafer by an electric field of high intensity and focused there by a uniform magnetic field with unity magnification.

The method was first proposed by researchers at Westinghouse in the late 1960s. But it was put aside in favor of the scanning system because of problems with the mask and mask-to-wafer alignment and layer-to-layer registration. Now Philips has solved these problems and has a prototype electron image projection in operation at Redhill. Other systems are being constructed. Philips aims to have some pilot products made with its image projectors available within three years.

The wafers processed by the Philips system have special metallic guides on them that emit X rays. These are used by an automatic aligner to align mask to wafer and layer to layer to an accuracy of ±0.1 micrometer. Devices that actually have been fabricated in this way include logic circuits with 2-μm details and bubble memories with 0.6-μm geometries. Minimum resolution of the Philips' electron-beam projector is 0.3 μm. Complete exposure of a wafer take as little as 30 seconds (120 wafers/hr).

Electron Beam Microfabrication is working on a system for the Naval Ocean Systems Center. Its objectives for its system are a throughput of 60 4-in. wafers per hour with 0.5-μm resolution and ±0.2-μm alignment accuracy. Delivery is slated for 1980.

P. J. Daniel of Philips' Research Laboratories is bullish on the chances of the image projector against direct electron-beam writing systems. These, he claims, will retain a specialized niche in the low-volume custom VLSI market, but the larger markets will be served by either X-ray lithography or electron-beam imaging systems, since high-volume production will justify the cost of the specialized masks either technique needs.

1977, p. 95]. This is a raster-scanning unit with a continuously moving table, full wafer alignment, and a laser interferometer to feed the table position back to the electron beam's computer.

ETEC's MEBES is the electron-beam system most heavily sold in the U. S. Twenty-four of the systems have been ordered, about 12 of which have already been shipped to firms like Fairchild, Motorola, National Semiconductor, General Instrument, and RCA.

ETEC's system, aimed specifically at mask making, has a 1-μm resolution and an overlay accuracy with ±0.25 μm. Exposure time for a 4-inch mask is 40 minutes. The present unit can take up to nine 5-inch wafers and developmental work is in progress for even larger wafer sizes. A 10-cassette airlock substrate feeder is now available. System price is $1.6 million.

Wafer confrontation

ETEC also has a program for an advanced system that could write directly on the wafer. Objectives of the program are a throughput of 10 to 20 3- or 4-inch wafers per hour with improvements in line resolution and alignment. The new unit will be built around vector rather than raster scanning [*Electronics*, May 12, 1977, p. 94].

Extrion has recently improved the data rate and software of its version of the Bell EBES. The 20-megahertz data rate of its older EBMG-20 has been uprated to 40 MHz in the newer EeBES-40, enabling it to process a 3-inch wafer in 20 minutes. Otherwise, the two machines are identical. The EeBES-40 still has 1-μm resolution and ±0.125-μm alignment error, and it can accommodate up to a 6-inch wafer. A vacuum loading system for 10 cassettes is standard.

Varian has delivered one of these $1.5 million machines to Ultratech Corp., a maker of masks rather than chips, and recently sold another to Fairchild.

John Dougherty, Extrion's manager of electron-beam lithography, says that making reticles for direct-step-on-wafer machines is becoming just as important as 1:1 mask fabrication. Because of this Extrion has developed special software to go from the optical pattern-generator format to one acceptable to the EeBES-40. For extremely large, dense reticles, the electron beam is faster than optical pattern generation. For instance, a reticle for a large LSI chip that took 1 million flashes to expose at 25,000 flashes per hour would require about 40 hours to produce with an optical pattern generator. The same reticle could be turned out in less than an hour on the new Extrion system. Dougherty also points out that the uprated machine turns out reticles four to five times faster than its earlier version.

Cambridge Scientific Instruments is producing two vector-scanning electron-beam systems, EBMF-1 and -2. The EBMF-1 is a low-cost system having a 75-mm stage and with full computer control and full precision pattern writing. The EBMF-2 is Cambridge's main system with a laser-controlled 105-mm stage and temperature-controlled chamber and airlock; a fully operational system costs about $700,000.

As mentioned before, the EBMF system is a laboratory tool for directly writing high-resolution features onto resist-covered silicon (although it certainly can also be used to make a mask). Minimum line width is about 0.25 μm and alignment accuracy is specified at ±0.15 μm (2-sigma value). The EBMF-2 can handle 4-inch wafers or 5-inch mask plates. Since it is intended for use in research and development, throughput is less important than resolution and layer-to-layer registration.

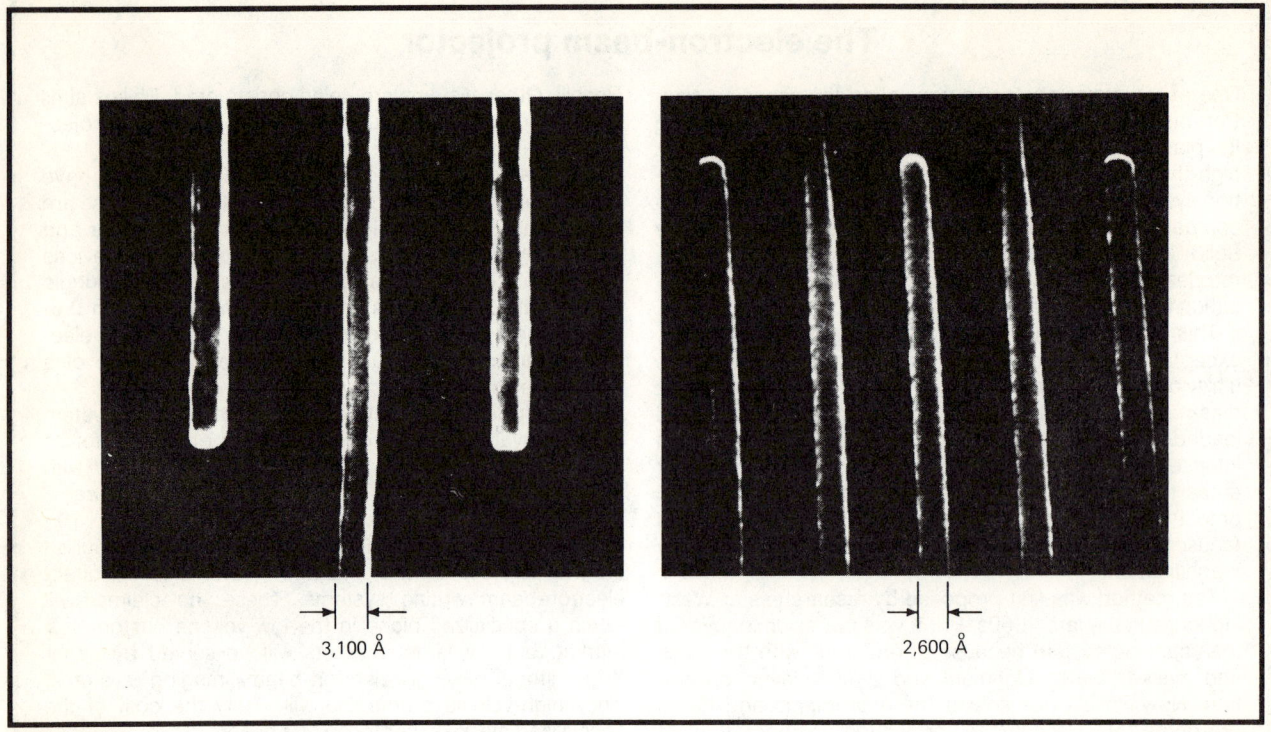

8. Fine lines. These are samples of submicrometer geometries exposed on a Hughes electron-beam system. The two views show the interdigitated fingers of the pattern of surface acoustic-wave delay lines. Even finer lines have been successfully exposed.

Cambridge Scientific Instruments has been delivering electron-beam systems and components for more than a decade. Its U.S. customers include Xerox, Rockwell International, Hughes Research, Motorola, Cornell University, and General Motors. A European competitor is Philips Gloeilampenfabrieken in the Netherlands, which makes the electron-beam pattern generator EBPG-3 [*Electronics*, Nov. 9, 1978, p. 68].

Versatile

The EBPG-3 is a vector-scanning machine for either making masks or generating patterns directly on the wafer. The typical minimum line width it produces is 0.4 µm. It takes 20 minutes to cover a 3-inch wafer with a pattern using lines less than 2 µm wide and exposing 30% of the chip area to the beam; for a line width of 0.8 µm it takes 65 minutes.

Philips has orders for six machines both from outside customers and from in-house users that make components, such as its Dutch Elcoma division and affiliates outside of Holland like Signetics in California. The $2 million price of the EBPG-3 includes installation, training of personnel, and operational support.

In Japan under the guidance of the VLSI Cooperative Laboratory [*Electronics*, June 9, 1977, p. 104], two commercial electron-beam systems have been developed. One is a raster-scanning system developed at the research and development center of Toshiba Corp. but now being manufactured by an affiliate, Toshiba Machine Co. The Toshiba system was followed by one developed by Jeol Ltd., which vector-scans rectangular areas of variable size.

A prototype of the Toshiba system and the first commercial unit are both in operation at Toshiba. The system is designed for beam diameters of 0.25, 0.5, and 1 µm, with minimum line widths of four times these values to prevent rounding of the corners. The two finer resolutions are designed for direct exposure of masks; the coarsest resolution is designed for making 10× reticles for direct stepping on the wafer. A lanthanum-boride emitter provides density almost an order of magnitude higher than tungsten emitters of other systems.

Toshiba engineers claim their system can write patterns easily for several different devices on a mask without loss of time—an especially helpful feature during development. Typical exposure times for a 100-mm-square mask are 208 minutes for a minimum line width of 1 µm, 58 minutes for 2 µm, and for a 100-mm-square reticle, 34 minutes for a 4-µm minimum line width. Price of this system is 400,000 yen.

The Jeol Ltd. variable-rectangle vector-scanning system differs from the Toshiba unit in that the exposure time is proportional to the complexity of the pattern. The sides of the rectangles can vary between 2 and 12.5 µm in increments of 0.1 µm, with a maximum area per shot of 100 µm². The normal exposure rate inside the 2-mm-square area within which the beam can be deflected is 50,000 shots per second.

Movement of the stage makes it possible to expose masks up to 5 inches square. Benchmark performance for a 4-inch hard chrome mask with a resist sensitivity of 10^{-6} coulomb per square centimeter and a pattern with about 200,000 rectangles/cm² is about 25 minutes.

The system can also be operated to generate lines as fine as 1 µm wide. Three of these units have already been delivered to Japanese customers for a price of 420 million yen each.

The company introduced its earlier vector-scanning

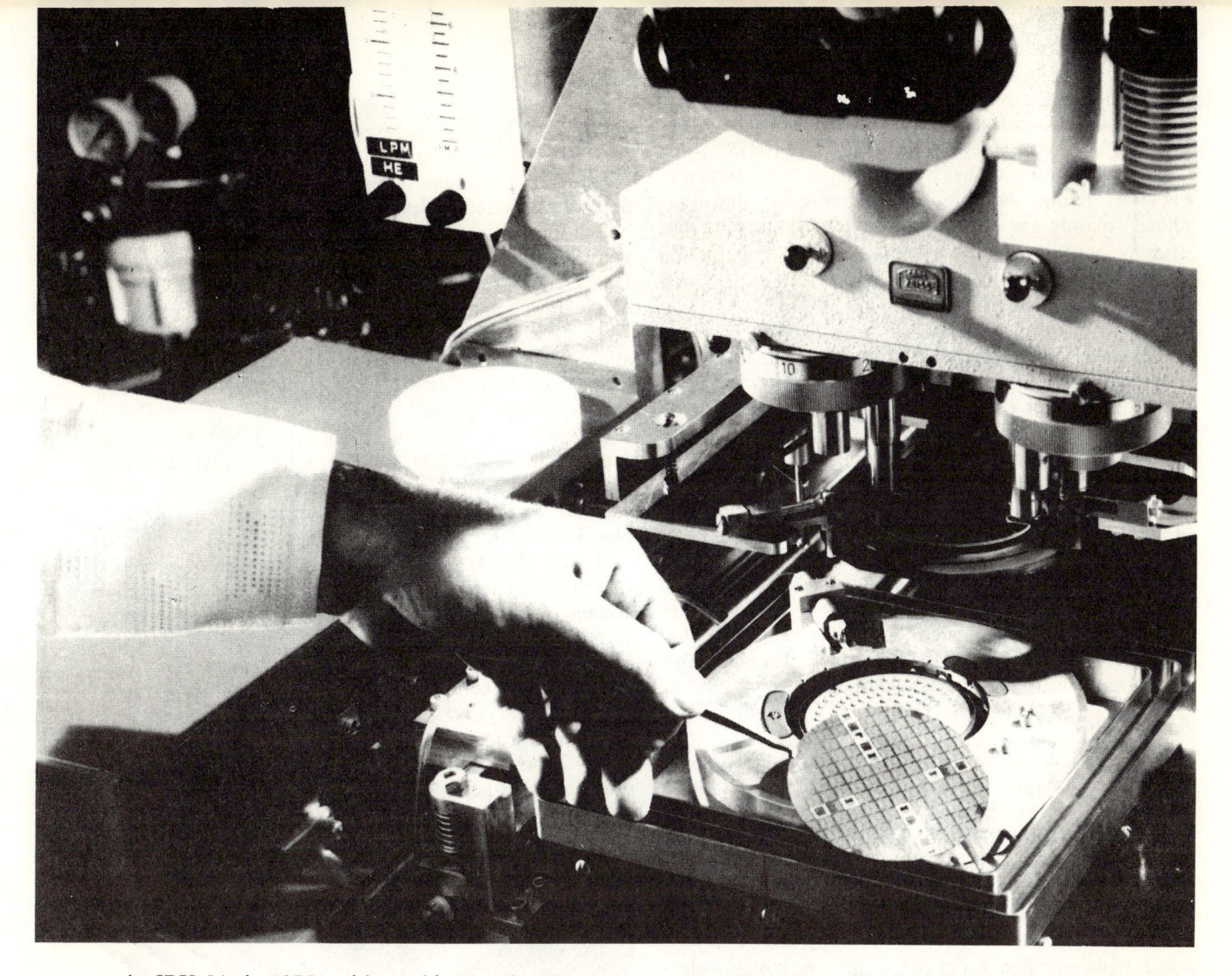

unit, JBX-5A, in 1975 and has sold more than 20 copies of it. The maximum wafer diameter the JBX-5A can handle is 3 inches, and the exposure time for sensitive resist is about 20 hours. The system is capable of submicrometer patterns. Some customers have used the system for direct wafer exposure, in particular for microwave transistors, which are small and expensive.

In-house electron-beam units

Some of the most sophisticated and highest-resolution chips are being exposed on sophisticated machines designed in house at firms like Hughes. IBM, TI, and Bell Labs. In fact, IBM is one of the few companies in the world to write directly on production wafers.

Hughes Aircraft Co.'s Research Laboratories has been writing circuit patterns directly on wafers for about 12 years. It specializes in building high-performance submicrometer devices like gallium-arsenide field-effect transistors and charge-coupled devices. GaAsFETs with 0.5-μm details have been routinely produced, samples of the same circuits with 0.3-μm details have been made, and a program for 0.25-μm devices is under way.

At its Malibu facility Hughes has built five vector-scanning electron-beam systems based on its modification of scanning electron microscopes from Cambridge Scientific Instruments (Fig. 7). These systems can resolve 0.1-μm details, as shown in the views in Fig. 8, and have an alignment accuracy of ±0.01 μm. Exposure

9. X-ray lithography. Bell Laboratories uses this experimental X-ray tool for exposures of 1-μm features on experimental integrated circuits. A palladium target provides 4.3-Å radiation for exposure of an organic, X-ray–sensitive resist through a gold-patterned mask.

time for a 4-inch wafer is about 57 minutes.

Being an R&D facility, Hughes has been less concerned with throughput than submicrometer ability. But now, with more demands from other Hughes facilities, it is becoming more production-oriented and is considering building or buying a system that could scan four wafers per hour at submicrometer levels.

IBM is one of the pioneers in electron-beam research, and its twofold effort backs both vector and raster scanning. Its Thomas J. Watson Research Center has developed the VSI system, a laser-interferometer–controlled machine that takes about 5 minutes to vector-scan a 3-inch wafer with 2½-μm details. For lines 1 μm wide, overlay accuracy is ±0.1 μm. The system is exclusively an R&D tool and has exposed 0.5-to-0.6-μm line widths routinely, 0.016-μm and 0.025-μm details with the aid of a special technique.

IBM's other track

At IBM's facility in East Fishkill, N.Y., the EL-1 raster-scanning system is producing 22 2¼-inch wafers per hour in a direct writing mode. Minimum line width is 2½ μm. The system is at present being used to expose patterns for customized IC layers on logic chips for IBM's System/38 [*Electronics*, March 15, 1979, p. 108].

Texas Instruments has developed its own series of vector-scanning, laser-interferometer–controlled machines, mainly for production photomask making but also for writing directly on the wafer for evaluation purposes. TI's goal is a machine that can scan 10 4-inch wafers with 2-μm details per hour.

William Holton, manager of research development and engineering for TI's Semiconductor Group in Dallas, predicts that by next year electron-beam systems will be making masks for 1:1 projection systems and reticles for wafer steppers as well as writing on chip directly in a few areas. By 1985 he sees them becoming TI's primary tool, with X-ray lithography a "maybe."

Bell Laboratories in Holmdel, N. J., is the originator of the EBES raster-scanning electron-beam lithography system on which both the ETEC and Extrion commercial systems are based. Bell uses the EBES mainly for mask making, but has set one machine aside to see if its parameters can be extended. At the present time this machine has been converted to vector scanning, and extensive software redesign is also in progress.

X marks the spot

Bell Labs and the other firms that use electron-beam systems all recognize that they are too expensive and too slow. This is why there is renewed interest in X-ray lithography. Such systems promise to combine electron-beam resolution with higher throughput for the price of a projection aligner. As Jerry Molitor, manager of the physical electronics laboratory at Hughes Research Laboratories, says, "Eventually we are going to need a batch replicator like X-ray lithography or possibly electron-beam projection to get the throughputs needed."

The first commercial X-ray lithography machine has been produced in Japan, and a U. S. commercial machine is in the works. Meanwhile home-built machines at Bell, Hughes Research, General Instrument, and elsewhere are successfully turning out 1-μm, multimask ICs with good yield and reliability.

X-ray lithography is simply a form of contact printing in which an X-ray source replaces the UV source. It appears suitable for lines 0.5 to 2 μm wide and, with appropriate power sources, resists, and masks, should be capable of throughputs of 10 to 60 wafers per hour.

Resists designed specifically for X-ray exposure are now becoming available. But the masks, which must be opaque to X rays, have been a special problem; made of gold deposited on a layer of silicon, Mylar, or polyimide only 2 to 10 μm thick, they are very fragile. Fortunately, despite their fragility, they can be aligned easily with wafers by modified commercial optical contact aligners. Other problems with X rays are distortion caused by ripple in a processed wafer and a lack of standardization of mask designs.

Bell's X-ray approach

Bell Labs, one of the leaders in X-ray lithography, is now using a second-generation machine (Fig. 9) that has a stationary 4-kilowatt palladium source and uses a gold-plated Kapton mask on a Pyrex support. It can expose a 3-inch wafer in 4 minutes and has achieved registration accuracies of less than 0.5 μm on fast metal-oxide-semiconductor devices with 1-μm details.

Bell opted for a stationary power source to avoid vibration that could blur details in the exposed wafer. Don Herriot, head of lithography systems development in advanced LSI at Bell Labs in Holmdel, N. J., says that line widths so far are limited by the resist, the optical aligner, and the electron-beam–fabricated mask. However, he feels 1-μm devices are possible. Figure 10 shows some samples of present Bell devices.

In an effort to evaluate the reliability of devices exposed to X-ray beams, Bell took similar optically exposed devices and irradiated them with X-ray doses equivalent to those experienced during X-ray lithography. Their performance was not degraded.

Hughes Research has already made 1-μm devices in a similar program in X-ray lithography. The Hughes system employs a rotating 10-kW aluminum source, a Kapton mask, and a Cobilt contact aligner. Hughes can expose a 2-inch wafer coated with copolymer resist in 8 minutes. Tests comparing X-ray and optical lithography complementary-MOS-on-sapphire shift registers produced similar results to those of Bell Labs.

Synchrotron radiation has often been discussed as a source of high-power radiation for X-ray lithography. In France, a half dozen firms are testing this concept on

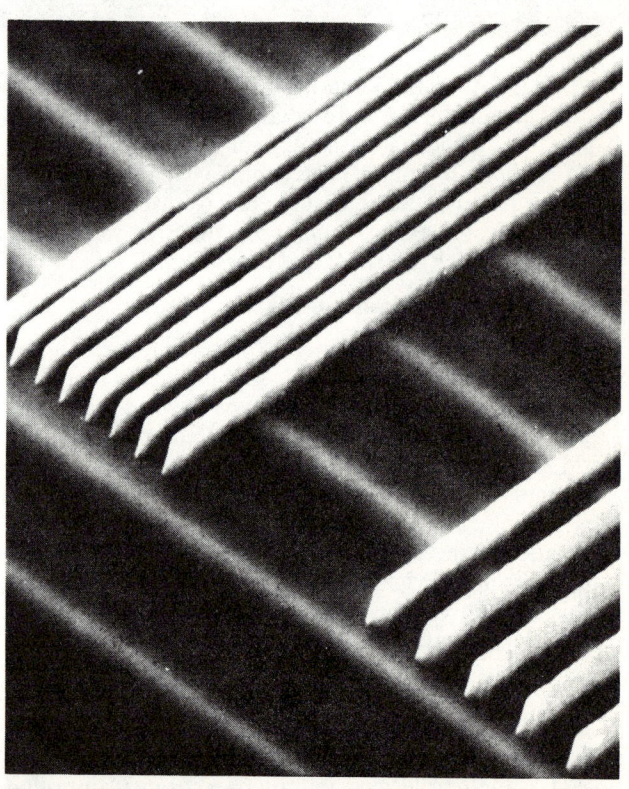

10. X-ray exposure. Shown above is a resist pattern of 1-μm-wide lines and spaces created by exposing a 2.6-μm-thick resist to X rays on a Bell Laboratories X-ray system. The steps shown are 1 μm high. Note the perpendicularity of the stepped features.

an electron storage ring at the Laboratoire d'Utilization du Rayonnement Electromagnétique at the University of Paris [*Electronics*, March 29, 1979, p. 74].

The firms share 18 hours a week on one of the six light lines out of the storage ring. Experiments being run include the effects of X-ray radiation on mask materials and silicon wafers. So far the damage to these materials is less than was expected.

Commercial rays

The single commercial X-ray aligner available to date was developed jointly by the Electrical Communication Laboratory of the Nippon Telephone and Telegraph Public Corp. and Nippon Kogaku (Nikon). Nikon has a prototype scheduled for completion in April and then will seek orders for delivery for 10 months or so later.

The system uses a rotating water-cooled silicon target with power inputs of 20 kW at voltages of 20 to 25 kilovolts. The system is relatively compact with a distance of only 350 mm between X-ray source and wafer. Mask-to-wafer spacing is 5 to 10 μm ± 1 μm.

The minimum incremental step of the work stage is 20 μm with a resolution of better than 0.1 μm. Pattern position detection is also better than 0.1 μm. Alignment accuracy is within 0.2 μm—more than sufficient for Nikon's stated goal of 1-μm-wide lines.

Several types of masks have been used with this machine. The most promising is a sandwich of silicon dioxide between two layers of silicon nitride on a silicon wafer, because its optical transparency of 80% facilitates optical alignment and the three layers minimize stress that could otherwise warp the thin substrate. Silicon is etched away only in the region corresponding to the actual active chip areas, being left untouched elsewhere for added strength. Thus the mask resembles a multipane window, having a gold pattern formed by electron beam lithography on its planar lower surface.

In-house X-ray systems at General Instrument, Hughes Research, and Bell Labs all use Cobilt contact aligners to align delicate X-ray masks to wafers. In view of this, it is no surprise that Cobilt is developing its own X-ray lithography system complete with source and aligner and due out in about two years.

What users can expect

Most of this report has been about the makers of lithography equipment makers and the electronic giants who generate and use their own in-house lithographic equipment. But the IC industry, especially in the U. S., is composed mostly of firms that go outside for their equipment. In general, at independents like Motorola, RCA, Fairchild, National Semiconductor, and General Instrument, most small- and medium-scale ICs are still being made with contact aligners. Large- and very large-scale ICs (down to 2-μm geometries) are produced on Perkin-Elmer's 1:1 projection systems with masks made on electron-beam machines. Step-and-repeat projection systems at these firms are still being evaluated.

Each of these firms has a scanning electron-beam system committed to full-time mask making, and a few have already ordered a second such unit.

Only General Instrument is actively pursuing an X-ray program. This firm has constructed its own machine [*Electronics*, Nov. 9, 1979, p. 106], as well as masks and aligner, and arrived at a satisfactory resist. It has successfully fabricated MOS LSI chips with 1- to 2-μm line widths.

Most other independents are either not considering X-ray lithography or are wary of it. For instance, Jim Dey, manager of the Image Technology Center of National Semiconductor Corp., Santa Clara. Calif., says: "X-ray potentially is the highest-resolution lithographic process currently under development. But there are many problems to be solved before it can become a production technique. It will probably come into prominence at about the 1-μm level. But it could be as long as 8 to 10 years before it becomes economical to use in high volume." Dey thinks that the gap between 1 μm and 2.5 μm will be filled by wafer-stepping techniques and the use of electron-beam wafer writing to design products and write on selected layers in volume production.

For the 1980s, Bill Howard of Motorola can see getting down to 1-μm size with a million components per die by incremental improvements of present technology. "For now, it's electron-beam below 1 μm," says Howard. "But after 1982–3, all bets are off."

Bob Fink, director of industrial engineering of General Instrument Corp. in Hicksville, N. Y., has an entirely different projection of the future from the last two IC people. By 1980 he sees 1:1 projection lithography sharing VLSI with the wafer stepper. And he sees X-ray lithography taking hold as a production tool in 1982.

That is the foreseeable future, for, of course, everyone could be wrong. There is always the chance that some completely new form of IC lithography—perhaps based on ion beams, step-and-repeat X-ray, or laser scanning—could appear and upset all predictions. □

Demands of LSI are turning chip makers towards automation, production innovations

by Jerry Lyman, *Packaging & Production Editor*

☐ The fight for a competitive edge in the semiconductor industry is taking place as much on the production line as in the design-and-development laboratory. Spurred by the explosion of applications brought on by the LSI revolution, the semi houses are making huge investments in automatic equipment in order to expand capacity and upgrade yield. What's more, increased circuit densities and higher performance are straining the limits of conventional production technology, so that integrated-circuit makers must adopt new procedures to keep up.

To meet the demands of large-scale integration, IC makers must make bigger chips and put more on them, while increasing throughput and yield. These rules tend to work against one another, because the number of defects on a wafer goes up as it gets bigger or as more components are crowded on.

As well as bigger wafers, smaller geometries are an answer. But traditional production methods are rapidly reaching their limits in both directions.

While there are significant improvements under way in the chip-layout and wafer-production processes outlined in Fig. 1, the most radical advances are taking place in the process that turns a raw wafer into finished IC chips. There are four major changes occurring in the loop that makes up steps 14 through 20 of Fig. 1:
- The beginning of microprocessor-controlled conveyor-belt operation of the lithographic steps.
- Replacement of contact pattern printing by projection printing and even more advanced methods.
- Substitution of plasma dry etching for acid wet etching when removing protective layers.
- Automation by minicomputer and microprocessor of the diffusion process and related activities.

Even the prosaic steps of probe testing and scribing and

Wafers get bigger and bigger

Today production lines of most U. S. semiconductor houses are in transition from 3-inch to 4-in. wafers. Already however, many of the firms are looking down the line at the 5-in. wafer.

For instance, General Instrument Corp. uses mainly 3-in. wafers but is beginning production with 4-in. ones. The transition has been relatively smooth, since 85% to 90% of the firm's 3-in. wafer processing equipment was purchased with convertibility to 4-in. production in mind. The firm's new plant in Chandler, Ariz., will have a basic 4-in. line that will be convertible to 5-in. wafers.

Texas Instruments Inc., Fairchild Semiconductor, and Plessey Semiconductors Ltd. are other IC firms setting up 4-in. lines convertible to 5-in. wafers. But the majority of the IC firms are content to be conservative and stop at 4-in. wafers for now.

Why are the IC companies going over to the larger wafers? "There are basically two reasons," says David Peterman of TI's semiconductor research-and-development laboratories. "One is to increase the equipment capacity for a given space and number of people, and the other is simply to get more output per person per hour. In short, productivity is the motivating force."

Marshall Wilder, manager of the 4-in. wafer line at Advanced Micro Devices Inc., Sunnyvale, Calif., puts it another way "A 4-in. wafer has 1.8 times the dice of a 3-in. unit and the labor cost is the same—making the dice from a 4-in. wafer cheaper."

However as many IC firms' process people point out, there are two disadvantages of larger wafers. First, wafer lithography becomes more critical with increased size. Second, breakage of a 4-in. or 5-in. wafer of complex large-scale integration can be an extremely costly loss. These two factors, plus the equipment limitations, will probably slow the general move into 5-in. wafers.

As the wafers get larger, equipment makers will either have to retrofit or completely redesign their machines. Practically all the critical machinery for IC processing can handle 4-in. wafers. But there is no projection aligner, for instance, that can expose a 4-in. wafer, and many planned 4-in. facilities will use projection lithography.

For 5-in. wafers, there are even more holes in the vital lineup of equipment. Henry Styskal president of Teledyne TAC points out, "Equipment makers can only afford to keep one jump ahead of the IC industry. We designed our 4-in. probers when IC people were making 3-in. wafers. We can't afford to go ahead too far on our own until we are sure the entire industry is changing its wafer size."

---

breaking are becoming semiautomated and automated under microprocessor control.

Turning 3- and 4-inch silicon wafers into semiconductor chips begins with a coat of silicon dioxide that protects the wafer from unwanted processing effects. Then wafers undergo lithography to print circuit patterns onto them. They are coated with a photoresist that acts as a selective mask for or against etching (for a fuller explanation of the technical terms, see "An integrated-circuit processing glossary." The wafers are baked to harden the resist, then aligned to a photomask and exposed to ultraviolet light to print the mask patterns on. After the patterns are developed, the wafer undergoes etching and diffusion. The entire process can be repeated a number of times to build different circuits on the dice.

Moving wafers

These lithographic steps take place in a work area that has come to be known as the yellow room, from the use of yellow lights that do not affect UV-sensitive photoresists. At first, yellow-room workers moved the wafers from one step to another with tweezers or vacuum pickups. But these inherently slow modes of transportation result in wafer contamination, damage, and breakage. Today, most yellow rooms are composed of modular stations, each with independent analog or digital control. The wafers are in cassettes, which feed into the modules' conveyor belts or tracks. The wafers move serially through a module where part of the printing process takes place and feed into an output cassette. After about 25 wafers have been processed, a worker carries the output cassette to the next station.

Although the cassette-to-cassette flow reduces contamination, damage, and breakage, it has deficiencies—some of them carried over from the earlier methods. One is a lack of uniformity in the processing caused by worker control of the process parameters. Another is the time and labor it takes to transfer the cassettes from one station to another. Also, the individual process modules take up a lot of floor space.

So equipment manufacturers have developed a new type of wafer-processing equipment for the yellow room. Using an in-line approach, the modules are mechanically linked with wafer transports. Moreover, a central microprocessor or a hard-wired digital device controls all process variables. The equipment is more compact than the separate modular stations. This compactness makes possible as many as four processing tracks where only one track of the older equipment fits. The equipment is designed for 3-, 4-, and even 5-in. wafers.

Now there are only single input and output cassettes for the worker to handle. The wafer wends its way through the various connected modules on an air track or on fully mechanized tracks. The tracks may branch, switch, and even make a complete U turn.

Down the tracks

Four California companies are supplying in-line equipment: the Sunnyvale division of GCA Corp., Kasper Instruments Inc., and Macronetics, both also in Sunnyvale, and the Cobilt division of Computervision Corp. in Santa Clara. GCA uses an air transport, which works on the principle shown in Fig. 2. Macronetics' wafer transport (Fig. 3) uses pneumatically driven Teflon supports to gently move the wafers through the modules. The Kasper and Cobilt systems move the wafers from module to module on belt drives.

A typical GCA Wafertrack system (Fig. 4) links spinners, ovens, scrubbers, and developers under micropro-

IC PROCESSING, STEPS

DESIGN AND ARTWORK

IDEA → 1. DESIGN CIRCUIT LOGIC → 2. LAYOUT CIRCUIT → 3. DIGITIZE LAYOUT → 4. CUT RUBYLITH → 5. PHOTO-REDUCE (RETICLE) → 6. PHOTOREDUCE, STEP AND REPEAT (MASTER MASK) → 7. DUPLICATE PHOTO-MASKS

SILICON PREPARATION

SAND → 8. REDUCE TO SILICON → 9. REFINE TO POLYCRYSTAL SILICON → 10. GROW SINGLE CRYSTAL SILICON → 11. SLICE INTO WAFERS → 12. POLISH WAFERS → 13. VISUALLY INSPECT

WAFER PROCESSING

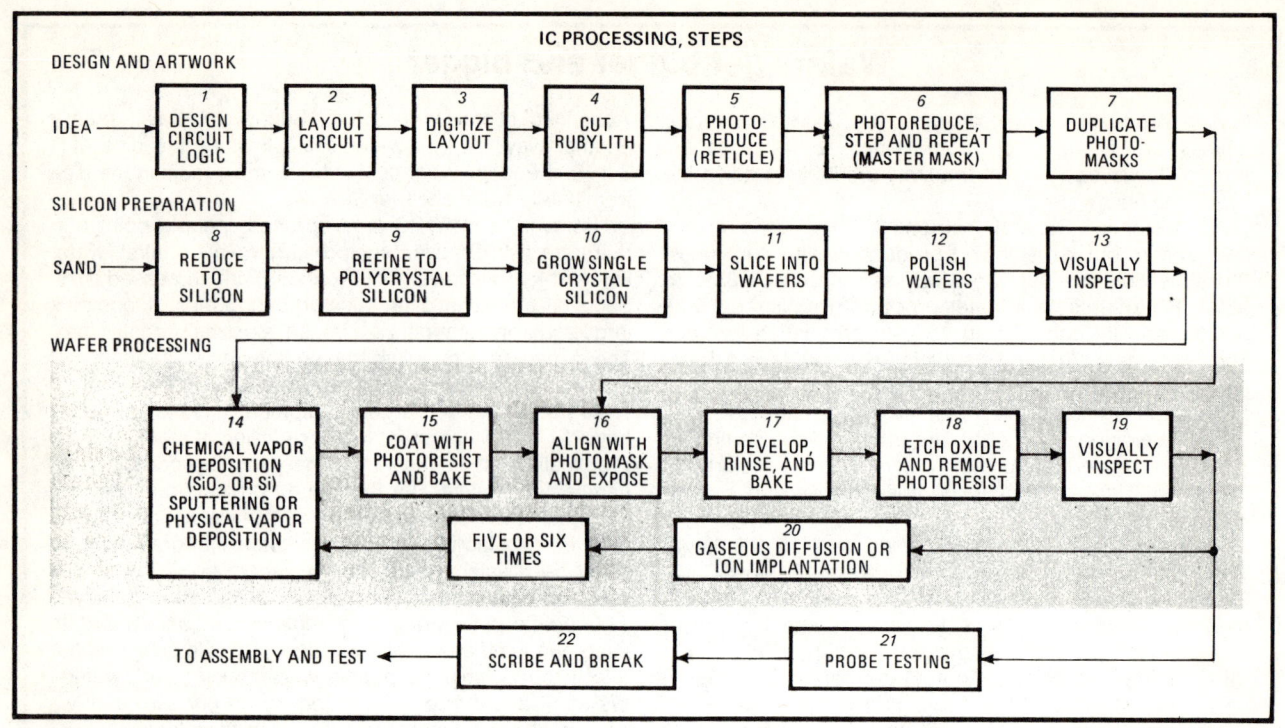

14. CHEMICAL VAPOR DEPOSITION (SiO_2 OR Si) SPUTTERING OR PHYSICAL VAPOR DEPOSITION → 15. COAT WITH PHOTORESIST AND BAKE → 16. ALIGN WITH PHOTOMASK AND EXPOSE → 17. DEVELOP, RINSE, AND BAKE → 18. ETCH OXIDE AND REMOVE PHOTORESIST → 19. VISUALLY INSPECT

FIVE OR SIX TIMES ← 20. GASEOUS DIFFUSION OR ION IMPLANTATION

TO ASSEMBLY AND TEST ← 22. SCRIBE AND BREAK ← 21. PROBE TESTING

1. Circuits from sand. Fabricating a wafer of ICs involves many techniques: circuit design, photoreduction, photolithography, wet and dry chemistry, materials handling, metal deposition, and gaseous diffusion. Last steps in the process are inspection and electrical probing.

cessor control. These modules (which may be used independently) are linked by a standard communications bus. Modules from other companies, such as semiautomatic or automatic mask aligners or probers, can be interfaced to this control bus.

Many tasks

The processor monitors the status of each module more than 10 times a minute. It also monitors and displays process parameters, controls the wafer transport, computes batch statistics, and generates alarms. In addition, it can control peripheral devices and can communicate with larger management computers, a feature that will expedite full-scale computer control of several Wafertrack lines (Fig. 5).

In general, a microprocessor-controlled in-line system has much finer control of process variables than does the independent-module setup. There is better application of photoresist and better control of baking temperatures, for example. The tight tolerances maintained on process variables can raise wafer yield at least 5% over the cassette-to-cassette approach, according to William Loveless, vice president of marketing for GCA.

An in-line system can do one batch of wafers to one microprocessor program and the next batch to another. Use of the modules also has a degree of flexibility. For example, a two-track system can temporarily split into three tracks to make use of three ovens. Also, one operator often can run two or three independent lines.

Eventually, in-line systems will change the yellow room to a fully automatic operation. Both Macronetics and GCA have proposed automated photomasking systems composed of one or two identical three-track lines. While each track would operate independently, a central microprocessor or minicomputer would control the entire system. GCA's Loveless says that systems incorporating many of the automated steps have already been ordered by some IC firms.

Even greater automation is found in a Kasper Instruments' double-track system that is available now. This processing line takes a wafer through a cycle of scrub, bake, spin, bake, align and print, develop, and bake. It has a fully automatic contact mask aligner, unlike the proposed Macronetics and GCA systems, which would require operators in place at each semiautomatic projection aligner.

Many IC makers are using the components of in-line systems in the stand-alone mode, with each module's resident memory programmed to set repeatable, consistent process parameters. Advanced Micro Devices Corp., Sunnyvale, Calif., operates a Wafertrack in a cassette-to-cassette mode. However, the modules for spinning and baking are in-line, and they are controlled by a single microprocessor.

While it is possible to move easily from the independent to the in-line mode with the components of these new systems, IC firms are approaching in-line production with caution. "I don't believe in a fully automatic yellow-room concept," says Robert Fink, director of industrial engineering at General Instrument Corp., Hicksville, N. Y. "I will automate a coat-bake section, and I like the idea of a scrub-bake controlled by a microprocessor. But I don't want my automation chain too long, because then I'll run into down-time problems."

Different approaches

Different circumstances call for different automation approaches, says David Peterman of Texas Instruments Inc.'s semiconductor research and development laboratories in Dallas. Volume production is suited for a highly automated line, he says, while a high-technology product that may require periodic process changes is more suited to the semiautomated approach.

Peterman sees the in-line approach attaining higher yields and throughputs, but he agrees with GI's Fink that one breakdown could stop an entire line. He also points out that the in-line process is relatively inflexible, in that special new machines cannot be added once the production line has been set up.

With product lines all of complex large-scale-integrated ICs, Intel Corp., Santa Clara, Calif., is committed to the cassette-to-cassette approach. Its production lines must be capable of quick changes for new products or modifications of existing parts. "In-line's lack of flexibility has stopped us from full automation," says Willard Kauffman, director of component production. "Intel won't go in-line until this type of equipment is fully proved out and reliable."

Unfamiliarity also slows the move to full automation, points out David Turcotte, Motorola Semiconductor Group's manager of n-channel MOS wafer fabrication in Austin, Texas. "We end up going through a 1½-to-2-year learning curve on almost any new piece of equipment we buy. There is a general reluctance to put new equipment in a manufacturing area until this learning process is over." All in all, fully automated yellow rooms are probably at least two years away.

Projection printing

Perhaps the greatest changes in wafer processing are taking place in the pattern-printing step. The well-established contact printing is being replaced by projection printing, and coming up fast are such new techniques as exposure of the pattern onto the wafer with electron beams.

In contact printing (Fig. 6), an emulsion or hard-surfaced pattern mask is aligned with reference points on a resist-coated wafer, then pressed directly to the wafer and exposed to UV light. Projection printing uses a complex refractive or reflective lens system to project the mask's image onto the wafer. Because mask and wafer do not touch, it has two obvious advantages over the contact procedure: extended mask life and elimination of damage or contamination to the wafer.

Most contact printing uses high-quality emulsion masks that cost between $5 and $10. These masks must be discarded after about 15 exposures. A hard-surfaced (chrome) mask, which costs about $25, can last for perhaps 150 contact printings—but it must be cleaned and inspected after every 15 exposures, so its actual cost is almost $50. With projection printing, the chrome masks can be used for at least a year before being discarded.

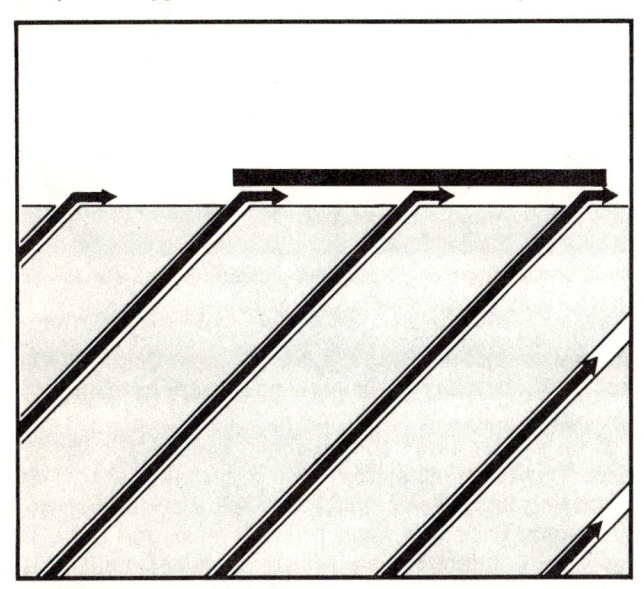

2. Jet transport. In GCA's Wafertrac, an air transport system gives gentle touchless wafer handling. The mechanism provides a rolling vortex of air that surrounds, moves, and supports the wafer.

3. Untouched by human hands. Pneumatically driven teflon supports transport wafers in Macronetic's in-line wafer-processing equipment. Wafers move from the left cassette to the transport, through a scrubber, to the right transport, and back to a cassette.

The leading supplier of projection-printing equipment, Perkin-Elmer Corp., Norwalk, Conn., says that one of its customers, operating around the clock, reports mask savings of about $200,000 a year for each projection system. Other suppliers of the projection equipment are Canon Inc., Tokyo, and Kasper Instruments, with Cobilt reportedly aiming to make its first gear available by year's end.

Substantial increases in device yields are an equally important selling point for projection printing. The procedure uses chrome master masks ordinarily used to contact-print production masks. Since these masters are the nearest thing to a perfect mask, imperfections are held down. Projection printing also eliminates yield losses from such factors as stripping of mask emulsion onto the wafer, sticking of the photoresist to the emulsion, and pressing onto the emulsion of epitaxial spikes, silicon pieces, and dust particles (which then print onto succeeding wafers).

Raising yield

Typically, projection printing is applied to such LSI devices as memories, microprocessors, and calculators. These chips are large dice with low yields, so even small yield increases offer big savings. The graph of Fig. 7 shows estimated yield vs die area for a typical contact procedure and a typical projection procedure. In addition, Perkin-Elmer data indicates that projection equipment can give as much as 50% yield improvement.

The table summarizes the theoretical advantages of replacing contact printing with printing for an n-channel metal-oxide-semiconductor 4,096-bit random-access memory. Taken from a report by Perkin-Elmer and consultants Ruddel Associates of Sunnyvale, Calif., it shows that for this complex IC, projection printing results in a higher wafer yield along with lower mask expenses.

The projection procedure does, of course, have some disadvantages. For one thing, current printers cost well over $100,000, compared to $17,000 to $20,000 for a good semiautomatic contact printer. Moreover, there are fully automatic contact printers, but none yet for projection printing.

No commercially available projection printer can expose the entire surface of a 4-in. wafer—a shortcoming that makes them poor candidates for future production lines with even bigger wafers. As well as better coverage, IC houses want simpler maintenance of the complex optics of the equipment and less sensitivity to temperature change.

Nevertheless, semiconductor makers are using projection equipment. Jim Day, manager of mask-making engineering for National Semiconductor Corp., Santa Clara, Calif., says his company uses projection for very

4. Wafer railroad. This GCA Wafertrac in-line system takes cassettes of wafers through a microprocessor-controlled sequence of scrubbing, baking, spinning, and developing. It needs only a single person to load and unload the input and output wafer carriers.

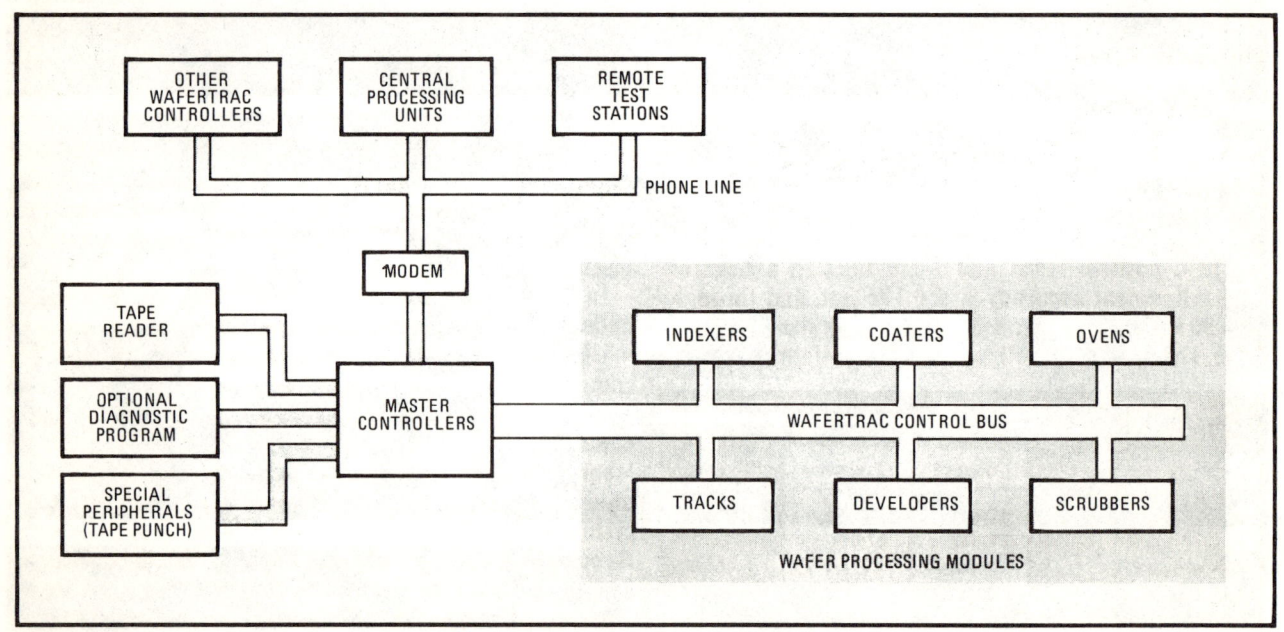

5. Dial-a-wafer. The Wafertrac system can be linked over a standard communication bus to other process modules and peripherals. With a modem, the system's microprocessor can communicate over phone lines with a master computer, other Wafertracs, or a remote test station.

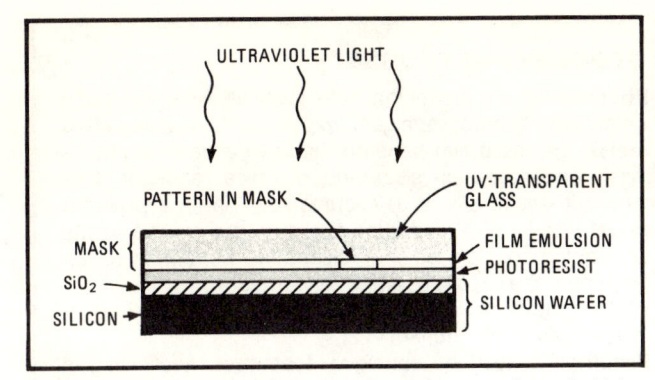

6. In contact. Contact printing of IC patterns consists of shining UV light through a chrome or film-emulsion mask that is pressed directly against a resist-coated wafer. Despite the method's deficiencies, it is still the dominant method of IC lithography.

complex large-area devices, especially those sensitive to surface damage. National uses semiautomatic equipment to expose 5-micrometer geometries. "With projection, we are getting between one and two years' mask life, and wafer yields are substantially higher," he says.

Day sees two reasons for the increased yield. Alignment registration from layer to layer is tighter—in a contact printer, alignment has a gap that causes a shift when the wafer and mask actually touch. Mask stability is assured, simplifying the finding of process problems.

The widely used Perkin-Elmer machines are 1:1 projection systems built around a reflecting optic system. The latest version, the model 140 (Fig. 8), has a high-performance condenser for shorter exposure times. It can expose 90% of a 4-in. wafer and can be loaded with a cassette of wafers. Earlier models must be hand-loaded. All the Perkin-Elmer machines can expose lines and spaces 2 μm wide with an alignment to 1 μm.

Industry sources report that Cobilt is developing a 1:1 machine that will automatically align wafers as large as 5 in. The machine reportedly will be interfaceable with either cassette-to-cassette or in-line equipment.

Stepping down

Besides the 1:1 equipment, there are systems available that project a reticle on to a wafer with a 4:1 to 5:1 ratio. Moving the reticle with a step-and-repeat mechanism exposes the entire wafer surface and gives significantly finer resolutions. Such a machine is the Canon 4:1 FPA-141 (Fig. 9) that costs $150,000 and can print 0.89-μm lines in a positive resist and 1-μm lines in a negative resist. Alignment accuracy is ± 0.125 μm, and throughput is 30 wafers an hour. Maximum wafer size is 3 in.

The Dutch firm, Philips Gloeilampenfabrieken, has

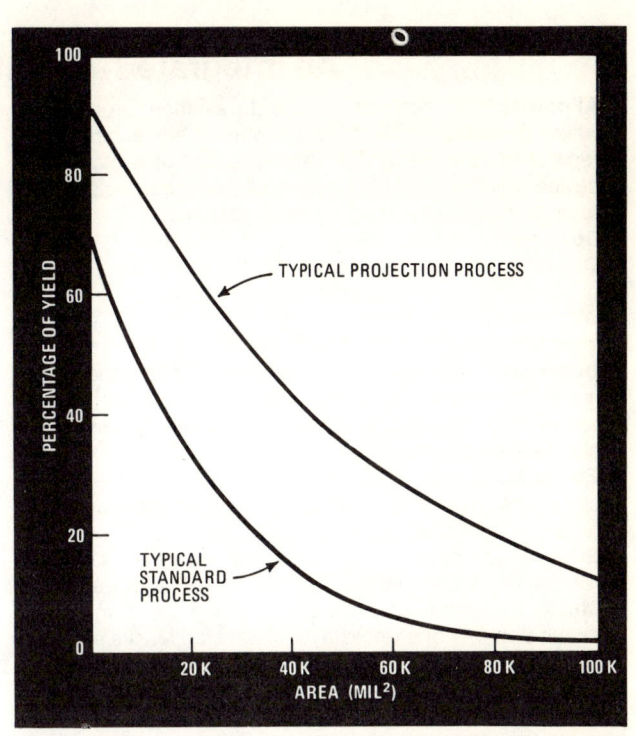

7. Big chips. As chip area increases, probe yield goes down. For a usable yield on LSI wafers, many IC firms are switching from contact lithography to projection alignment and printing.

developed a 5:1 projection machine with 2-μm line resolution but with automatic alignment accurate to 0.1 μm. Maximum wafer size is 4 in.

In the U. S., GCA's Burlington, Mass., division and Ultratech Corp., Santa Clara, Calif. are developing reduction projection systems. Improved versions of such systems with 1-μm line resolution, automatic alignment, and 4-to-5-in. wafer capability could rival the existing examples of the electron-beam pattern-exposure systems—while doing the job at less cost and with higher throughputs, says Warren Moore, project marketing manager of Kasper Instruments. He predicts such machines before the end of 1977.

Electron-beam lithography

Optical projection methods just about reach their resolution limit around 1 μm. For geometries smaller than this, IC makers must look to either direct electron-beam exposure or X-ray exposure through a contact mask manufactured with an electron beam.

In the direct-exposure method, an electron beam under computer control exposes patterns onto a wafer or mask coated with an electron-sensitive resist. The computer determines the pattern.

The technology can hardly be said to have emerged from the laboratory. IBM's East Fishkill, N. Y., laboratory has produced wafers at the rate of 22 2¼-in. units an hour, and Japan's Cooperative Laboratory has a system with about the same throughput for 3-in. wafers. Those U. S. semiconductor firms that have an electron-beam capability are concentrating on making high-resolution, low-defect masks for use with optical projection systems. In Europe, firms like France's Sescosem division of Thomson-CSF, Germany's

TABLE: THEORETICAL ADVANTAGES OF PROJECTION LITHOGRAPHY FOR N-MOS RAM			
	Contact	Projection	Difference
Rework rate	12%	6%	6%
Finished wafers/year	81,432	87,672	6,240
Finished dice/year	3,909,000	6,312,000	2,403,000
Annual mask cost	$332,000	≈$15,000	≈$317,000
Product manufacturing cost	$4.17	$3.55	$0.62*

*Without adjustment in direct overhead rate for savings in masks.

An integrated-circuit processing glossary

Alignment: a technique in the fabrication process by which a series of six to eight masks are successively registered to build up the various layers of a monolithic device. Each mask pattern must be accurately referenced to or aligned to all preceding mask patterns.
Boat: a wafer holder used in a diffusion furnace.
Cassette: an open metal or plastic carrier used on IC production lines for moving groups of wafers.
Die: a portion of a wafer bearing an individual IC, which is eventually cut or broken from the wafer.
Diffusion: a high-temperature process involving the movement of controlled densities of n-type or p-type impurity atoms into the solid silicon slice in order to change its electrical properties.
Emulsion: a suspension of finely divided photosensitive chemicals in a viscous medium, used in semiconductor processing for coating glass masks.
Etching: a process using either acids or a gas plasma to remove unwanted material from the surface of a wafer.
Mask: a chrome or glass plate having the transparent circuit patterns of a single layer of a wafer. Masks are used in the defining of patterns on the surface of a resist-covered wafer.
Master mask: a chrome mask of a complete wafer's multiple images. It is used either in projection printing on a wafer or to contact-print additional masks.

Photoresist: a material that selectively allows etching of a wafer when photographically exposed. With a **negative resist,** the resist film beneath the clear area of a photomask undergoes physical and chemical changes that render it insoluble in a developing solution. In a **positive resist,** the same areas after exposure are soluble in the developing solution, so they disappear, permitting development of the exposed pattern underneath.
Plasma etching: an etching process using a cloud of ionized gas as the etchant.
Probing: a testing technique that uses finely tipped probes to make electrical connections to a sample chip.
Reticle: a glass-emulsion or chrome plate having an enlarged image of a single IC pattern. The reticle is usually stepped and repeated across a chrome plate to form the master mask.
Step and repeat: a method of positioning multiples of the same pattern on a mask or wafer.
Stripping: a process using either acids or plasma to remove the resist coating of a wafer after the exposure, development, and etching steps.
Yield: the number of usable IC dice coming off a production line divided by the total number of dice going in. Yield tends to be reduced at every step in the manufacturing process by wafer breakage, contamination, mask defects, and processing variations.

Siemens AG, and Holland's Elcoma division of Philips have purchased or are building electron-beam equipment for mask making. Sescosem's equipment was developed by Masktechnique, also part of Thomson-CSF, to make the 10:1 reticles that are required in the fabrication of master masks.

The leader

Texas Instruments Inc. most probably leads the other semi houses in electron-beam technology and application. The Dallas-based firm is using its EBMIIs for mask making, machine development, and experimental direct wafer exposure. "By 1978, all masks for new designs will be electron-beam produced," says Turner Hasty, manager of semiconductor research and engineering laboratories for the company. "By 1980, sample quantities of ICs directly exposed by electron beams will be available."

Most U.S. IC firms are considering buying the machines of Etec Corp., Hayward, Calif.; Varian Associates' Extrion division, Gloucester, Mass., and Cambridge Instruments Ltd., Melbourne, Royston, North Hertford, Great Britain. Also, GCA Burlington has started development of an electron-beam system.

The machines available all can expose patterns on either masks or wafers, but their limited throughput makes them cost-effective only for mask making. Etec's $1,600,000 MEBES and Extrion's $1,000,000 EBMG-20, both improved versions of Bell Telephone Laboratories Inc.'s EBES, are intended for mask making. They can expose patterns on masks for 5-in. wafers in about an hour.

Extrion recently sold its first system to mask maker Ultratech, while Etec has already delivered MEBES units to Fairchild Camera and Instrument Corp. and two other IC makers. In about a year or so, about a dozen MEBES will be in the field, predicts Nelson Yew, technical director of Etec.

Fairchild has already used its machine to make a master mask for a 65,536-bit charge-coupled-device memory. The Mountain View, Calif., firm bought the expensive system because of "the low mask-defect rate and quick turnaround time," says Donald Brettner, vice president of manufacturing. "Also, it can be used for larger chip sizes than can optical pattern generators, which are limited to a chip size of 400 square mils."

A major impetus for development of electron-beam machines is their effortless handling of a big problem with bigger wafers. All lithographic systems require an optically flat wafer; yet the larger the wafer, the more vulnerable to distortion it is. For instance, the high-temperature processing steps can distort a 3-in. wafer to the extent of several micrometers by growth or shrinkage along its planar surface or by bowing. For LSI devices with 1-to-2-μm geometry, any one-step masking process will register properly over only a small percentage of the total wafer area.

Since the beam of electrons is infinitely and minutely adjustable, exposure can be tailored to the shape of the wafer. No rival production technology can do this. However, wide-scale use of electron beams awaits the development of machines with throughputs of at least 30 wafers an hour.

X-ray lithography

Another lithography possibility is X-ray exposure of a wafer in contact with a mask produced by electron beams. It is a simple and inexpensive approach, although

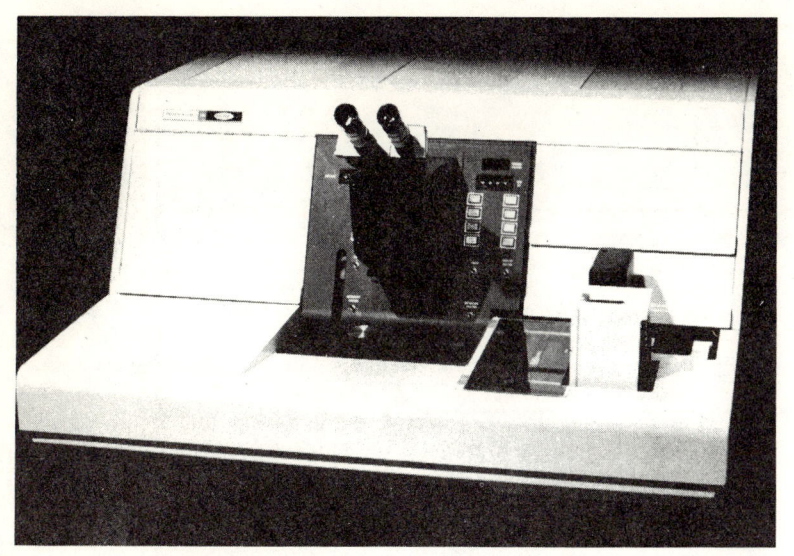

8. Projection aligner. Perkin-Elmer's Micralign model 140 direct-projection mask aligner can expose 90% of a 4-inch wafer and is equipped with automatic wafer loading. Older versions of this unit are used for much of the current projection lithography.

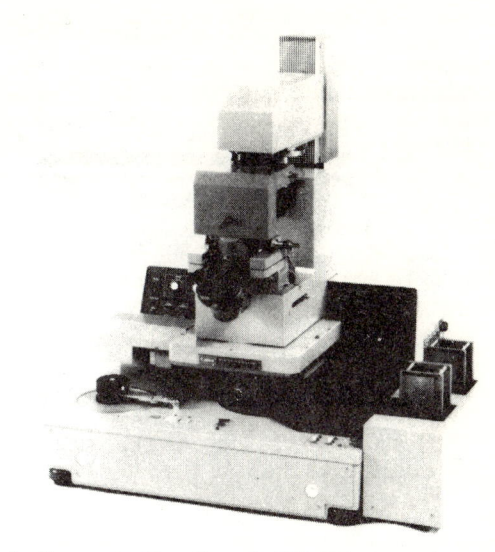

9. Reduction projection. Canon's FPA-141 step and repeats a 4:1 mask image onto a resist-covered silicon wafer. The unit can print 0.89-μm lines in positive resists and 1-μm lines in negative resists.

the masks are costly and alignment is difficult with a 3-μm-thick mask which no commercial mask aligner is equipped to handle. Also early systems had excessively long exposure times — 20 wafers a day.

General Instrument Corp. is working with the Massachusetts Institute of Technology to develop a system of X-ray lithography. Submicrometer lines have routinely been exposed on prototype ICs, the firm says.

An early example of a GI system uses gold-on-Mylar photomasks in contact with wafers coated with a special positive photoresist. X-ray exposure is from a source generated by a 20-kiloelectronvolt electron beam of 1.5 kilowatts. A modified Cobilt contact aligner is used to align successive mask levels. Exposure times are less than an hour, comparable to those of commercial electron-beam machines.

Plasma processing

After lithography comes a step equally critical for yield: removing the photoresist and etching off the protective layers of silicon dioxide or silicon nitride (which prevents unwanted passivation). Almost from the beginning of the IC industry, the standard method of stripping and etching has been with sulphuric, hydrochloric, or phosphoric acids, which have given the procedure its name of wet etching. Now IC firms are switching almost completely to a dry procedure that is called plasma etching.

A plasma essentially is a volume of ionized gas atoms capable of supporting a current. Some fraction of the atoms are ionized; positive and negative ions are present in roughly equal numbers. The plasma contains a substantial group of free radicals—electrically neutral atoms that can form chemical bonds. The free radicals react with the photoresists and the protective layers for an etching effect.

Figure 10 is a simplified representation of a plasma etching system. The plasma is generated in a gas contained in a cylindrical reactor by energy supplied by a radio-frequency generator. Free radicals (CF_3 and F) reach a wafer also in the reactor. The radicals react with the wafer's surface layer of SiO_2 to produce SiF_4 and oxygen.

There are a number of reasons for the turn to plasma etching. Its reagent gases, such as freon, oxygen, and argon, are safer to handle and are less expensive than wet etching's corrosive chemicals, which also produce dangerous fumes and wastes. It provides finer resolution, sharper etching, and less undercutting of that part of the protective layers that is to remain.

Plasma etching makes possible sequential etching and stripping operations on the same machine. Most important, it raises yields because of its inherent cleanliness, few process steps, and elimination of wafer handling.

The flow chart of Fig. 11 illustrates the steps in etching a silicon-nitride layer on a wafer with the plasma process. Wet processes require an added layer of silicon dioxide on top of the nitride layer to make the resist adhere, and they need six or seven rinse and dry steps.

Plasma throughput

One serious drawback that must be solved before the plasma procedure takes over completely from the wet process is the relatively slow throughput of 40 to 50 wafers an hour. The semiconductor makers want a rate of 150 wafers an hour.

Another problem is unwanted etching of the underlying silicon substrate when the protective layers are etched. Varying gas pressure and rf power makes it possible to create free radicals that will favor the silicon-dioxide/-nitride layers. The principle is found in plasma etchers from LFE Corp., Waltham, Mass., and DW Industries, Sunnyvale, Calif.

The cylindrical reactors that have been used present problems also. The plasmas they produce cannot etch aluminum, which is the interconnection material evaporated onto the wafer in one of the final processing steps. Thus this layer has had to be chemically etched. The

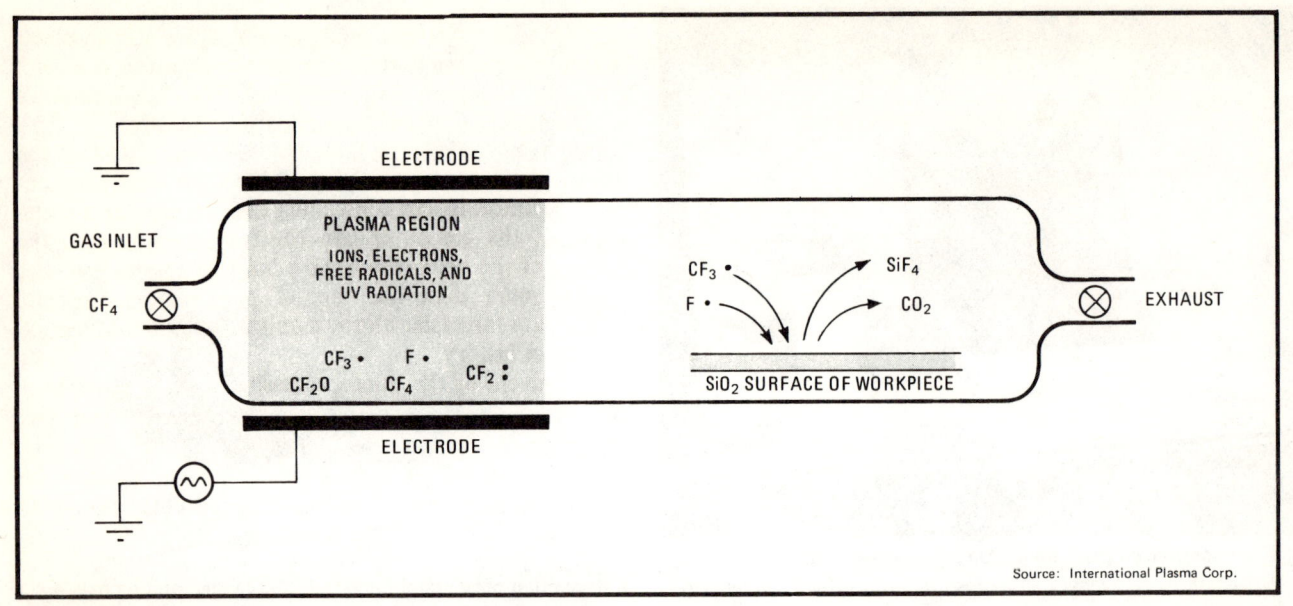

10. Plasma. In a plasma reactor during an etching process, plasma is generated in the space between two electrodes. The wafer to be processed lies outside the plasma region, but free radicals survive to reach the wafer and react chemically with the silicon dioxide.

nonuniform nature of these plasma fields leads to uneven etching, which can cause across-wafer variations and wafer-to-wafer variations in the same run. Differences in temperature between runs can cause run-to-run variations. These three etching variations can cumulatively make a difference of 10% to 20% in the depth of the material etched.

The solution to these etching problems appears to lie with a new type of reactor, the parallel-plate machine. The cylindrical reactor generates plasma remote from the wafer to be etched, while the parallel-plate models generate it directly over the wafer. This location permits the relatively unstable free radicals to spend more of their lifetime on the wafer. Among other benefits, this permits the etching of aluminum.

Minimizing plasma variations

The uniform plasma field produced by these reactors minimizes most of the across-wafer and wafer-to-wafer variations. The parallel plates can be water-cooled to keep wafer temperatures constant, thus minimizing many run-to-run variations and eliminating a preheat operation necessary in the cylindrical reactor.

All the major U.S. makers of plasma-etch machines are working on parallel-plate models. Two such machines from DW Industries and International Plasma Corp. Hayward Calif., can etch into aluminum at the rate of 2,000 angstroms a minute. The DW machine can process 35 3-in. wafers at a time, while the International Plasma machine can do 10 3-in. wafers at a time.

A third new machine from Applied Materials Inc., Santa Clara, Calif. is used for low-temperature plasma deposition of silicon-nitride films for device passivation (Fig. 12). Tegal Corp., Richmond, Calif., also has a new parallel-plate reactor for depositing silicon nitride on a processed wafer to act as a moisture-inhibiting passivation layer.

Parallel-plate reactors still are relatively small machines, but much larger models are on the way. They will be able to handle bigger wafers and to interface with cassettes or with in-line production systems.

Ion-beam milling

Another dry-etch method to which IC manufacturers may turn is ion-beam milling. In this technique, a collimated beam of ions is focused onto a resist-covered wafer in a vacuum chamber. The beam selectively mills out unmasked material by displacing ions of the wafer. Unlike plasma etching, milling will etch any material, including garnet and nickel iron. Therefore it has been extensively used with magnetic-bubble memories, which are built on substrates of these materials.

Right now, ion milling machines from Veeco Instruments Inc., Plainview, N. Y., and Technics, Alexandria, Va., are mostly confined to laboratory use, since their wafer holder is a plate only 3 to 5 in. in diameter. However, a new Veeco system has a wafer holder 10 in. in diameter that can hold 38 2-in. or 14 3-in. wafers. This unit could move ion-beam milling out of the lab, especially if bubble memories take off.

Automated diffusion

Among the various procedures wafers undergo after the yellow room, the most common and most important is diffusion of impurities onto the wafer to provide the semiconductor properties. To raise product yields and cut operation costs, considerable automation of this procedure is taking place.

In the diffusion process, a batch of wafers is placed in a furnace and subjected to specified conditions of temperature and gas flow for a set period of time. A typical cycle is plotted in Fig. 13. To get uniform results, each load of wafers must experience the same environmental conditions for the same length of time. The critical process parameters are time, temperature, and gas composition.

The original control approach was simple analog elec-

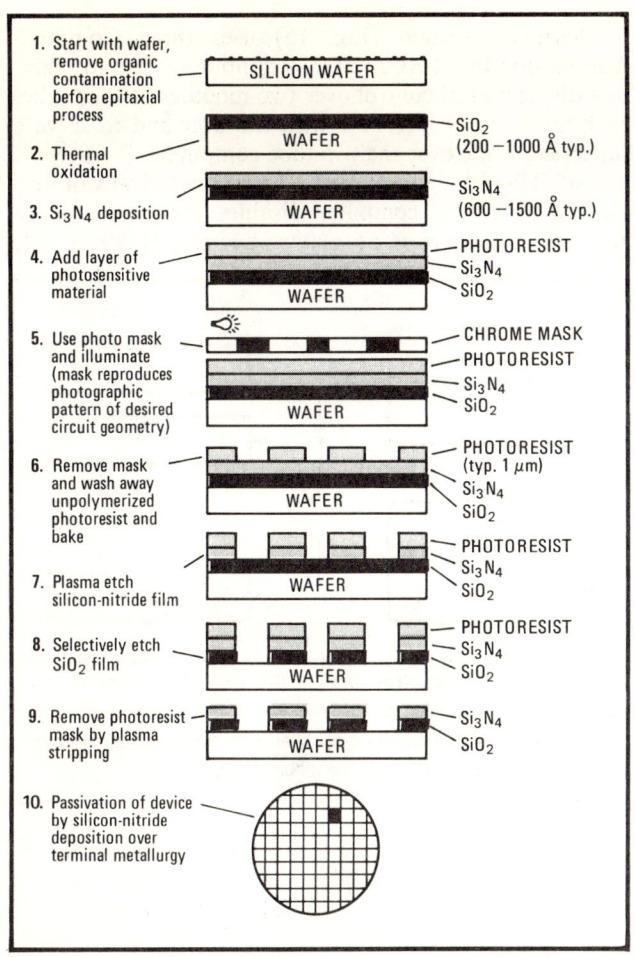

11. Plasma processes. A typical process for a silicon wafer shows that plasma processes may be applied for wafer cleaning, oxide etching, resist stripping, and silicon-nitride deposition (passivation).

tromechanical systems, many of which are still in place. However, this semiautomatic instrumentation does not give the required uniformity, so the minicomputer approach in Fig. 14 has been tried successfully.

One minicomputer can control as many as 64 of the tubes in which diffusion takes place. As well as accurately monitoring and controlling time, temperature, and gas flow, the computer can control insertion into the furnace of the boat (the fixture holding the wafer), can provide more than one mix of parameters, and can record data for easier alarm monitoring and compilation of product history.

The advent of the microprocessor made it possible to have the equivalent of minicomputer control at less than the price of some of the older semiautomatic electromechanical units. It also is leading to a solution for excessive down time when the minicomputer or multiplexer fails, putting all the tubes out of business.

Using the microprocessor

Single microprocessors control eight-tube diffusion furnaces built by Thermco Products Corp., Orange, Calif., and Sola Basic Industries Inc.'s Tempress Microelectronics division, Los Gatos, Calif. Thermco's furnace control is built around National Semiconductor's IMP-16C/400, while Tempress uses an Intel 8080.

Both furnaces are hybrid systems. The microprocessor controls the time sequences, but the temperature and gas-flow controllers are analog. However, the microprocessor does control the set points (desired temperature and gas flow) through digital-to-analog converters. If the microprocessor fails, either of these furnaces may be run manually with each tube's analog controller.

To overcome the down-time problem, Bruce Industrial Controls Inc., North Billerica, Mass., has developed a

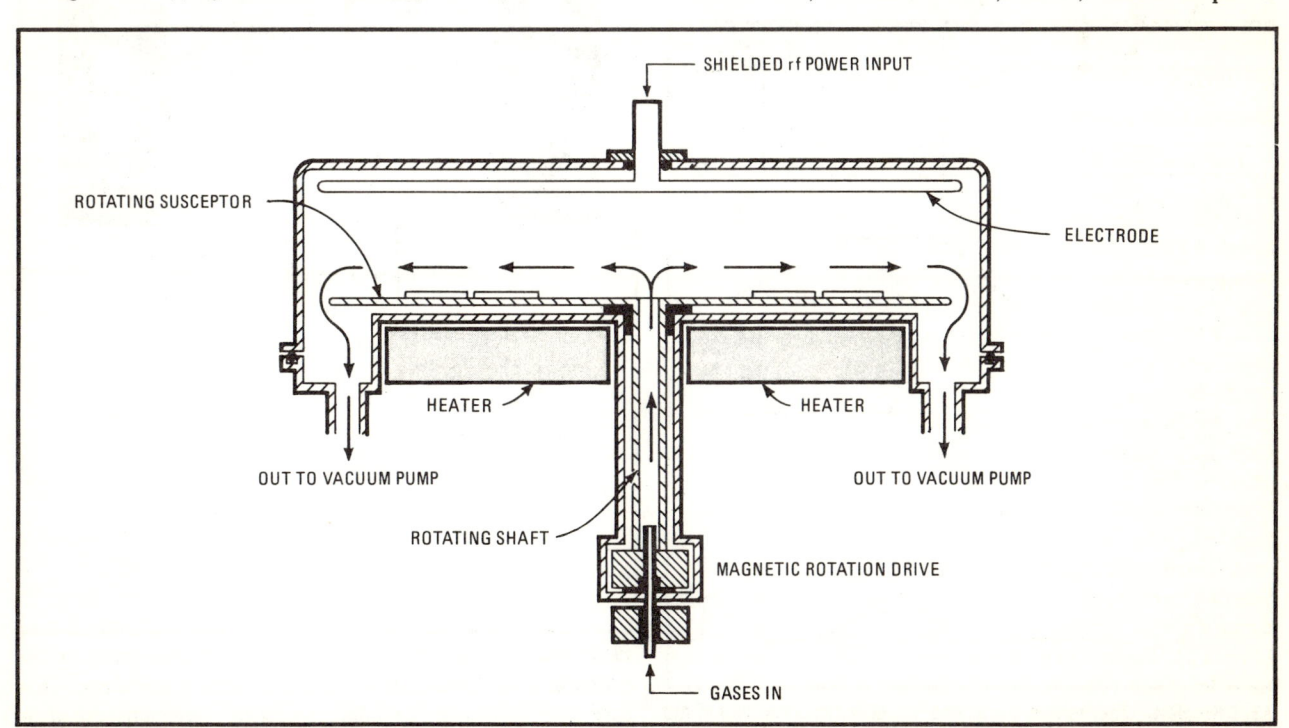

12. Parallel plate. A new type of reactor using parallel-plate construction, rather than the older barrel method, is appearing for both etching and deposition. Applied Materials uses the parallel-plate reactor, shown in cross section, in its Plasma 1 silicon-nitride deposition unit.

furnace with a microprocessor assigned to each tube. Again, the tubes have analog controllers for temperature and gas flow.

Since each MOS Technology Inc. 6502 is dedicated to only one tube, its failure affects only that tube, which can be operated in a manual backup mode. As well as performing the functions performed by the Thermco and Tempress furnace controls, the Bruce unit scans analog outputs, comparing them against values in memory and identifying discrepancies. It also can link with digital peripherals and has provision for expansion to direct digital control of gas flow and temperature.

This versatility does have an obvious drawback. At least for the time being, the processor-per-tube approach is more expensive than the time-shared processor method.

Direct digital control

Even more precise than the hybrid fully automatic approach is direct digital control of all three critical parameters. It gives freedom from the long-term drift, overshooting, and low resolution of analog controllers and has the computing capacity to provide more control functions at a substantial reduction in initial cost. Such an eight-tube diffusion furnace has been available for the past 3½ years from Thermco.

"We went to DDC primarily for very tight control of temperature gas flow and boat speed," says Dick Dunn, marketing manager of Thermco. "In this type of system, we are not limited by the performance of analog controllers. Now we can calibrate an entire system—every one of the eight tubes is exactly alike. With analog controllers, recovery rates from tube to tube aren't necessarily identical. With DDC, all tubes have identical recovery rates and identical reactions."

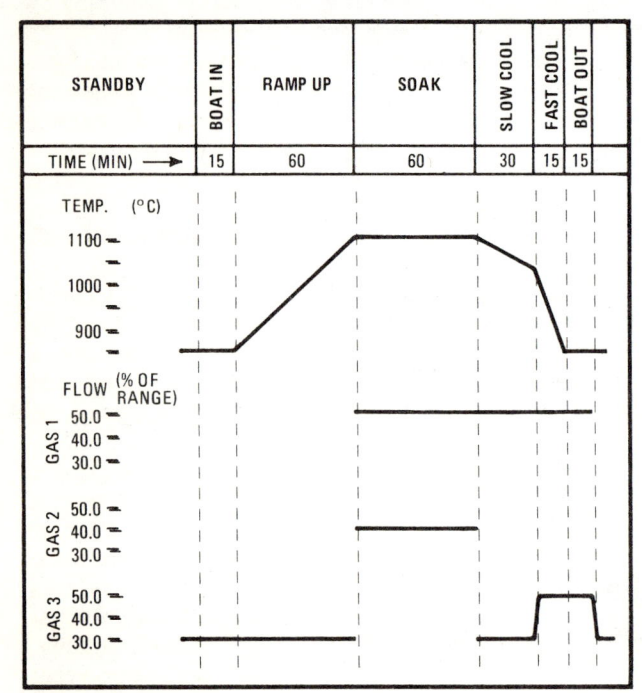

Thermco's system (Fig. 15) uses three Computer Automation Inc. LSI 2/20 minicomputers: two to maintain direct digital control over two modules of four tubes each, and one to interface with the user and to serve a supervisory role over the two tube computers.

The Thermco setup does have one disadvantage. Failure of a tube computer disables the four tubes it controls, and there is no analog backup. However, the

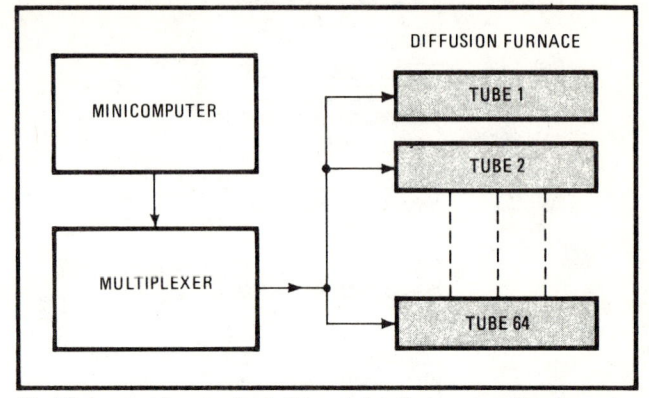

14. Minicomputer control. One of the first approaches to automating multiple diffusion tubes was to let a minicomputer with a multiplexer act as the master controller. However, failure of either of these components disables all the diffusion tubes.

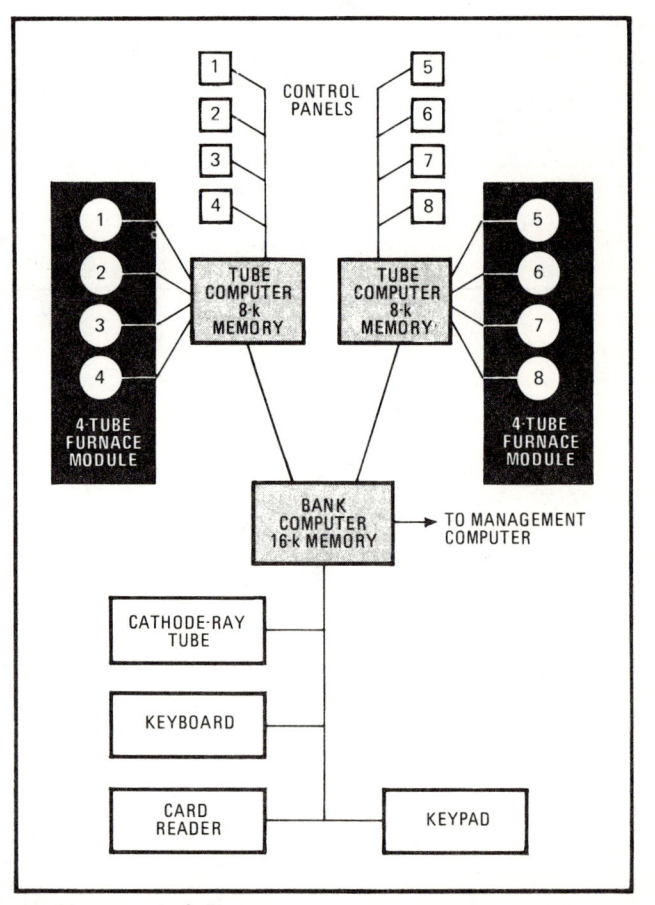

13. Diffusion. Time, temperature, and gas flow are controlled in the diffusion process. Note that temperature vs time functions use ramps rather than step functions to maintain process control.

15. Direct control. Thermco's direct digital control system for an 8-tube diffusion furnace uses three LSI 2/20 minicomputers—one bank computer and two tube computers. Each tube computer controls four diffusion tubes and eliminates all analog controllers. The bank computer supervises the other two computers.

17. Microprocessor control. The Electroglas Model 1038 reflects the improvements that IC firms wanted in probers. It is microprocessor-controlled and has a cassette-to-cassette mechanical interface and Z-axis compensation for wafer warpage.

minicomputers have proved extremely reliable in use, Dunn maintains.

The next step is to introduce microprocessors. Bruce Industrial Controls has already demonstrated in a developmental model that it can adapt its processor-per-tube approach to direct digital control, and Thermco says that its next generation will use microprocessors.

After a wafer takes five to eight trips through the wafer-processing steps, it emerges for its first electrical tests at a probe station interfaced with a large automatic device tester. It is placed on a special staging plate and aligned with the tester. Then tungsten-tipped probes on a card fixed to a ring are placed in contact with each die's input/output pads.

Probing the future

Most IC producers are committed to probers that are manually loaded and semiautomatically aligned. An operator makes an initial alignment, and then the machine probes rows of dice sequentially.

This semiautomatic probing does increase yield by reducing operator error, but IC makers would prefer a cassette-to-cassette or in-line feed to the prober to further reduce damage. In fact, they would prefer to eliminate the operator, or at least to to reduce staffing to one person supervising as many as five machines. Such automation calls for microprocessor control of the aligning and probing steps.

Above all, prober users want Z-axis control. This feature is needed to compensate for the surface warpage of large processed wafers—which can be as much as 5 mils. With warped wafers, the prober can either damage dice or fail to contact them. Some semiconductor houses find that dice at the edge of wafers are rejected needlessly because of this effect.

The solution is Z-axis control, which senses the height of the wafer at each die and compensating by moving the wafer platform up or down. This feature is available on probers that are being field-tested: the Cobilt CP-4400, the Tac PR-100 from Teledyne Inc.'s Teledyne TAC division, Woburn, Mass., and the model 1038 from Electroglas Inc., Santa Clara, Calif.

In the Teledyne and Electroglas units, Z-axis sensing and compensation takes place with each die. The Cobilt prober maps wafer topography with capacitive sensors and then applies to each die a Z correction generated by an algorithm in an on-board microprocessor.

Only the CP-4400 has fully automatic wafer alignment, but Electroglas expects to have this important feature fairly soon. All three machines have cassette-to-cassette interfaces for wafer handling, and the Electroglas machine can be interfaced with GCA's in-line Wafertrac.

The 1038 (Fig. 16) uses a Motorola 6800 to store 10 separate test programs that can be loaded by a teletypewriter or an RS-232 communications interface. Coordinates of bad chips can be stored in the 6800's memory to help in discarding unusable dice. Cobilt also uses a microprocessor, but the Teledyne unit has hardwired logic for its programming. All three units can be controlled by an external computer.

Automation may eventually bring about a complete hands-off operation, all the way from sand to packaged chip. The manufacturers of the production equipment feel that advance will come soon, but IC firms are more conservative.

Certainly the yellow room's steps and the probing stations are approaching full conveyor-belt automation. But it is a fairly safe guess that it will take some years before the etching diffusion, ion implantation, and metalization procedures catch up, which leaves many steps still to be automated. □

Plasma processes set to etch finer lines with less undercutting

Wet chemical etching is starting to give way
to plasma-assisted techniques that now can etch selectively
any level on a semiconductor device with greater dimensional control

by C. J. Mogab, *Bell Laboratories, Murray Hill, N. J.*
and W. R. Harshbarger, *Bell Laboratories, Allentown, Pa.*

☐ If there were a Rip van Winkle of semiconductors and if he were to fall asleep today, he would wake up to an industry as vastly changed in manufacturing technology as it will be in design advances. One important difference he would note right away would be plasma etching's displacement of wet etching.

Based on the use of reagent gases, plasma etching is safer and less expensive than wet etching, which is based on sulphuric, hydrofluoric, or phosphoric acid. Also, it provides finer resolution and less undercutting, so it is better suited to the era of very-large-scale integration.

As interest in plasma etching has grown over the past several years, new equipment and new processes have been developed that overcome certain drawbacks. For one thing, it is now possible to etch selectively any level on a metal-oxide-semiconductor or bipolar device. Furthermore, it is now possible to etch thin films of semiconductor material anisotropically (leaving a perpendicular edge), as well as isotropically (leaving a concave edge). This advance over wet etching greatly improves the dimensional control of integrated-circuit patterns, leading to ever-higher circuit density.

Physical and chemical phenomena

A plasma is an ionized gas, composed of nearly equal quantities of positive and negative charge (electrons and positive and negative ions) Plasma etching techniques employ weakly ionized gases produced by low-pressure electric discharges. Inelastic collisions between gas molecules and energetic electrons created by the electric field produce the ions and free radicals that can cause etching by physical and chemical interactions with surfaces.

An important characteristic of these discharges is that the gas temperature is much lower than the free-electron temperature. The low gas temperature means that it is possible to directly expose thermally sensitive materials, such as photoresists, to these plasmas.

Plasma-assisted etching had its beginning in the 19th-century work of Michael Faraday and Sir William Crookes on gas discharges in vacuums. W. R. Grove reported the erosion of matter by sputtering processes in a plasma more than 100 years ago.

Sputtering is a wholly physical process, in which the plasma's ions are accelerated by an electric field and made to hit a solid's surface. They transfer their momentum to the surface, ultimately causing atoms of the solid to be ejected. The physical phenomenon of sputtering forms the basis of two plasma-assisted etching techniques: sputter etching and ion-beam milling, also called ion milling.

To achieve useful etch rates, that is, to direct many ions toward the solid, these processes require relatively low pressures (less than 0.1 torr), as well as high particle energies that make them prone to causing damage to semiconductor devices.

Even worse, these etching techniques provide very little selectivity in etch rates among the materials used in silicon IC fabrication. Selectivity is a key word in the delineation of patterns, signifying that the etching does not go below the layer for which it is intended.

However, there is also a chemical plasma-assisted technique, called simply plasma etching, in which an electrically induced gas discharge provides active free radicals that combine chemically with a solid to produce volatile products. Generally it is done at higher pressures than the physical etch techniques, and since it needs lower-energy particles, it is less damaging to the semiconductor solid.

In 1971, S. Irving, K. E. Lemons, and G. E. Bobes of Signetics Corp. reported that carbon tetrafluoride (CF_4) can be dissociated in a plasma to produce fluorine atoms and other reactive free radicals. Fluorine atoms react vigorously with silicon and its compounds to produce

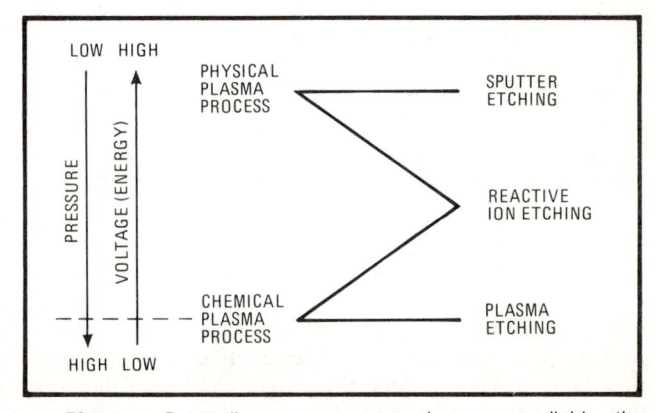

1. Plasmas. Depending on pressure and energy available, the spectrum of plasma-assisted etching can go from chemical to physical. The techniques are becoming the processes of choice for LSI and VLSI manufacturing and are gradually displacing wet etching.

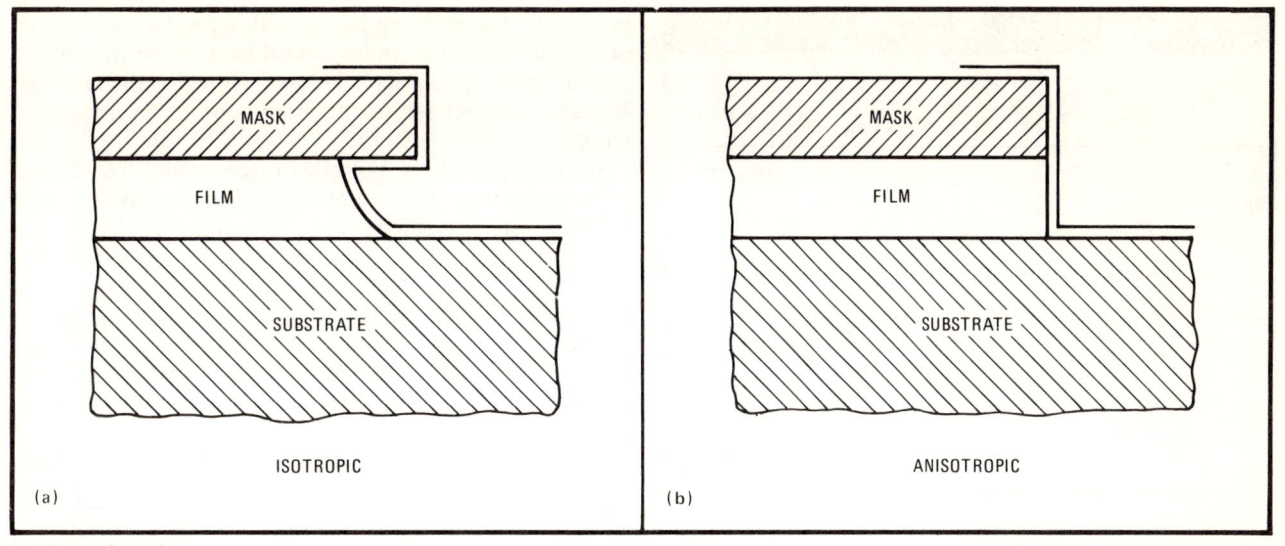

2. Along the edge. A fully isotropic edge profile film (a) undercuts the photoresist mask by an amount equal to the film thickness. An anisotropic edge (b), on the other hand, has no undercut. Anisotropic etching is an absolute must for the high-resolution patterns of VLSI.

SiF$_4$, which is a gas at room temperature. The reaction is close to a purely chemical process; indeed, if argon is simply substituted for CF$_4$, the etch rate of silicon is insignificant, indicating negligible sputtering.

Between the extremes of physical sputtering and plasma etching, there exists a broad spectrum of plasma conditions (Fig. 1). In 1974, N. Hosokawa and his co-workers at Nippon Electric Varian Ltd. discovered a way station between the two extremes: reactive sputter etching, also called reactive ion etching. Their experiments demonstrated that the etch rates and selectivity of physical plasma sputtering could be improved by substituting halocarbons for noble gases. As Fig. 1 shows, this substitution involves lower energy requirements than does straight sputter etching.

In fact, reactive sputter etching suggests that there is no sharp demarcation between physical and chemical techniques. In general, both may occur, perhaps synergistically, in plasma etching. Of course, there is more chemical etching and less physical etching as pressure increases or, more generally, as the average energy of the plasma decreases.

Still, at just what point, if ever, physical ion bombardment becomes unimportant is not known precisely. For instance, R. A. H. Heinecke of Standard Telecommunications Laboratories Ltd. in England has demonstrated that plasma etching under high-power conditions can improve selectivity and feature profiles in silicon dioxide.

Good performance

Wet chemical etching techniques would be expected to etch isotropically, giving rise to edge profiles of the kind shown in Fig. 2a. The concavity of the etched feature undercuts the mask (on top of the circuitry for illustration only; in actual production, it would have long since been removed). Thus the etching does not faithfully reproduce the dimensions of the mask.

This undercutting is not a serious problem until the ratio of feature width to layer thickness drops below about 5 : 1, as it inevitably must for VLSI. Imagine trying to produce 2-micrometer lines and spaces in a 1-μm-thick film with isotropic etching that would produce the edge profile of Fig. 2a.

However, the nature of the boundary layer between a plasma and a solid surface is such that the charged particles have trajectories perpendicular to the surface. With many semiconductor materials, such directional bombardment can lead to the edge profiles of Fig. 2b. Clearly, under conditions that permit such anisotropic etching, the resolution limits on pattern transfer depend only upon the lithographic technique used to generate the mask and not at all upon the plasma-assisted etching technique. This high-resolution capability, of course, is the single most compelling reason for development of such etching techniques.

The ever-increasing scale of integration on silicon chips has led to extensive development of techniques that can handle the handful of materials used in IC fabrication. Thus, single-crystal silicon, polycrystalline silicon, thermally grown and deposited silicon dioxide (doped and undoped), silicon nitride, and aluminum metalization all have been plasma-etched.

Moreover, these materials can be etched selectively in the presence of masking layers, such as organic photore-

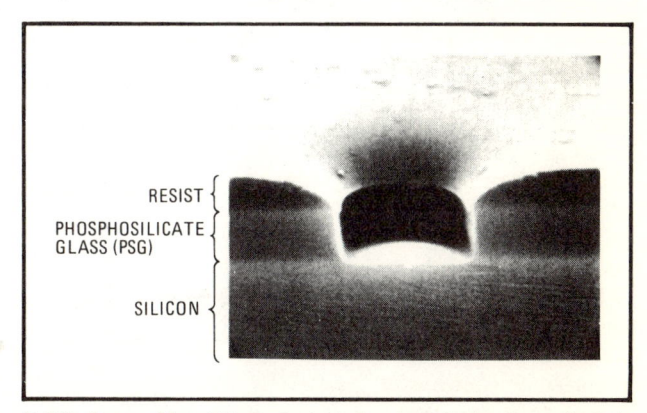

3. Window making This contact window is etched anisotropically in a 1.7-μm thick phosphosilicate-glass layer on silicon. It has been opened in the glass dielectric to allow electrical contact between the polysilicon substrate and the aluminum conductor lines.

TABLE: PLASMA-ASSISTED ETCHING FOR IC FABRICATION				
Film	Substrate	Film etch rate (Å/min)	Selectivity (etch rate ratio of film to substrate)	Resolution (edge profile)
Si_3N_4	SiO_2	300	4 : 1	isotropic
Si_3N_4	Si	700	7 : 1	anisotropic
Poly-Si	SiO_2	1,000	30 : 1	isotropic
Poly-Si	SiO_2	1,100	8 : 1	anisotropic
SiO_2 thermal	Si	600	6 : 1	anisotropic
PSG	Si	1,300	13 : 1	anisotropic
Al	SiO_2	1,000	6 : 1	anisotropic
Si N*	Al	2,000	∞	isotropic

* Plasma-deposited silicon nitride

sists, and of underlying layers that are a permanent part of the multilevel IC. An important example of this selectivity is etching contact windows, typically the smallest feature to be defined in the production of an integrated circuit.

In the case of a polysilicon-gate MOS device, for instance, an intermediate phosphosilicate-glass dielectric insulates the polysilicon from the aluminum conductor lines deposited later. Windows must be opened in the dielectric to permit electrical contact to the polysilicon and to the underlying silicon substrate. Making such a window (Fig. 3) with dimensions at the resolution limit of the mask-making technique requires anisotropic etching of the phosphosilicate glass relative to the polysilicon and the silicon.

The selectivity required for the windows depends upon the thickness of the polysilicon layer and the depth of

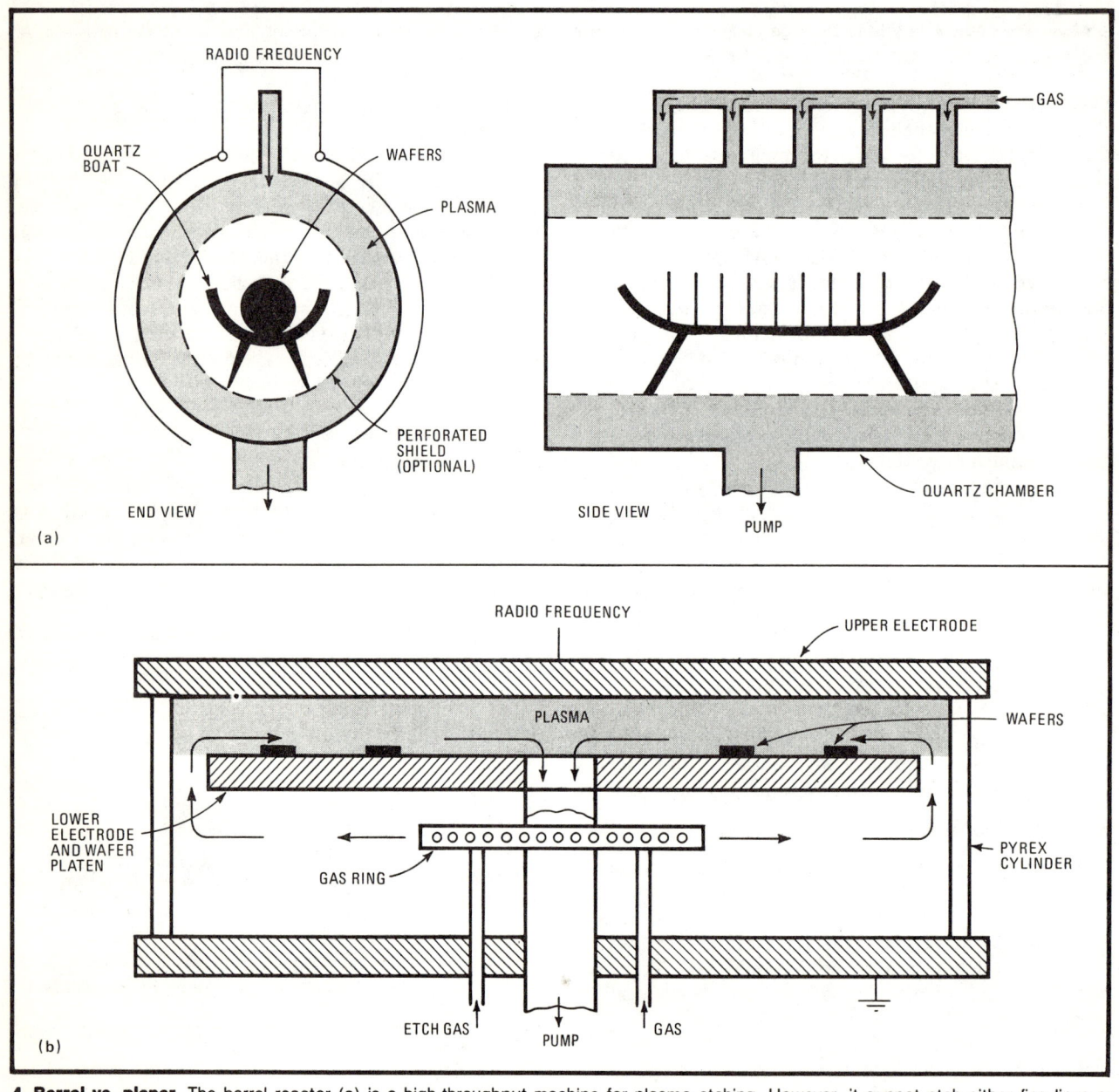

4. Barrel vs. planar. The barrel reactor (a) is a high-throughput machine for plasma etching. However, it cannot etch either fine lines or aluminum. The parallel-plate reactor (b) overcomes these deficiencies but has a low throughput in the present models.

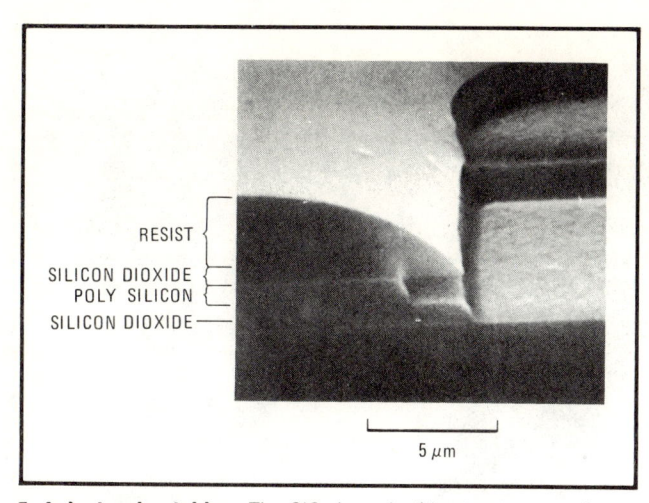

5. Anisotropic etching. The SiO₂ layer in this structure was intentionally undercut by overetching with a wet chemical etch in order to demonstrate the anisotropic character of plasma etching of the polysilicon layer in a Bell Systems production-line process.

junctions in the substrate. For very-large-scale integrated work, selectivity requirements naturally will be even more stringent.

The table summarizes some of the features in etching IC materials with plasma-assisted techniques. Actual etch rates depend on such system parameters as gas composition, pressure, power density, and type of plasma-assisted etching. They generally range from a few hundred to a few thousand angstroms per minute. Many materials not listed, including metals such as titanium, tantalum, tungsten, molybdenum, gold, and plantinum, and semiconductors and insulators such as germanium, gallium arsenide, carbides, nitrides, and oxides, also can be etched in plasmas.

Commercial reactors

Commercially available reactors for plasma-assisted etching are generally of two types, both generating plasma with radio-frequency power (usually 13.56 megahertz). The first type consists of a chamber with external electrodes into which wafers are stacked vertically in a suitable carrier (Fig. 4a). The chamber usually is cylindrical, accounting for the designation given this design: barrel reactor. The second type has internal electrodes and the wafers are loaded horizontally (Fig. 4b). The electrode arrangement usually is planar, accounting for this design's designation: parallel-plate reactor.

The large batch capacity of the barrel reactor is its most attractive feature and makes it ideal for such noncritical uses as photoresist stripping. However, such a reactor generally cannot be used for anisotropic etching, for etching aluminum, or for selective etching of silicon dioxide over silicon at practical rates. Thus this type cannot be used for most of VLSI's high-resolution, exacting etch operations.

Despite smaller batch capacities, parallel-plate reactors are finding increasing use in semiconductor production because they can perform high-resolution anisotropic etching. Moreover, they allow greater selectivity and can do certain operations impossible with barrel reactors, such as etching contact windows and aluminum.

A particularly effective parallel-plate design is what is called the radial-flow reactor, shown in Fig. 4b. It was designed by A. R. Reinberg of Texas Instruments Inc. for uniform plasma deposition of insulator films over a large area. Its flow geometry tends to compensate automatically for the effects of gas-phase depletion and electron-density gradients.

The radial-flow principle will work equally well for etching. Parallel-plate reactors recently have become available from several commercial suppliers; however none employ the radial-flow principle exactly as developed by Reinberg.

Advantages and limitations

While plasma equipment is justly extolled by its manufacturers as dry, clean, and safe, some of the volatile reaction products are toxic and corrosive gases, which must be handled respectfully. On the other hand, the reaction products from stripping of photoresists in an oxygen plasma are primarily water, carbon dioxide, and carbon monoxide, which are much easier to handle than are the phenols or heavy-metal acids present in the commercially available wet chemical stripping agents.

Also, plasma-assisted etching is fully compatible with present photoresist technology, and resist adhesion is seldom the problem it can be with wet chemical etching. Thus it often is possible to eliminate intermediate masking layers that compensate for the lack of adhesion.

Another bonus with most commercial plasma reactors is their fully automated operation. The wet chemical processes usually are manual or semiautomatic.

For the design and processing engineer, of course, the main advantage is the high resolution obtainable in pattern transfer. If 1-to-2-μm features can be reliably imaged in the resist, plasma-assisted etching can replicate them in silicon, polysilicon, silicon nitride, silicon dioxide, aluminum, and so on. The isotropic physical processes involved in wet chemical etching means there is now no way to obtain comparable results.

Plasma-assisted etching is not without its limitations, such as the restricted batch size in parallel-plate reactors. Moreover, with some IC materials, etch selectivity compares poorly with wet chemical etching. For example, plasma etching of silicon dioxide over silicon generally results in selectivities on the order of 12:1, depending on the doping levels of the constituent materials. In contrast, buffered hydrofluoric acid can give a nearly infinite etch-rate ratio.

Another problem with plasma-assisted etching can be the presence of unwanted particulates on the wafer. Wet chemical etching often is accompanied by some type of agitation, so it is less susceptible to such contamination.

Finally, plasma-assisted etching may result in damage to ICs through photon or charged-particle bombardment. This type of damage tends to vary from one level of the device to another and has not been fully characterized yet. Fortunately, it often can be annealed in the standard furnace operations found in production.

Plasma-assisted etching at Bell Laboratories and Western Electric Inc. generally reflects use throughout the electronics industries. G. K. Herb of Bell Labs has developed production processes for chip separation,

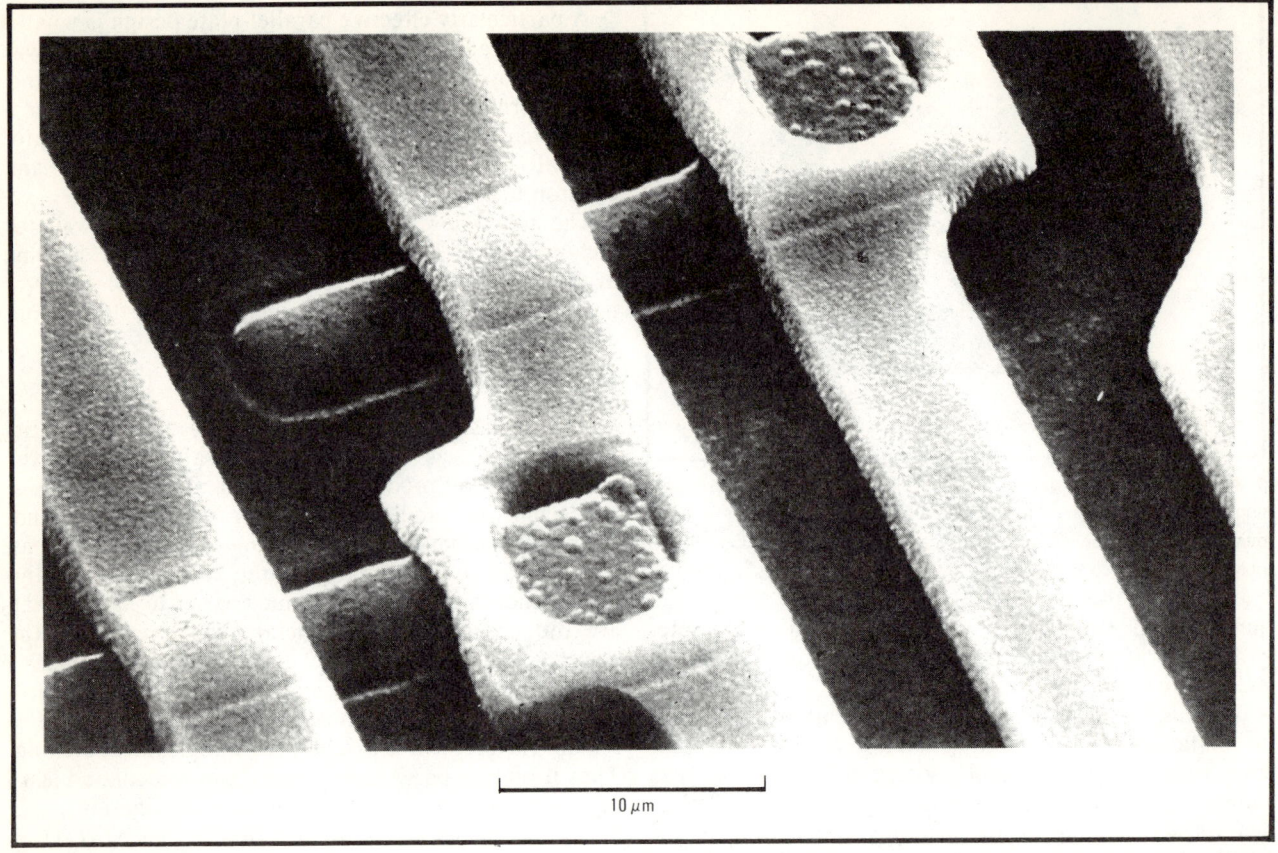

6. Etching aluminum. A recent development in plasma etching is the introduction of plasma reactors that can etch aluminum. These anisotropically plasma-etched aluminum conductors are on a 16-K MOS dynamic random-access memory used in Bell System equipment.

using commercially available barrel reactors. They also are used to strip photoresist and clean substrates during processing. Parallel-plate reactors have been used for several years to isotropically etch silicon nitride and polysilicon.

Anisotropic etching at Bell

More recently, Western Electric has begun anisotropic plasma etching in IC manufacture. For the past year, plasma etching has been used to define contact windows in PSG on a 16-K n-channel MOS random-access memory manufactured for Bell System applications.

Plasma-assisted etch technology also has been extended into new areas. R. A. Porter of Bell Labs has developed an anisotropic etching process for polysilicon (Fig. 5), now being introduced on production lines. D. B. Fraser of Bell Labs has developed anisotropic etching of aluminum films, also being introduced into manufacture (Fig. 6).

Evaluation of these new etching processes generally discloses that, for devices with large design rules of 6 to 8 μm, yields and reliability are not significantly different than with wet chemical etching. However with 2-to-5-μm design rules, with their tighter dimensions and tolerances, there is no comparison between the two. Devices and structures beyond the capabilities of wet chemical processes are routinely manufactured with plasma-assisted techniques.

Several developments may be expected in plasma-assisted etching for semiconductor processing. First of these are improvements in commercial equipment, such as scaled-up parallel-plate reactors for bigger batch sizes. Also coming are novel reactor designs with high throughput capability. In-line wafer handling with serial, rather than batch, wafer handling is the next step in this evolutionary process and should appear soon.

Improvements in the etching of some materials will be made to increase marginal selectivities. Control over resist erosion for particularly sensitive materials such as electron resists will improve, making plasma techniques fully compatible with the most advanced lithographies. Moreover, plasma-assisted etching will expand into technologies other than silicon: III-V-compound semiconductors, magnetic-bubble devices, integrated optics, etc.

Flexible edge profiles

Better understanding of the influence of charged particles on the development of edge profiles during etching in reactive gases may ultimately permit complete flexibility in controlling these profiles, producing variable tapers as well as straight walls.

Finally, as designers and process engineers become more familiar with plasma processing, they will evolve new sequences to exploit its advantages. To date, plasma etching has been used merely as a one-to-one replacement for wet chemical etching in the etching and stripping of thin films. However, the ability of plasma-assisted etching to produce anisotropic plus isotropic etch profiles should lead to the fabrication of device structures impossible with wet chemical etching. □

Part 2
Thick- and Thin-film Hybrids

Hybrid-circuit technology keeps rolling along

Riding high on a surge of technological innovation, improved hybrids are triumphing in new fields as a cost-effective circuit-assembly technique

Hybrid techniques

Advances in materials, components, processes, ensure hybrid prosperity in the LSI age

by Jerry Lyman
Packaging & Production Editor

☐ The billion-dollar hybrid-circuit industry is starting to transcend its old military/aerospace limits and expand into the more lucrative, high-volume fields of computers, business machines, telecommunications, and industrial equipment. In these areas, the attraction of the technology is its cost-effectiveness and its reliability as a circuit-assembly technique. These two factors have, for example, made hybrid circuits the No. 1 assembly method in the mass markets of automotive electronics and digital watches.

Additionally, new developments in IC chip packaging, which get away from the use of bare chips, are allowing almost complete automation of the hybrid process—an automation that is lowering costs while raising yields and reliability. Other cost-cutting developments in the process are the availability of new nonceramic substrates, the development of lower-cost conductive and resistive thick-film inks, and the use of computer-aided design (CAD) to generate complex artwork.

Monolithic semiconductor technology is both the hybrid's greatest competitor and an additional stimulus to the success and expansion of hybrid technology. There have always been predictions that, as IC chips get even more complex, the need for hybrids will be eliminated. And in many cases it's true that the new LSI has eliminated older hybrid designs.

But overall the new LSI has simply become another component for still larger and still denser hybrids. For example, extremely large memories have been built with standard 4,096-bit random-access-memory chips on large multilayered substrates. Hybrids seem to be evolving into a multilayered interconnect package with only active chips on the surface—a sort of super chip carrier.

While LSI will always be denser and less costly than hybrids, the older technology can still boast lower tooling costs and quicker turnaround, the ability to mix different logic families plus discretes, higher power dissipation, reparability and easy modification, better performance, and the capability of matching pairs of active or passive discrete components.

LSI, a mainly digital technique, has forced the hybrid industry into the use of large multilayered hybrids. The multileaded LSI chips call for an extremely dense interconnection pattern that cannot be achieved on one level. This has led to hybrid thick-film packages with up to 15 conductive layers, having chip densities of 20 to 40 units per square inch. This density cannot be achieved by any other packing technique.

Thick-film technology is going off in all sorts of directions to meet the demand for more multilayering, while also responding to a level cost-reduction call. The technology is in the process of adopting new inks, new substrates, and new multilayering methods, while new lower-cost methods of component attachment are becoming available for both technologies. In general, the thin-film process has become a mature technology, with few changes seemingly in store for the future except possibly a method of multilayering or use of radically new substrates.

Hybrids today

About 70% of all hybrids produced are thick-film types, the rest being thin-film. Thick-film hybrids are circuits or systems built by screening and firing alternate conductive, resistive, and dielectric pastes onto a ceramic substrate. Thin-film hybrids have resistors and conductors vacuum-deposited onto a flat substrate. Active components are attached to both types.

The basic advantages of hybrid technology are:
- Small size and weight.
- Wide operating frequency range because of short lines and low capacitance.
- Increased reliability because of a reduction in interconnections.
- Functional trimmability.
- Compatibility with all types of active and passive chips and packages.
- Efficient thermal dissipation.

1. Nonceramic substrates. Specially plated, epoxy-glass boards manufactured by Dyna-Craft can be used for hybrid applications like digital watches, calculators or displays, where IC chips are epoxy-die-attached and ultrasonically wire-bonded to substrates.

- Easy conversion of thick film to automated operation for cost reduction.
- Economical small production runs.
- Ease of achieving close component tolerances.

Thick or thin

The two hybrid technologies—thick and thin film—are not competitors. In fact, each has carved out separate application areas defined by the properties of each method. This is why most hybrid manufacturers have both a thin- and thick-film hybrid capability. Since thin film can pack interconnecting lines (2-mil lines and 2-mil spaces) more densely than thick-film (10-mil lines and 10-mil spaces), it is ideally suited for high-density single-sided packages. The properties of thin-film resistive metals (nichrome or tantalum nitride) make for resistors with lower temperature and ratio-tracking coefficients than is possible with thick-film types. For instance, thin-film resistors have temperature coefficients in the range of ±25 parts per million per degree celsius. A pair of these resistors will track at ±1 ppm/°C. Thick-film types have a temperature coefficient in the 100-to-250-ppm/°C range, and pairs will track to ±20 ppm/°C. The low noise level of thin-film resistors makes them suitable for the preamplifier applications.

Thin-film resistors have a limit of 5,000 to 1 in the range of resistor values that can be fabricated on a given substrate. The same parameter for thick film is as high as 10^7 to 1. In addition, thick-film capacitors are widely available, whereas thin-film technology has mainly an interconnective and resistive capability.

In microwave applications thin-film technology has earned a dominant position over thick-film. This is due to the requirement for fine line widths where interconnections are part of a transmission line or path.

Thick-film hybrids are less costly to make since the basic equipment needed (inks, screens, screeners, ovens) is a smaller investment than complex evaporation or sputtering equipment for thin films. Another great advantage of thick films is that they can be multilayered for extremely dense applications. Thin-film hybrid multilayering is still in the laboratory stage.

At times, the two technologies are blended. It is quite common to find thin-film resistor networks on glass or silicon substrates used as chip components on thick-film hybrids on ceramic substrates. This often occurs in data converters, voltage regulators, or any thick-film application requiring precision resistors.

In microwave applications at RCA's Government and Commercial Systems Missile and Surface Radar division, Moorestown, N.J., for instance, a thick-film resistor is screened and fired on an ultrasmooth ceramic substrate. Then the fine-line gold interconnections are put on with standard thin-film techniques.

Another microwave application of mixed hybrid technology takes place at Applied Technology, a division of Itek in Mountain View, Calif. The bottom of a hybrid substrate is screened as a ground plane, and the top side has a thin-film interconnect sputtered on. A thin-film line is sputtered around one edge of the substrate to tie the ground plane to the top of the substrate. The sputtering process deposits a thin metal layer on a surface in a vacuum, using a cathode made of the metal.

In general, thin film's accurate and stable resistors allow the manufacture of high-accuracy circuits such as sample-and-hold, precision voltage references, digital-to-analog and analog-to-digital converters, instrumentation amplifiers and preamplifiers, and active filters. Thick-film hybrids are suited to almost any application (especially if they use thin-film resistors). As far as the high-volume consumer, industrial, and automotive fields are concerned, it is a thick-film world on the basis of cost alone. Also, thick-film hybrids have a much higher power capability (up to hundreds of watts) than thin-film hybrids, which basically must operate at a low power level because of the low current-carrying capability of their fine lines.

Hybrid thin-film processes

Thin-film hybrid circuitry encompasses three approaches. Resistors may be deposited directly on a monolithic chip and laser-trimmed on the chip to provide improved performance, or they may be deposited on an oxidized silicon or ceramic wafer. Typically, 60 to 500 resistor networks are delineated by photoresist techniques on a single 2-inch wafer. The wafer is subsequently probed, scribed, diced, and inspected. The individual resistor-network dies are then treated like semiconductor or IC chips.

Thin-film resistors also may be deposited by evaporation or sputtering on a ceramic circuit substrate. Photoetching can lay out the resistor network, as well as the interconnecting path and chip pads. Following final assembly of the IC and capacitor chips, the circuit can be completed by trimming the resistors.

Usually nickel chromium (nichrome) or tantalum nitride is picked as the resistor material, and gold is used for the conductors. In the process used by Micro Networks, Worcester, Mass.—it's representative of today's techniques—nichrome is deposited on a substrate of silicon, glass, or ceramic under computer regulation of pressure, deposition rate, and the oxidation of nickel and chromium. Without breaking the vacuum, a layer of nickel is evaporated, followed by a layer of gold. The gold is used for interconnections, while the nickel is

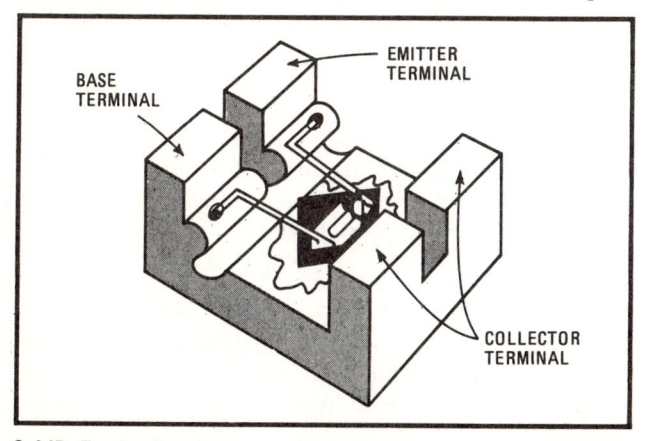

2. LID. The leadless inverted device (LID) is a ceramic body with elevated terminal pads. Here, a transistor has been die-attached and wire-bonded. The entire unit is sealed with a drop of epoxy, making the device almost immune to damage from handling.

Hybrid techniques

used to prevent interaction between the nichrome and gold. Then photolithographic maskings etch the resistors and conductor paths on the substrates.

Thin-film-hybrid manufacturing resembles the manufacture of ICs and semiconductors, without the diffusion steps. Thick-film manufacturing, on the other hand, is a process closer to printing. Substrates are manufactured by a screen-and-fire process. The precision, stainless-steel screens have 80 to 400 openings per inch. A precise image of a hybrid's connection patterns is projected onto a photographic emulsion on the screen, sensitized, and developed. A simple one-sided hybrid with screened-on resistors might only require two to four screens; a multilayered hybrid would require perhaps two screens per layer.

The thick-film process

Usually a squeegee forces conductive ink through the screen onto a ceramic substrate. After air drying, the substrate is fired in a furnace with a carefully controlled temperature profile. Next, a dielectric or resistive ink is screened onto the ceramic, and the procedure is repeated. Unlike the thin-film process, this method lends itself to automated screeners feeding substrates on a conveyor belt into automated ovens.

The result is the 10-mil-wide conductive patterns that have been accepted as a standard by most U.S. thick-film-hybrid manufacturers. This spacing has been found to give much better yields than the narrower lines that can be fabricated (as low as 2 mils). On the whole, the thick-film hybrid is getting away from thin lines. Many manufacturers would rather add a layer of interconnections than go to finer lines on one layer. Several companies are getting finer conductive lines by first screening and firing on conductive inks and then etching the width of the lines down.

In many ways the hybrid thick-film process resembles silk-screening. The ceramic substrate is equivalent to canvas or paper, the stainless-steel screen is equivalent to a silk screen, and the various hybrid inks or pastes are equivalent to paints.

On the substrate, the screenable thick-film materials look rather like artwork, but they serve very practical purposes—ending up as conductors, bonding pads, vias, resistors, insulating layers, capacitors, adhesives, solders, and surfaces to which later on to attach lead frames or covers.

Hybrid inks

Any paste, or ink, has three components: a functional phase, a binder, and a vehicle or carrier. The functional phase is the chemical compound that will serve as conductor, resistor, dielectric, or capacitor. The binder cements the functional phase to the substrate. The vehicle or carrier is a liquid, containing the first two elements, that makes the ink suitable for screening.

For a conductive paste, for instance, the functional phase may be a powdered metal or powdered metal alloy, while suitable resistive or dielectric materials are introduced for resistive and dielectric pastes. The binder may be glass or occasionally a reactive oxide, and the vehicle is composed of resins and solids.

Conductive pastes available are pure gold, silver, and nickel, compounds like platinum-gold, platinum-silver-gold, palladium-gold, palladium-silver, palladium-copper-silver, and platinum-silver. For military, aerospace, or avionic hybrid hardware where active and passive chips are eutectically die-bonded and ultrasonic or thermocompression wire-bonding is used, gold is the preferred conductor. Most of the other materials are more suited for reflow-soldering of components.

Resistive pastes available are mostly based on ruthenium oxides and cover a resistance range of 3 ohms per square to 10 megohms per square, with temperature coefficients as good as 50 ppm/°C. Capacitor dielectrics of barium titanate are available with dielectric constants ranging from 8 to 1,500 along with a whole range of temperature coefficients.

Non-noble inks

Copper is the only high-conductivity material that is both bondable and solderable, the two methods of attachment used with hybrids. In addition, it has good adhesion to alumina, good thermal conductivity, good solder leach-resistance, good microwave properties, and good radiation resistance. Moreover, it will not migrate into ceramic.

These important properties are why the laboratories of almost all thick-film-hybrid manufacturers are evaluating new nitrogen-fired copper and dielectric inks. Sample quantities are at present becoming available

3. Flip chip. Solder bumps on the chip metalization are the key to economical assembly of the reflow-soldered flip chip. Advantages of this leadless package are lower assembly costs, greater yield than the chip-and-wire process, and an ability to be easily tested.

from Du Pont, Wilmington, Del., Electro Materials Corp. of America (EMCA), Mamaroneck, N.Y., Electro Science Laboratories Inc. (ESL), Pennsauken, N.J., and Thick Film Systems, Santa Barbara, Calif.

Even though the basic cost of copper is much lower than gold, manufacturing costs have made copper paste only slightly less expensive than gold pastes and more expensive than silver. So there will be no great savings in using copper unless the price of gold skyrockets again. Adrian Rose, technical service manager at ESL, predicts the copper pastes will penetrate into industrial and computer applications and then into automotive and consumer use. Most of these applications use reflow soldering in which heat is applied to two materials that have already been coated with solder.

Copper does have several disadvantages. It requires a nitrogen-fired furnace, while the hybrid industry is based on air-fired types. Also, it offers no screenable resistor material, although this is in the research and development stage at several ink companies.

Another non-noble ink in fairly heavy use is nickel, for screening interconnections on the ceramic boards of plasma displays. Unlike gold, nickel conductors do not sputter from the plasma's action—something that would contaminate the display.

Resistors on top

The surface of the dielectric layer in hybrids is neither smooth nor stable enough to support a thick-film resistor. The stress and strains of the dielectric can change the value and temperature coefficient of the screened resistor. The problem is to match the dielectric to the resistor, but the solution usually is to use discrete passive components.

However, ESL and EMCA now have dielectrics designed for screened-on resistors, and Du Pont will come out with a screenable dielectric fairly soon. Thick Film Systems' engineers have come up with a different approach. They have a special ruthenium-based resistor ink that can be screened on top of a mating dielectric.

Various hybrid companies are evaluating both approaches, and it will probably take at least a year for any decision. At any rate, there is another problem with screened-on passives: how to trim a resistor on the top of a multilayered substrate without causing a short or open in the next layer. Jason Provance, vice president of Thick Film Systems, suggests lowering the power on the laser trimmer or using abrasive trimming as possible solutions to the difficulty.

Ceramic and nonceramic substances

Most hybrids are constructed on aluminum-oxide ceramic substrates, with occasional use of beryllium oxide for a substrate that must dissipate high power (BeO has about six times the thermal conductivity of alumina). Thick-film circuitry is usually screened onto a 96% pure alumina (Al_2O_3) substrate, which has a surface finish in the 25-microinch range. Thin films, on the other hand, are sputtered or evaporated onto a 99.5% alumina substrate with a smooth surface finish ranging from 10 to 1 microinches.

To house the large hybrids being produced, alumina substrates are being supplied up to a 6-by-6-in. size. For minihybrids and passive networks, substrates as small as 0.05 by 0.05 in. are available.

In the small runs of typical custom military-aerospace hybrids, the 10-cents-per-square-inch cost of the ceramic is negligible, compared to the cost of semiconductor chips and labor. However, in applications that use ICs in chip form, like calculators or digital watches, where product runs go to at least 50,000 units, cost of the substrate is important. So the Mica Corp., Culver City, Calif., and others have developed rigid plastic substrates clad on one or both sides with a thin copper foil (typically 200 microinches thick).

Mica's new copper-coated plastic laminates range in price from 3 cents per square inch for glass-epoxy to 6 cents per square inch for a Xylok type (a high-temperature resin) to 10 cents per square inch for types based on Triazine A or polyimide.

Aside from the lower cost, the metalized substrates provide panel sizes up to 18 by 24 in. (vs 4 by 4 in. for ceramic). Extremely fine lines are possible because, unlike thick film, these circuits do not require screen printing (production line widths are 4 mils compared to 10 mils using thick-film technology).

High line density is achievable and repeatable at high production yields at much lower cost than thin-film processes on ceramic. The substrates may be multilayered, and active and passive chips can be bonded to the laminates using standard techniques.

For hybrid substrates that are to have active chips attached, gold-nickel fine-line conductor patterns are plated on, using the copper cladding as a plating electrode. A single immersion in a suitable etchant quickly removes the remaining thin copper.

With the proper platings on the etched copper pat-

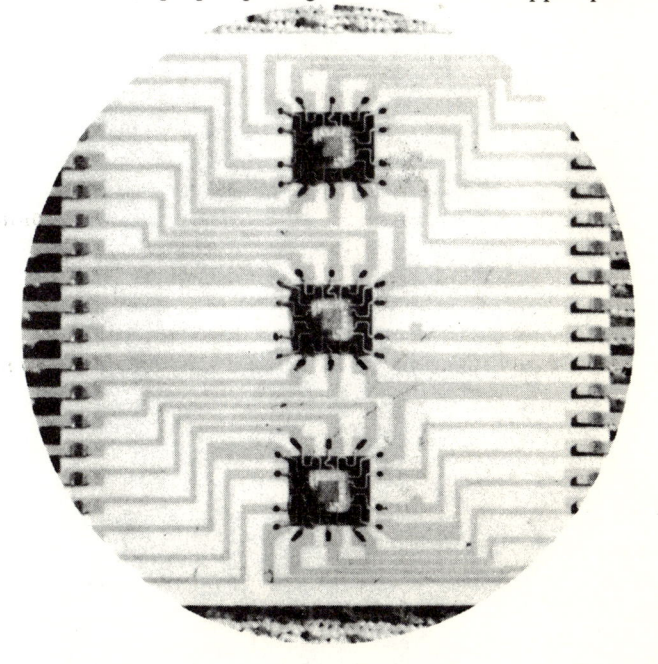

4. Chips on film. A prototype multichip thin-film assembly, built at RCA's Solid State division, uses pretested IC chips on tape carriers. The film-carrier packaged chips are wire-bonded to the thin-film conductive pattern by semi-automated machinery.

Hybrid techniques

terns, it is generally possible to use ultrasonic aluminum wire bonding on glass-epoxy substrates; ultrasonic gold-ball bonding on polyimide- and Triazine-A-based substrates; pulsed thermocompression bonding and solder reflow on all of the four Mica substrates, and thermocompression on the Xylok and Triazine-A substrates.

Other boards

Manufacturing wire-bondable substrates out of glass-epoxy laminate is not exclusive to Mica. Dyna Craft Inc., Santa Clara, Calif., a subsidiary of National Semiconductor, produces large numbers of fine-line specially plated glass-epoxy substrates (Fig. 1) for its parent's displays and calculators.

Most of the hybrid developments at Johns Hopkins University's Applied Physics Laboratory, Laurel, Md., are in the thin-film field. However R.E. Hicks and D. Zimmerman are using hybrid substrates made from Pyralin, a copper-clad polyimide glass-epoxy laminate [*Electronics,* June 12, p. 36]. This approach was successfully used in an experimental navigational receiver, and the technique has since been applied to other systems at the lab.

Packaging engineer Fred Muccino, of APL, prefers to use a copper-coated glass-epoxy board as a basic substrate instead of alumina. With glass-epoxy, he points out, all fabrication uses standard photoetching processing; ordinary tools are adequate for drilling and shaping, and there is no loss of material during modification or parts replacement.

Components attached to Muccino's nonceramic substrate are chip capacitors and resistors. The semiconductors, diodes, and transistors are leadless inverted devices. Solder cream is used to attach all chip components to the substrate, with a hot plate reflowing the solder at the interface. With this attachment process, components could be changed as many as 20 times—there is no loss of metallic contact as with thick- or thin-film techniques after only a few chip removals.

Getting attached actively

In the customized hybrid fields of avionics, military, aerospace, and medical implants, where relatively few high-reliability items are turned out, there are two main techniques of attaching active components to the hybrid substrate. They are the chip-and-wire method and, to a much lesser extent, the use of beam-leaded chips.

In the chip-and-wire method, all active components are in the chip, or die, form. Inspection is a significant problem, because the microscopic probing equipment often damages the chip under test. After inspection, chips are connected to the substrate by the eutectic or epoxy method.

Eutectic bonding brazes the silicon of the die to the gold of the pad at about 400°C, creating a gold-silicon alloy. Even though this process is highly reliable, it is expensive and does not permit replacement of a chip.

This has brought on the use of organic adhesives such as epoxy as an alternative to eutectic attach. Parts attached with epoxy can be replaced fairly easily. However, the military still tends to favor eutectic bonding since it feels there isn't enough long-term experience with epoxy bonding. In commercial hybrids using chip-and-wire construction, all die attachment is with epoxy.

After die-bonding, fine metal wires are thermocompression-bonded or ultrasonically bonded between the chip pads and the substrate. Thermocompression bonding consists of applying concentrated heat and pressure to the points where wires are attached to both chip and substrate. Ultrasonic wire-bonding consists of producing sufficient heat at the junction between the wire and pad to create a weld. Ultrasonic bonding usually uses aluminum or gold wires, while thermocompression uses gold wires.

Beam leads

To avoid the difficulties of the chip-and-wire approach, the beam-leaded chip was created. This device replaces the separate wire leads with gold beam leads, which are grown apart of the chip production process. The leads are basically extended input/output pads 4 to 5 mils wide. Instead of mechanically scribing and breaking the individual die from the wafer, an etching operation is used.

During assembly, the beams are aligned with the substrate pads. A wobble-bonding process is used, which heats each of the beams sequentially so that they make eutectic bonds with the pads. The width of the beams means the bonds are strong enough for separate die bonding *not* to be required as it is in the chip-and-wire method of attachment.

5. Ceramic chip carriers. The ceramic chip carriers, used to house integrated-circuit chips on this multilayered hybrid substrate, make active chips easy to test and replace. Small hybrid substrates may also be enclosed in the carriers for the same reasons.

Many hybrid companies are successfully using beam-lead chips. However, availability is a major problem, since only Texas Instruments, Raytheon, and Motorola manufacture them. Also, only a limited number of IC types is available, which often leaves manufacturers no choice but to mix beam-leaded and conventional chips.

Even with automatic bonding, both chip-and-wire and beam-lead devices are bonded serially. And the great amount of handling in wafer probing, scribing, chip probing, several optical inspections, and bonding combine to keep the overall yield down. So four different approaches—the LID, the flip chip, the film carrier, and the chip carrier—have evolved to produce chip packages for hybrids that can be batch-processed by the reflow-soldering process, are testable without damage, and can be handled easily.

LIDs and flip chips.

The LID (leadless inverted device) is basically a chip in a ceramic body with elevated terminal pads made with a three-layer metalization (molymanganese, nickel, and gold). An IC or transistor chip is eutectically die-bonded to a cavity on the body. Then the chip's pads are wire-bonded to the LID's pads as shown in Fig. 2. After inspection, the chip and leads are sealed with a drop of epoxy, resulting in a strong ceramic package.

Sole supplier for the device is Amperex Electronic Corp., Slatersville, R.I., which offers a good range of active devices in LIDs. The smallest—discrete transistors, diodes, and field-effect transistors—are 43 by 75 by 35 mils. Medium-size LIDs, used for 14-and 16-pin ICs, are typically 190 mils square. A 32-pin, 300-mil-square package is Amperex's largest LID.

Amperex can put a customer-specified chip in a LID, but tends to shy away from handling proprietary devices because it has no control over the chip screening, probing, and inspection.

Perhaps the main advantage of the LID is that it can be reflow-soldered, eliminating the need for the skilled operators required for wire bonding. "LIDs can be attached to a substrate in 10 to 15 seconds in a batch operation," says Ronald Goga, marketing manager of Amperex. "Since the LID is completely encapsulated in epoxy, it can take a tremendous amount of shock. We've had devices survive 60,000 g."

Another big advantage is increased hybrid-circuit yields. Goga says that, in Amperex' own thick-film hybrid operation, yields of 90% are realized the first time through on a finished unit, vs about 50% for a chip-and-wire hybrid. And he says that any parts that do fall out are easily replaced.

The complete encapsulation of the chips eliminates the breakage problems inherent in testing and handling. In addition, LIDs may be tested for such parameters as switching time, high- and low-current operation, matching transistor pairs, special selections at odd biases, and other parameters—in contrast with bare chips, which usually are probed only for dc static parameters.

LIDs have two disadvantages. Not every digital logic or linear IC is available off the shelf from Amperex. Also, for high-density hybrid packages, the 14- and 16-pin ICs take up much more real estate than a bare chip.

Nevertheless, the LID is in fairly heavy use. For instance Motorola Communications, Fort Lauderdale, Fla., uses them in a pager application, General Electric in Lynchburg, Va., is applying LIDs in mobile communications, and Applied Physics Laboratory in Silver Springs, Md., is using LIDs in a pacemaker hybrid module and communication hybrids.

Flip chips

Another alternative to the bare chip is the flip chip. This is a semiconductor device designed for face-down reflow-soldering on any type of hybrid. (Chip-and-wire devices require face-up wire bonding.) This technology, developed by IBM in the 1960s, is known there as the "controlled-collapse technique." The main difference between them and conventional IC chips is that all flip-chip contacts are made on the active (front) side of the chip. The chip itself is protected by a glass layer over the active area and its metalization. Solder bumps are formed on the chip for the reflow connections.

Flip-chip bumps are nominally 95% lead and 5% tin and, on transistor chips, are on 0.018-inch centers (Fig. 3). The bump is typically 6 mils in diameter and 4 mils high. The chips are 30 by 30 mils in area and 11 to 13 mils thick.

Micro Components Corp., Cranston, R.I., and Motorola Semiconductor Products group, Phoenix, Ariz., are the only open-market suppliers of flip-chip devices in the United States. IBM, Delco, and Fairchild Semiconductor Products, Mountain View, Calif., manufacture their own devices, but they are not for sale. Both Micro Components and Motorola offer only a limited number of bumped ICs and discretes—mainly discrete transistors, diodes, FETs, and linear ICs. Both companies will "bump" a customer's own chip. The largest IC from ei-

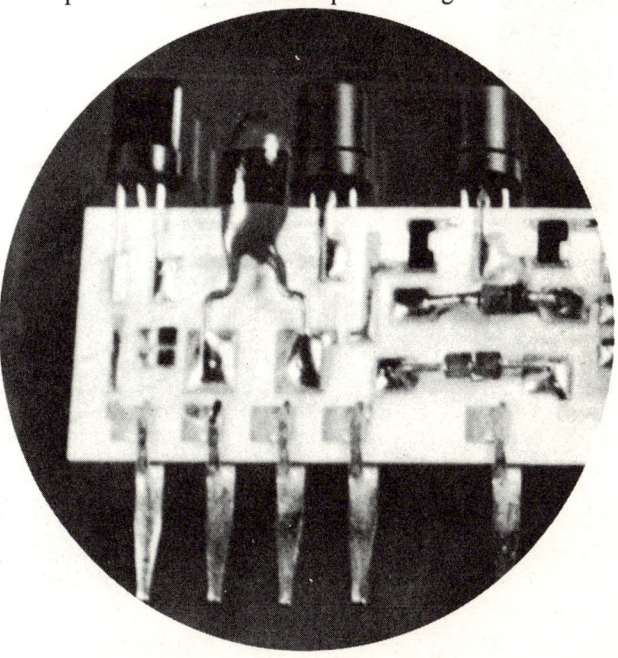

6. Hybrids with discretes. Much hybrid business involves flow soldering of discrete components to ceramic substrates, rather than bonding of bare chips. Shown is a Centralab tachometer driver using discrete components and screened-on resistors.

Hybrid techniques

ther company has 16 leads. However, devices in the 40-bump range have been demonstrated experimentally.

"The chief advantage of solder-bumped chips is their cost-saving potential for the hybrid customer," says Egons Rasmanis, vice president of sales for Micro Components. He estimates that the cost to a hybrid maker is 3 cents per wire bond. For a 16-pin IC then, the manufacturer saves 96 cents per IC position by using solder bumps (16 wires by two bonds per wire by 3 cents).

In general, the cost of assembling a flip chip into a hybrid circuit, regardless of the number of leads involved, is even less than the cost of eutectic diebonding. The labor and material costs of wire bonding are completely eliminated. In addition, savings are realized because of higher yields in assembly, lower capital-equipment requirements, and excellent rework capability.

The glass-coated chips also enhance reliability and eliminate requirements for expensive hermetic encapsulation of individual components. Also, the bumped devices are cut to very accurate dimensions and can be automatically handled and tested without damage. Probing or testing is through the solder bumps, and scratches fortunately just disappear during the reflow-solder operation.

The chief disadvantage of flip-chip devices today is that there just isn't enough variety of device types—particularly in digital ICs. "There is no reason why most digital logic families can't be bumped, but we won't do it until there is a demand from our customers," says Dave Cavanaugh, product engineering manager for flip chips at Motorola.

Both Micro Components and Motorola Semiconductor have been approached by companies who wanted to apply the solder-bumped chips to film carriers—which is another method for eliminating lead bonding on hybrids.

The film-carrier system uses a sprocketed, nonconductive film with a copper surface etched into IC micro-interconnections [*Electronics*, Dec. 25, 1975, p. 61]. Specially bumped IC chips are automatically gang-bonded to the inner leads of the tape interconnects. In a hybrid operation, the chips would be tested on the tape and then cut out, placed on a substrate, and either soldered or wire-bonded down.

This method lends itself to automatic attachment of the chip to the hybrid. However, it isn't possible to buy chips already on tape; bumped chips with the proper metallurgy aren't available, and low-cost equipment for bonding the tape is lacking. So this method is still very much in the future. Honeywell Information Systems, Phoenix, Ariz., has been applying film carriers to large thick-film multilayer hybrids, for some time. RCA in Somerville, N.J., has bonded film carriers to thin-film hybrids (Fig. 4).

Ceramic carriers

Still another method of avoiding lead bonding to a substrate is the ceramic chip-carrier: small, leadless, square, cofired ceramic gold-plated packages, in a variety of sizes. Lead metalization is connected internally to gold solder pads on the bottom face of the units.

These packages allow a user to employ standard die-attachment and to wire-bond a chip to the cavity of the carrier. In this way he dodges the problem of limited availability of LIDs, flip chips, and film-carrier hybrids.

Once in the carrier, the encased chip can be tested in a computer-controlled IC tester with a special socket for accepting the chip carrier. Then it is sealed and reflow-

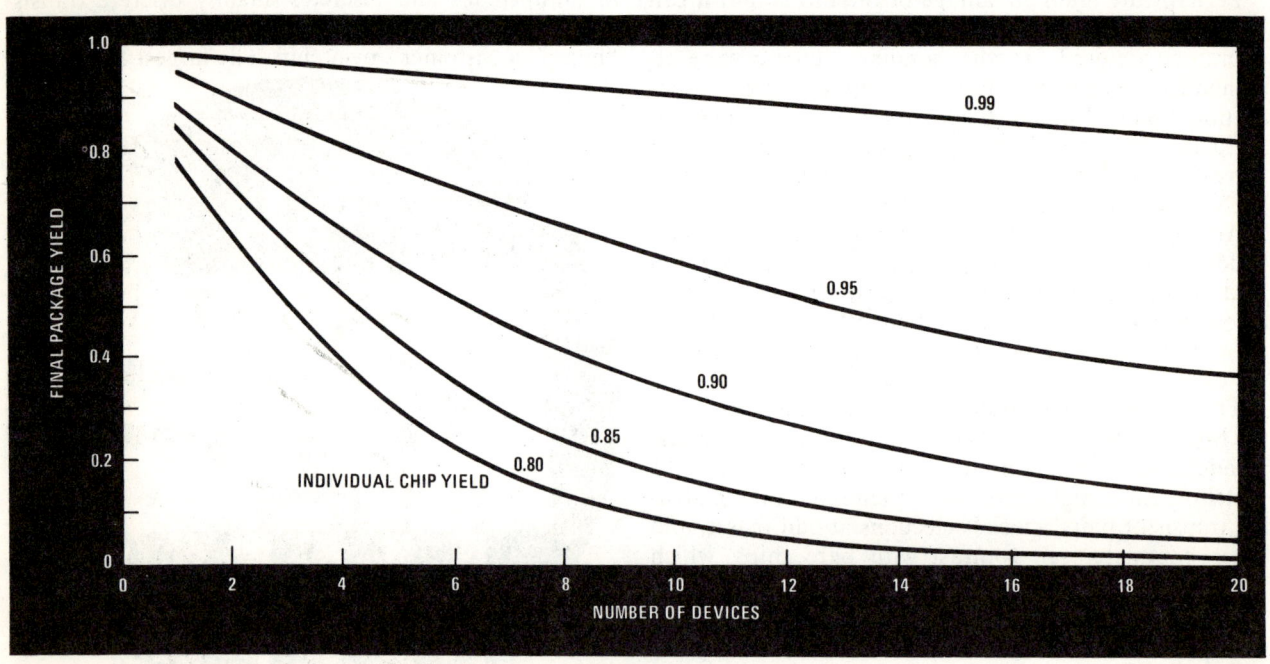

7. Chip yield vs hybrid yield. Active-chip yield may drastically affect overall hybrid yield. For instance, a chip yield of 0.8 gives a hybrid yield of about 0.15 for a 10-chip package. A chip yield of 0.95 gives a hybrid yield of 0.62 for the same unit.

soldered to a ceramic substrate. Even with the carrier sealed, it is still testable because the leads extend through the carrier. Also it is relatively easy to remove a ceramic carrier from a substrate and substitute a new chip in a carrier.

About the only disadvantage of the carrier is that it is relatively large. However, it is extremely well suited for expensive LSI, like memories and microprocessors, where the large hybrid substrate often has very little passive circuitry.

RCA, Moorestown, N.J., is using the ceramic chip carriers in a series of plug-in programable waveform modules housed on standard plug-in ceramic multilayer boards. An RCA fast-Fourier-transform memory module with 12 chip carriers and 2 flatpacks is shown in Fig. 5.

RCA has built up special test substrates with square Zebra elastomeric conductors resting on each carrier's substrate pads. A Zebra conductor is a frame of alternating vertical conductive and nonconductive strips. It is a solderless interconnection between carrier and substrate. To test a chip carrier, it is simply pressed against its own Zebra on the motherboard. This saves the trouble of soldering a bad chip in a carrier to the motherboard substrate.

Attaching discretes

While ultraminiature hybrids use chip and wire and the alternatives to it, a great part of the worldwide hybrid industry is committed to reflow-soldering standard packaged devices onto ceramic substrates with screened-on resistors and conductors.

An example is the Centralab thick-film tachometer driver shown in Fig. 6. It has three epoxy-cased low-level transistors, a power transistor, a discrete tantalytic capacitor, discrete diodes and screened-on resistors. Unlike in chip-and-wire hybrids, there are more discrete components on the back of the substrate.

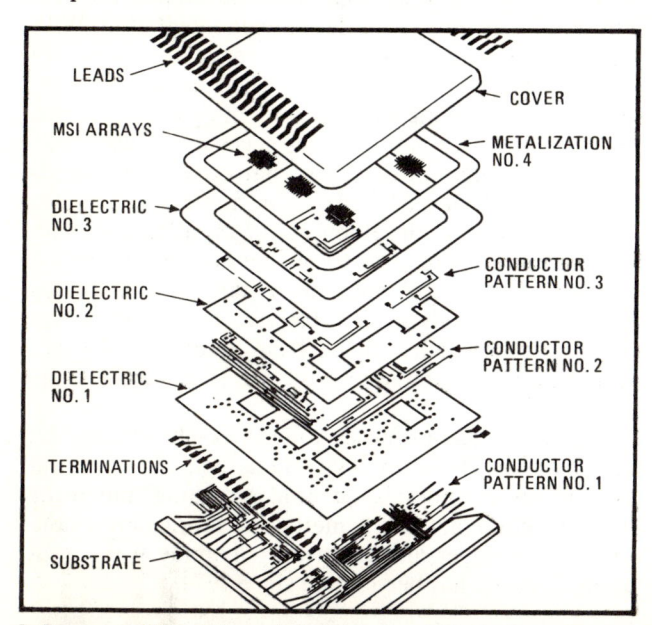

8. Screen and fire. This exploded view shows the buildup of a multilayered thick-film hybrid made by the sequentially screen-and-fire process. Screen printing can be repeated many times, and substrates with as many as 15 conductive layers have been fabricated.

Two years ago, Centralab, a Milwaukee, Wis., firm, had both chip-and-wire and solder-reflow capability, but it found the commercial market didn't want the more sophisticated packaging. "Automotive, data-processing, industrial-control, consumer-electronics, and telecommunications hybrid parts needed low-cost, high-yield, automated production, rather than high parts density," says Duane Kobs, the firm's general manager for thick-film circuits.

Centralab, where thick-film was first used in a proximity fuze during World War II, has a highly automated operation in which silver palladium or silver is the basic interconnection. Carbon or cermet resistors are screened on, and anything that may be reflow-soldered is attached to a substrate. Since this system is tied to off-the-shelf components, delivery is no problem. Cost is low because all discretes can be attached in one batch-soldering operation. Yet the usual advantages of hybrids—substantial size and weight reductions, better performance, and increased reliability—still exist.

Other companies in the discrete-component hybrid field are Sprague Electric Co., Nashua, N.H., Sprague Electromag, Belgium, and CIT-Alcatel in France. CIT-Alcatel, a telephone equipment producing company in the Compagnie Générale d'Electricité group, has a micro-electronics division that will turn out some 2.5 million hybrids this year. Most will be thin-film types from a fully automated production line.

The company's attitude toward active-component attachment, which reflects the opinion of most European hybrid manufacturers, is not quite the same as Sprague and Centralab in the U.S. "We use bare chips only when there's no other possibility. Whenever we can, we persuade customers to accept prepackaged chips with provision for connections by reflow soldering," says Claude François, sales manager.

"Sometimes—mainly for military equipment—the package has to be so tight that chips are necessary. Under these conditions, the active chip can't be fully tested until the circuit has been fabricated. As a result costs double or triple compared to a circuit having prepackaged chips. Many circuits have to be repaired or discarded."

François doesn't think chips are worthwhile unless the density of the package makes them mandatory. He'd like to see more beam leads, flip chips, or—best of all—chips on tape (film carriers). The film carriers would be ideal for the tape-based automatic production equipment that his firm has put into service.

Like CIT-Alcatel, all the other manufacturers in the U.S. and Europe who attach components by reflow-soldering are looking toward the flip chip to increase component density without losing the advantages of batch processing. Meanwhile, many European hybrid companies, CIT-Alcatel included, are reflow-soldering transistor, diode, and IC chips prepackaged in a small, leaded, plastic package onto their hybrid substrates. These components, which have several names, being known as Timtors (TI's version), SOT-23 (Philips, Amperex, Siemens), Micropax (SGS-ATES), and Minimold (Nippon Electric Corp.), are not manufactured in the U.S. and are just starting to become available here.

Hybrid techniques

Centralab is already evaluating these devices.

Now, mainly transistors and diodes in three-leaded packages (3 by 3 by 0.85 mm) are available in the U.S. and Europe. However, SGS-ATES, an Italian firm, is about to come out with a line of linear ICs in its Micropax package. Medium-power transistors are available from Amperex in a slightly large SOT-89 (4.6-by-2.6-by-1.6-mm) package.

As with the flip chip, there is a complete lack of digital types in this small plastic chip carrier. However these units, about the size of capacitor chips, can easily be handled by automated equipment for testing and placement and soldering.

Hybrid testing

Testing of hybrids, especially the chip-and-wire types, can be broken into three phases: active-chip testing, bare-substrate testing, and functional testing. Most hybrid manufacturers agree that malfunctioning active chips are the biggest cause of low hybrid yields.

Figure 7, from RCA, Moorestown, shows how chip yield affects hybrid yield. For a 10-chip hybrid, a chip yield of 0.8 gives a hybrid yield of about 0.15. A chip yield of 0.95 gives a hybrid yield of about 0.62. These results suggest that 100% inspection of IC and semiconductor chips is mandatory to ensure a good yield.

But most hybrid companies do not completely test every incoming chip and wafer because of the cost of inspection and the potential damage from handling of chips. On the assumption that they will pick up the bad chips in functional test, many companies order chips to their specification and test only samples.

However, several hybrid manufacturers do inspect incoming chips and wafers 100%. Typical of these quality-control-conscious companies are Teledyne Microelectronics, Circuit Technology, Raytheon Quincy, Lockheed Missiles and Space Co., and Autonetics. At Teledyne, a staff of 35 subjects each incoming wafer or package of chips to two visual inspections and electrical tests at 25°C and 125°C. "In the dice test area, the chip yields fluctuate like a yoyo from week to week" says Les Sutton, manager of application engineering at Teledyne. "Yields from the same vendor vary from 95% to 65%. If the yield kept constant, we could let up on inspection and set an acceptable quality level . . . Our average hybrid has 30 chips, so with 95% chip yield, we'd be in the rework business. We have made hybrids with 100 chips with no rework."

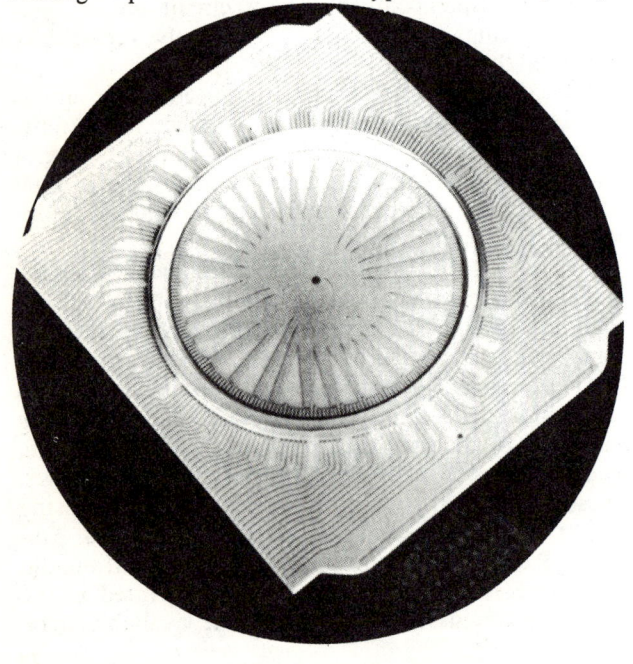

Substrate testing for simple one- or two-sided hybrids is usually confined to complete optical inspection and no electrical inspection. Larger samples of complex multilayer substrates are checked with IC probe cards, the resistance bridge of laser trimmers, and, in one instance, by a miniature bed of nails feeding into an automatic continuity tester. (A bed of nails is a test fixture with many spring pins on top.)

Most hybrids are functionally tested by only three types of automatic testers: the extremely large and costly ($500,000 and up) LSI testers like the Fairchild 5000 or Tektronix 3260, large functional testers like the HP 9500, and for certain jobs, calculator-controlled testers. The Fairchild and Tektronix testers are used extensively throughout the hybrid industry, since they can test either hybrids or ICs. The HP 9500 is especially suited for testing both analog and digital hybrids.

Multilayer hybrids

Multilayer, or multilevel, hybrids consist of conductive layers connected by thin vertical metal channels (vias) through layers of isolating dielectric. They have evolved as the original transistors developed into ICs with small-scale and, later, large-scale integration. With only transistors and diode chips, it was relatively easy to screen or evaporate the circuitry used in the 1960s on one side of a ceramic substrate. As the circuits became more complex, it was sometimes necessary to use crossovers: thin bridges of dielectric that isolate two successively screened conductors from one another.

Finally, with the development of the digital IC with 14-, 16-, 22-, 32- and even 40-pin connections, there just wasn't any way to put complex digital or digital-analog circuits on one side of a substrate. The solution was multilayering, which evolved from avionics and space work, where an extreme reduction in space and volume is vital.

For all practical purposes, only thick-film circuits are multilayered. There are two main methods of multilayering on ceramics: sequential screening and firing and the cofired method. A method of fabricating a nonceramic multilayer substrate with polyimide as the insulating dielectric has been in use since 1974.

9. Cofired multilayer. A complete silicon wafer (master slice) will be housed within the circular portion of this cofired multilayered ceramic substrate. The back surface of the substrate has a heat sink attached to dissipate the heat of the wafer.

The oldest and most common of the layering methods is the so-called sequential-screening process. The first step is to screen a conductive layer of ink on an alumina substrate and then fire it. The second step is to screen and dry a dielectric-layer with windows for vias. This step is repeated, and the two layers are fired. Then a second conductive layer is screened over the insulating layer, filling the vias. This layer is fired, and the process is repeated as required by the circuit or system interconnections. Figure 8 shows the steps in the sequential method of fabricating a four-layer thick-film hybrid.

Sequential screening

Hybrids with as many as 15 conductive layers have been made by this method, but most types in large production runs have two to six layers, since yield is reduced as the number of layers is increased. The advantages of this process are that it uses readily available equipment, it can be inspected at each step, it is cost-effective, and it is conducive to quick delivery.

The main disadvantage is that the process requires many steps—typically three screenings per conductive layer. Moreover, dielectric and resistive inks are not compatible, although several ink manufacturers are on the verge of solving this problem. But it is not yet possible to screen resistors into multilayer substrates—although by using partial multilayers, an area with resistors can be left as a single layer.

Cofired ceramics

The cofired multilayering process, based on the use of uncured (green) ceramic tapes, can only form interconnections, since no resistor system now exists. The process also is used for construction of LSI packages and ceramic chip-carriers. Metalization, typically tungsten or molybdenum, is screened on the ceramic tape and interconnected through vias. The screened layers are laminated together and then fired in a furnace with a hydrogen atmosphere at 1,600°C.

This process forms a one-piece structure, in which the customer—a hybrid manufacturer—sees only vias in the top and bottom surfaces. He must evaporate or sputter a layer of thin-film gold on the top and bottom surfaces and etch on his own interconnects and pads.

Advantages of the cofired multilayer system—supplied by the Electronic Products division of the Minnesota Mining & Manufacturing Co., St. Paul, Minn., and Ceramic Systems, San Diego, Calif.—include the monolithic all-ceramic structure with no glass interface, high strength, and high temperature stability. It also offers increased thermal conductivity and high circuit density. The one big disadvantage is the extensive tooling and lead time required for the type of substrate—12 to 20 weeks, depending on complexity of the job.

According to 3M engineers, who have already made 15-layer cofired substrates, there seems to be no limit to the number of layers that can be stacked. One 3M 15-layer substrate has 220 chips attached over its 3½-by-3½-in. surface and 3,000 vias. The cofired substrate of Fig. 9 has a complete wafer (master slice) on one side and a heat sink on the other.

At least two companies that have been using the sequential screening process—Raytheon Industrial Components Operation in Quincy, Mass., and the Electro-Optical and Data Systems Group of Hughes Aircraft Co., Culver City, Calif.—are trying to develop universal cofired substrates. With such a product, their customer would order a cofired substrate with specified standard inner conductive layers and vias but he would customize the top and bottom to suit his application.

Nonceramic multilayering

Pactel Corp., Westlake Village, Calif., has developed a nonceramic multilayer substrate for complex high-density circuits. This system uses polyimide for the insulating layers.

The process starts on a conductive temporary support (15 by 18 in.) with an ultrasmooth finish. A photoresist is put on, and the desired conductive pattern and vias are exposed onto the surface. After this, a conductor (typically gold, nickel, or copper) is plated on by the additive method. Then a layer of polyimide is put over the conductive pattern everywhere except the vias.

The polyimide, with the pattern locked on, is then peeled away from the support, and the process is repeated for other layers. The layers are then laminated to an aluminum substrate to provide mechanical strength and heat-transfer capabilities.

Using this technique, Pactel's engineers can create substrates with as many as five layers with 3- to 4-mil lines on the conductive layers—a resolution nearly as good as that of thin-film conductors. Because of the fine line widths, Pactel claims, it can eliminate 30% to 40% of the layers required for a ceramic sequentially screened substrate. There are no limitations on the size of the substrate, and any method of chip attachment can be used because of the excellent thermal properties

10. Hybrids on hybrids. Since thick-film resistors normally cannot be screened on the top layer of a multilayered surface, many manufacturers instead screen the resistors onto small ceramic substrates, trim them, and mount them on the main substrate.

Hybrid techniques

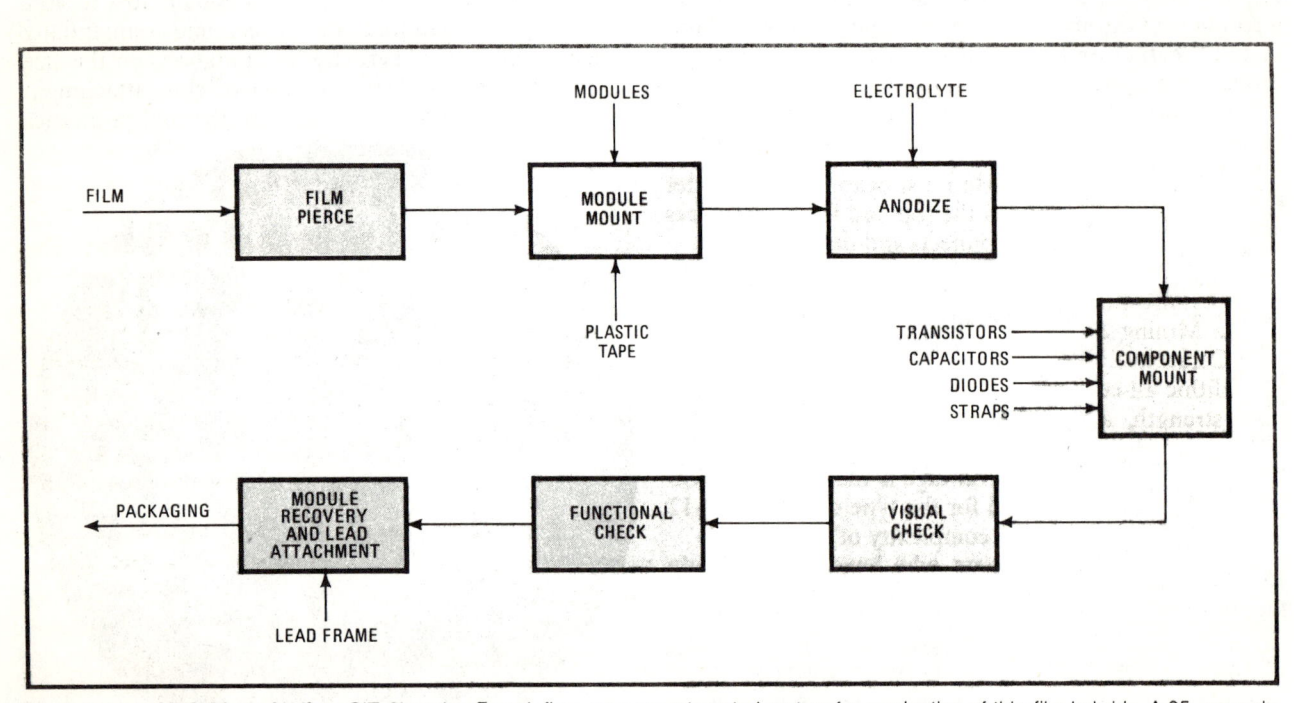

11. Automated hybrid production. CIT-Alcatel, a French firm, uses an automated system for production of thin-film hybrids. A 35-mm polyester film carries substrates throughout the entire process. The machine shown at left attaches modules to the tape. The upper right machine trims the thin-film resistors, while the lower right machine mounts and solders all active and passive components.

of the substrate with its built-on heat sink.

Advantages of this type of multilayering are lower cost, higher density, and better resistance to shock and vibration than are possible with other methods.

Hybrid tricks

Even without multilayering, the hybrid designer has many options for increasing circuit density. By screening both sides of a thin-film substrate, it is possible to mount all active chips on top and all discrete and screened components on the bottom. Interconnections can be wired or made through the leads brazed to the substrate.

Another method is to stack circular substrates on risers, making interconnections through the risers. By far the most popular technique is to use hybrids on hybrids. Most motherboards have only the interconnections plus a few discrete passives. Finished hybrids are mounted on the ceramic motherboard. This approach is extensively used for the Navy's NAFI boards.

The hybrid in Fig. 10, manufactured by General Instrument, uses four small substrates because testing in sections keeps the yield of completed hybrids high. Also, if this were a multilayer single substrate, no thick-film resistors could be screened and laser-trimmed on top layer.

Another version of the hybrid-on-hybrid can be generated with a ceramic chip-carrier. The largest ceramic carriers can accept small hybrid substrates. In one such application, a small ceramic-packaged hybrid, serving as an interconnection, is being used to change the circuit function of a large complex hybrid substrate.

Hybrid automation

Many independent hybrid manufacturers in the U.S.—both thick- and thin-film—are committed to the chip-and-wire method, since it suits tightly packed, complex military hardware. The relatively small quantities of units involved do not warrant such equipment as automatic screeners, conveyer belts, and parts placers used on automated production lines. Often the only automation is laser trimmers and automatic test equipment. Usually the finished substrate ends up in a room full of manually operated bonding equipment.

Centralab, whose hybrid output is based on soldering discrete parts to a substrate, has a completely automated line in Lafayette, Ind., where substrates are screened, fired, trimmed, tinned, tested, and coated. Typical machinery installed includes magazine-fed automatic screeners, probe fixtures, and automatically positioned laser heads for thick-film-resistor trimming.

CIT-Alcatel's microelectronic division in Montrouge, France, has an interesting fully automated line for thin-film hybrids that have thin-film resistors sputtered on and discrete components soldered on (Fig. 11). Other automatic equipment sputters on connections, pads and resistors before the substrates enter at the point labeled "modules" in Fig. 11. A 35-millimeter opaque polyester film is used as the medium for moving substrates through the system.

The first machine punches windows 20 mm in diameter on 38-mm centers on the film. The second machine (Fig. 11b) feeds substrates loaded in magazines down to the moving polyester tape, where they contact adhesive pads in the windows fastened to the film. The pads are automatically cut from a special plastic tape that will handle temperature as high as 250°C.

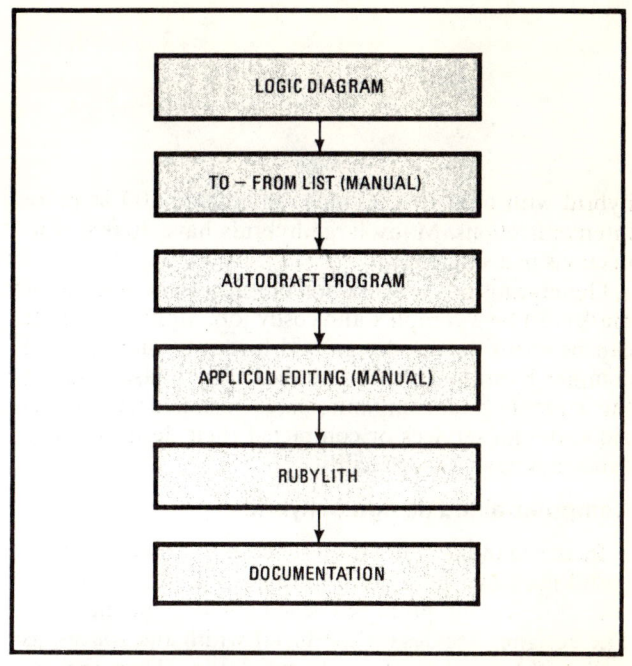

12. CAD. Cost-effective fabrication of extremely dense multilayered hybrids can really only be done with the assist of computer-aided design. The system shown has cut the cost of manual hybrid artwork and documentation at RCA, Moorestown, by a factor of three.

Next, the film with the attached substrate is fed into a resistor-trimmer that uses the anodizing principle as shown in Fig. 11 (upper right), where the film-attachment machine is in the background. The trimmer can adjust as many as 10 resistors in one step or as many as five in two steps. At the end of the machine, a programable scanner checks all resistor values, including those not subject to trimming.

The component-mounting machine (Fig. 11, lower right) is set up to serially attach as many as three Timtors, one chip capacitor, one diode, and one strap. The machine orients each component, picks it up, transports it, deposits flux, and solders components to the substrate. Next, there is a semiautomatic visual check. Throughput is 200 modules per hour, and yield is 99.5%

The next position is a functional check by computer-controlled automatic test equipment. Finally, the last machine detaches good modules from the film and attaches them to lead frames at the rate of 240 modules per hour. Bad modules are left on the film.

CIT-Alcatel is working on a "double-film" assembly method. This uses a 35-mm film to carry the substrates and a 16-mm film at right angles to the first for carrying all discrete components. This would eliminate the need for separate magazines for each component type.

As was mentioned earlier, the trend in hybrids is toward extremely large, all-digital, multilayered units carrying hundreds of LSI devices. The number of interconnection points on such devices is immense—a small hybrid with only 10 LSI chips can contain 500 or more interconnections. Many large hybrids have 10,000 connections in a single layer.

Generating the artwork for the necessary screens or masks can be a complex and costly job. Months of effort can be spent on the design and layout of one of these complex hybrids. To speed up the design phase, some of the large hybrid companies have installed computer-aided-design services or contracted their designs out to firms that have a CAD facility.

Computer-aided design of hybrids

In the manual method of generating artwork, which is still used by the majority of hybrid producers worldwide, the starting point is a schematic or logic diagram. The designer decides on conductor width and spaces, as well as the number of conductive layers. Then the design is laid out on graph paper with a scale 10 to 20 times final size.

When the design is satisfactory, a circuit layout is detailed, using the proper line widths and spaces on accurate graph paper. In highly complex designs, this step is repeated for each conductive and nonconductive material. After the layout of each layer is completed, the rubylith artwork is cut on a light table, using the various layout drawings as masters.

A fully automated hybrid-design system at RCA, Moorestown, is shown in flow graph form in Fig. 12. This system is built around an Applicon graphic terminal, which is used interactively to edit or redesign the hybrid created by a large data base from digitized information from a logic diagram. RCA, Moorestown, which uses this system to generate artwork for thick-film screen and to create drawings for hybrid assembly and inspection, has cut the time and cost of generating hybrid artwork and documentation by 300%.

So far, CAD is used only by a small number of custom-hybrid manufacturers, where the cost of the automation is offset by the savings in manhours. Many large independent hybrid producers that still prefer to lay out all their hybrids manually say CAD is not cost-effective because the system is not used often enough.

It is interesting to note that Centralab, a company that deals with mostly one- and two-sided commercial and industrial hybrids, is awaiting delivery of an Applicon to automate its artwork generation. For Centralab, the 25%-to-50% savings on the sheer volume of different designs each year will justify the cost.

For hybrid manufacturers without CAD capability, design services such as Algorex Data Corp., Syosset, N.Y., or Automated Systems Inc., El Segundo, Calif., are available. Both companies have large computer facilities with programs committed to design of hybrid artwork.

Most hybrids, whether for space, military, avionics, industrial, automotive, or consumer use, are built around linear and digital ICs, discrete transistors, and FETs. However, there is no limit to the circuits hybrids can handle.

13. Hybrid heater. Thick-film resistors are used as heating elements in a 25-watt thermistor-controlled heater on a ¾-by-½-inch beryllia substrate. The heater, built by Integrated Microsystems Inc., is used in a solid-state oscillator manufactured by Watkins-Johnson.

An unusual thick-film hybrid built by Integrated Microsystems Inc., Mountain View, Calif., for Litton's Data Systems division, Van Nuys, Calif., carries a LED matrix that acts as an alphanumeric display. The top of the three-layer, ceramic, 1½-by-3-in. substrate has more than 32 lines per inch of LEDs. A total of 144 holes and 191 vias connect the substrate to electronic circuitry (drivers and logic) that programs the display.

A hybrid heater

The relatively uncrowded hybrid of Fig. 13 is a 25-w thermistor-controlled heater on a ¾-by-½-in. beryllia substrate. Thick-film resistors on the substrate are used as heater elements for external, solid-state oscillators on this hybrid, built by Integrated Microsystems for the Watkins-Johnson Co. An external resistor programs the temperature of the hybrid heater element.

Itek's Applied Technology division has designed a group of specialized optoelectronic hybrids for systems that warn pilots that their planes are being tracked by a laser ranging system. These hybrids have special optical windows in their lids. Mounted on the hybrid under the lids are an optoelectronic sensor and electronic circuitry. A laser beam passes through the window and activates the hybrid's warning circuitry.

A data-keyboard-switch matrix mounted in a 3-by-3-in. space has been built in a ceramic substrate by Centralab. Alphanumeric keyboard symbols are screened onto the top of the substrate. The bottom has two custom ICs, screened-on resistors, and a conductive thick-film interconnection. The entire unit acts as a capacitively coupled keyboard. □

Why the design nod goes to resistors made as thin-film monolithic networks

Process adapted from IC making gives microminiature resistors that offer precision performance, convenience, and low cost

by Donald B. Bruck and Allen L. Pollens, *Hybrid Systems Corp., Bedford, Mass.*

☐ Monolithic thin-film resistor networks are rapidly winning affections away from matched assortments of wire-wound and metal-film discrete resistors. Capitalizing on advances in integrated-circuit manufacturing, these microminiature resistors are attractive, not just for small size, but for precision performance, convenience, and low cost, also. By and large, they owe these lures to wafer batch processing adapted from IC manufacture—including vacuum deposition, etching, photolithography, and computer-controlled laser trimming.

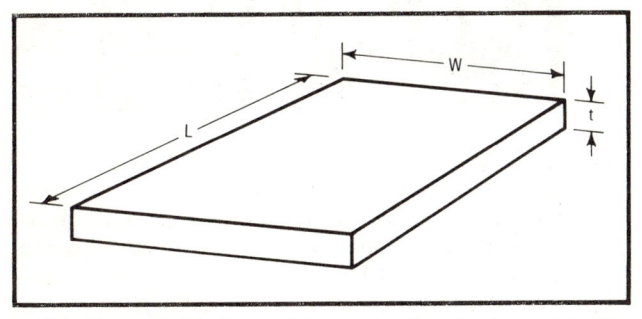

Today's thin-film resistor networks are routinely characterized by a ratio match within 0.01%, an absolute accuracy within 1%, a tracking temperature coefficient of 2 parts per million per degree centigrade, noise of less than 0.1 microvolt per volt, absolute drift of less than 0.1% per 1,000 hours (at 125°C), and ratio drift of less than 0.01% per 1,000 hours (at 125°C).

Furthermore, designers find it simpler to specify, purchase, and test a monolithic resistive network than to work with one built with discrete parts. In addition, a thin-film network eliminates all assembly of matched components. Standard and custom configurations are available in TO cans and dual in-line packages, and, if standard IC packages are not small enough, the networks are available in chip form for use in thick- and thin-film hybrids.

Monolithic thin-film networks are now used in sample-and-hold circuits, precision voltage references, digital-to-analog and analog-to-digital converters, instrumentation amplifiers and preamplifiers, active filters—in fact, wherever resistor accuracy and stability

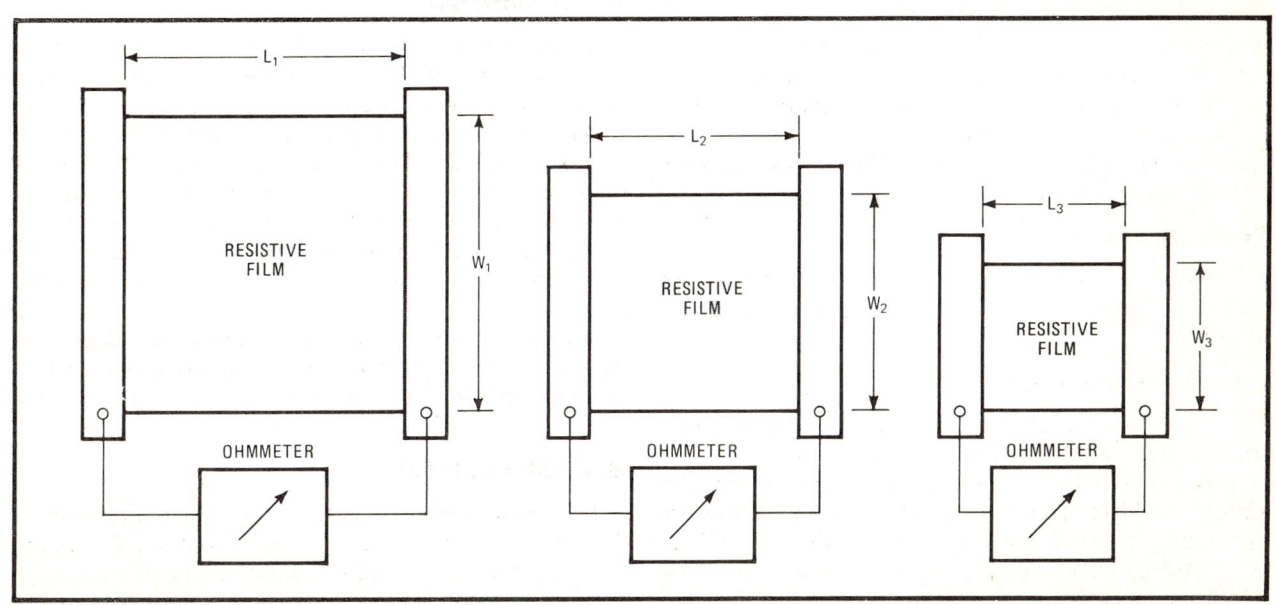

1. Thin-film resistor. The resistance of a thin film is given by the equation $R = (\rho/t)(L/W)$ where ρ is resistivity, t is material thickness, and L/W is the aspect ratio. Metallic films are usually Nichrome or tantalum nitride, depending on the resistance value required.

2. Ohms/square measurement. The ohms-per-square parameter (ρ/t) can be measured by connecting an ohmmeter across the contacts of a film of known exact dimensions. In the case shown, the constant-aspect ratio would result in equal values for each resistive film.

3. Film deposition. Thin films may be deposited on a substrate in one of two ways. In a sputtering system, below, an ion beam hits a metal target causing metallic atoms to land on the substrate. An evaporator, right, is a heat source that boils off metal atoms.

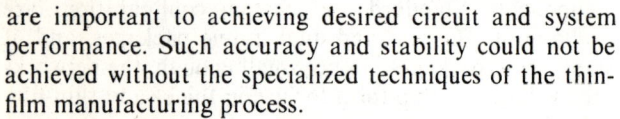

are important to achieving desired circuit and system performance. Such accuracy and stability could not be achieved without the specialized techniques of the thin-film manufacturing process.

The thin-film process is based upon deposition of a metal film, typically of 100-angstrom thickness, on a flat, nonconductive substrate. From Fig. 1, the resistance, R, is equal to the sheet resistance times its aspect ratio, or:

$$R = (\rho/t)(L/W)$$

where ρ is material resistivity and t is film thickness. The ratio of ρ/t is known as the ohms-per-square resistance, and the L/W relationship of the sheet's length and width is the aspect ratio. For a constant t, the value of R is a function only of the aspect ratio, so that theoretically the resistor can be any size.

Sheet resistance

Figure 2 shows how the Ω/sq sheet resistance can be measured. Since the aspect ratios of (a), (b), and (c) are all equal:

$$L_1/W_1 = L_2/W_2 = L_3/W_3$$

Therefore, the resistances of each of the different squares of material should measure the same.

There should be no limit to how small a thin-film resistor could be. In practice, minimum line widths are dictated by power-dissipation requirements, the difficulty of etching excessively narrow lines, and so on. Line widths are typically 1 mil and are seldom less than 0.5 mil. For maximum temperature tracking, all the resistors in a network should be of uniform width.

Alumina, glass, and silicon are the three commonly used substrates for thin-film deposition. Substrate choice is a function of the special demands of the application, plus cost and ease of manufacture. Alumina (ceramic) offers the advantage of low substrate-to-network capacitance but must be laser-cut or sawed. Glass can be scribed and broken apart but has comparatively poor heat-transfer capability. Silicon is frequently the material of choice.

Silicon substrate

Silicon wafers with an oxide insulating layer scribe exceedingly well and have excellent thermal-conduction properties. Handling techniques for the types of silicon wafers used in the manufacture of semiconductors are now well established and can be applied to processing of thin-film networks. Moreover, substrates of silicon can be packaged and bonded by thermocompression or epoxy to another substrate in the same way semiconductor chips are bonded to an alumina substrate in hybrid manufacture.

Silicon's network-to-substrate capacitance does limit its operation in high-frequency, submicrosecond switching applications. Here, glass is the usual choice for a substrate material.

Metal-film materials

The most common materials for thin-film resistor deposition on a substrate are nickel chromium (commonly called Nichrome) and tantalum nitride. Others, such as chrome silicon, are occasionally used for high-resistivity applications.

But Nichrome and tantalum nitride are most often

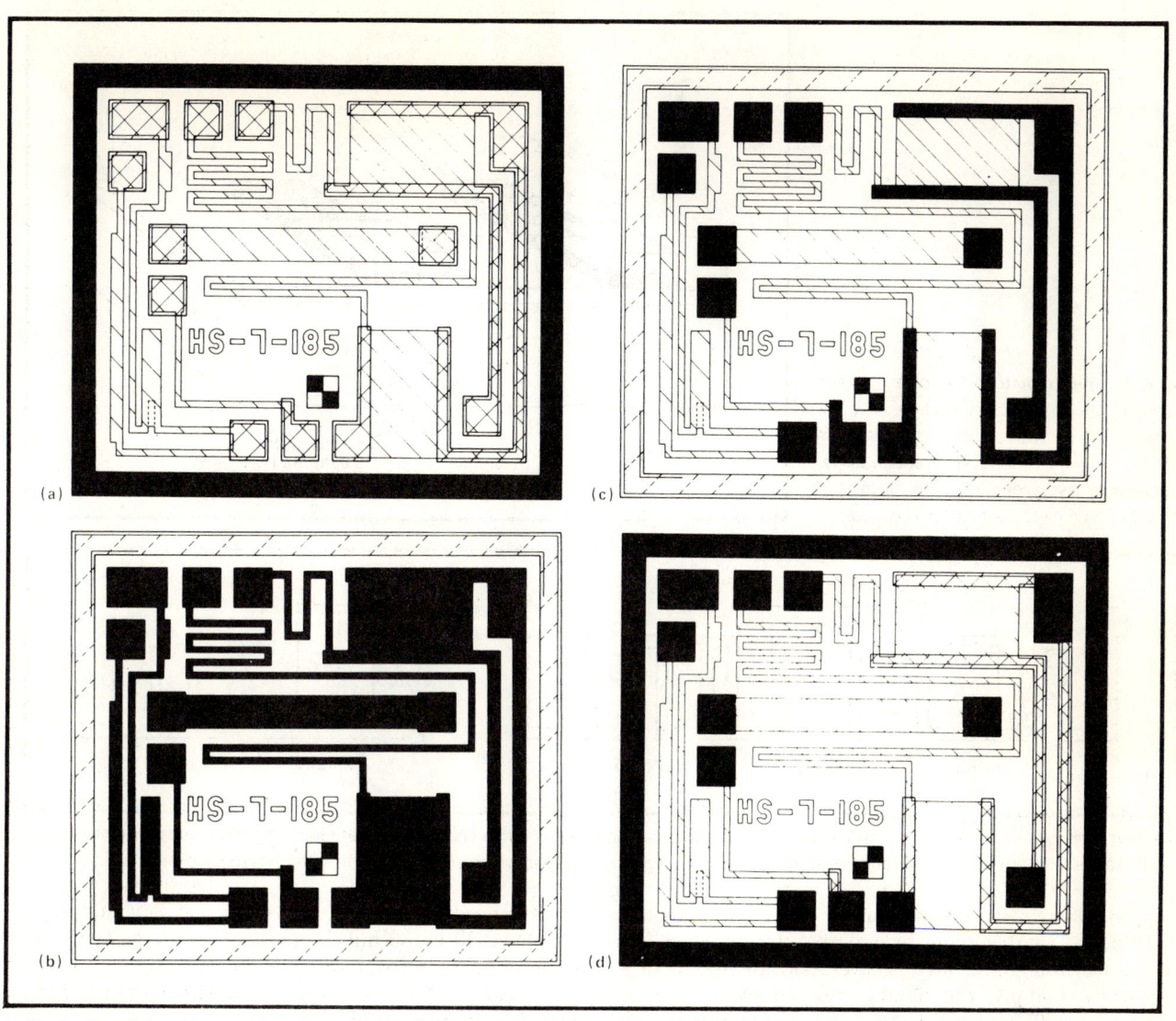

4. Masked resistor. Four masks are needed to fabricate a thin-film resistive network: (a) the grid mask, (b) the Nichrome mask, (c) the aluminum mask, and (d) a glass passivation mask. All are the size of the resistor and are stepped and repeated across the wafer.

used at a sheet resistivity of approximately 300 Ω/sq. At this level, Nichrome has absolute temperature coefficients that are in the range of 0 to +50 ppm/°C with 30 ppm/°C being typical. At 300 Ω/sq, tantalum nitride produces coefficients from −50 to −200 ppm/°C with −75 ppm/°C being typical.

Above 400 Ω/sq, Nichrome resistor values tend to become unstable. However, with tantalum nitride, 1,000 to 1,200 Ω/sq can be obtained before values become unstable. On the other hand, worsening of the absolute temperature coefficient to perhaps −200 ppm/°C may render the high resistivity useless. Variables that affect sheet resistivity include the ratio of the metals in the alloy to one another, film thickness, substrate temperature during deposition, postdeposition heat treatment, and so on.

Tantalum nitride was the first metal to be used, and it continues to be specified for older, on-going resistive requirements. Nichrome is the more usual choice for new requirements. Earlier reservations concerning its ability to withstand moist environments have generally been eliminated by the use of passivation techniques such as glassivation and ultra-dry packaging (vacuum baking and hermetic sealing).

Deposition of these metal films on wafer substrates is accomplished by either of two vacuum techniques: sputtering or evaporation. Generally speaking, tantalum nitride and chrome silicon are sputtered, and nickel chromium is sputtered or evaporated.

Film deposition

In sputtering, an ion beam bombards the metal target, causing the metal atoms to disperse and land on the substrate material. Figure 3 shows an open tantalum-nitride sputtering chamber. The wafers are loaded onto the four circular pallets, six per pallet, with a seventh position occupied by a thickness sensor.

The chamber sputters the metal film onto one pallet of wafers at a time. Upon completion of the sputtering onto a given pallet, the interior table is rotated to place a new pallet under the target. This procedure is repeated two more times, whereupon the metal film will have been deposited on the wafers of all four pallets.

An open nickel-chromium evaporator is also shown in

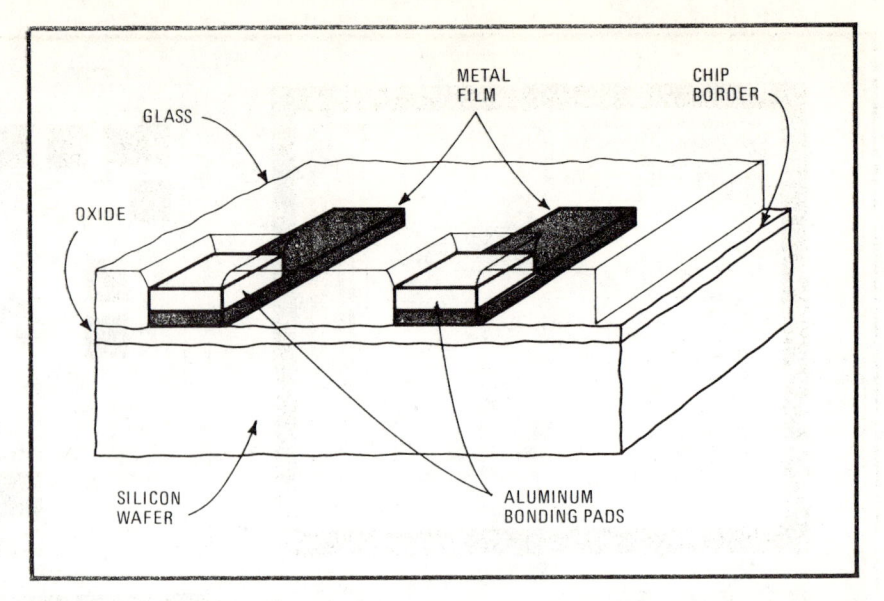

5. A layered resistor. A typical thin-film resistor, shown in cross section, is composed of five layers. The bottom layer is silicon, followed by a layer of silicon oxide. Successive layers of metal film, aluminum, and glass complete the resistor structure.

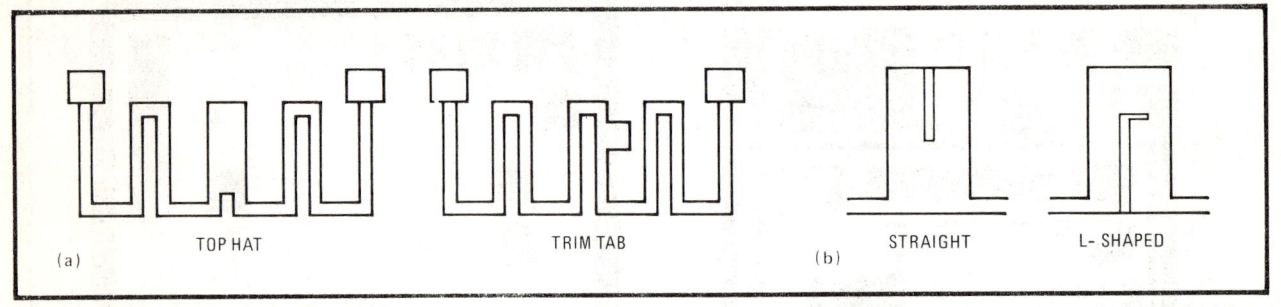

6. Laser trim. Monolithic thin-film resistors may be trimmed to their final values with a laser. To facilitate trimming top hats or trim tabs (a) are added to the resistor pattern. The actual laser trim may be a straight or L-shaped cut (b), but usually the latter has much greater effect.

Fig. 3. Here, 36 silicon wafers are on a planetary fixture that rotates about several axes during deposition to ensure film uniformity. The evaporator uses a heat source to "boil off" the metal atoms at typical temperatures of 1,200° to 1,500°C. A single evaporator of this type can complete three deposition runs per 8-hour shift, producing 108 wafers per shift, or as many as 200,000 40-mil-by-40-mil chips from 2-inch wafers.

No matter what choice is made among substrates, resistive materials, and deposition methods, the overall manufacturing process is pretty much the same. Usually the first step in thin-film processing is to create the 1:1 masks needed for the various steps of this process.

Resistor masks

Several masks (Fig. 4) are required in the manufacturing sequence of a resistive network. The first defines the boundaries between the many networks on the silicon substrate. These boundaries are used for scribing and breaking apart the individual chips after trimming and testing.

The second mask is the basis for the resistor pattern, and a third puts down aluminum bonding pads and conductor runs. The final mask removes glass from the bonding pads and chip boundaries subsequent to glassivation, a sealing process that deposits a layer of silicon dioxide (glass) over the entire wafer.

A resistor network of average complexity would require chip area of 30 by 60 mils. A chip with a center-tapped resistor, on the other hand, would typically have 30-by-30-mil dimensions. Mask patterns for either of these resistor designs are typically drawn at 200× scale, and a rubylith (cut and stripped film) of the pattern is cut on a coordinatograph.

Then the rubyliths are photographically reduced to the actual size of a resistor chip. Each chip's image is reproduced on the wafer surface on an automatic step-and-repeat basis as many times as can be accommodated by the size of the wafer. A 2-in.-diameter wafer, for example, has enough area for nearly 3,500 30-by-30-mil chips. Since the photographic plate is square rather than round, the mask pattern is actually repeated nearly 4,500 times.

Once all masks for a typical monolithic thin-film resistor network have been created by photoreduction, the actual production process flow (assuming it starts with a silicon substrate that has an oxide insulating layer) is straightforward.

Initially, a layer of photoresist (a photosensitive lacquer-like film) is uniformly applied to the oxidized silicon and allowed to dry. The grid mask is positioned over the photoresist surface and the light-sensitive surface is exposed. Chemical processing leaves photoresist only in those areas not exposed. The oxide layer, in the light-exposed borders between the chips, can then be etched away to facilitate scribing and separation.

Next, the metal film is coated on the silicon wafers by one of the vacuum-deposition methods. Another layer of

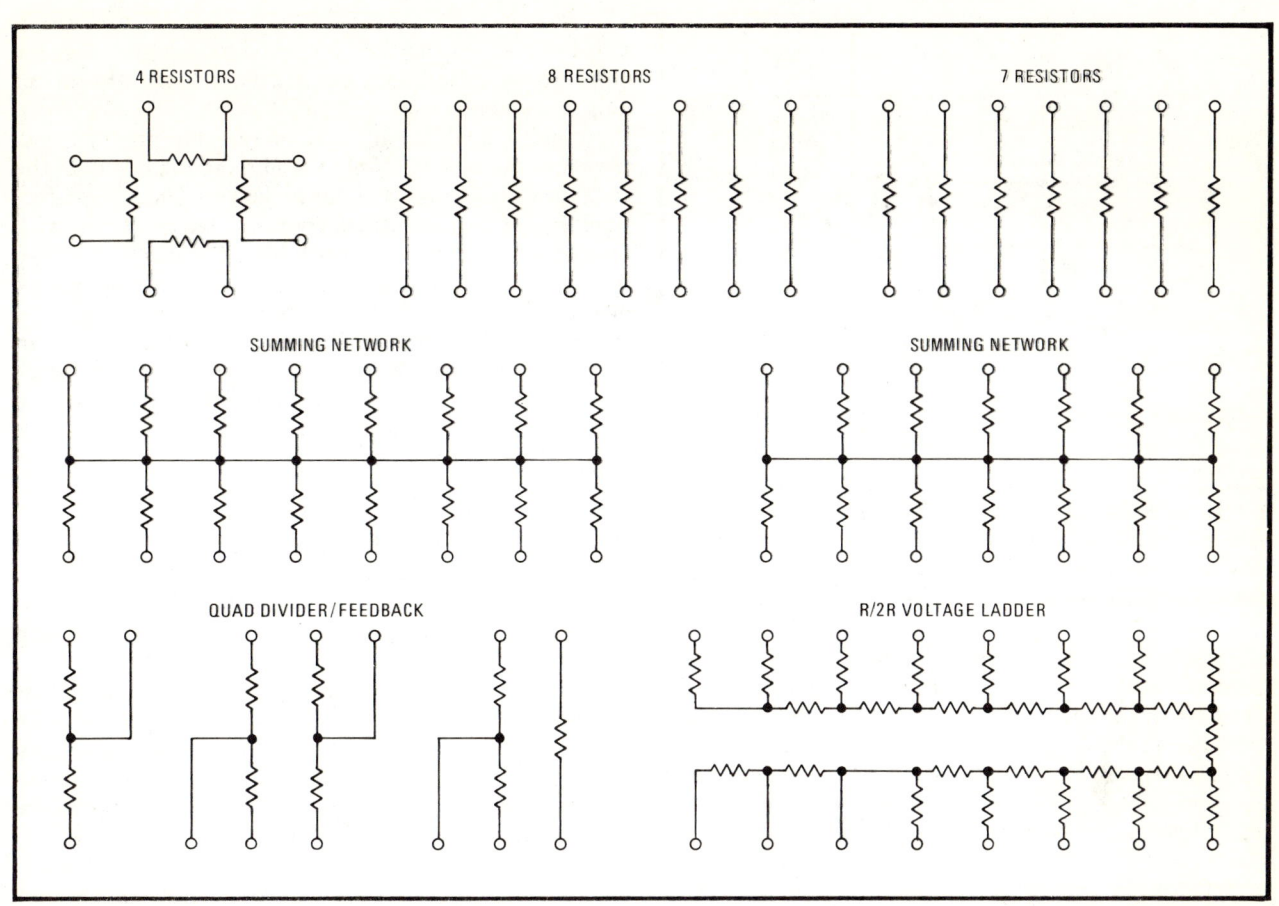

7. Off-the-shelf. Thin-film resistive networks are supplied in the seven standard forms shown. Most popular are the summing networks that are used for operational amplifiers and the R/2R voltage ladder that is used for analog-to-digital and digital-to-analog conversion.

photoresist is applied, and, with the resistor mask in position, the wafer is exposed to light. As before, the light-exposed photoresist is removed, and an etchant removes thin-film material from between resistor runs and other places where it is inappropriate.

A layer of aluminum next is vacuum-deposited on the wafers. In similar fashion to the prior two operations, photolithography and etching leave the aluminum only where desired for bonding pads and conductor runs and do not disturb the thin-film material or the oxide layer.

After the glassivation, the fourth and final use of photolithography and etching removes the glass seal from atop the bonding pads and grid borders. The resulting layered structure is shown in Fig. 5.

Trimming and testing chips

With the manufacturing techniques discussed, films can be within 10% of target resistance values. If greater precision is desired, wafers may be subjected to a trimming procedure.

Early in the history of thin-film resistor manufacturing, when tantalum nitride was the metal most often used, networks were trimmed by an anodizing process. Now it is far more common for resistance values to be adjusted by a laser beam.

With a laser, active trimming can take place. The laser beam is used to burn away portions of the resistor film, thereby altering its physical dimensions and the resistance value. The resistor can be probed and its value can be continuously monitored to determine when the laser has cut sufficiently to produce the value and precision required.

To facilitate such trimming, features called top hats and trim tabs (Fig. 6a) are added to the layout of a thin-film resistor network. The trimming is performed on them. The top hat has greater effect and is therefore used for coarse trim. The trim tab has significantly less effect and is used for fine trim.

An L cut

It may appear that considerable latitude exists for both the shape and size of the cut made on the top hats and trim tabs. However, it can be demonstrated theoretically and proven in practice that trimming methods can have a significant impact on resistor stability and reliability. In fact, an L-shaped cut has much greater effect than does a straight cut (Fig. 6b).

Laser trimming is usually computer-controlled and automatic, a big advantage considering the large number of networks on each wafer. In the automated operation, defective chips (those that cannot be trimmed to the desired values) have an X burned into them. This mark allows them to be readily recognized and culled later during visual inspection.

A useful byproduct of computerized operation is the tabulation of measurement and yield statistics. If trimming is not required, automatic probing, measurement, and inking of defective networks is done instead. After

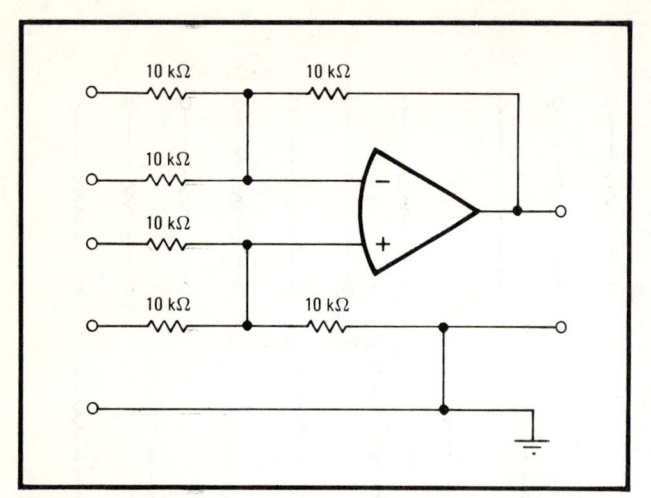

8. Thin- vs metal-film. For this typical operational amplifier resistive network with its 0.01% ratio accuracy and its 2-ppm/°C tracking, bulk- or metal-film discrete resistors would cost about $35 versus $10 or less for a standard thin-film monolithic resistive network.

laser trimming and testing, the borders between the many networks on each substrate are scribed, and the chips are broken apart by pressure. The individual chips are loaded into waffle packs for microscopic inspection, typically by 100× magnification. Those chips with scratches or other blemishes are removed. For military requirements, the inspection is performed in accordance with MIL-STD-883A, method 2010.

Networks in unpackaged chip form are often used by hybrid-IC manufacturers. The trimming is frequently done after the chip has been incorporated into the IC. It can be based on the performance requirements of the completed circuit, thereby taking variations in other components into account.

For those users that require packaged networks, chips can be mounted in TO cans, flat packs, or DIPs by eutectic or epoxy die attachment. Wires are connected between the chip pads and the interior terminations of the package leads. Aluminum ultrasonic or gold thermocompression bonding of the same type used in the packaging of monolithic ICs is employed. After vacuum bake to eliminate any traces of moisture, the package is hermetically sealed.

Manufacturing precautions must be observed and procedures followed to avoid network damage such as permanent resistance shifts that can result from the high temperatures encountered in bonding and other packaging operations. Labor and materials constitute the largest share of packaged-network manufacturing costs.

Cost factors

Wafer-manufacturing costs are essentially independent of the particulars of the mask patterns. Consequently, the cost of resistor chips is primarily a matter of how many can be contained on the wafer and what kinds of yields can be obtained. A good rule of thumb is: the more chips per wafer, the lower the cost per chip.

Networks of smaller total resistance require less metal film, use less chip area, pack more chips per wafer, and so cost less. Similarly, more terminations per network mean more bonding pads per chip, a larger chip, fewer chips per wafer, and therefore produce a higher cost per chip. Power rating and required trim range also affect chip size and cost.

Actually, doubling the size of the chip produces less than half the number of chips and more than doubles the cost per chip. A major factor is the very small imperfections in the original silicon wafers. Larger chips with fewer total chips per wafer will have a higher percentage of defective, less usable wafer area. Another way to look at this is to realize that the larger the chip, the greater the likelihood of an imperfection being included in it. In the extreme, a one-chip-per-wafer design that occupies the whole wafer would have a 0% yield because the imperfections would have to be part of the chip.

Custom charges

Another cost factor is nonrecurring charges for custom networks. Although a customized network can be produced for a nominal one-time charge (typically $2,000 to cover special masks, the trim program, probe cards, and so on), standard designs (Fig. 7) can satisfy many requirements.

Furthermore, it is frequently possible to adapt an existing design, particularly where the manufacturer's original judicious inclusion of trim tabs and top hats allows alteration of resistor values to meet a wide range of requirements. In some cases, it may only be necessary to modify the bonding-pad mask in order to tap the network at a different location. Thin-film resistor manufacturers now have substantial libraries of mask sets, and it behooves a prospective user to discuss his custom requirements with manufacturers before freezing his resistor specifications.

The economic benefits of using monolithic thin-film resistor networks can be demonstrated by considering a thin-film net versus discrete bulk-film resistors in the construction of the typical operational-amplifier circuit shown in Fig. 8, which is based upon 0.01% ratio accuracy and 2 ppm/°C tracking. In hundreds, standard thin-film monolithic resistors for this circuit would cost about $10 versus $35 for a bulk-film network.

Thin vs bulk-film resistors

With the thin-film network, the circuit's gain accuracy can be controlled by simply specifying the ratio tolerance. In contrast, with discrete resistors, the gain precision can only be assured by specification of tight absolute tolerances, which costs extra. In addition, the monolithic network, by virtue of being packaged in a single DIP, can be assembled into a printed-circuit board with less labor and, in fact, may be used with automatic insertion equipment.

The future may be expected to bring improved materials and processing that will facilitate the creation of networks with better absolute accuracies, reduced temperature drift, and increased long-term stability. Higher sheet resistivity will make possible greater total network resistance and smaller chip size which, in turn, will reduce chip costs.

Further work will reduce costs of packaging networks. Finally, costs will come down as quantities continue to grow and further economies of scale are realized. □

Thin-film layers shrink rf inductors to chip size

Getting a lot of turns of "wire" into a thin-film inductor is not easy, especially if the part is to remain small. More turns—and inductance—usually means more surface area, but not when the inductor is made with a new technique developed by the Thinco division of Hull Corp., Hatboro, Pa.

Instead of adding turns by depositing film on a flat surface, Thinco fabricates the turns vertically, depositing alternate layers of metal and silicon-dioxide dielectric on an alumina substrate. Furthermore, the components—Thinco has produced capacitors, too—are batch-produced in a vacuum chamber during only a single vacuum pump-down.

Harold de Palma, Thinco's general manager, calls his company's "picominiaturization" of passive components "a great advance." But he does not think that Thinco, organized some five years ago to develop the production process, will stop there. "Ultimately, we will build passive filters on a chip."

Two types. Presently, Thinco sells two types of inductors designed to be included in hybrid radio-frequency circuitry. The XL-30 can be made with inductances specified between 3 and 56 nanohenries. It has a self-resonant frequency of about 2 gigahertz and measures 85 by 50 mils. The XL-55, with inductances between 15 and 210 nH and a self-resonant frequency of about 700 megahertz is somewhat larger at 67 by 67 mils.

Chip inductors in these ranges are currently available, but they are made with thick-film materials in cubic packages about 100 mils on a side. Both of the new inductors are, at most, only 2 mils high, and are about the same size as a transistor chip. The quality coefficient, or Q, for both units are up in the 40 to 50 range at 400 MHz.

Thinco's engineers use substrates about 8 to 10 mils thick, sturdy enough to be handled in a production environment. Thus, the volume of an inductor is only about 4% of that of the commercial thick-film type. Using an electron-beam gun in a vacuum chamber, the thin-film materials are deposited through precision masks onto the substrate. The gun beam vaporizes material stored in any of four crucibles contained in a "lazy susan" arrangement within the chamber. The chamber also has a library of masks and a mask changer and indexer.

Each mask pattern is stepped and repeated in a grid, enabling thousands of identical components to be put down on a 2-by-2-in. substrate. Thinco will soon add a mask-changing mechanism that can accommodate a 6-by-6-in. substrate, enough for 6,000 XL-55-size inductors in a single pump-down.

Capacitors, too. Thinco has also built capacitors using the technique. One prototype, measuring 30 by 50 mils and 2 mils high has been made with capacitances between 2.7 and 100 picofarads. "With higher dielectric constant materials, we expect to obtain capacitances up to 1,000 pF in a 1-mil-high structure measuring 40 by 50 mils," says de Palma.

Film thicknesses, which range between 5,000 and 50,000 angstroms, can be controlled to several hundred angstroms, according to de Palma. Lot-to-lot repeatability is high because of this close film control, as well as the precision mask indexing that is possible, he says. A computer program helps calculate layer thicknesses to tailor inductances, capacitances, and self-resonant frequencies.

Future plans call for the fabrication of passive filter networks incorporating combinations of inductor, capacitor, and resistor elements. A preliminary design of a very high-frequency pass-band filter has dimensions of 295 by 60 mils by 12 mils high; another is 160 by 110 mils by 12 mils high.

Samples of the new inductors are already being evaluated by companies including Raytheon, General Electric, TRW, and Watkins-Johnson, de Palma says. The parts sell for between 70 and 90 cents each and de Palma hopes that further automation of the process could bring this down to 50 cents. □

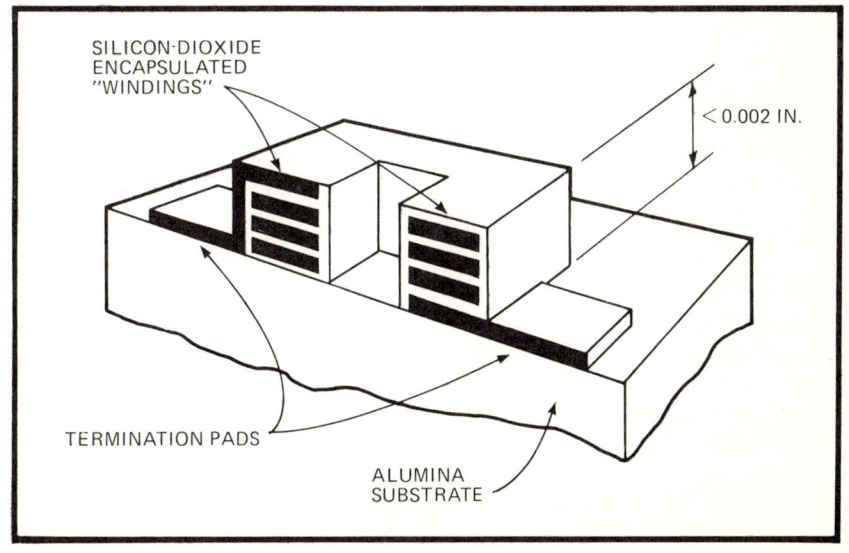

On a chip. Radio-frequency inductors built of vacuum-deposited thin films have ranged as large as 210 nanohenries. Two chip types are about 4,250 and 4,490 mil².

Part 3
Printed Circuit Board Technology

Part 3
Fisheries Management Technology

New methods and materials stir up printed wiring

Changes are cutting processing and chemical costs, increasing board density, giving designers less expensive substrates

by Jerry Lyman, *Packaging & Production Editor*

☐ After several years of relative technological quiet, the printed-circuit industry is now in the midst of introducing new and in some cases radically different methods and materials. These changes promise to be extremely valuable to pc-board users as well as to provide significant benefits for the manufacturers.

Among the most important are:
■ A transfer of thick-film techniques to conventional pc substrates and to enamel-coated steel. This merger of technologies could cut pc production costs drastically.

■ Other new but more conventional fabrication methods are shrinking printed-circuit conductor widths, allowing packaging engineers to increase pc component density.
■ New types of dry-film resists are reducing chemical costs and easing the environmental considerations of manufacturing printed-circuit boards.
■ New general-purpose and high-frequency laminates are giving the designer a whole family of low-cost, high-performance alternatives to more expensive laminates now in general use.

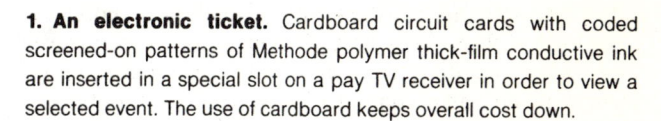

1. An electronic ticket. Cardboard circuit cards with coded screened-on patterns of Methode polymer thick-film conductive ink are inserted in a special slot on a pay TV receiver in order to view a selected event. The use of cardboard keeps overall cost down.

2. Burn-in. This is a small portion of a large printed-circuit board used for burning in several hundred ICs in DIPs. The black areas are screened-on PTF load resistors. They cost less than discrete types and eliminate the need for automatic component insertion.

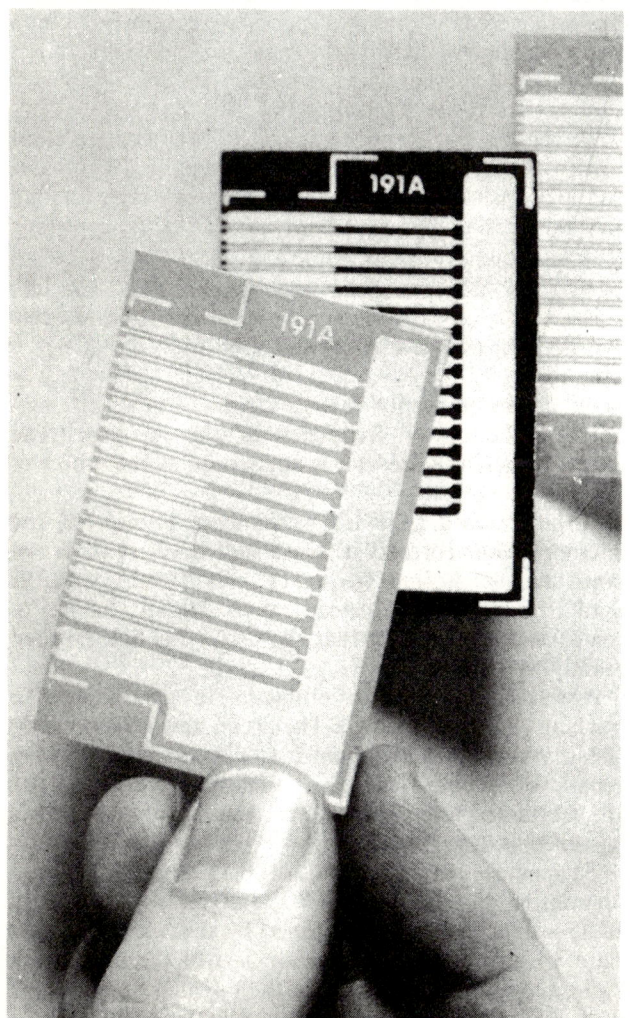

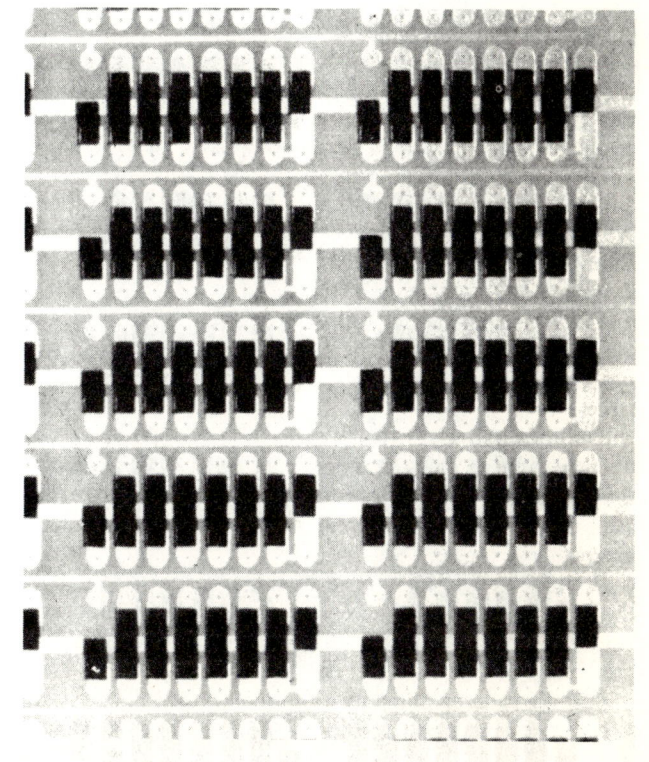

Printed-circuit manufacturers have long envied the simple processes used for fabricating thick-film hybrids. Instead of employing the complex etching (subtractive) or plating (additive) methods of the pc industry, the hybrid manufacturer simply screens and fires conductive, resistive, and dielectric thick-film inks on an alumina substrate. However, the 1,000°C firing temperature of these inorganic inks rules out the use of the low-cost plastics that form the bulk of pc substrates.

Printed printed circuits

Now, two parallel developments have made it possible to print pc boards with components as well as conductors. The first is the creation of a family of polymeric thick-film inks that can be screened onto conventional pc substrates and cured. The second is a rival to the electroplated board that employs extremely cheap enamel-coated steel as a substrate and modified inorganic thick films for resistors, conductors, and insulators.

Polymer thick-film inks emerged only in the last two years. They are now available from companies like Methode Development Co., Chicago; Electro-Science Laboratories Inc., Pennsauken, N. J.; Electro Materials Corp. of America, Mamaroneck, N. Y.; and the Electronic Materials division of E. I. du Pont de Nemours & Co., Niagara Falls, N. Y.

Conductors and resistors are formed of a polymer combined with a silver or a carbon paste, respectively. Crossovers and dielectrics are simply screenable polymers with no paste content. All these materials fire at about 160°C—well within the maximum temperature range of standard pc laminates.

Aside from the major advantages of eliminating etching and plating and the use of copper-clad substrates, this new system offers several other important benefits. For one thing, printing resistors directly on the circuit board cuts labor costs, eliminates the need for automatic component insertion, lowers component costs (PTF resistive ink costs about 1¢ per square foot), and increases component density. For another, the screened-and-cured resistors resemble improved carbon-composition types, with a temperature coefficient of under ±300 parts per million, and the surface resistivity of the inks ranges from 1 ohm per square to 1 megohm per square. Thus the screened-on resistors are more than sufficient for all but the most demanding circuitry.

Also, since a complete family of conductors, resistors, and dielectrics is available in PTF, it is possible to use thick-film techniques to build up successive layers of conductors insulated from each other. This procedure is much less costly than the standard multilayering process, since with PTF inks the circuitry of a two-sided printed circuit with plated-through holes can be placed entirely on one side of the board. Another advantage is that literally anything that can stand the curing temperature without shrinking more than 2% can be a substrate. Besides conventional materials, wallboard, vinyl siding, and even cardboard have proved successful.

The PTF process uses the same type of stainless-steel-mesh screens as conventional thick-film systems for hybrids. As a result, the conductor resolution of PTF boards resembles that of alumina boards with thick-film inks screened on—lines and spaces are routinely 10 mils wide and can be as narrow as 3 mils. Such fine resolution cannot be achieved with standard pc processing, for reasons that will be discussed later.

Wayne Martin, general manager of Methode Development, which is both an ink and a pc supplier, points out that PTF inks are being applied in two ways. In one method, resistors, conductors, and crossovers are screened onto a conventional board with etched or plated copper conductors. In the more up-to-date method, everything is screened onto a bare plastic laminate. The first method was used heavily initially, but "as people gain more confidence in PTF," Martin notes, "they go to an all-thick-film pc approach."

The possibilities of this system can be seen in some of the ways in which it has already been applied. For example, one of Methode's customers for PTF inks made an intrusion alarm in three versions—a standard line and two models with added functions. Originally, the customer went to a conventional multilayer pc board for the top-of-the-line unit. However, the top-priced alarm did not sell in sufficient quantities to justify the use of a multilayer board. The company's solution was to build its standard unit on a conventional two-sided board with plated-through holes. By screening on additional circuitry with stock screens, this board could easily be converted for use in the more complex alarms. Now only one standard board is needed.

A novel application of PTF is shown in Fig. 1. Here a pattern of PTF conductors is screened onto a piece of cardboard. The finished board is a "ticket" for a pay TV system now being developed. In this system, in order to view a special event a customer inserts his ticket into a slot on an electronic accessory added on to his TV receiver. The use of low-cost cardboard rather than a standard pc substrate keeps the price of the ticket down.

Appliance manufacturers have screened PTF control circuitry right onto the plastic housing of a blender. Moreover, with PTF resistive inks, as many as 1,000 resistors have been screened onto large integrated-circuit burn-in boards (Fig. 2); in addition, such inks can be used to form potentiometer windings. These two examples represent large savings in component costs.

Sidney Stein, president of Electro-Science Laboratories, gives another example of the economies possible with PTF inks. A German customer of his was manufacturing some 1 million pc boards, each with 12 discrete carbon resistors costing about 1 cent apiece. The total cost for the resistors plus installation came to almost $250,000. Replacement of the discrete resistors with screened-on ones eliminated installation and made the price per resistor so low that it could be ignored as a factor in the manufacturing costs.

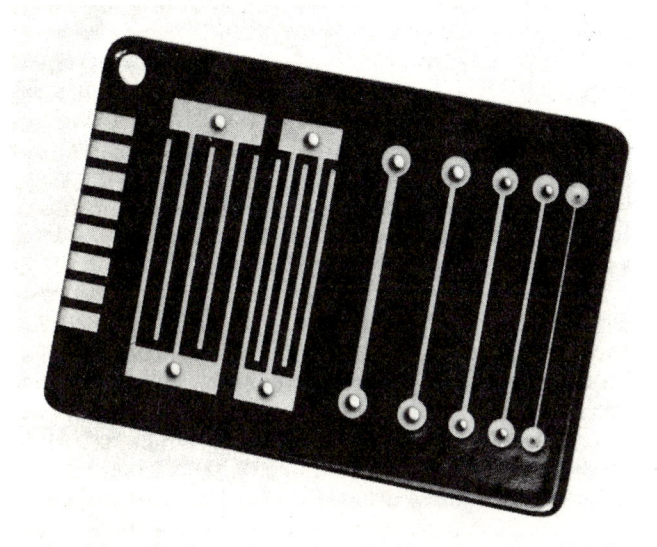

In general, PTF is being used in the United States in watches, calculators, automotive controls, intrusion alarms, appliances, electronic games, and TVs and stereos. Electro-Science Laboratories, for instance, supplies PTF inks to hundreds of customers. According to both Stein and Martin, it is also an accepted technique for making consumer electronic items in Japan. But in Europe, it is only just starting to be designed in.

Overall, the future of PTF processing looks bright both here and abroad, although it will probably share its market with enameled-steel printed circuits.

Steel pc boards

Enamel on steel, normally used on appliances like refrigerators or freezers, is hardly a material one would expect to consider for a pc substrate. However, its cost is extremely low, it can be made in large sizes, and it has good thermal and mechanical properties.

Within the last few years, thick-film ink companies like Electro Materials Corp. and Du Pont have come up with inorganic thick films for conductors, resistors, and dielectrics that can be screened onto an enamel-coated steel sheet and fired at 600°C to 700°C. The substrates are available from two sources, Erie Ceramic Arts Co., Erie, Pa., and Alpha Advanced Technology, a division of Alpha Metals Inc., Jersey City, N. J. Boards of up to 20 by 20 inches can be supplied at a price of 1 to 2 cents per square inch.

The enameled-steel technique has several advantages over the PTF process. For one thing, a ruthenium-oxide ink with a lower temperature coefficient than that of

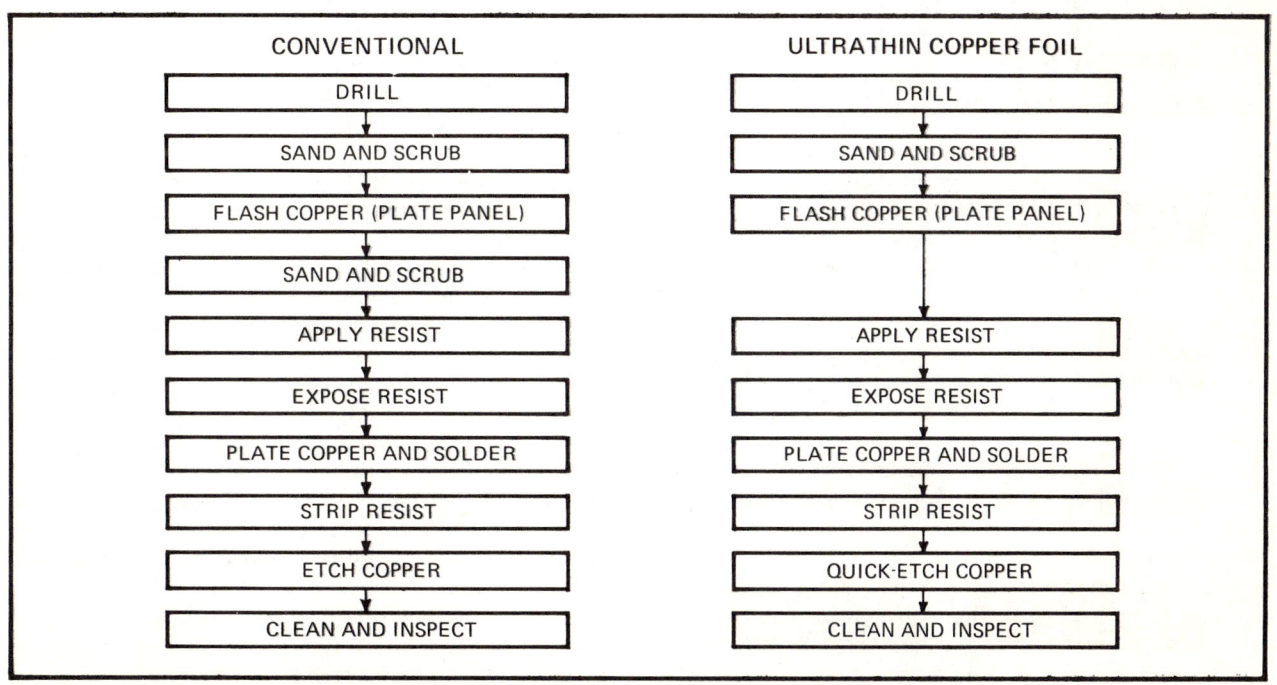

3. Steel boards. Enamel-coated steel with screened-on thick-film conductors and resistors is a low-cost, high-heat-dissipating alternative to plastic pc substrates. Shown is a test pattern of copper thick-film conductors on enameled steel from Alpha Metals.

CONVENTIONAL	ULTRATHIN COPPER FOIL
DRILL	DRILL
SAND AND SCRUB	SAND AND SCRUB
FLASH COPPER (PLATE PANEL)	FLASH COPPER (PLATE PANEL)
SAND AND SCRUB	
APPLY RESIST	APPLY RESIST
EXPOSE RESIST	EXPOSE RESIST
PLATE COPPER AND SOLDER	PLATE COPPER AND SOLDER
STRIP RESIST	STRIP RESIST
ETCH COPPER	QUICK-ETCH COPPER
CLEAN AND INSPECT	CLEAN AND INSPECT

4. UTF vs subtractive. Making printed circuits with ultrathin copper foils has many similarities to a conventional etching process, as the two flow charts show. UTF etching may require the additional step of peeling a protective carrier away from the microfoil.

TABLE 1: COMPARING THE SUBTRACTIVE, ULTRATHIN-FOIL, AND SEMIADDITIVE PROCESSES FOR FINE LINE PATTERNS

	Subtractive 1-oz/ft² foil	Subtractive ultrathin foil	Semiadditive CC-4 Unclad
Fine lines	difficult to achieve: slivers, undercuts	easier to achieve: no slivering or undercutting	easier to achieve: no slivering or undercutting
Processing	difficult, costly	easy: no tin-lead needed	easy: no tin-lead needed
Substrates	all NEMA grades	only FR-4 at present	all NEMA grades
Reworking	no	no	yes
MIL spec approval	yes	pending	under study
Special steps	none	removal of carrier for some laminates	oxidation
Cost (1-oz/ft² foil = 100)	100	103	93
Burrs produced by drilling	yes	some	no

SOURCE: PHOTOCIRCUITS DIVISION, KOLLMORGEN CORP.

carbon-based PTF ink can be used. Secondly, it is easier to wire-bond or wave-solder to this substrate than to those used in the PTF method. If necessary, copper thick-film inks, easily solderable, can be screened onto the coated steel and fired, as shown in Fig. 3.

So far, production examples of boards made of enameled steel are few. (Two are used in Sylvania and General Electric flashbulb arrays.) Erie Ceramic Arts, however, has samples out to over 100 potential users, of which three are close to going into production.

Both the PTF and enameled-steel techniques will likely coexist and be applied mainly in automative and consumer applications. The thermal and mechanical properties of the coated-steel substrate make it ideal for high-temperature–high-shock environments. On the other hand, the high dielectric constant of enamel (6.5) may rule it out for applications covered by high-frequency boards easily made with PTF.

Shrinking conductor widths

At the present time, the majority of the pc boards made for computer, telecommunications, and instrument equipment are made subtractively by etching plastic substrates clad with 1-ounce-per-square-foot copper foil. These boards typically carry 10- to 20-mil-wide copper conductors etched from the 35-micrometer-thick foil.

Over the last two to three years, however, there has been a constant demand from pc users, particularly in the computer field, for higher circuit density. The result is three different processes, each with the same aim—to produce pc boards with fine lines of 5 to 10 mils. The most widely used is a slightly modified form of etching of epoxy-glass laminates clad with ultrathin copper foils. The other contenders for fine-line production are the semiadditive and resistless additive processes.

The success of ultrathin foils (UTF) in fine-line etching is due to the fact that they undergo a minimum of undercutting. Undercutting occurs because the etchant penetrates at about the same rate laterally as vertically. The amount of undercutting is therefore proportional to the thickness of the copper foil. Excessive undercutting limits line widths to about 10 mils on 35-μm-thick (1-oz/ft²) copper-foil-clad boards.

Ultrathin-foil laminates, on the other hand, are clad with copper foil approximately 9 μm (¼-oz/ft²) or 5 μm (⅛-oz/ft²) thick. These foils etch proportionally quicker than the 35-μm types, drastically reducing the amount of undercutting and allowing for finer lines. They are supplied either with a peelable-copper or etchable-aluminum protective carrier foil or with no carrier at all.

The application of ultrathin copper foils was originally held back by material shortages and developmental troubles. Today, the material and development problems have all but been eliminated, and perhaps 4% to 10% of all printed circuits produced in the U.S. use such foils.

Conventional etching of a 1-oz/ft²-clad board and the fairly similar process for etching a UTF-clad board are compared in Fig. 4. As can be seen, fewer steps are needed with ultrathin foils, and conversion from the former to the latter is relatively simple because of the similarities of the two processes.

In addition to its capability for producing fine lines, the UTF process has other advantages over conventional etching. First, etching times and chemical costs are less. For a ¼-oz/ft² foil, times and costs are approximately 25% those incurred with 1-oz/ft² foil. Also, since the amount of etchant is reduced, waste treatment of the etchant is obviously less of a problem.

Despite the higher cost of UTF laminates compared with 1-oz/ft²-clad types, processing UTF-clad boards can still cost less. For example, according to information furnished by Fortin Laminating Corp., San Fernando, Calif., processing costs for a ¼-oz/ft²-clad, carrierless pc board are 54¢/ft² less than for a comparable one clad with 1-oz/ft² foil. In fact, this savings suggests that ultrathin copper foil could be economical for boards with 10- to 20-mil lines.

UTF fine-line processing does have two disadvantages, however: the clad substrates are expensive, and they require an added process step to remove the carrier. The trend, though, is now to bare foils.

With the increasing demand for fine-line circuits, the

advantages of the 20-year-old semiadditive process are being reevaluated by at least two major independent U.S. printed-circuit houses. This process, heavily used in Europe and Japan but not in the U.S., is based on electroless plating of a 0.10-mil-thick layer of copper on a bare substrate. This layer is even thinner than ⅛-oz/ft² copper foil, the actual thickness of which is 0.17 mil. The flashed electroless copper will therefore etch faster than the thinnest available UTF, with less undercutting, making possible even finer lines.

Semiadditive processing

The semiadditive technique has three additional advantages over the subtractive UTF method. With it, any type of plastic substrate can be used. UTF has been limited to clad FR-4 as a substrate (see below) simply because until recently only FR-4 was supplied with ¼- and ⅛-oz/ft² foil. However, Atlantic Laminates is now just starting to supply its CEM-3 material, AL-910, with ¼-oz/ft² foil.

The second advantage of the semiadditive process is that copper patterns can be erased (that is, the copper can be chemically removed), which is not possible with an all-subtractive method. Furthermore, since drilling in the semiadditive process is done on a copperless substrate, deburring is eliminated. A comparison of conventional subtractive, UTF subtractive, and semiadditive methods is shown in Table 1.

The semiadditive process starts with bare substrate on which conductor patterns are built up by electroplating copper over a thin film of electroless copper. The production flow of a typical semiadditive process (that of the Photocircuits division of Kollmorgen Corp. for making its CC-4 Unclad) is shown in Fig. 5. Note that after the electroless-copper flashing, all succeeding steps resemble those for etching UTF.

Litton Industries Inc.'s Advanced Circuitry division, Springfield, Mo., has recently developed its own semiadditive process, called Semi Plus, which was shown a couple of months ago at Nepcon West in Anaheim, Calif. According to Robert Schutz, vice president of marketing at Advanced Circuitry, "In mass production, our process gets lines down to the state of the graphic arts, which is in the 5-to-6-mil range."

Although semiadditive processing has undeniable advantages over the subtractive UTF technique, it does have one serious drawback: to convert a subtractive production line to a semiadditive one requires new equipment and new chemical processing. In contrast, the conversion of a standard subtractive production line to handle carrier-protected ultrathin foils is simple. In fact, if the UTF is carrierless, no conversion is needed.

Fine lines without resists

The ideal method for producing fine-line circuitry would be a fully additive plating process that requires no resist, masking, or etching. Two such processes do exist. N V Philips Gloeilampenfabrieken has one it calls physical development by reduction, or PD-R, which it introduced in early 1977 and is now using at its Elcoma division in Eindhoven, the Netherlands, for prototype telecommunications boards. In the U.S., Photocircuits

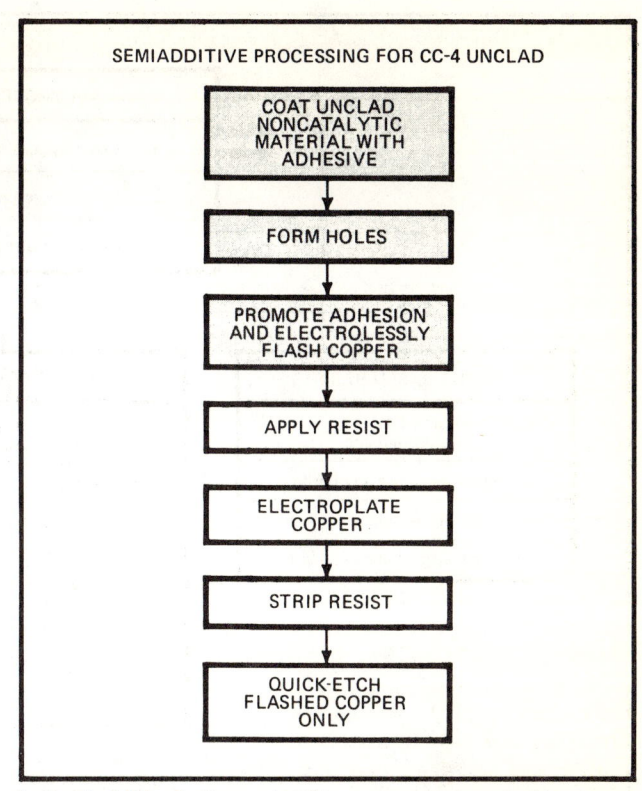

SEMIADDITIVE PROCESSING FOR CC-4 UNCLAD

COAT UNCLAD NONCATALYTIC MATERIAL WITH ADHESIVE
↓
FORM HOLES
↓
PROMOTE ADHESION AND ELECTROLESSLY FLASH COPPER
↓
APPLY RESIST
↓
ELECTROPLATE COPPER
↓
STRIP RESIST
↓
QUICK-ETCH FLASHED COPPER ONLY

5. Semiadditive. In the semiadditive process, extremely thin copper is electrolessly plated on a bare pc substrate. The resulting layer of copper, even thinner than ultrathin foil, allows quick etching of fine conductor widths comparable to those of UTF.

has been working on its Photoforming process since 1975. It expects to have a line for prototype production applying Photoforming by late 1978.

Philips' process starts by coating an FR-4 substrate with a layer of titanium oxide. Holes are drilled in the board, which then goes through a swell-and-etch step to ensure a firm surface for conductor patterns. Next, the roughened laminate is prepared for photoexposure by a dilute solution of palladium chloride, which provides a source of palladium ions to react with the titanium oxide. Conductor patterns are then contact-printed onto the substrate with ultraviolet light. After exposure, the remaining palladium chloride is washed off, leaving a pattern image of metallic palladium.

A special electroless process deposits a thin layer of copper over the palladium image. At this point, flawed patterns can be "erased" and the laminate reused. A second electroless plating brings the copper up to its final thickness of about 25 μm. The final step is to clean the board of process chemicals.

Photoforming resembles PD-R but with some important differences. It can start with any adhesive-coated substrate, unlike PD-R, which can only be plated onto FR-4. Also, it uses a recently developed noncatalytic film adhesive, rather than the titanium-oxide coating of the newer method, and a rolled-on activator based on copper salts instead of the noble-metal palladium solution used in Philips' system.

Both processes look promising for fine-line boards. Philips has found PD-R to be 30% cheaper than a straight subtractive approach, while Photocircuits says that with

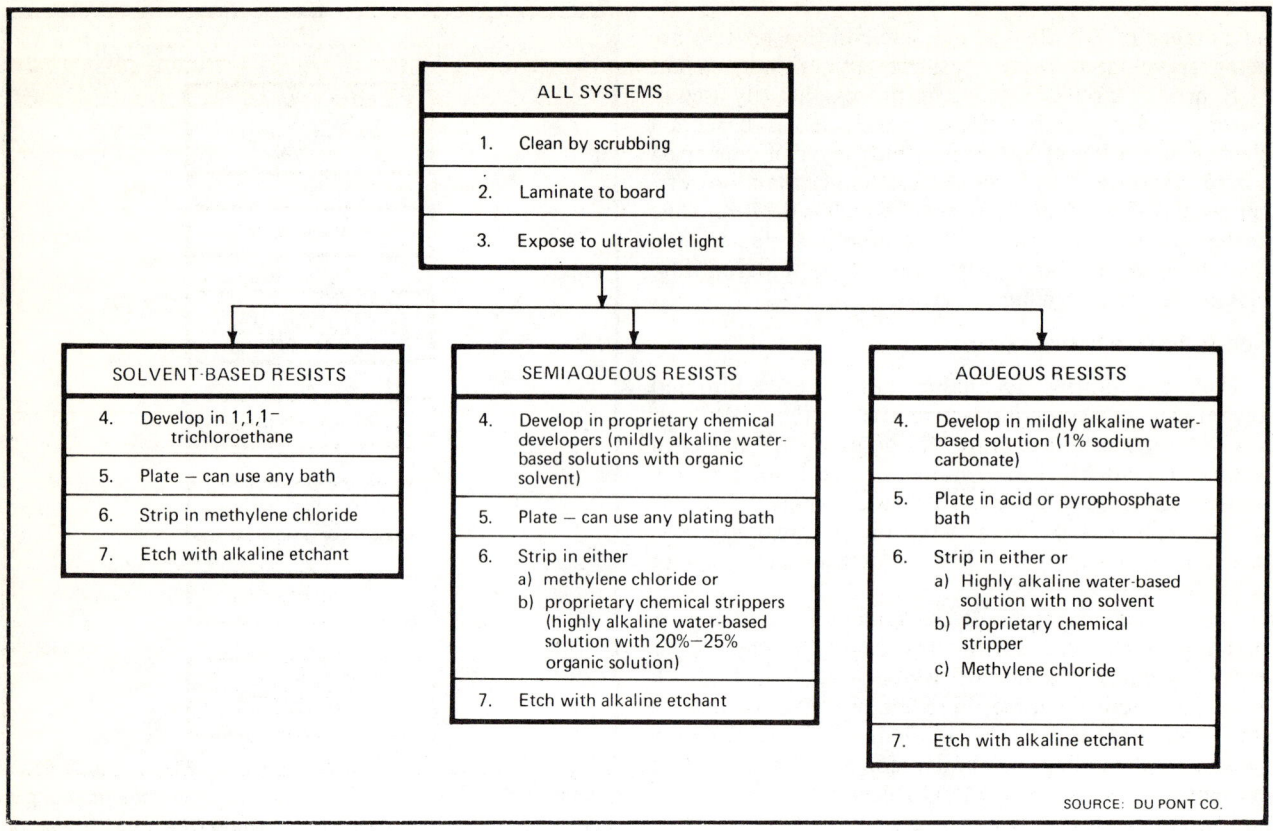

ALL SYSTEMS
1. Clean by scrubbing
2. Laminate to board
3. Expose to ultraviolet light

SOLVENT-BASED RESISTS
4. Develop in 1,1,1–trichloroethane
5. Plate – can use any bath
6. Strip in methylene chloride
7. Etch with alkaline etchant

SEMIAQUEOUS RESISTS
4. Develop in proprietary chemical developers (mildly alkaline water-based solutions with organic solvent)
5. Plate – can use any plating bath
6. Strip in either a) methylene chloride or b) proprietary chemical strippers (highly alkaline water-based solution with 20%–25% organic solution)
7. Etch with alkaline etchant

AQUEOUS RESISTS
4. Develop in mildly alkaline water-based solution (1% sodium carbonate)
5. Plate in acid or pyrophosphate bath
6. Strip in either or a) Highly alkaline water-based solution with no solvent b) Proprietary chemical stripper c) Methylene chloride
7. Etch with alkaline etchant

SOURCE: DU PONT CO.

6. Film resists. Normally, dry-film resists are both developed and stripped in organic solvents. But semiaqueous and aqueous films are now available to cut solvent costs and disposal. Aqueous dry films use no solvent, while semiaqueous films require only a small amount.

photoforming it can produce photoprinting-quality boards at screen-printing costs. Because there is no etching, neither process will undercut or sliver conductors, and therefore conductor resolution is limited only by the quality of the photographic master used in contact printing of the treated substrate. Furthermore, disposal problems are greatly reduced.

Both processes bear watching. David Frisch, manager of development engineering at Photocircuits' Technology group, says: "Formerly, Photoforming was tied to a resin-rich FR-4, which limited potential users to one source of laminate. Now, we have developed a low-cost dry-film adhesive that can go on any standard pc substrate, and this should help make Photoforming price-competitive with other fine-line methods."

New directions in film resists

All the well-established pc processes—subtractive, semiadditive, additive—have a step in which a photosensitive polymer is placed either on a copper-clad or -plated board (subtractive and semiadditive) or on a bare board (additive). The image of the printed wiring is then exposed on the resist and developed. Next, the sensitized image is either selectively plated or etched.

Today, there are two main types of resist, screenable and dry-film. Seventy-five percent of all resists used are screenable. However, the properties of this type are not consistent enough to produce fine lines. In fact, without dry-film resists, high-resolution pc boards (from 13-mil conductors down) would be impossible. To achieve high optical resolution and high product yield, the optical and photochemical characteristics of dry film are needed.

The first dry films were introduced by Du Pont in 1958. Films of this type require processing with organic solvents. They are developed in 1,1,1-trichloroethane and stripped with methylene chloride. However, chemical costs are high, stainless-steel tanks are needed to handle the fluorocarbons that the films contain, and effluent-disposal problems have appeared as a result of the recent concern over environmental pollution.

During 1971–72, the pc industry grew tremendously. At that time, there were as many as 600 to 800 firms involved. Many were small and could not afford the investment needed for processing solvent-based films. To meet this need, Du Pont's Riston Products division, Wilmington, Del., and Thiokol/Dynachem Corp., Santa Ana, Calif., introduced semiaqueous dry-film resists. These films require solvent only for stripping. The procedure for processing these resists is given in Fig. 6.

Semiaqueous resists represents a considerable improvement over solvent-based dry-film resists. They use proprietary chemical developers with a mildly alkaline water-based solution plus some organic solvent, and they can be stripped either with methylene chloride or with a highly alkaline water-based solution.

The new resists lowered the cost of dry-film processing. For example, a typical solvent-based resist processor costs about $25,000, whereas a comparable unit for a semiaqueous film costs $12,000 or less, and stripping could be done in plastic tanks. In addition, the chemical-disposal problem is much smaller. For large installations (greater than 100,000 ft²), though, the cost of a good

TABLE 2: LAMINATE COMPOSITION, APPLICATIONS, AND COST

	Composition	Applications	Cost ($/ft²)
XXXP	paper-based, impregnated and bonded with a phenolic resin	consumer, automotive	1.00
FR-2	paper-based, impregnated with a flame-retardant phenolic resin	consumer, automotive	NA
FR-3	paper-based, impregnated and bonded with an epoxy resin and incorporating a flame-retardant additive	consumer, automotive	NA
FR-4	woven glass-cloth impregnated and bonded with an epoxy resin and incorporating a flame-retardant additive	computer, military, telecommunications, instruments	2.10
CEM-1	a composite incorporating an epoxy resin and a flame-retardant additive. The core is a nonwoven cellulosic felt similar to FR-3 sandwiched between cover sheets of woven glass similar to FR-4	consumer, automotive	1.40
CEM-3	a composite incorporating an epoxy resin and a flame-retardant additive. The core is a nonwoven glass felt sandwiched between cover sheets of woven glass similar to FR-4	computer, telecommunications, instruments	1.85
Polyester-random-glass	random-glass matte combined with a polyester resin with or without a flame-retardant additive	consumer, automotive	1.28

SOURCE: PHOTOCIRCUITS DIVISION, KOLLMORGEN CORP.

closed-loop solvent-recovery system is not much more than the cost of a semiaqueous system, and conversion is therefore not profitable.

Semiaqueous resists caught on particularly in the Northeast, where both Digital Equipment Corp., Maynard, Mass., and Data General Corp., Westboro, Mass., adopted them for their large in-house facilities. However, it was still desirable to have a family of film resists that required no organic solvent at all in processing. The first fully aqueous resists were put on the market by Du Pont in 1975. The processing steps with this type of resist are also given in Fig. 6. Here, developing is in a low-cost solution of 1% sodium carbonate in water at 100°F to 110°F. Stripping may be done by a highly alkaline water-based solution with no organics, by a proprietary chemical, or by methylene chloride.

This first family of fully aqueous resists was successfully applied to simple printing and etching operations on the inner layers of multilayer boards. On the overplated lines of double-sided boards, though, stripping was difficult with nonorganic compounds. Many firms using these resists simply went to organics for stripping—really a semiaqueous procedure. Still, overall chemical costs were cut compared with a production line for a solvent-based dry-film resist.

The 1975 generation of resists is now being superseded by newer aqueous dry films from Du Pont called Riston 3000 and 3300, which were introduced recently at Nepcon West. Also, later this year, Dynachem will offer improved versions of their aqueous resists.

Du Pont claims that the new Ristons will have more processing latitude and will eliminate the acid dip after developing required with older aqueous film. In addition, the newer material will strip better in an all-aqueous solution. Brian O'Conner, product marketing manager for the Riston Products division, says: "The primary application for Riston 3000 is the inner layers of multilayer boards. Since this is a straightforward print-and-etch operation, it is easy to match its chemical processing to an aqueous resist. We're working with many pc firms to apply aqueous resists to all forms of pc fabrication. Eventually, we hope to make the processing costs of these resists competitive with those of screened types." But although they are fine for subtractive and semiadditive processes, O'Connor points out that aqueous films will not work for fully additive pc plating because the pH of the additive solutions is too high.

At present, aqueous processing of dry-film resists is mostly being done by new or small firms, since, as was mentioned, it is not worthwhile for big companies with heavy investments in large closed-loop solvent-recovery systems to change over to the newer system. But for a firm about to start a pc facility or add a new production line to an existing facility, aqueous film resists can result in great savings.

For instance, Mike Busby, president of Cirtel Inc., in Anaheim, Calif., began using fully aqueous processing at the start-up of his pc company 5½ years ago—about the time fully aqueous dry films came out. At first, Cirtel's engineers had trouble with the new Du Pont and Dyna-

TABLE 3: MECHANICAL AND ELECTRICAL PROPERTIES OF LAMINATES						
	Flexural strength (lb/in^2)		Impact resistance (IZOD) (ft-lb/in)		Water absorption (% of maximum value)	Insulation resistance (MΩ)
	X direction	Y direction	Lengthwise	Crosswise		
XXX P	12,000	10,500	0.45	0.40	0.75	2×10^4
FR-2	12,000	10,500	0.45	0.40	0.75	2×10^4
FR-3	20,000	16,000	0.55	0.50	0.65	1×10^5
CEM-1	40,000	30,000	2.5	1.5	0.35	5×10^5
CEM-2	40,000	30,000	2.5	1.5	0.35	5×10^5
FR-4	50,000	40,000	7.0	5.5	0.40	5×10^5
Polyester-random-glass	18,000	18,000	3.0	3.0	0.40	2×10^4

SOURCE: PHOTOCIRCUITS DIVISION, KOLLMORGEN CORP.

chem resists, particularly with brittleness. The troubles were solved by the vendors, and Cirtel's production line has run smoothly ever since, turning out etched fine-line and multilayer printed-circuit boards.

As predicted, Cirtel has had lower equipment and solvent costs. In addition, its disposal problems have been simplified. Cirtel is now adding another production line for prototyping, and it, too, will be fully aqueous.

Above board

The keystone of any printed-circuit process is the substrate itself. It must have adequate flexural strength and impact resistance, a small amount of water absorption, and good electrical properties like low dielectric constant and dissipation factor. Table 2 gives the composition, applications, and costs of the most popular laminates currently used, and Table 3 lists their mechanical and electrical properties.

Until fairly recently, XXXP, FR-2, and FR-4 were the most heavily used, with paper-phenolic boards filling consumer slots and epoxy-glass boards dominating the computer and military markets. As Table 3 shows, the mechanical and electrical properties of FR-4 are far superior to those of the less expensive XXXP and FR-2. What was needed were materials with properties approaching those of FR-4 but at a lower cost. Two new types of substrates, one made of random-glass matte and polyester resin and the other of composite epoxy materials, have now appeared to fill the gap.

The polyester-based material has excellent impact resistance (about half that of FR-4), but its flexural strength is only about as good as that of paper-phenolic materials. Because of its high impact resistance, it is displacing XXXP and FR-2 substrates in automotive and consumer applications where more complex circuitry requires a laminate with better electrical properties than paper-phenolic types. But because of its low flexural strength, it is pretty much ruled out for applications that use FR-4. However, Cincinnati Milacron Co.'s Molded Plastic division, Blanchester, Ohio, which produces a polyester–random-glass laminate called Cimclad, now has a newer version called Milclad with double the flexural strength of the older material. Milclad could conceivably be applied as a substitute for epoxy-glass.

The composite epoxy materials, which were developed a little later, come in two basic types: CEM-1 is composed of an internal core of cellulose paper encapsulated between two layers of woven glass, and CEM-3 has an internal core of a nonwoven glass.

CEM-1 (General Electric's PC-75 is an example) is superior to paper-phenolics and polyester and is proving popular in automotive electronics, video games, and smoke detectors. As Table 2 shows, it is price-competitive with XXXP and Cimclad. But according to J. E. White composites project manager of GE's Laminated and Insulating Materials Business department, Coshocton, Ohio, "polyester–random-glass material is competitive with CEM-1 in price only. PC-75, for instance, has a longer die life and higher yields." However, the cellulose-paper construction of CEM-1, with its attendant high water absorption, prevents it from replacing epoxy-glass boards in demanding applications.

Of the new materials, CEM-3 is the one that could have a large impact on the pc field, since on the whole it has the same specifications as FR-4 at a lower cost. In addition, it can be used in any pc process that employs FR-4, and it can be drilled or punched more easily than its epoxy-glass counterpart or other competitive plastic substrates. Moreover, there is no difficulty in making plated-through holes, as there is with CEM-1 and polyester–random-glass.

Atlantic Laminates, Franklin, N. H., a division of Oak Industries, was the first to come out with a successful CEM-3, called AL-910. Derek Russel, product manager for punchable substrates for the company, sees CEM-3 replacing FR-4 in 85% to 95% of the boards in the near future. He points out that CEM-3 meets or exceeds FR-4 in every specification except flexural strength. For instance, 1/16-inch-thick AL-910 has an flexural strength of 61,000 pounds per in.^{2}, whereas an FR-4 laminate of the same thickness has a 79,400-lb/in.^{2} flexural strength.

Russel believes that very few applications need the

	Material	Copper clad	Dielectric constant (at 10^6 Hz)	Dissipation factor (at 10^6 Hz)	Continuous-use temperature °F
Low-cost materials	ABS	no	2.4 – 3.8	0.007 – 0.015	180
	Epoxy-paper	yes	4.0	0.018	250
	Polyester-random-glass	yes	4.5	0.020	290
	Noryl	no	2.7 (at 60 Hz)	7×10^{-4} (at 60 Hz)	220
	Epoxy-glass	yes	4.5	0.020	290
	Polysulfone	no	3.1	3.4×10^{-3}	345
	TPX	no	2.1	2.5×10^{-5}	320
	Polycarbonate	no	2.9	0.010	250
Medium-cost materials	PPS-glass	no	3.9	4.1×10^{-3}	400 – 500
	PPS	no	3.2	4×10^{-4}	400 – 500
	Epoxy-polyimide-glass	yes	5.1	0.017	425
	PPO	no	2.6	7×10^{-4}	220
	Tefzel	no	2.6	5×10^{-3}	300
High-cost materials	PPO	yes	2.6	7×10^{-4}	220
	Teflon-glass	yes	2.5	8×10^{-4}	500
	Teflon	yes	< 2.1	$< 2 \times 10^{-4}$	500
	X-linked polystyrene-glass	yes	2.6	4×10^{-4}	190

TABLE 4: HIGH-FREQUENCY PROPERTIES OF PC LAMINATES

SOURCE: TEKTRONIX INC.

flexural strength of a glass-epoxy board and that most boards are overdesigned and could use a CEM-3 material. Only military or aerospace boards really need the structural and impact strength of FR-4, he says.

As the operating frequency of a circuit board moves up, dielectric constant and dissipation factor suddenly become more important. Table 4 lists the properties of plastics suitable for circuit boards. The materials are arranged roughly according to cost. For example, 1/16-in.-thick ABS is less than $1.00/ft², whereas the materials at the bottom of the third group—clad Teflon and X-linked polystyrene—cost $30 to $50 a square foot. Note that the general-purpose materials previously discussed—epoxy-paper, polyester–random-glass, and epoxy-glass—all have relatively high dielectric constants and dissipation factors that exclude them from high-frequency applications for all practical purposes.

High-frequency boards

Tektronix Inc., Beaverton, Ore., has a large in-house pc facility and makes many high-frequency boards. Its engineers considered polysulfone, ABS, and TPX as hf substrates, but the last two were discarded because they softened under soldering. Bare polysulfone, though, did have the electrical and thermal properties required.

Tektronix now makes high-frequency boards from relatively low-cost unclad polysulfone using its own semiadditive process. These boards are fulfilling all their requirements, says Jerry Jacky, senior chemical engineer at the company. In addition, the use of polysulfone has cut raw-board costs by about $10,000 a month.

The next development in this area will be copper-clad polysulfone. With clad polysulfone, straight etching could be done, thus cutting the price of processing by utilizing available etching equipment. Steve Nelson, marketing manager for Union Carbide Corp. in New York, says, "We are working with several laminators to develop sources." One of the companies is the Norplex division of Universal Oil Products, La Crosse, Wis., which is now supplying such samples.

Union Carbide, a larger supplier of polysulfone, sees two markets for this material. One is for high-frequency boards, either bare or clad. The other is for general-purpose boards with premolded features like holes and connector housings. Premolding makes it possible to combine structural and circuit-board functions, and the printed wiring can be plated or screened on with conductive inks. The resulting boards would most likely be employed in consumer items.

In general, all the developments discussed above are in their formative stage. Polymer thick-film inks and enameled-steel boards have a good chance of moving from consumer to industrial applications.

Practically every independent pc supplier is now either developing or using ultrathin copper foils. In fact, in the future, standard etching may shift to ½- rather than 1-oz/ft² foil as a result of the influence of fine-line boards. Also, copper foil suppliers are now experimenting with even thinner foils, and boards with 1- to 3-mil-wide conductors may appear in a few years. ☐

Packaging & production

New substrate causes a stir

Porcelain on steel could be key to size and cost barriers
of conventional alumina material in thick-film hybrids

by Jerry Lyman, Packaging & Production Editor

For several years, porcelain on steel with a simple conductive pattern screened on has been used in hundreds of thousands of flashbulb arrays made by the General Electric Co. and GTE Sylvania Inc. Now, it is beginning to look like the large-circuit substrate of the future.

This rugged, easily manufactured material overcomes the size and cost limitations of the alumina substrates used for thick-film hybrids. At the same time, it presents an attractive alternative to present printed-circuit laminates, with better power dissipation and a built-in ground plane.

Applications in sophisticated electronic circuitry are just beginning to undergo examination. For example, both E. I. du Pont de Nemours & Co. at its Electronics Materials division at Niagara Falls, N. Y., and Singer Co.'s corporate research and development laboratory at Fairfield, N. J., are in the midst of evaluations of porcelain on steel with glass-based thick-film materials to determine whether it is a usable circuit-board medium.

Early results from both programs do look promising, but there are still problems to be solved. These include minimizing the alkali ion content, eliminating a damaging phenomenon known as brown plague, and a determination of the long-term reliability of circuits built with this technique.

Despite its problems, Trevor Allington, senior product specialist at du Pont, says, "The answer to whether porcelainized steel is a bonafide substrate is a qualified yes. However, the task of proving this is by no means complete."

During both du Pont's and Singer's testing of fired conductive materials on the substrates, a strange and deleterious effect—brown plague—occurred. It showed up as a discoloration of the silver conductors. More serious than the discoloration was the fact that the plagued material affected resistor values and was difficult to wire-bond or solder. Du Pont researchers found that the brown plague could be eliminated by keeping the conductor pattern from coming within 10 mils of the substrate's edge.

Edge cause. Early research by Murray Spector, general manager of materials supplier Alpha Advanced Technology Inc. of Newark, N. J., indicates that the plague may be caused by a strong electric field generated at unrounded edges of a porcelainized steel substrate. The electric field generates silver sulfide in the thick-film silver conductors. To prevent this, the rounding of the edges must be controlled carefully.

Singer's tests turned up interesting data on a comparison of the electrical properties of porcelainized steel and alumina that indicates the new material is not best for all applications. For example, alumina's lower dielectric constant makes it the more attractive material for microwave applications.

Daniel Wicher, a staff scientist at Singer, says, "We have a positive but qualified answer to whether porcelain on steel technology is viable. The qualifications come in two major areas. One is that porcelainized steel will not replace alumina in high-frequency, high-performance, or high-stability systems. The other is that porcelainized steel-based technology will find numerous applications, particularly in the consumer product area, where large area substrates with higher performance and lower cost than standard pc systems are needed." ☐

ELECTRICAL CHARACTERISTICS OF PORCELAIN ON STEEL VS. ALUMINUM			
Property	Conditions	Porcelain on steel	Alumina
Volume resistivity (Ω·cm)	200 Vdc, 25°C	> 10^{11}	> 10^{11}
Dielectric strength (V/mil)	25°C	900	800
Dielectric constant	1 MHz, 25°C	6.4	9.3
Dissipation factor	1 MHz, 25°C	0.008	0.0003
Loss factor	1 MHz, 25°C	0.07	0.0028

SOURCE: SINGER CO.

Porcelain-on-steel boards can launch a thousand chips

Thermal and mechanical properties of popular appliance material cut processing steps in hybrid and pc-board manufacture

by Murray Spector, *Alpha Advanced Technology Inc., Newark, N. J.*

☐ Porcelain-coated steel, the material found in every kitchen, comes close to being the ideal substrate long sought by manufacturers of both thick-film hybrid circuits and printed-circuit boards. It has excellent mechanical and thermal properties and is low-cost in large volumes.

Its ruggedness will enable the hybrid makers to move from their present small, fragile alumina substrates to ones at least the size of pc boards, having built-in heat sinks and ground planes, and promising eventually to carry over a thousand chips. Its good thermal properties will enable the pc-board manufacturers to print and fire conductive and resistive inks directly onto it (Fig. 1), cutting their costs tremendously by eliminating all the plating and etching now needed, as well as much of the component assembly.

Still other benefits will be shared by both groups. For instance, the metallic substrate can be bent and shaped easily. After enameling, it survives environments inimical to both ceramics and plastics. Finally, it is possible either to wave-solder discrete components or wire-bond chips to the new steel boards.

A porcelain-coated steel substrate has a core of low-carbon steel and a coating of fired-on porcelain enamel. The porcelain enamel is a ceramic with a firing temperature high enough to allow hybrid components to be fired onto it subsequently at 650°C. Conductors may be of copper, silver, palladium-silver, platinum-silver nickel, gold, or any other metal available as a screenable ink. Resistors are ruthenium-oxide–based, and the smooth surface of the porcelain often makes their fabrication so uniform that trimming is unnecessary. When necessary, however, it may be done abrasively or by laser—and with the circuit energized. This kind of active trimming is possible because of the thermal mass of the steel, which is so great that it prevents the resistor from heating up much during the process.

From heaters to circuits

Way back in the 1930s, heaters were being made of copper on porcelain-coated steel. So in the 1940s, when printed wiring boards were invented, there were immediate attempts to make boards with a metal core. But success came only in 1965, when steel boards were insulated with epoxy. Epoxy, though, cannot take the firing temperatures needed for hybrid circuits, so in 1967

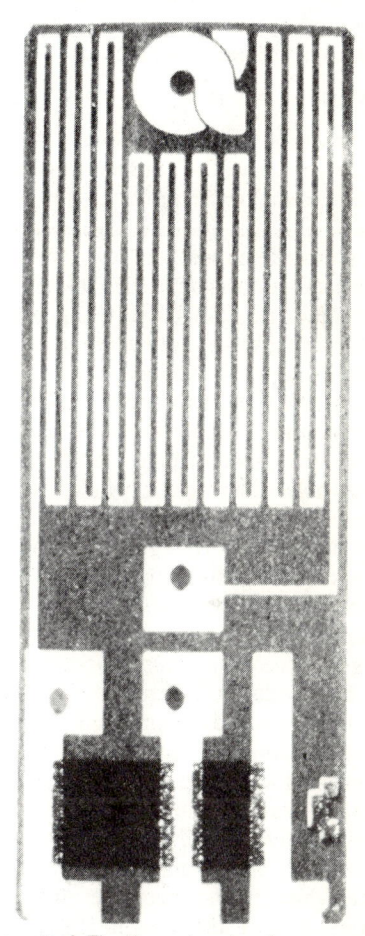

1. Porcelain on steel. The thermal and surface properties of porcelain-coated steel allow modified thick-film inks to be screened and fired onto its surface at a relatively low temperature. A sample, shown above, has screened-on resistors and conductors.

the first serious attempts were made to use porcelain-coated steel in electronics. Some manufacturers of thick-film inks made inks that would fire at the right temperatures for it, but additional engineering problems with the inks and the coated steel held up the full emergence of the new substrate until about 1977.

Today porcelain-coated steel represents a new substrate technology. It has its limitations, which will be discussed later, but evaluations currently under way at several large potential users will probably result in its use

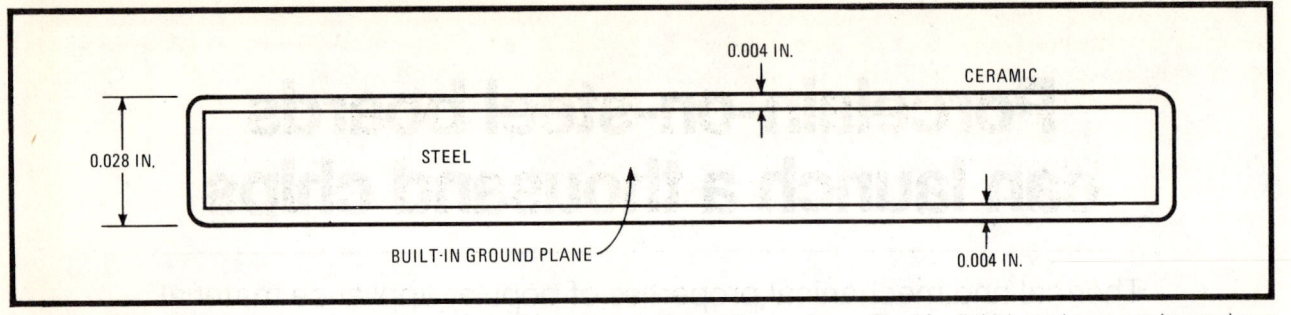

2. Big boards. After processing, a porcelain-on-steel substrate has a cross section as shown. The 28-mil-thick steel core can be used as a built-in ground plane. Boards of this composite material, aimed at large-scale circuitry, can be fabricated in sizes as large as 12 by 18 inches.

this year in both military and consumer electronics—automobiles, telephones, telephone switching circuits, television sets, and appliances. Still other possible areas of application are computers and medical electronics.

The metal-working step

The manufacturing process starts with the fabrication of the steel core by any of a number of common metal working techniques. For prototype runs, it is most economical to chemically mill the steel core. For larger quantities, a stamping process is best, being very inexpensive on a per-piece basis once the tooling is in place.

The tooling, however, can be very expensive, because it needs to be unusual. Simply drilling circular holes straight down through the steel substrate is not enough. Picture one such hole in cross section. It would have a square shoulder, and a square shoulder causes a porcelain coating to pull away from it and create a ridge encircling the hole. The result would be a point of minimum insulation and of extremely high electrical field density—in other words, an insulation breakdown hazard bad enough to defeat the use of this design in many electronic applications.

The solution to this problem is either to flare or to bevel the edge of the hole. This rounding of the sharp corners has the additional benefit of facilitating the automatic insertion of leaded components.

Once the steel is formed into its final shape, it starts through the processes leading to enameling. It is cleaned and then pickled in an acid solution, which roughens the surface and promotes the adhesion of the enamel. Pickling is followed by a rinsing operation and then by nickel-plating, which further promotes enamel adhesion. After additional rinsing the steel is ready to be dipped into the enamel.

Electronic-grade porcelain

The enamel used is a formulation designed for continued operation in an electric field at elevated temperature. It is applied in a water solution to the part, which then must be very carefully dried before firing; otherwise the water will vaporize explosively and pit the enamel surface. The firing yields the finished substrate—a 0.028-inch-thick core of enameling-grade steel coated on both sides with a black ceramic layer 0.004 inch thick (Fig. 2).

Metalization is next applied to the porcelain. This involves screening on a metal paste and firing it. The Electronic Materials division of E.I. du Pont de Nemours & Co., Niagara Falls, N.Y.; Cermalloy, Cermet division of Bala Electronics, West Conshohoken, Pa.; Electronic Materials Corp. of America (EMCA), Mamaroneck, N.Y.; Electro-Science Laboratories Inc., Pennsauken, N.J.; and Thick Film Systems, Santa Barbara, Calif., all manufacture precious-metal conductors, resistors, and dielectrics formulated for porcelain adhesion and having firing points between 600° and 650°C. These inks resemble those used on alumina substrates except for their lower firing temperatures. Additionally, DuPont and Cermalloy both have copper inks available. Where through-hole printing is required on double-sided boards or substrates, a special ink formulation is necessary to provide the requisite kind of plastic flow, or rheological properties.

Wave-soldering to copper

Copper inks are especially suited to printed-wiring-board applications since they allow the use of 60/40 tin-lead solder. Precious-metal inks, preferred by hybrid-circuit manufacturers, have the disadvantage of requiring a more expensive silver-bearing solder to inhibit the precious-metal ink from being leached off the board during soldering.

Wave-soldering a porcelainized steel board with copper conductors is not the same as wave-soldering a conventional epoxy-glass printed-circuit board. Preheating temperature and soldering speed must both be higher than those necessary with the epoxy board.

The first adjustment must be made because of the high heat capacity of the steel core. The core acts as a heat sink to the copper and prevents it from reaching soldering temperature unless the board has been sufficiently heated beforehand.

Conversely, at the time of application of the solder, when the board is over the solder wave, the core acts as a heat source and could encourage the solder to flow too far up the component leads. This can be prevented by speeding up the belt.

The steel core of the substrate may be used as a ground plane. First the bare steel must be exposed; then contact is made to it by conductive epoxy or by deposition of a thick film.

Comparing board materials

The biggest difference between steel and plastic boards lies in their thermal conductivity. None of the plastics used in standard printed wiring boards is good at conducting heat—a problem there have been many attempts to alleviate. One approach uses heavier copper laminates, but then the thickness of the copper demands that all lines and spaces be widened. Another approach is to back the board with aluminum or steel, but this is awkward and expensive and has gained relatively little favor with the industry.

A porcelain-coated steel board is far superior to a plastic type in thermal conductance, even when no direct contact is made to the steel core. By way of comparison a 2-watt precision resistor operating at rated power on an epoxy-glass board is 25.2°C hotter at its surface than an identical resistor mounted on an Alpha porcelainized steel substrate, trade-named Alphamet.

Even this performance can be enhanced by making ground contacts to the steel core and putting the core in metal-to-metal contact with a system's chassis. This is made possible by masking the steel so as to prevent porcelain from being deposited on any area to be directly contacted. The procedure often eliminates the need for heat-sinking components.

A second difference is that plastics all expand much more than steel with increased temperature. It is because of this characteristic that the pad areas around the holes on plastic boards have to be large, to compensate for any misalignment of the final etch pattern with holes. On porcelain-coated steel boards, the pad areas can be smaller because the much lower thermal expansion of steel allows more precise alignment.

Finally, standard printed-circuit board materials cannot be used reliably in high-temperature environments, as in an automobile engine, oil drilling apparatus, and thermal printing heads. These are all applications where porcelainized steel boards are found to be highly satisfactory substitutions.

Steel vs alumina

All-ceramic substrates, in contrast, are inferior to steel not only thermally but in their electrical and above all their mechanical properties.

Indeed, mechanical fragility has historically been the greatest drawback of alumina and beryllia wafers. It precludes their use entirely in many harsh mechanical environments, or only if equipped with expensive steel backing plates. The same fragility keeps yields low in both substrate fabrication and finished circuit assembly and creates a cost structure that increases exponentially with substrate size. Porcelain-coated steel, on the other hand, is virtually indestructible, and its cost per square inch is unaffected by size.

To illustrate comparative thermal performance, identical resistors were printed on alumina and Alphamet substrates. Then the increase of temperature at the resistor surface was plotted as a function of power in the resistors. As can be seen in Fig. 3, alumina heats up faster and to a higher temperature than the Alphamet.

It is worth noting that a simple conductivity concept is not enough for a proper theoretical analysis of a composite material like porcelain-coated steel. Instead, the conductance of all five layers of the structure must be considered—the two outer porcelain layers, the innermost steel core, and the two transition layers between the porcelain and the steel. In addition, unlike other substrates, porcelainized steel radiates a significant amount of heat. The emissivity of its black porcelain is a measure of its efficiency at this form of thermal transfer. A mathematical model taking all these factors into account has been found to generate experimentally verifiable results.

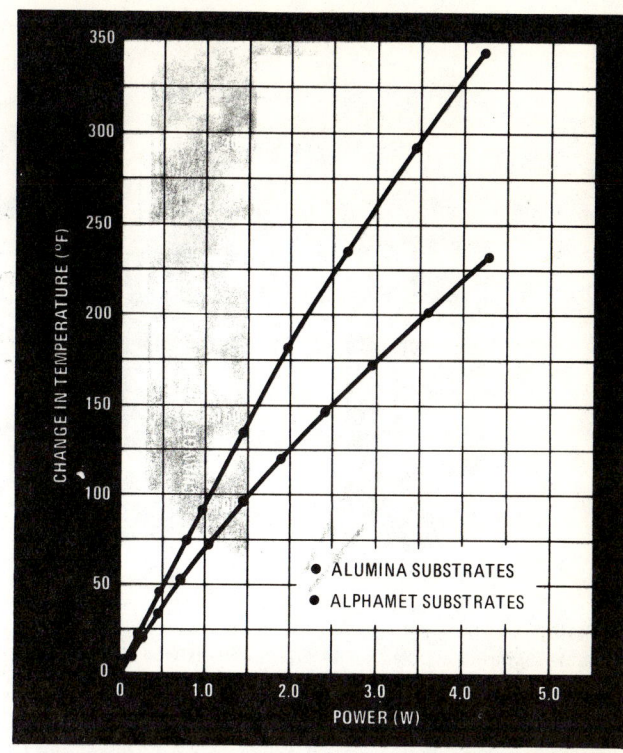

3. Hot plates. A comparative plot of temperature change versus power shows porcelainized steel has a lower temperature rise per watt than similarly sized alumina substrates. The new substrate's steel core contributes to its excellent thermal conductance.

As already noted, the thermal performance of porcelain-coated steel may be further enhanced by supplying metal-to-metal thermal paths away from the substrate. This is typically done by providing the substrate with unporcelainized mounting pads for bolting it to a chassis.

Finally, the steel core serves as an electrical plane, useful either as a ground plane or sometimes as a power distribution plane. Either use eliminates the need for one conductor and one dielectric layer.

Electrical performance may also be enhanced by the availability of two-sided construction. Conductor-filled holes can be provided as vias, or plated through-holes, connecting the two sides of the substrate.

For multilayer circuits the number of firings is halved by using a cycle of print-dry-print-dry-fire. In applications where crosstalk is a factor, circuits can be separated and then shielded from each other by their common steel substrate.

Bending steel

The presence of the steel core creates two other useful properties that are not to be found in other rigid substrates. First, the steel can be worked in three dimensions; on a simple level, the edges of a board can be bent so that it becomes its own chassis. Second, the porcelain surface absorbs no water, and the ceramics used in

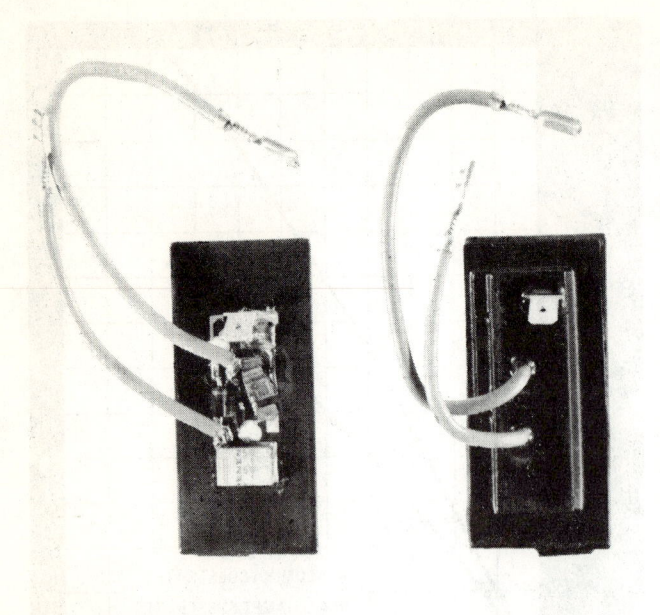

4. Potted power. Porcelain-coated steel replaces alumina in substrate for power-supply circuitry. The same material also replaces the aluminum heat sink. Hybrid thick-film techniques were used in both cases. An unpotted module is pictured to the left.

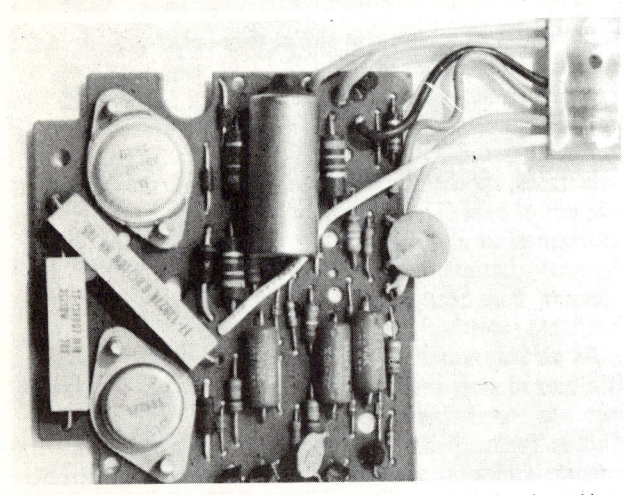

5. Printed wiring on steel. This porcelain-coated steel pc board has been substituted for a plastic type in an electronic ignition system designed at Motorola. Note the discrete components. Final version of this circuit will be a thick-film hybrid on a steel board.

hybrid substrates allow the porcelain-coated steel actually to be the outer wall of any package.

For all its advantages, porcelain-coated steel does present some problems, both technical and otherwise. The first is ion migration.

Porcelain enamel is a complex material. From an electrical point of view, it is generally considered to be simply an insulator. However, at a thickness of 4 mils, it is far from ideal in that role. The metallic oxides that comprise the porcelain ionize under the influence of heat and voltage. In a dc field, the ions move and constitute a leakage current to a working circuit. To minimize this problem, the current under an 0.080-in. pad (the size of a typical integrated circuit) should be less than 10^{-13} ampere. Also, the rate of ion generation should be so low that the adhesion of the metalization to the porcelain will not be affected by ion accumulation for 40 years.

The other technical problem is brown plague—the oxidation of silver alloy conductors at the edge where they accidentally come in contact with the steel core. The mechanism is not well understood, but can be avoided by using copper for ground contacts and keeping silver 0.060 in. away from the substrate edges.

The third problem is economics and concerns the cost-volume relationship. Tooling for steel substrate manufacture is so expensive that the process becomes cost-effective against epoxy-glass printed wiring boards only at a volume of over 100,000 substrates.

The last, and perhaps the largest, problem is psychological. The unavailability of all-ceramic substrates in large sizes at a reasonable cost has conditioned designers of hybrids to think in terms of circuit functions rather than systems. Designers now have to reorient themselves to today's reality in which 12-by-18-inch substrates are available, along with screen printers, drying ovens, and firing furnaces big enough to handle them.

Packaging power

Despite these obstacles, enameled steel substrates are being evaluated for many tasks. Figure 4, for instance, shows an application in a small power supply. The porcelainized steel substrate replaces an alumina substrate and an aluminum heat sink. Figure 5 shows a pc-board-like application—an automotive electronic ignition system—in which screened-on metallic conductors are combined with discrete components on porcelain-coated steel. It is a transitional version, meant to be followed by a completely hybridized version.

Steel substrates are eminently suited to the rough under-the-hood environment of cars and indeed in any high-shock environment. For example, ceramic hybrid boards are often backed with metal in such shock environments. This is an expensive procedure since both materials require extremely precise machining if they are to mate well enough to withstand the shock. A ceramic-on-steel substrate eliminates the need for a two-piece part as well as the high cost of machining surfaces in the first method.

Similarly, a circuit board, when it is required to act as a mechanical member of a mechanical assembly, often needs a metal backing. In one system a large plastic pc board had to be stiffened with a piece of aluminum before it was able to serve as a panel for rack-mounted equipment. A porcelain-coated steel board, on the other hand, could replace the aluminum plate and the epoxy-glass board.

A look to the future

Applications for porcelain-coated steel are moving in two major directions, toward combining existing functions and toward altogether new functions.

Its potential for serving as both circuit board and chassis or even package has already been mentioned. Another possibility is for the same substrate to support thin and thick films and discrete components.

As for the jobs that only porcelain-coated steel can do, thermal sensors have already been designed on such a substrate for direct insertion into a car's engine block. □

Mass molding a boon to fast logic

Makers of computers and peripherals are using newly popular method of making multilayer boards that is faster, simpler, and cheaper

by Jerry Lyman, Packaging & Production Editor

A nine-year-old technique that offers a simpler, low-cost, fast-turnaround alternative to the standard multilayer board has suddenly hit the big time. And what has propelled mass molding to its new heights of popularity is the growing interest being shown by makers of computers and peripheral devices in multilayer printed-circuit boards.

The usual multilayer board, with up to 30 conductive layers, is pretty much a high-density packaging approach suitable only for military and space applications. But with the advent of high-speed logic, a simplified four-layer form has begun to find its way into computers and peripherals because it is so much denser than single-layer printed-circuit boards. Moreover, mass molding, a method developed by Fortin Laminating Co. of San Fernando, Calif., can turn out these simpler boards almost on an assembly-line basis.

Comparison. conventional multilayer-board making amounts in effect to a custom process, providing the user with a ready-to-go pc board. The mass molding process provides semi-complete boards, made by stepping and repeating circuit patterns on the inner layers, each for a different board, on laminate sheets that are up to 48 by 24 inches in size. The first method turns out a board at a time, manufactured layer by layer

With mass molding, the system manufacturer receives a multilayered laminate from pc-board firms like Fortin, Lamination Technology Inc. of Santa Ana, Calif., and the Mica Corp. of Culver City, Calif. It has two conductive layers—usually a ground plane and a power plane—and two outer layers covered with copper foil. The customer then need only subtractively etch the patterns he needs from the two outer copper surfaces (see diagram).

Mass molding uses much the same materials as those found in conventional board manufacturing—cores, or epoxy-glass layers with copper foil on both sides, and prepregs, or thin sheets of partially polymerized epoxy resin. For the two internal layers, circuit details are first subtractively etched on the core material. Then several insulating prepregs are placed over the top and bottom, and standard copper-covered pc epoxy-glass material make up the final two outer layers. Lastly, the entire sandwich is laminated together in a molding press.

That contrasts strikingly with the conventional method of making multilayer boards. Most standard multilayer presses can handle only material that is 18 by 24 in., which limits the number of similar boards that can be turned out simultaneously. A conventional four-layer board is put together from a top circuit core with copper foil on both sides (layers 1 and 2) followed by prepregs and a bottom circuit core (layers 3 and 4).

Distortion a threat. In the manufacturing process, layers 2 and 3 are etched, then pins are used to ensure exact registration of the cores and prepregs, and the assembled layers are laminated together in a press. After this step, layers 1 and 4 have their circuit patterns subtractively etched onto them. The whole process uses much more labor than mass molding and requires many additional steps. What's more, any distortion of the laminates or pins due to the heat and pressure of the laminating press can knock the layers out of registration.

On the other hand, mass molding requires fewer steps and there are fewer opportunities for error. Registration of the outer to inner layers is simpler: pilot holes are drilled into the outer layers through which registration targets on the etched inner layers are found. Layers 1 and 4 are matched to these targets. This method virtually eliminates the registration defects that can crop up in the standard way of manufacturing multilayer boards.

Ordinarily, the

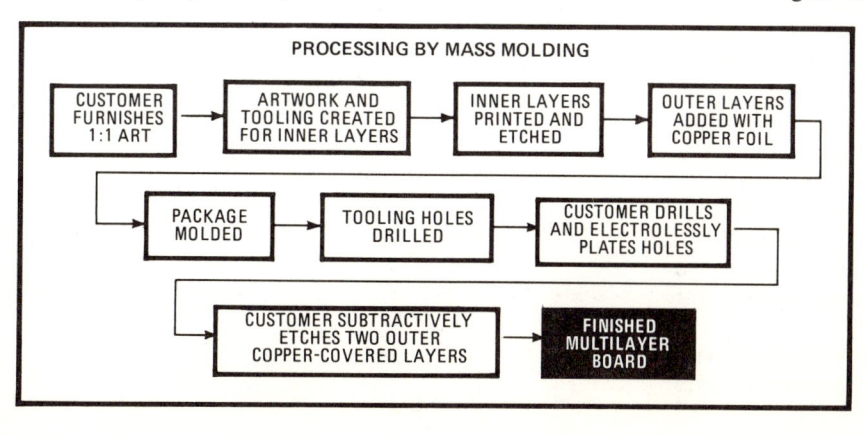

```
                    PROCESSING BY MASS MOLDING

    CUSTOMER          ARTWORK AND         INNER LAYERS        OUTER LAYERS
    FURNISHES   →     TOOLING CREATED  →  PRINTED AND     →   ADDED WITH
    1:1 ART           FOR INNER LAYERS    ETCHED              COPPER FOIL
                                                                   ↓
    PACKAGE           TOOLING HOLES       CUSTOMER DRILLS
    MOLDED      →     DRILLED          →  AND ELECTROLESSLY
                                          PLATES HOLES
       ↓
    CUSTOMER SUBTRACTIVELY
    ETCHES TWO OUTER          →   FINISHED
    COPPER-COVERED LAYERS         MULTILAYER
                                  BOARD
```

mass-molded board is fabricated with 70-micrometer (2-ounce-per-foot) copper foil on its inner layers and 35-μm (1-oz/ft) copper foil on its two outer layers. Now many customers want a thin copper foil (9 to 5 μm thick) on the two outside layers in order to be able to etch densely packed, fine-line circuitry from it. In fact, many makers of computers and peripheral devices have already standardized on mass-molded four-layer boards with 5-to-7-mil conductors etched into the outer layers.

While most mass-molded boards now are four-layer types, the industry will not stop there. Some six-layer boards are being supplied commercially, and Lamination Technology can supply boards with eight layers. Over 90% of all mass-molded boards are digital types, but the technique also should be ideal for high-speed analog circuitry and backplanes.

Advantages. Mass molding has many important advantages over standard multilayering. For one thing, the huge capital investment and engineering staff required to set up an in-house multilayer board-manufacturing line is eliminated. Also, the relatively long 6-to-10-week turnaround needed with the conventional multilayer method is cut to three to four weeks.

On the other side of the equation, mass molding is good for the pc-board manufacturer because it means instant added capacity without the addition of equipment or people. Finally, mass molding can be used to create a whole family of different circuit boards, particularly for boards with inner ground and power planes. New designs simply require new circuit patterns on the two unetched outer layers.

Paul Benke, general sales manager of Mica Corp., says most mass-molded boards are going to computer, minicomputer, and peripherals makers. Another big customer is the end user who has substantial in-house manufacturing capability. Digital Equipment Corp. and others have been using this technique to make high-speed logic boards for some time.

However, pc-board makers with strong multilayer capabilities are actually the heaviest exploiters of mass-molded boards. Such firms as National Technology Inc. of Santa Ana, Calif., and Metropolitan Circuits Inc. of Costa Mesa, Calif., mass-mold four-layer boards and save their multilayer capacities for more complex boards.

Mass molding now represents 10% to 15% of the total multilayer market—for example, Benke says Mica's sales grew 40% to 50% in the past year. While the number of layers is creeping up, it does appear that for the highly complex applications, where painstaking albeit slower production is desirable, the conventionally manufactured multilayer boards will continue to dominate the field. □

Fine-line printed circuits catch on

Technique uses ultra-thin copper foil and yields much greater density than can be obtained with conventional packaging processes

by Jerry Lyman, Packaging & Production Editor

A printed-circuit process that produces much denser circuits than the current norm is beginning to win a following. In the fine-line process, 5- to 10-mil-wide conductors and spaces are subtractively etched from 5- to 10-micrometer-thick copper foil. Conventional techniques use 40-micrometer-thick foil to get elements 10 to 20 mils wide.

Sales figures from two of the largest producers of thin copper foils, divisions of Gould Inc. in Cleveland and Yates Industries Inc. in Beaumont, Calif., put such fine-line pc boards at 5% to 10% of the total pc market now. Most of these end up as densely packaged, high-speed digital boards in computers and computer peripherals. Eventually, the technique may also dominate specialized boards.

While the fine-line approach also offers processing advantages, it is the increased density that brings a gleam to users' eyes. "This technique provides 20% more packaging density and capacity than a standard IC-socket wire-wrapped panel and achieves twice the packaging density of conventional two-sided boards," says Jeff Waxweiler, president at Algorex Corp., a Syosset, N. Y., firm that produces both the fine-line boards and computer-designed artwork for them. What's more, he adds, "the densities can equal or exceed those of more expensive techniques, like multilayer boards or multiwire. It's possible to use two-sided fine-line pc boards and still get more than two dual in-line packages into the square inch." With such a board, two to three conductors could be run between DIP leads—an impossibility with conventional pc traces.

At many independent pc firms like Multi-Circuits Inc., Manchester, Conn.; Metropolitan Circuits Inc., Franklin Park, Ill.; and Cirtel Inc., Irvine, Calif., 20% to 30% of the boards being produced are already fine-line types. In fact, Mike Busby, president of Cirtel, expects a third to a half of his company's output will be fine-line by 1980.

At Memorex Corp.'s operation in Eau Claire, Wis., where all of the firm's own and custom pc work is done, product control manager Jim Berry, says, "Fine-line is the way of the future." Memorex has already committed itself to the fine-line geometry in all of its new products with boards for high-speed logic. IBM Corp. and Western Electric are also heavily involved with it.

The process. The fine-line technique resembles the older, conventional process, based on 40-μm foils, in most of its steps, but eliminates several of them (see chart). Its ability to produce finer lines is due to the speed with which the thin foil can be etched—fast enough not to undercut the final circuit pattern.

Additionally, the microfoil's thinness means less etching time is required, so costs of chemicals are reduced: process users indicate that time and costs are about 25% of those rung up with 40-μm foils. Less etchant also means that waste treatment of the liquid is less of a problem than in standard production work.

Also indirectly responsible for the fine-line process's advantages is a copper or aluminum protective carrier about 2 mils thick that can be supplied with the copper foil. The carrier decreases drilling time, thus extending drill life, and the drill

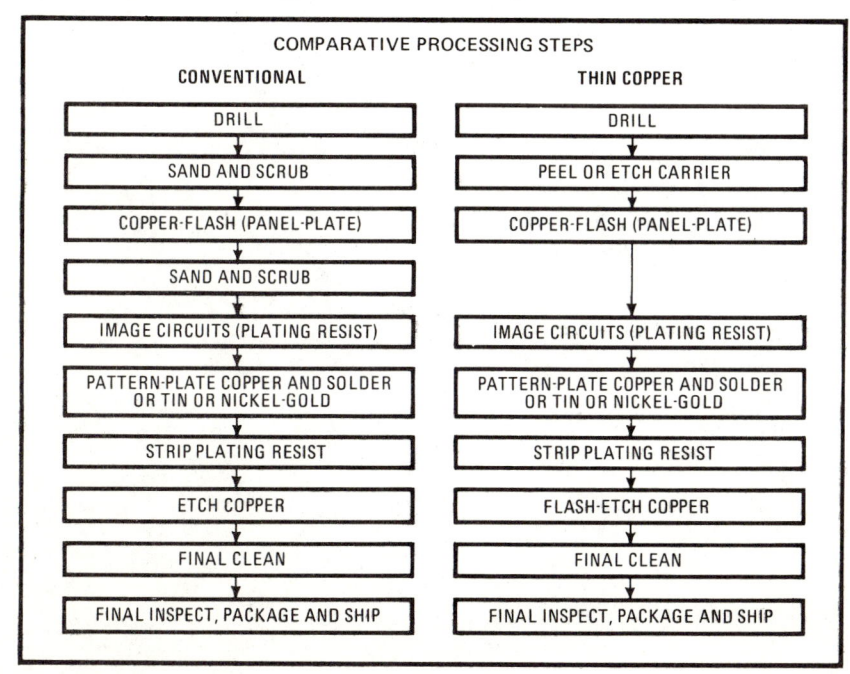

COMPARATIVE PROCESSING STEPS

CONVENTIONAL	THIN COPPER
DRILL	DRILL
SAND AND SCRUB	PEEL OR ETCH CARRIER
COPPER-FLASH (PANEL-PLATE)	COPPER-FLASH (PANEL-PLATE)
SAND AND SCRUB	
IMAGE CIRCUITS (PLATING RESIST)	IMAGE CIRCUITS (PLATING RESIST)
PATTERN-PLATE COPPER AND SOLDER OR TIN OR NICKEL-GOLD	PATTERN-PLATE COPPER AND SOLDER OR TIN OR NICKEL-GOLD
STRIP PLATING RESIST	STRIP PLATING RESIST
ETCH COPPER	FLASH-ETCH COPPER
FINAL CLEAN	FINAL CLEAN
FINAL INSPECT, PACKAGE AND SHIP	FINAL INSPECT, PACKAGE AND SHIP

holes need not be deburred.

Finally, any circuit manufacturer can use thin-foil copper laminates with present subtractive equipment. He need only add machinery for carrier removal.

Dissenters. Not all pc-board companies are sold on the new process, however. One of the larger independents, Circuit-Wise Inc. of North Haven, Conn., is sticking to conventional subtractive and additive processes. "Our pc lines are highly automated—it simply would not be cost-effective for us to invest added capital in redesigned automated pc lines for fine-line," says vice president John Mettler,

Another independent resisting the trend is the Photocircuits division of Kollmorgen Corp., Glen Cove, N. Y. Its engineers use a semi-additive rather than thin-foil process for boards with fine-line conductors. John Dennis-Browne, its manager of technology sales and licensing, says that despite undeniable improvements in the subtractively etched thin-foil method, the semi-additive process is the less costly and also produces boards that can be reworked instead of having to be scrapped. He points out, too, that the thin-foil process can only be used with FR-4 (glass-epoxy) boards, while the semi-additive process is usable with many types.

The cost of the thin copper laminates is undoubtedly one factor holding back fine-line pc boards. Another is the added process step needed for carrier removal, and the fact that pc manufacturers have not standardized on any one method for it. The copper carrier/copper foil furnished by Yates must be mechanically stripped and the carrier is salvageable. An aluminum carrier/copper foil supplied by Gould is also mechanically strippable. A third material supplied by the Swedish firm Perstorp AB has an aluminum carrier that can be etched away.

Still others—pc firms like Cirtel and the Collins Radio Group of Rockwell International Corp., Cedar Rapids, Iowa—are successfully working with unprotected foil, eliminating the peeling step.   □

Designer must plan early for flat cable

Matrix enables interconnections to be optimized in initial stages to achieve cuts in design and fabrication costs

by James A. Henderson, Westinghouse Electric Co., Baltimore, Md.

Although spawned out of military necessity to save weight and space and to augment equipment reliability, planar interconnection techniques have now developed to the point that their advantages should no longer be ignored by designers of consumer and commercial equipment. And indeed they are not, as a glance into modern computer and communications systems coming off the line will attest.

Planar wiring is attractive to system designers because it offers cost savings in many situations where traditional round-wire cable interconnections must be forced into small spaces, or where a lot of costly hand labor is required to form the harnesses needed in a particular design.

There are two major approaches to planar interconnections—flat cable and flexible circuit wiring. Flat cable, basically a planar version of a circular cable containing round wires, is usually fabricated from either flat rectangular or round conductors, made of bare or plated copper and laminated or molded in flexible plastic insulation. The cable is available in a wide range of wire sizes, conductor spacings, and configurations.

The other major category of planar wiring—flexible circuits—is a first cousin to the printed-circuit board, except that a flexible plastic, like polyester or polyimide, is substituted for the rigid materials of pc boards. Laminated to this plastic base is a thin layer of copper that is etched to form the required circuit configuration, just as with a pc board. A flexible circuit can be designed to fit exactly into a limited space and reliably make the required interconnections, thus eliminating the unwieldy harnesses, lacing terminal lugs, and other inconvenient and time-comsuming appurtenances of conventional round wiring.

In the following pages, three experts describe some of the new techniques being used to exploit the unique properties of flat and flexible cable interconnects. First, Westinghouse Electric Corp.'s Jim Henderson shows how he realized economy with flat cable by applying good design principles early in the design phase of an engineering project.

Then Joseph Marshall of Ansley Electronics Corp., a subsidiary of Thomas & Betts Inc., shows how flat cable can now handle the ultrafast pulses of modern computer systems.

And finally, Peter Maheux of Bell-Northern Research presents a use of flexible circuitry that produces extraordinary savings in assembly time.   —Stephen E. Grossman

☐ Electrical interconnections are seldom given much thought in the initial design phase of a sophisticated electronic system. Often they are ignored until system partitioning and packaging designs are completed, and then the spaces available for the cables are sometimes so small and cramped that costly redesign may be required merely to incorporate the necessary signal and power conductors.

Flat cable changes all this, for it imposes a requirement for careful planning and system integration of electrical interconnections. Because of its planar construction, the address of each lead falls into a given position, which cannot be altered later as in a conventional round-wire harness. On the other hand, because the cable is designed in and not added on as an afterthought, much costly branching can be eliminated and significant labor savings achieved.

Figure 1 pictures a flat-cable assembly, and Fig. 2 indicates the labor savings possible. In quantities of 100, the 26-connector conventional round-wire bundled cable shown represents $680 worth of materials and 100 hours of labor. On a branch-for-branch, connector-for-connector basis (Fig. 2b), material costs for a comparable flat cable rose to $785, but labor dropped to 67.5 hours. With repartitioning, the system material costs fell to $720, and labor dropped markedly to 38 hours.

Thus, the extra cost of flat cabling materials is more than offset by the elimination of individual wire laying, shaping, and conductor-terminating techniques associated with round-cable assemblies. They are replaced with one set of design steps that determine the form and fit of the cable as an integral unit.

Interchanges

While conventional wiring employs break-outs and branch points to enable each wire to exit from the bundle and seek its termination, flat cable employs interchanges. These interchanges may consist of a single-sided printed-circuit wiring board with crossovers accomplished by judicious location of cable inputs on the board.

Alternatively, the connections may be automatically (or semiautomatically) wired matrices or multilayer printed-wiring matrices (Fig. 3a and 3b). In the latter,

1. Challenger. Typical wiring harnesses are being replaced by new flat-cable techniques, which eliminate expensive prototype and production pitfalls.

the cables are usually terminated at a junction box and interconnected within it by the same techniques. Finally, the interchanges may be a welded matrix (Fig. 3c), in which one layer of flat-conductor cable is laid on top of another, and conductors are welded together through the insulation as desired. This last technique lends itself best to thermoplastic insulations.

Interchanges are expensive, however, not only in dollars, but also in lowered reliability, increased requirements for quality control, and higher system weight. A good designer will eliminate them whenever possible by careful planning and partitioning.

A case history

As an example of the design procedure, a weapons system was originally interconnected by hand-formed harnesses, but it was failure-prone because of wire breakage at the short lead-ins to the connectors from the main branches.

To optimize the design, a functional flow diagram of the system was prepared, outlining initial system partitioning and subassembly identification (Fig. 4). The signal and power-connection requirements were entered in an elementary interconnection matrix, with the matrix sequenced in order of planned subassembly location—that is, the subassemblies were listed on the left side of the chart in the order of their planned physical placement in the system. This list was repeated in identical order from left to right along the top of the chart, forming a square matrix. The number of interconnecting leads required between subassemblies is shown at the proper intersections in the matrix—for example, three twisted pairs and nine single conductors were needed between subassemblies D and E.

This interconnection matrix was developed as a practical offshoot of a shortest-path algorithm for interconnecting cables. With the matrix ordered in the physical sequence of the subassemblies in the system, it is easy to see where interconnections must take long paths, by virtue of their being relatively far removed

2. Evolution. Merely replacing a traditional harness (a), on a branch-for-branch, connector-for-connector basis as in (b), will yield significant savings in labor, with a slight increase in material cost. However, repartitioning the system can optimize labor and material costs (c).

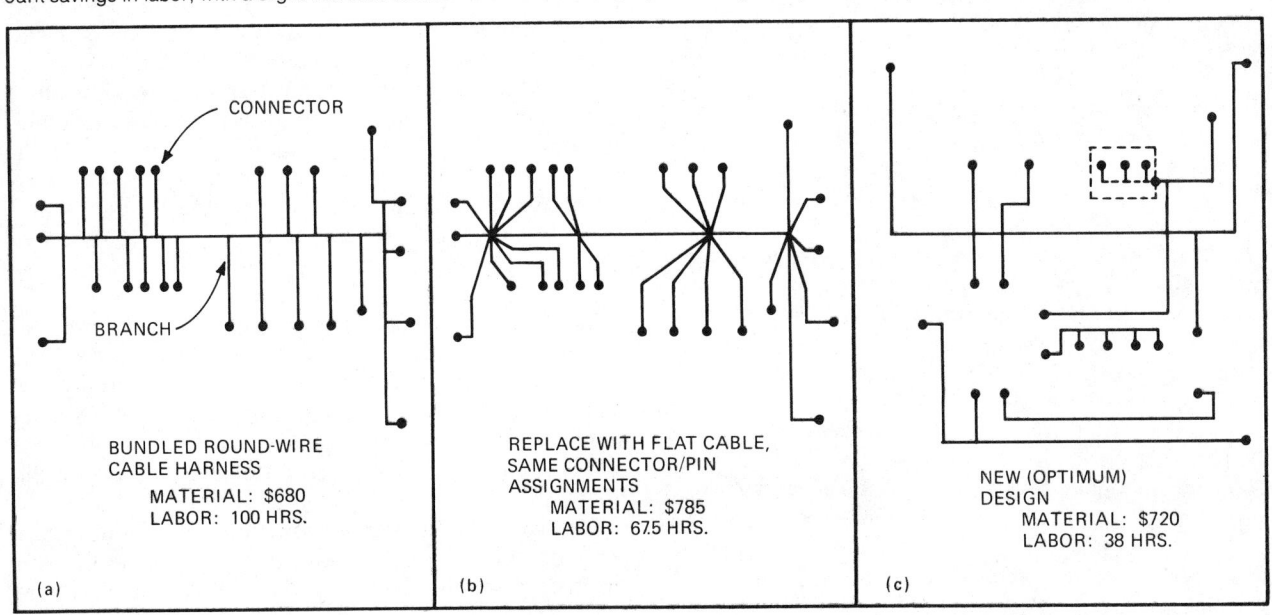

(a) BUNDLED ROUND-WIRE CABLE HARNESS
MATERIAL: $680
LABOR: 100 HRS.

(b) REPLACE WITH FLAT CABLE, SAME CONNECTOR/PIN ASSIGNMENTS
MATERIAL: $785
LABOR: 67.5 HRS.

(c) NEW (OPTIMUM) DESIGN
MATERIAL: $720
LABOR: 38 HRS.

(a) WRAPPED-WIRE MATRIX

(b) MULTILAYER PRINTED CIRCUIT MATRIX

(c) WELD-THROUGH MATRIX

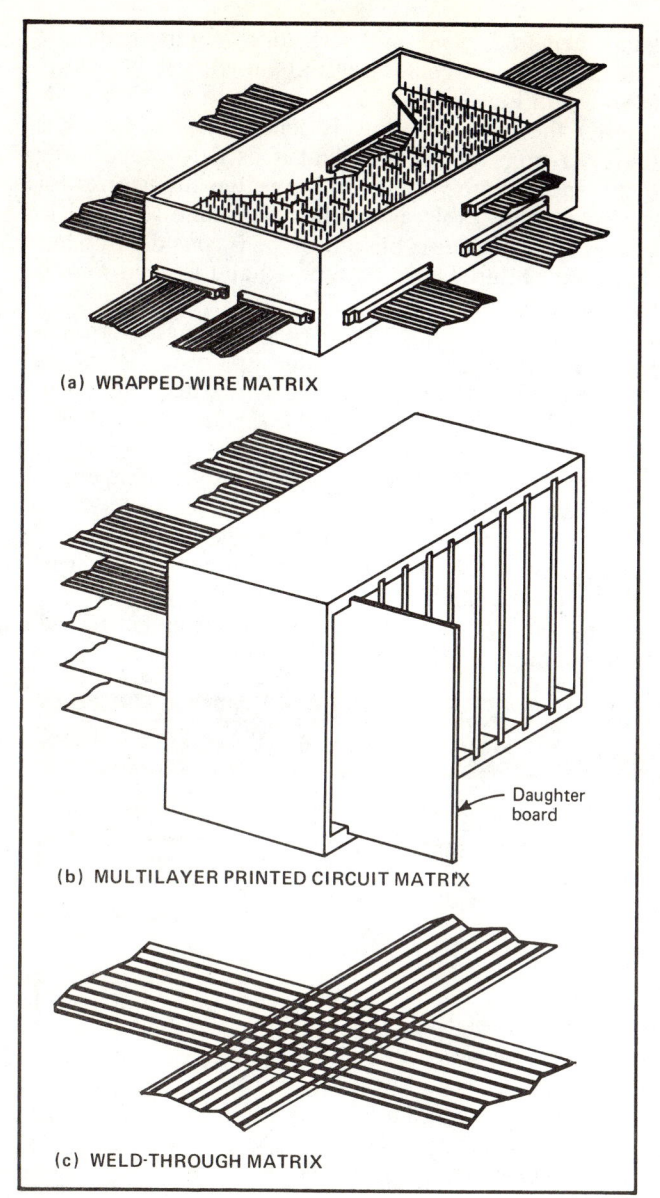

3. Complex. The wrapped-wire matrix (a) is desirable when changes are anticipated. In large systems with complex wiring, a multilayer back panel (b) is programed by daughter boards. The compact welded-through matrix (c) is fine for a planar interface.

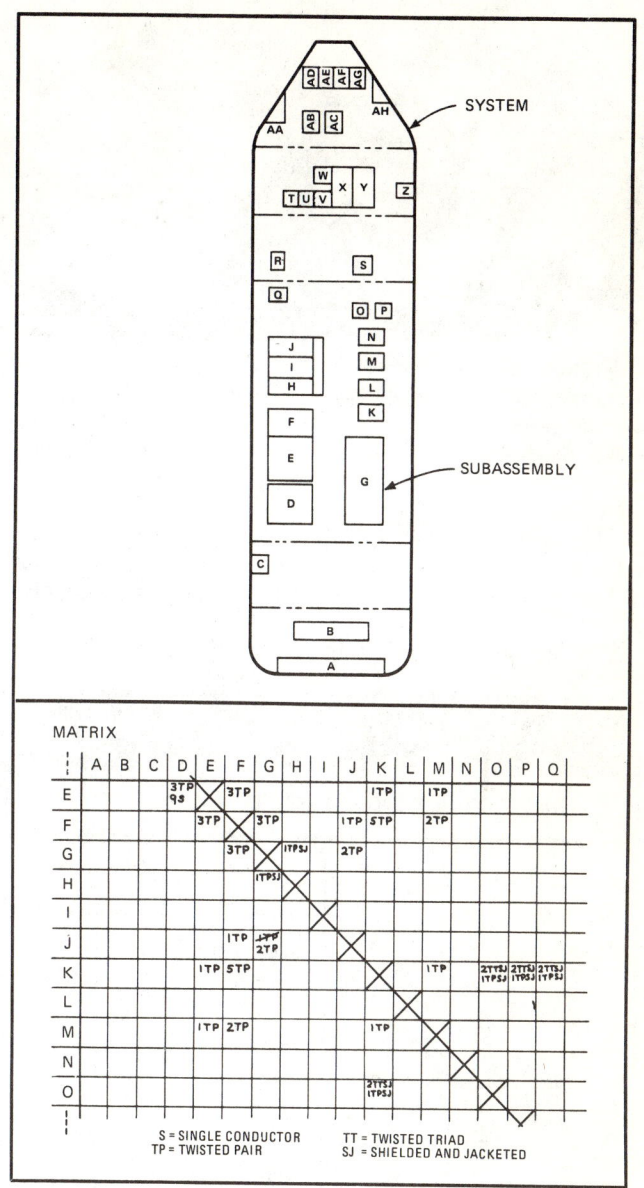

4. Tool of the trade. Interconnection matrix is the key to optimizing a flat-cable design. Letters are assigned to the various subassemblies in the system. Interconnections are located as close as possible to the diagonal line.

from the diagonal "zero-distance" axis (see, for example, row E, column M in Fig. 4).

To alleviate this situation, one or both of two possible courses of action may be taken: one of the subassemblies may be relocated to shorten the path, or the circuit function requiring the stray pair of wires may be moved to a different subassembly. Both of these techniques are used to optimize the system in an early stage in the design phase—a simple task, thanks to the matrix. The matrix also enables early resolution of wiring discrepancies, as in the example shown in boxes JG and GJ. There is little likelihood that a wiring requirement will be overlooked.

After subassembly input/output requirements have been established, the signal and power cables are defined, and conductors chosen within the cables so as to obtain the greatest degree of separation possible between noisy emitter and sensitive receiver circuits. This essential step departs from the practice of allowing the individual subassembly designers to select connector/pin assignments. For if the designers were free to select them, this would probably result in random assignments of signals to conductors and compromise cable design by requiring added interchanges. Instead, once the flat-cable designer has completed the conductor/signal assignments, he assigns signals to connector pins at all interfaces to complement the conductor assignments.

The mockup

Each cable section has to be designed with enough spare leads to accommodate minor changes and additions to the circuitry, since extra conductors cannot be added one at a time, as in round cable. For major changes, requiring more than the spares complement, new cables would have to be added.

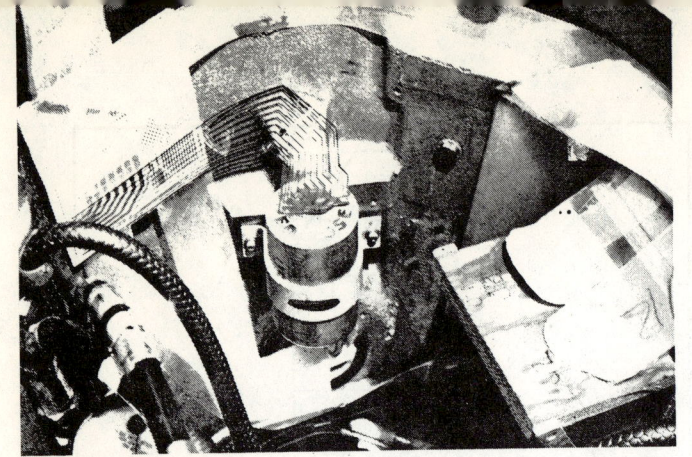

5. Mockup. Mylar film serves as a model for flat-cable wiring in developing prototype of cable assembly. Note the model for the planar interchange on the wall. At the right are patterns for the flat-cable interface with the connectors.

The next step is to develop a cable mockup using clear Mylar film. For this purpose, an accurate mockup of the component layout of the actual system must be available as a basis for determining how and where the cables are to run, how they are to be mounted, and what their fold, bend, and retraction requirements are. The actual system enclosure can be used as the mockup, as shown in Fig. 5.

After the Mylar mockup is completed, the Mylar is removed from the mockup and is used to prepare the detail and assembly drawings, as well as the fixtures for the production cable assemblies. The time required to assemble such a flat-cable prototype is considerably less than that required to prototype a hand-formed round-wire harness.

These factors, coupled with the initial fit and form development using the Mylar mockup, result in cables that fit the first time. What's more, the need for documentation change to relieve poor-fit areas is virtually eliminated.

Table 1 demonstrates the power of prototyping with a Mylar mockup. It compares the labor costs for engineering a 300-wire cable system employing traditional cabling techniques with the costs of the new design techniques employing flat cable. Although both engineering and drafting costs are somewhat higher for flat cabling, the manufacturing time for the prototype is

TABLE 1: NONRECURRING LABOR COSTS FOR 300-WIRE ROUND AND FLAT CABLES

TRADITIONAL BUNDLED-ROUND-WIRE DESIGN				FLAT-CABLE DESIGN			
	HOURS				**HOURS**		
	ENG.	DRAFTING	MFG.		ENG.	DRAFTING	MFG.
Basic cable design: select connectors and wire and prepare wire	80	40 (Note 1)		Prepare interconnection matrix and resolve discrepancies	60		
				Define cable and connector-pin assignments and tabulate	80		
Develop cable, wire by wire			600	Verify cable run lengths, make Mylar layup, and prepare drawings	40	360	Note 2
Mount prototype in unit			32	Mount prototype in unit; no recycle necessary[2]			8
Rework, correct errors, and relieve tight areas	40	20					
TOTALS: Eng. Drafting Mfg.	120	60	632	TOTALS: Eng. Drafting Mfg.	180	360	8

Notes
1. Or Electronic Data Processing.
2. Large manufacturing and rework effort is eliminated by engineering design employing Mylar mockup.

drastically reduced—down from 632 hours to only 8 hours.

Retraction techniques

As part of over-all cable design, retraction and extension must be considered. Some of the common techniques are shown in Fig. 6. The one to be used depends primarily on the available space.

The window-shade form of flat cable may be either retractor-controlled (as with an automobile seat belt) or formed into the memory of some insulation systems. The controlled fold may be rigid at the bends and flexible at the straight sections, or flexible throughout. It is better not to allow flexing at the cable bends because the work-hardening encountered during bend formation may reduce the life of the copper significantly. Controlled-roll flat cable provides a slide-rolling action within a narrow enclosed conduit. The design includes selection of the conductor and insulation systems as required for the electrical and environmental conditions.

The last step is to produce a prototype of the cable to validate both mechanical and electrical design. The final cable assembly produced from preformed detail parts is the one illustrated in Fig. 1. This assembly is easily mounted into place, using adhesives and support brackets, as shown in Fig. 7. ☐

6. Retract. Flexible cable is usable with sliding drawers, doors, and hatches—window-shade (a) and controlled-fold (b) versions used where motion is straight pull-out and return. The controlled-roll type (c), confines the flexing cable to a shallow conduit.

7. Installed. Jacketed cables, with potted connectors that are costly to assemble, are replaced by flat cabling, which eliminates expensive fabrication procedures. The flat-cable assembly is clean and compact and provides short thermal paths for all conductors.

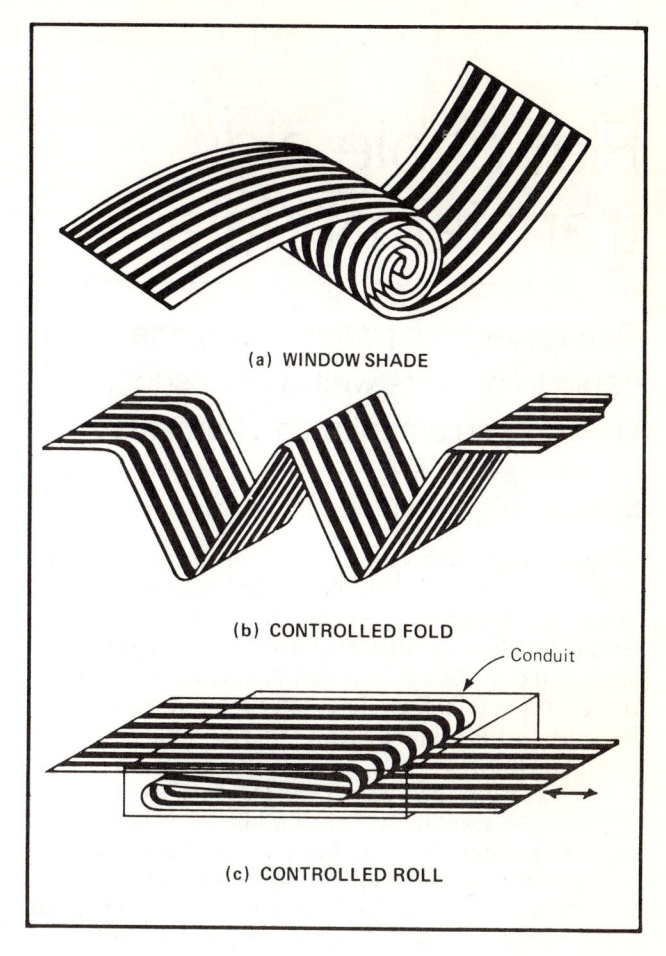

Flat cable aids transfer of data

Two dielectrics permit multisignal cable to operate well at nanosecond speeds with a minimum of crosstalk

by Joseph B. Marshall, Ansley Electronics Corp., Doylestown, Pa.

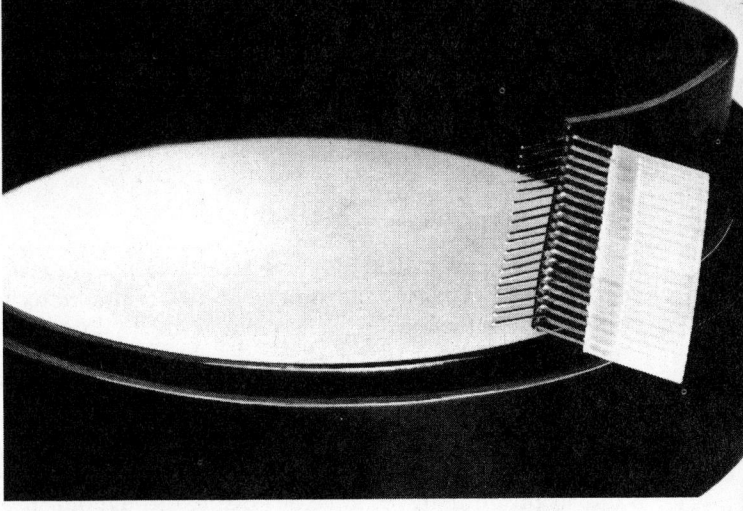

Boom to high-speed data. Twenty-channel flat cable handles 1-nanosecond data signals with negligible crosstalk. White polyethylene dielectric is folded back along with the 21 rectangular ground leads to expose the 20 signal lines.

☐ To transfer high-speed data from one point to another with minimal distortion, conventional cabling is not good enough. The system designer must think in terms of high-frequency transmission lines and their attendant parameters of impedance, attenuation, and particularly crosstalk. Such lines must often interface satisfactorily with printed-circuit boards, and their cost should be low.

Coaxial cable has been used for data transmission, and so have stripline techniques, but, besides requiring expensive connectors, and being expensive to fabricate, they are not really flexible enough. Nor, until recently, was flat cable able to perform where switching speeds were faster than 4 to 5 nanoseconds, because its far-end crosstalk was too high.

A recently developed dual-dielectric flat cable, however, provides the necessary transmission-line characteristics and, most significantly, has drastically reduced crosstalk at nanosecond switching speeds. Since its signal leads are spaced on 50-mil centers, it satisfies the density requirements and, compared with coax, it is also economical.

A flat-cable version of a transmission line is the culmination of a long-established trend. Cabling that interconnects pc boards should ideally be planar, like flat cable. Also, it should exhibit high wire density, as does flat cable, because the wide, rectangular body of insulation, covering a multitude of conductors, is physically strong, the conductors in such cable can be very thin—no thicker, in fact, than is necessary to fulfill attenuation requirements. Also, automated tooling is available to rapidly strip and terminate such cable, whereas twisted pairs or grouped coaxial cable cannot be terminated nearly as fast. However, conventional flat cable's serious shortcoming is its susceptibility to crosstalk at high switching speeds.

Crosstalk

Generally, when conventional flat cable is used to transfer data, it develops high crosstalk whenever the signal propagation delay along its length approaches the rise time of the pulse being propagated. The propagation delay is related to the length of the line and the square root of the effective dielectric constant of the insulating material. Some representative dielectric constant values for commonly used insulation materials are shown in Table 1.

A handy rule of thumb for adequate crosstalk control is that the principles of transmission lines should be applied when the delay time between connections exceeds 1/25 of the transmitted pulse's rise time (t_r). To get an idea of what this means dimensionally, these conditions are met by a 10-nanosecond pulse rise time on a polyethylene-insulated cable 3.2 inches long, or in air over a path only 4.8 in. long. Since data-processing equipment now commonly operates at two or three times this speed, and new designs employ subnanosecond risetime devices, it's obvious that most system and device interconnections must be made by high-frequency transmission-line techniques. At the same time, mechanical compatibility with pc boards must be maintained.

A worst-case crosstalk condition occurs when, say, four out of five adjacent signal transmission lines are pulsed simultaneously, with the quiet line in the center. The interference level on the quiet line is expressed either in decibels of isolation between the active pulse-carrying signal lines and the quiet line, or as a percentage of the signal level on the active lines.

Relevant equations[1] are:

Crosstalk (dB) = $20 \log (V_a/V_q)$
Crosstalk (%) = $(V_q/V_a) \times 100$

Figure 1 represents two adjacent uniform signal lines, one of which carries a pulse-type signal. The coupling between the two lines is represented by mutual inductive and capacitive components, L_{12} and C_{12}, respec-

TABLE 1 SIGNAL PROPAGATION DELAY IN VARIOUS DIELECTRICS	
Media	Propagation delay (ns/foot)
Air	1.016
FEP (Teflon)	1.47
Polyethylene	1.545
Polyvinyl chloride	1.8 – 1.9
Polyester	1.85
G–10 epoxy-glass-cloth printed-circuit board	2.0 – 2.2

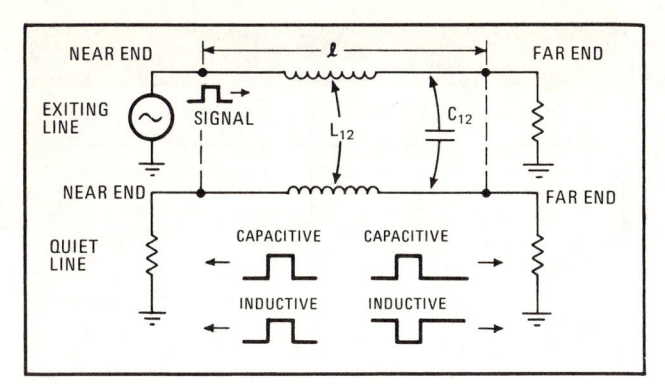

1. Crosstalk. Unshielded, signal lines that are adjacent and parallel are vulnerable to crosstalk. The distributed capacitive C_{12} and inductive components L_{12} couple the spurious crosstalk-signal components that propagate in the directions shown in the schematic.

tively. Signals introduced into the active line will be coupled into the quiet line by the electrostatic and electromagnetic fields that accompany the traveling signal.

The induced voltage and current propagate in both the forward and backward directions at the same speed as the exciting signal, but there are significant differences in the polarities of the components of the backward and forward crosstalk. The capacitively coupled component propagates in both directions with the same polarity as the signal in the exciting line. The inductively coupled component propagates backward with the same polarity, but forward with the opposite polarity.

The net effect of all these interactions comes to this. The near-end crosstalk will be the sum of both the capacitive and inductive effects; hence it will be an attenuated replica of the pulse on the active line. This pulse will reach a maximum amplitude when the effective propagation delay of the quiet line corresponds to $t_r/2$. (For example, a pulse with 1-nanosecond rise time will generate maximum backward crosstalk on a quiet air-dielectric line, 6 inches long or longer.) The far-end crosstalk component is the net difference between the inductively and capacitively coupled effects because they are opposite in polarity. Hence, if these components are equal, forward crosstalk is zero.

Coupling coefficients are expressed by the following formulas:

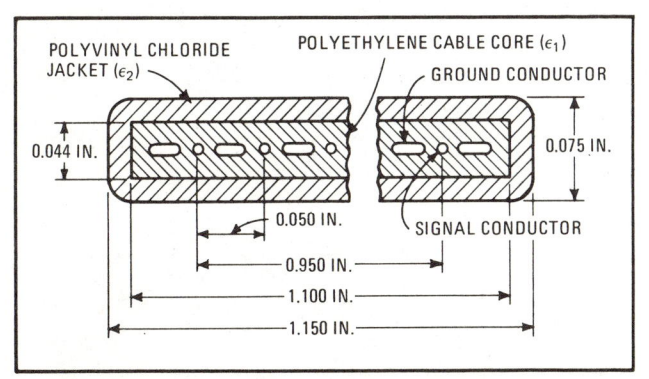

2. Two dielectrics. Inner polyethylene and 15-mil outer polyvinyl chloride dielectrics have differing dielectric constants (ϵ_1 and ϵ_2) that in combination minimize far-end crosstalk. Dense construction enables 20 signals to be carried in a cable a little over 1 inch wide.

$k_L = L_{12}/(L_{11}L_{22})^{1/2}$
$k_C = C_{12}/[(C_{11}C_{22})^{1/2} + C_{12}]$
where
k_L = inductive coupling coefficient
L_{11}, L_{22} = inductances of the independent signal lines
L_{12} = inductive coupling between adjacent signal lines
k_C = capacitive coupling coefficient
C_{11}, C_{22} = line capacitances of the two signal lines to the ground plane
C_{12} = capacitive coupling between two adjacent signal lines

The fact is that uniform matched impedance lines in a homogeneous dielectric medium exhibit $k_L = k_C$. The result is zero differential far-end crosstalk. But the dielectric medium of multiconductor cables, twisted pairs and conventional flat cables does not present a homogeneous cross-section to the propagating electrostatic and electromagnetic fields.

Portions of these fields extend beyond the solid dielectric into the surrounding air, altering the capacitive coupling coefficient considerably but not affecting the inductive coupling coefficient. The resulting differences in coupling coefficient polarities create far-end differential crosstalk. Far-end crosstalk increases by the length of cable and also by the faster pulse rise time; in the worst case could be as high as 40% of the transmitted signal.

A dual dielectric

The main reason for the development of high crosstalk in conventional flat cables is that their mutual capacitance, C_{12}, is much lower than the self-capacitances, C_{11} and C_{22}, compared to a transmission line where the fields are in a uniform dielectric. This deficiency can be compensated for, and crosstalk considerably reduced, by a dual-dielectric cable configuration in which conductors are surrounded by a core of low-dielectric constant material and that core in turn is clad by a material with higher dielectric constant, as shown in Fig. 2. This principle has been embodied in a line of flat cables designated Black Magic, manufactured by the Ansley Electronics Corp.

The concept was the side-effect of an attempt to produce a cable at reasonable cost that has an insulation with better electrical and self-extinguishing characteristics than was then available. The prototype cable with a polyethylene center core and vinyl jacket not only exhibited the desirable characteristics, but it unexpectedly appeared to virtually eliminate far-end crosstalk.

It turns out that the dual-dielectric cable corrects the imbalance between C_{12} and the self-capacitance because the higher dielectric constant of the vinyl jacket over the low dielectric constant of the polyethylene core equalizes the inductive and capacitive coupling coefficients. The configurations are shown graphically in Fig. 3.

Performance

Just how well does the dual-dielectric cable stack up against its competitors? Figure 4 illustrates the magnitude of the crosstalk on 20-foot lengths of both standard

TABLE 2: CROSSTALK IN TWISTED PAIR VS. DUAL-DIELECTRIC CABLE

SINGLE CABLE LAYER	SIX TWISTED PAIRS BUNDLED	DUAL-DIELECTRIC CABLE	IMPROVEMENT RATIO
Length of cable	10 ft	10 ft	
Pulse rise time	1 ns.	1 ns.	
No. of Active lines	4	4	
Near-End crosstalk, peak	18.8%	3.5%	5.4
Far-End crosstalk, peak	14.0%	2.5%	5.6
Length of cable	30 ft	30 ft	
Pulse rise time	1 ns.	1 ns.	
No. of Active lines	4	4	
Near-End crosstalk, peak	22%	4.2%	5.2
Far-End crosstalk, peak	21%	2.5%	8.4

SINGLE CABLE LAYER	TEN PLANAR TWISTED PAIRS	DUAL-DIELECTRIC CABLE	IMPROVEMENT RATIO
Length of cable	30 ft	30 ft	
Pulse rise time	5 ns.	5 ns.	
No. of Active lines	4	4	
Near-End crosstalk, peak	25.8%	8.1%	3.2
Far-End crosstalk, peak	21.5%	3.3%	6.5
THREE LAYERS OF CABLES			
Length of cable	30 ft	30 ft	
Pulse rise time	5 ns.	5 ns.	
No. of Active lines	8	8	
Near-End crosstalk, peak	26.7%	8.0%	3.3
Far-End crosstalk, peak	24.6%	6.0%	4.0

and dual-dielectric flat cables. Pulse rise time in both cases is 0.18 nanosecond.

Tests employing two twisted-pair geometries yielded the results shown in Table 2. The six pairs in the cylindrical bundle exhibited a characteristic impedance ranging from 50 to 65 ohms, whereas the 10 pairs of wires arranged side by side in the single plane were 80-ohm pairs. The results, as shown in the table, indicate that dual-dielectric cable affords an improvement in both near-end and far-end crosstalk. The most dramatic improvement occurs for a 30-ft cable carrying 1-nanosecond rise-time signals. The improvement ratio is over 8:1.

A dual-dielectric version of solid-logic-technique cable was also compared with conventional SLT cable. The far-end crosstalk improvement is five times for a 1-ns signal (Fig. 5). What is paramount, however, is that these tests show the dual-dielectric cable reduces crosstalk below a troublesome level for all the fast digital-logic families.

In summary, in regard to electrical performance, the polyethylene inner core determines the characteristic impedance, the propagation velocity, and the attenuation. The polyvinyl jacket, with its higher dielectric constant, limits the crosstalk.

With an over-all thickness of 55 to 77 mils, the dual-dielectric cable is well suited for intra- and intercabinet applications. Flexibility is excellent, in that the cable may be bent into a radius equal to its thickness. It may also be bent over at 45° to form right-angle turns—and all with neither physical damage to the cable nor detriment to the electrical performance. What's more, the

3. Boost the epsilon. In (a), the dielectric region between signal lines is largely air. Adding the polyvinyl chloride outer cover (b) raises the dielectric constant of the region, which tends to equalize the coupling coefficients. The result: lower far-end crosstalk.

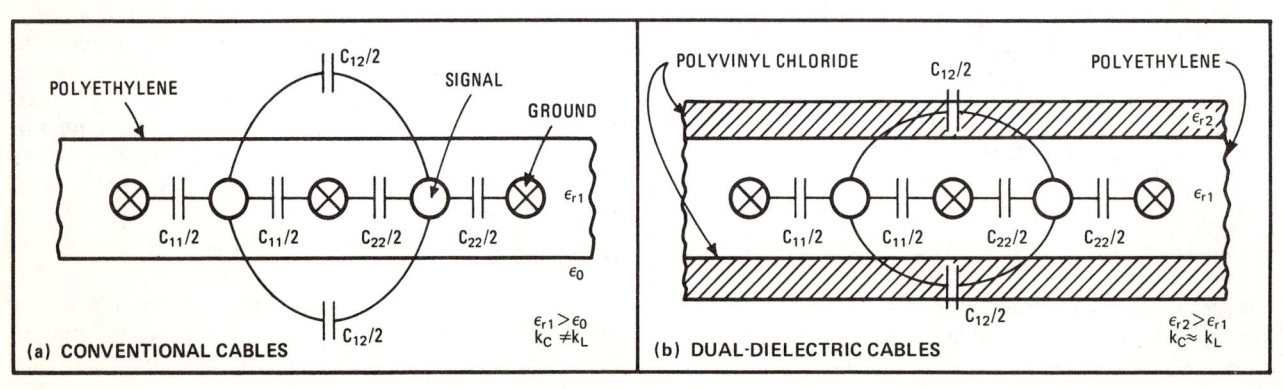

(a) CONVENTIONAL CABLES — $\epsilon_{r1} > \epsilon_0$, $k_C \neq k_L$

(b) DUAL-DIELECTRIC CABLES — $\epsilon_{r2} > \epsilon_{r1}$, $k_C \approx k_L$

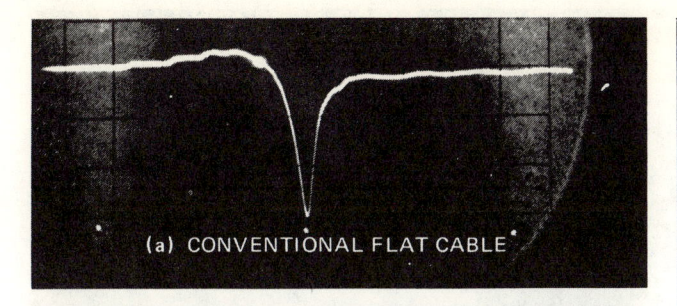

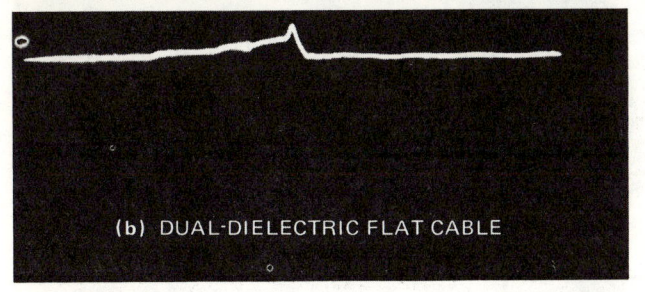

4. Performance. Dual-dielectric cable shows five-fold improvement in crosstalk over conventional flat cable when excited by 0.18-ns rise-time pulse. Cables are 20 feet long. Horizontal scale is 2 ns/division; vertical scale is 5% crosstalk/division.

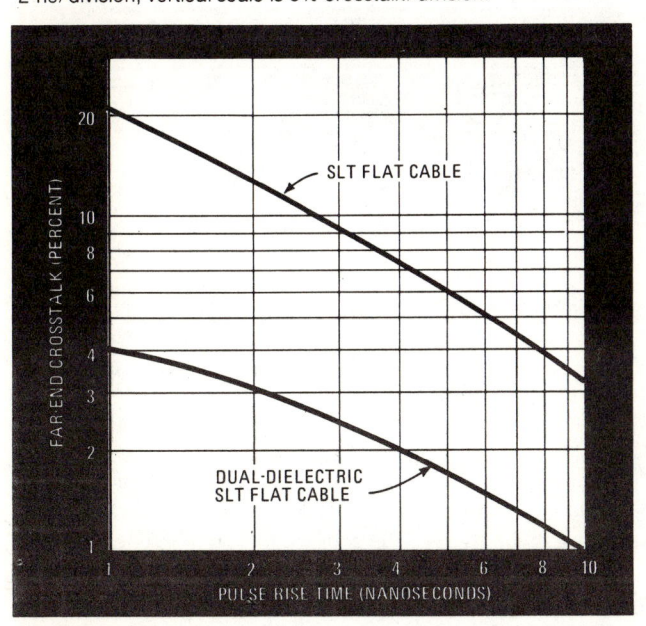

5. Comparison. The graph depicts far-end crosstalk for both the dual-dielectric and conventional SLT cable. Each is 10 feet long, with the quiet line between four sets of active lines. At 1 nanosecond, the dual-dielectric cable exhibits a five-fold improvement.

cable fulfills Underwriters Laboratories' requirements for flame retardancy.

There is only one ground conductor to terminate between two adjacent signals (in conventional round-wire flat cables, two or more grounds have to be used). With flat and round conductors alternating in the cable, identification of signal and ground conductors is easy. This feature is particularly helpful when signals and grounds will terminate to two different conductor planes—such as a printed-circuit board used as microstrip. □

Planar transmission lines

Planar transmission lines first appeared as stripline, which was introduced by R. M. Barrett for use in microwave components such as hybrid junctions, directional couplers, power-divider networks, and filters. This configuration evolved from coaxial geometry and made use of the etched printed-circuit board. As shown below, the cylindrical outer conductor of coaxial geometry is exchanged for two parallel ground planes. This promotes the propagation of a true transverse electromagnetic mode (TEM) along the line, just as in coaxial cable.

Microstrip is simpler in structure than stripline, having only one ground plane. It is characterized by an impure TEM-wave propagation, due to the air dielectric above the single strip. Still, with careful design, it will transmit high-speed data signals with fidelity. An even simpler geometry may be achieved by locating the ground and signal conductors in the same plane—a coplanar transmission-line configuration that achieves a performance comparable with that of microstrip.

Each of these geometries can be applied either to pc boards or to flat cables. In practice, though, stripline and Microstrip are more often applied to pc boards, and the single conductor plane is usually favored in flat-cable design.

A 16-conductor flat cable was used by IBM in 1963 in the 1440 series SMS computers. Individual signal-transmission-line properties were not considered, so it was designed for optimum mechanical and fabrication qualities: flexibility, foldability, simple strain relief, small size, and ease of termination, including stripping and soldering.

The logic speed of IBM 360-series computers (pulse rise time in the neighborhood of 5 nanoseconds), however, required high-frequency transmission lines for interconnections. In 1964, IBM developed the solid-logic technique (SLT) cable. Sixty conductors carry 20 signals, since each signal conductor is centered between two ground conductors. The dielectric is self-extinguishing polyethylene, and the nominal characteristic impedance is 95 ohms. Signal-propagation time is 1.57 ns/foot, attenuation at 75 megahertz is 15 decibels per 100 ft maximum, and the worst-case cross-talk is 10% with a 2.6-ns rise-time pulse.

Because the geometry of the SLT design can be adapted to various impedance values, conductor sizes, and insulation materials, it is still a favorite today for conventional signal-transmission-line flat cables. But, as rise times approach 1 ns, there's a need for an even better cable. Such an improved flat cable is the multisignal dual-dielectric cable, described in the accompanying article, in which crosstalk can be kept below 10% even in the region of picosecond switching speeds.

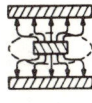

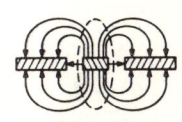

COAXIAL   STRIPLINE   MICROSTRIP   CO-PLANAR TRANSMISSION LINE

111

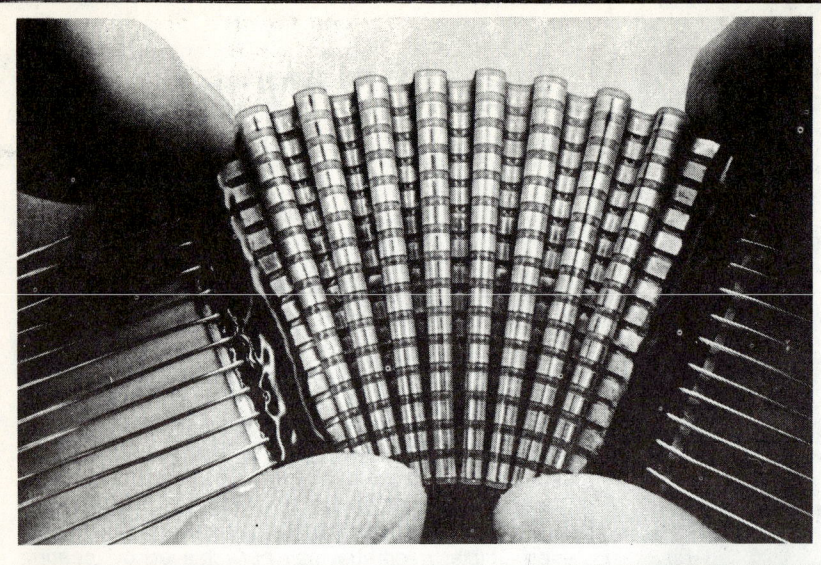

Accordian. Flexible circuit, such as one made by ITT-Cannon Electric (left), provides a dense yet flexible interconnection between two circuit assemblies. Pleat enhances flexability. Insulation is usually a polyimide, which is more stable over wide temperature ranges than competing Teflon and Mylar.

Zip. Flat cable (below) consists of parallel insulated conductors, which can be readily separated by hand or machine. Cable, such as this from 3M Co.'s Electro-Products division, which makes it in widths of up to 50 conductors, is fine for computer wiring, readouts, control panels, and electronic test equipment.

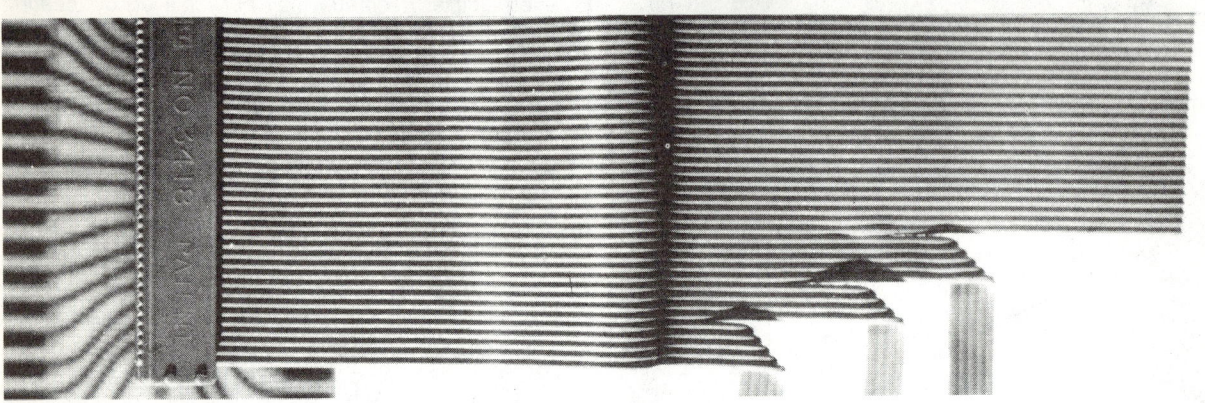

Multilayer. Flexible circuit (left) has a multilayer construction and mounts several connector types. This assembly was manufactured by the Parlex Corp. Note the simulated twisted-pair conductor pattern. Planar-wiring provides good thermal dissipation.

Shielded. Flat cable (below left) has 28-gage stranded conductors located on 50-mil centers and a mesh-shield ground plane. Made by 3M Electro-Products division for computer-mainframe and peripheral applications, it is available in 16-, 34-, 40-, and 50-conductor versions.

Woven. Planar cables (below right) can be custom woven with Teflon, Nylon, or other fibers to suit customer requirements. These cables, fabricated by Woven Electronics, illustrate both single-conductor and twisted-pair cables.

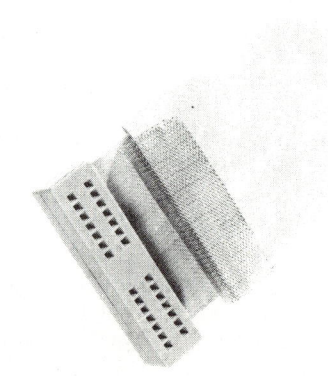

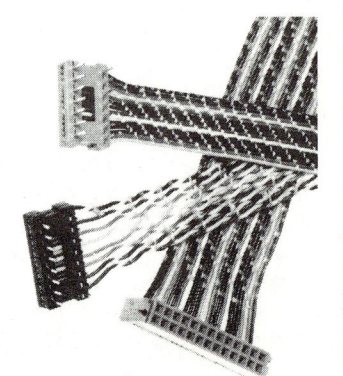

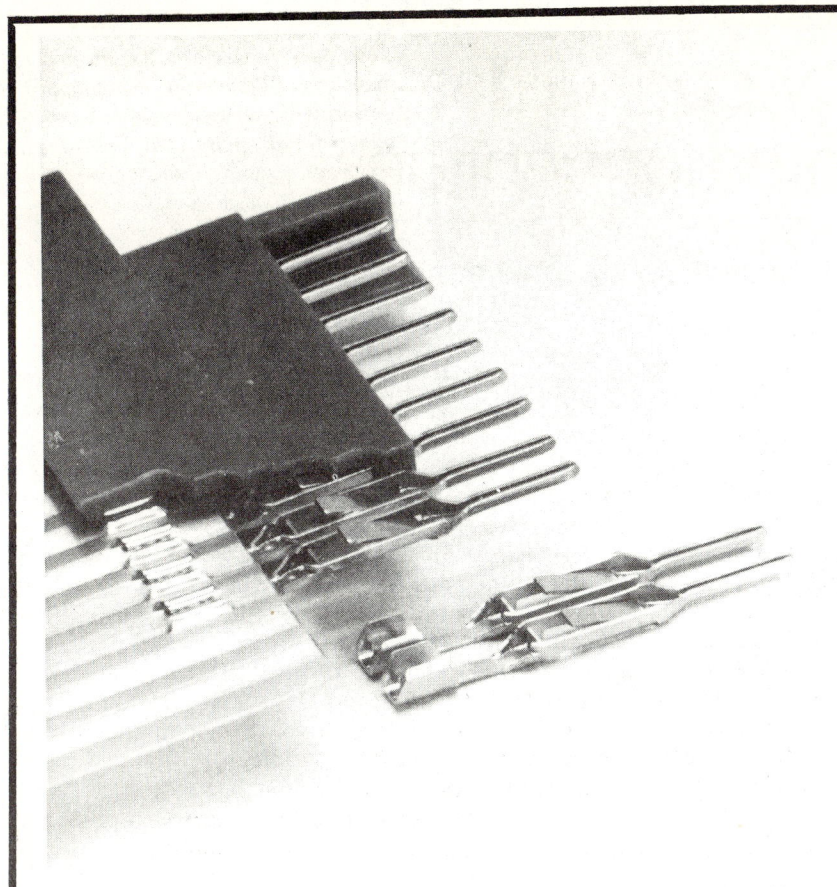

Tailored. Scissors makes fast work of separating wires of woven flat cable (below). Woven wiring employs textile techniques to combine wires into flat configurations required by a customer. This cable is made by Woven Electronics.

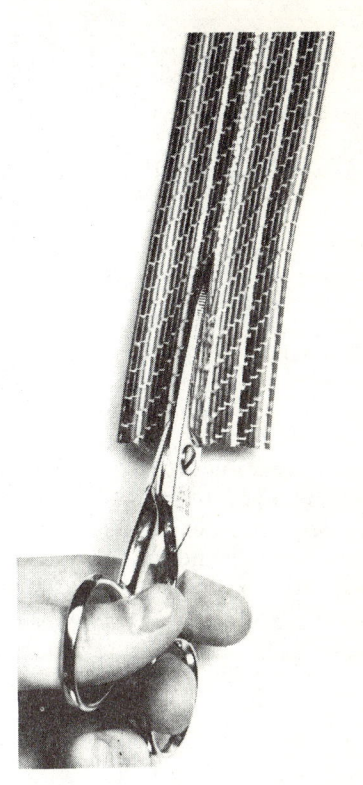

Pierce and crimp. These Amp Inc. flat-cable connector pins can be attached by automatic equipment at the rate of two per second. Designed for standard flat cable with 63-mil-wide conductors spaced 100 mils apart.

Quick Strip. In two-step automated stripping technique (below, left and right) tool, developed by the Spectra-Strip Corp. and Tektronix Inc., strips and separates end of the 10-conductor flat cable in 3 seconds.

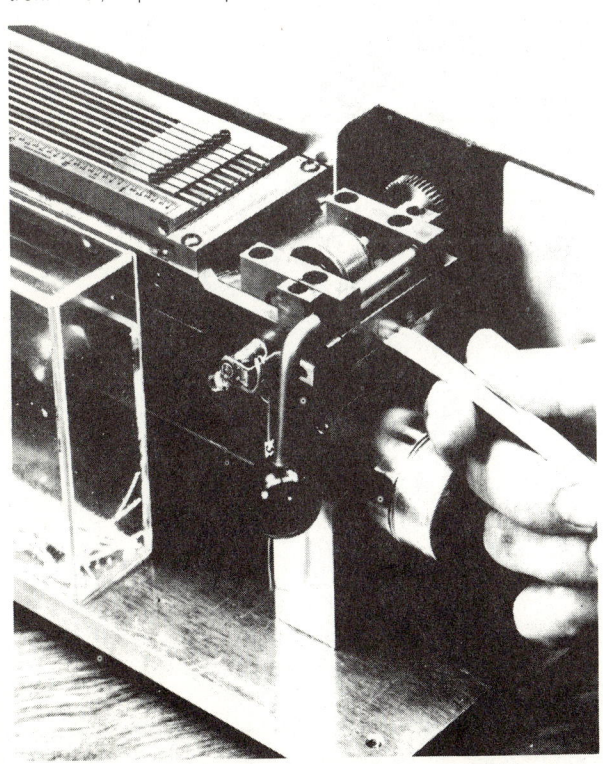

Flexible circuitry consolidates hardware for interconnections

Prefabricated wiring, which eliminates several conventional parts, allows simplification of assembly process, as well as the design of other components; it is especially suitable for solid-state displays

by Peter Maheux, *Bell-Northern Research, Ottawa, Canada*

☐ One of the greatest attractions of flexible circuitry is that it can entirely replace the individual rigid printed-circuit boards, hard wiring, and the piecemeal, cumbrous, and expensive interconnection techniques that boards and wire require.

A system that best illustrates these advantages is a telephone set. Figure 1 illustrates how the traditional arrangement in a telephone set can be reduced to three basic elements: flexible circuitry, connectors, and a housing. The key and the circuit contacts are portions of the foil exposed in the circuit's plastic covering. This does away with a set of contacts and the conventional multicomponent connector—often complex assemblies by themselves.

The circuit is fabricated and assembled before being installed in the housing, much as a ship model is assembled before it is inserted in a bottle. The housing, besides providing mechanical stiffening and a shield against the environment, also has cavities molded into its wall to mount connectors. The flexible circuit is then formed about the lips of the connector interfaces; retaining clips hold the flexible circuit securely in place about the male and female mating parts.

Figure 2 demonstrates how the conjunction of circuit

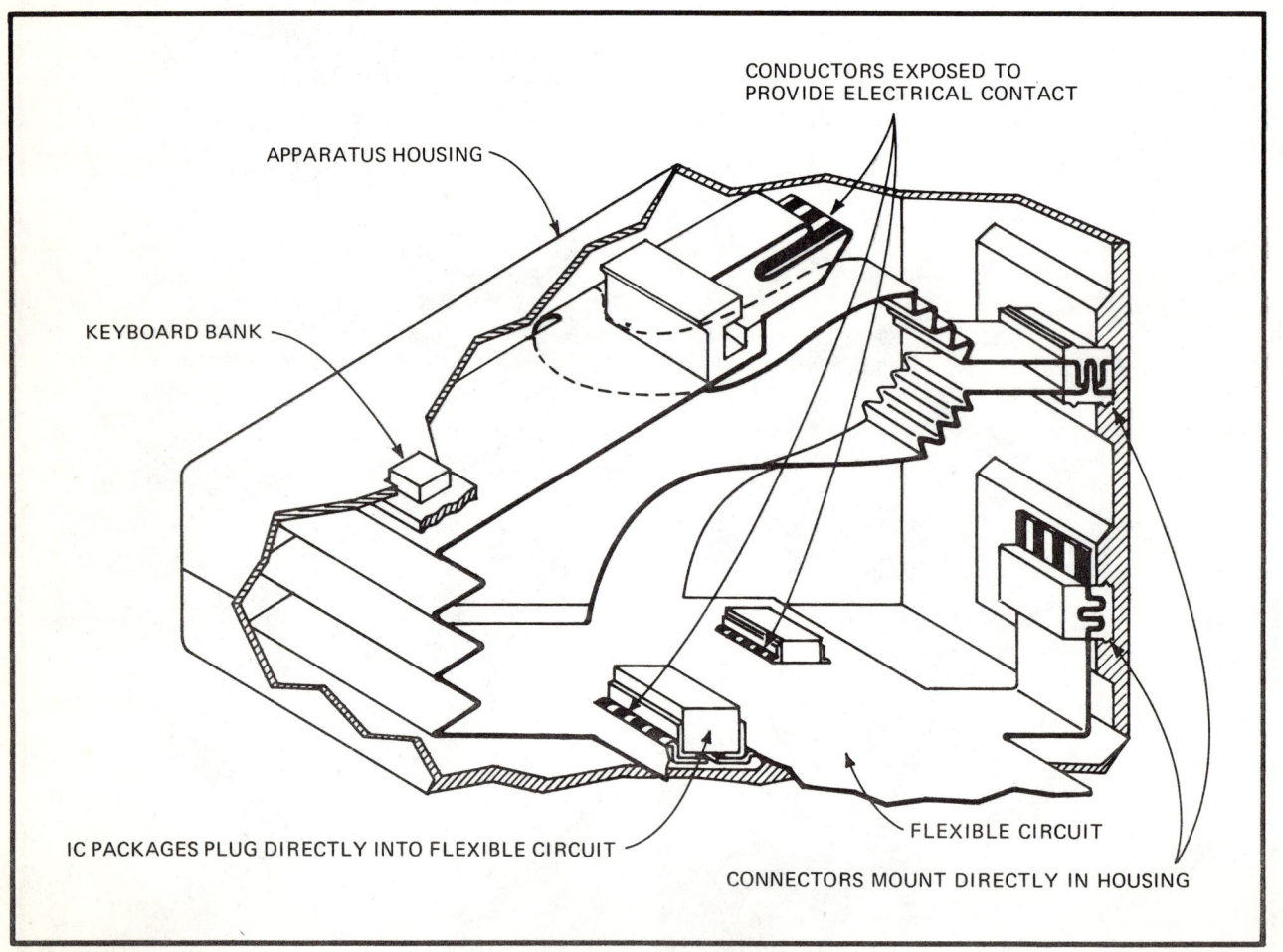

1. Package revolution. Flexible circuit banishes traditional discrete-component mounting methods and laborious point-to-point wiring, offering instead pressure interconnections that are quickly assembled. Terminals, soldering, and conventional connector hardware are gone.

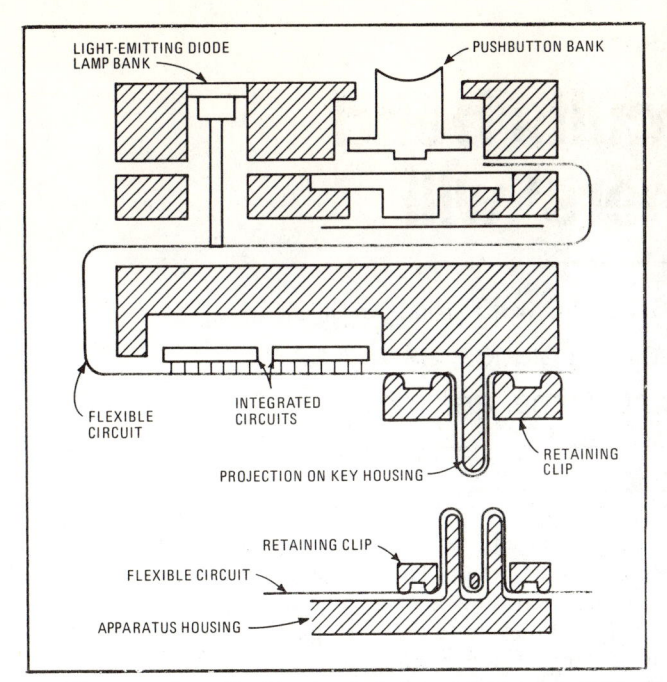

2. All wrapped up. Flexible circuit starting at the push-buttons of telephone-handset subassembly is connected to the light-emitting-diode lamps, secures and is connected to several ICs, and finally is wrapped round a projection to form a connector.

and components can provide electrical interconnection and mechanical placement. Here a light-emitting-diode bank and a push-button bank, like the one used in a telephone handset, become a single structure. The plastic projection on the key housing turns part of the flexible circuit into a male connector component. The whole assembly is plugged into the mating receptacle on the apparatus housing. (Not shown in the figure are the exposed areas of the circuit's conductive foil that form one set of contacts for the push-button bank.) The flexible circuit also makes contact with the pins on the integrated circuits and holds each IC in place, allowing the design of other components to be simplified and reducing the labor required—the assembler merely inserts the flexible circuit in the over-all assembly, adds the ICs, and presses the retaining clips in place.

Both the housing and the connector components are molded from acetal copolymer, which has good mechanical properties and dimensional stability, changing

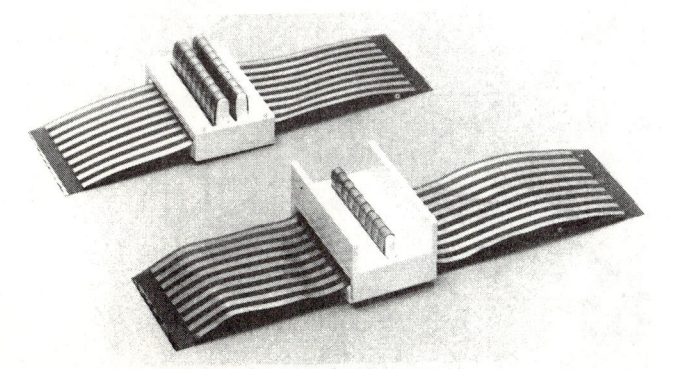

3. How it mates. The flexible circuit is wrapped about the male and female portions of the plastic shapes to form the contact interfaces. With this configuration, no separate connection step is required because the flexible circuit is both wiring and connector contact.

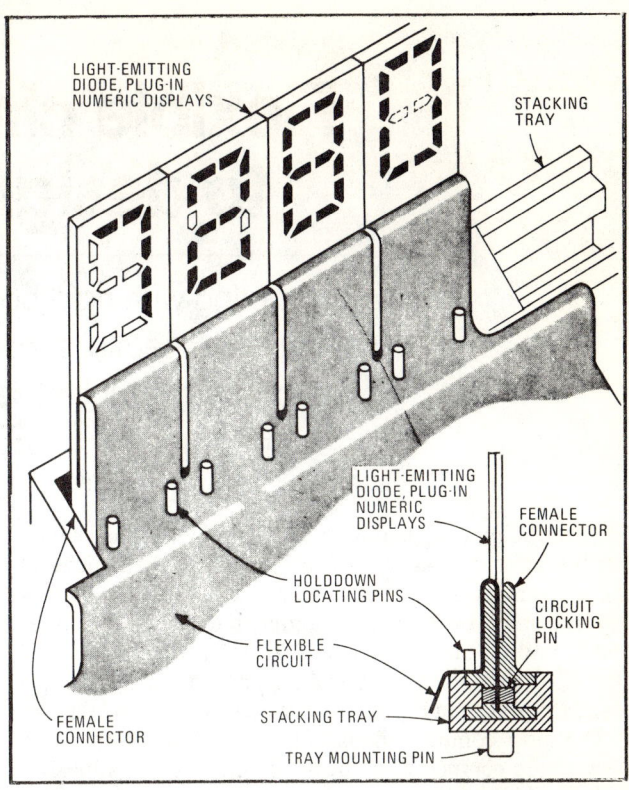

4. Ideal for LED displays. Flexible circuit teams with tee-shaped plastic connectors to form a versatile socket system for digital displays. Any number of connectors may be slid into stacking tray to mount the number of displays desired.

little in humid environments up to 85° C. It molds readily and resists cold flow after extended periods of sustained stress. A connector made up of these components and a flexible circuit is shown in Fig. 3.

The flexible circuit is fabricated from 3-mil polyimide and a 1.4-mil (1-ounce) copper foil. Gold plate 100 microinches thick is applied to the contact surfaces to provide reliable interconnections.

Flexible circuitry is especially good for mounting solid-state displays because it does away with an entire set of interconnections. Conventionally, the chips that make up the display would be wire-bonded to pads on a ceramic substrate. Then Kovar pins would be bonded to the pads, and the entire device encapsulated in plastic. Finally, the package would be either plugged into a printed-circuit board and soldered, or plugged into a dual in-line socket already soldered to the pc board.

With flexible circuitry, however, though the chip is wire-bonded to the ceramic substrate as before, the interconnect leads are run directly along the substrate to its edge. When the device is plugged into an associated female interconnect shape (Fig. 4), the substrate leads are forced against the exposed foil contacts on the flexible circuit. The tee-shaped base on the socket permits any number of sockets to be arrayed in a tray or other support structure. Again, the result is lower hardware costs and fewer assembly hours.

This kind of interconnect scheme has no metal parts, other than the conductive foil in the flexible circuit. Also, the connectors have but one interface, while many conventional connectors on the market today may have from three to five interfaces. □

Flexible circuits bend to designers' will

Now competing with rigid boards, flexible printed circuits can form connections in several planes, as well as saving weight, unit assembly time, and money

by Jerry Lyman, *Packaging & Production Editor*

☐ After being overshadowed for years by rigid circuitry, flexible printed circuits are moving out of their niche in the military and aerospace industries and into all areas of electronics, especially computer, industrial, consumer goods, and automotive systems. In fact, because of their cost effectiveness as an interconnect technique, it is now mandatory at many firms to try a flexible design first.

The most common manufacturing method starts with a substrate composed of sheets of thin copper, adhesive, and flexible insulating film laminated together. After lamination, circuit patterns are etched on the copper side of the substrate. The resulting device can be used as a harness, a pc board, or a combination of the two. Sanders Associates Inc., Manchester, N. H., originated flexible circuitry in 1952, when it developed this technique for a military application in which space and weight were at a premium.

For example, Fig. 1 shows a military system that has been converted from rigid boards plus wiring harnesses to a flexible circuit. These photographs illustrate one of the main advantages of flexible circuitry—its ability to be shaped into more than one plane or to conform to an irregular package. Such circuits can also be folded up to save space—say, in a small module—and can branch off in many directions, as shown in Fig 2.

Flexing the circuit

Another advantage of this kind of circuitry is its ability to be continually flexed from a small folded or rolled-up configuration to its full length. This ability accounts for the popularity of flexible-circuit harnesses in the moving members of plotting boards and magnetic disks.

Flexible circuits are also extremely thin. Their thickness runs from 4 to 11 mils on average, whereas a typical two-sided rigid board is 62 mils thick. And such thinness,

1. Flexible vs rigid. A control box from a military system is shown in both its rigid (left) and flexible circuit versions. Note the clean, uncluttered package of the flexible-circuit system. Converting a system of this type to flexible circuitry cuts size, weight, and assembly time.

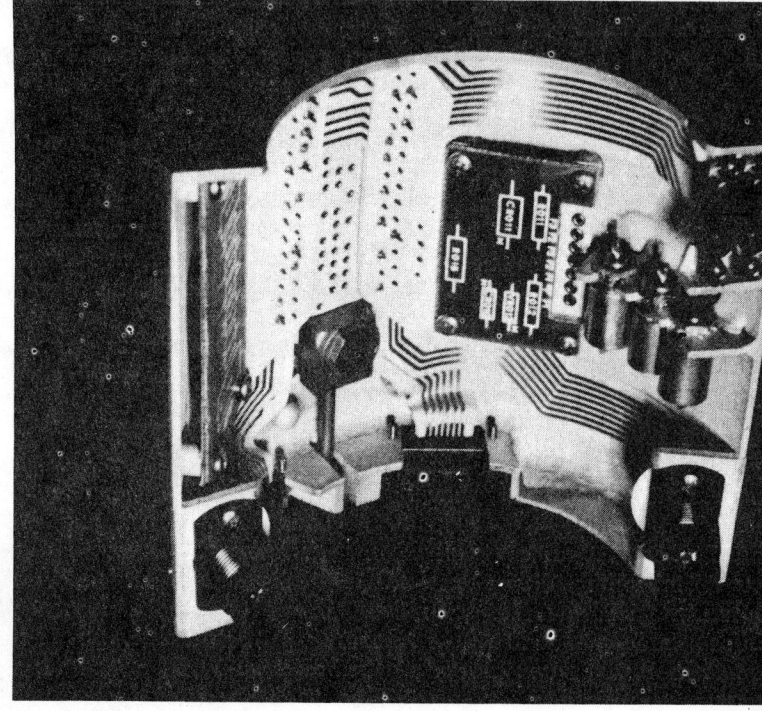

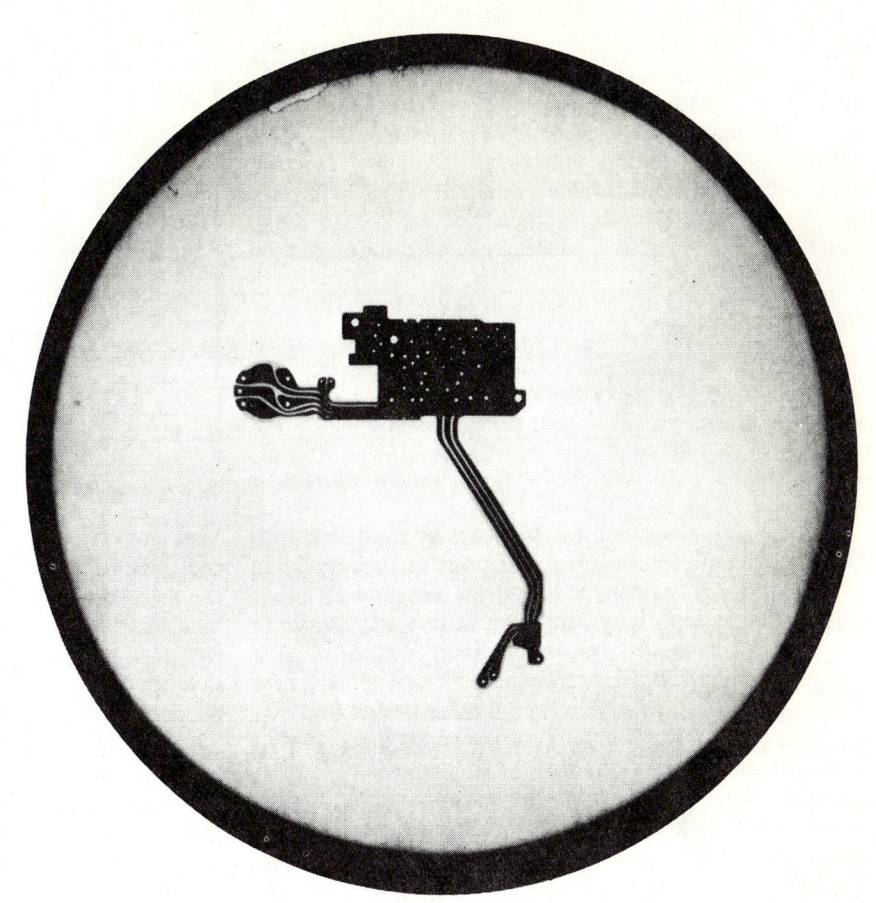

2. Multiplanar. Flexible printed circuits can be shoehorned into spaces where rigid circuitry could not possibly fit. This flexible printed circuit is bent into four planes and has two 180° fold-over sections to provide circuit reversal. End use of the board is in a portable movie camera.

plus the lightness of the insulating film, automatically brings with it a drastic reduction in weight.

Nowadays, most flexible circuits are replacing complete interconnection systems rather than individual hardboards. In these cases, in addition to saving space and weight, they eliminate wiring errors and cut testing time, rework, and assembly costs. Vic St. Amand, marketing director of Teledyne Electro-Mechanisms, Nashua, N. H., gives these examples: an avionics black box redesigned by his company's engineers with flexible circuits saved 129 hours per box in assembly time and cut weight 29%. A similar conversion of another military system cut 140 hours off unit assembly time and reduced weight by 50%.

If pins, wires, or plated-through holes are placed at its terminations, flexible circuitry can also eliminate the need for a connector. Still another advantage is shock-resistance. Vibrations and shock that would crack a rigid board have little or no effect on flexible circuits. That is why they appear in such diverse equipment as missile electronics and watch modules.

Drawbacks

Of course, flexible circuitry does have its disadvantages. One is that for high-frequency work it is difficult to control the characteristic impedance of transmission lines formed by the laminated system used in flexible designs, because of the many variations in the thickness of the layers of copper, adhesive, and film. Also, many users of flexible circuitry claim that it is difficult to use with automatic component-insertion equipment without the addition of hardboard stiffeners. However, Teledyne Electro-Mechanisms routinely inserts components automatically on unsupported flexible circuits.

What held back the growth of flexible circuitry in spite of its obvious advantages over the rigid-board-plus-harness technique? One factor was problems with materials. Early insulating films were unstable during processing, causing poor yields, a limited capability to hold tight dimensions, and poor solderability. Another problem was a lack of connectors and termination methods to interface flexible printed circuits with other parts of a system.

The introduction of polyimide laminates in 1965 provided the industry with a high-temperature, solderable film that solved the materials problem. Now there are four satisfactory flexible laminates that are available. In addition, many interconnection schemes have been designed for interfacing with flexible circuitry.

Today, the biggest problem flexible-circuit manufacturers face is the resistance of packaging engineers accustomed to designing with hardboards. As Steve Gurley, director of sales and marketing at Sheldahl Inc.'s Electrical Products division, Northfield, Minn., says, "Our biggest problem is educating people to use flexible circuitry. Many companies are just not willing to take a chance on a technique unproven to them."

In spite of the resistance, total sales of flexible-circuit boards will be $128 million in 1977 and will grow to $177 million in 1980, according to Steve Grossman, director of interconnection studies for Gnostic Concepts Inc., Menlo Park, Calif. The breakdown of available and captive flexible-circuit production is shown in Fig. 3.

Flexible insulating base materials are literally and

figuratively the backbone of flexible circuitry. In the early years, materials like Vinyl, Kel-F, Teflon, and glass-reinforced Teflon were tried and found lacking. Today, four insulation materials — Kapton, Dacron-epoxy, Nomex, and Mylar — dominate the field (see table). (Kapton, Dacron, Nomex, and Mylar are registered trademarks of E. I. du Pont de Nemours & Co., Wilmington, Del.)

Base materials

Kapton, a polyimide-based film, is perhaps the most widely used, particularly in military and space projects. It has good dimensional stability, electrical characteristics, and high-temperature properties, and it withstands temperatures produced by wave soldering. However, it is the most costly material of the four.

Dacron-epoxy, used extensively by Western Electric and ITT, consists of nonwoven polyester fibers embedded in an epoxy resin. It has excellent dimensional stability, high moisture and tear resistance, and good electrical characteristics. Wave-solderable and flame-resistant, it has a cost close to that of Kapton.

Nomex, a low-priced nylon-and-paper insulator, is wave-solderable but extremely moisture-absorbent. It is used in commercial applications — especially in cameras and cars — where humidity is not critical.

A low-cost polyester-based film with good electrical properties and good dimensional stability at room temperature, Mylar has poor high-temperature characteristics and limited solderability. It is heavily used for the flexible circuitry of automotive dashboards (mainly as a harness), where soldering is often eliminated.

In addition, a lower-cost insulating film with properties similar to Kapton may soon appear. Exxon Chemical Co. USA, Houston, Texas, has been developing a film made with polyparabonic acid (PPA) for some time. Called Tradlon, it has properties approaching that of Kapton at about 65% to 75% of the price. Tradlon is being evaluated by several manufacturers. However, Exxon is manufacturing it so far only on a pilot basis.

Manufacturing

Flexible printed circuits, like rigid ones, can be made by either the additive or the subtractive process. In the additive process, electroless copper is selectively plated onto a substrate. The subtractive process selectively removes copper by etching a copper-covered substrate. Practically all flexible circuitry is made by the subtractive process, which is shown in Fig. 4a for a simple single-sided board and in Fig. 4b for a two-sided board with plated-through holes.

As the flow chart of Fig. 4a shows, resist (a material that resists etching) can be either screened on through a fine mesh (similar to the process used in thick films) or photographically exposed and developed. Normally, for lines and spaces 10 mils wide or less, most flexible-circuit manufacturers shift from screened-on to photographic resists.

Fine lines

As with rigid pc boards, the great majority of flexible circuitry is based on 10- to 20-mil lines. However, almost all flexible-circuit manufacturers can supply circuits with 3-mil conductors and spaces; moreover, in an engineering model, Sanders Associates has now achieved 1-mil conductors with subtractive etching. Fine-line (3 mils or less) circuits are of course much more expensive than ones with the normal line work, because processing is more complicated and yields are lower. They are being used to connect to magnetic recording heads or directly to integrated-circuit chips.

At this point, most flexible-circuit manufacturers are just beginning to investigate the use of additive plating. Only Buckbee Mears Co., Nashua, N. H., and Pactel

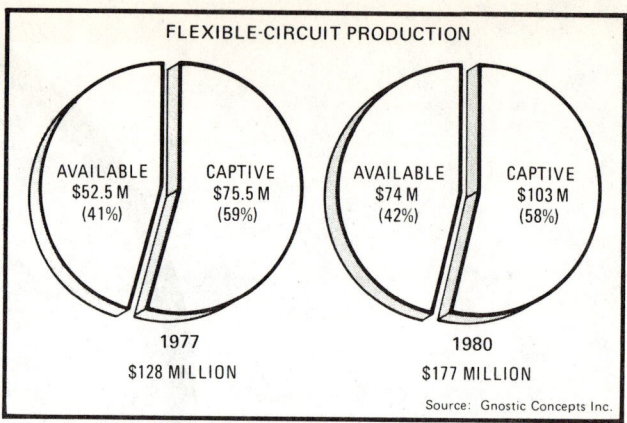

3. A growing market. According to Gnostic Concepts, the flexible-printed-circuit market will increase 38.2% by 1980. Captive production will retain the major share of this growing field, particularly in automotive and telecommunications systems.

	Sample thickness (in.)	Tensile strength (psi)	Tear strength (gm/mil)	Folding endurance (cycles)	Ultimate elongation (%)	Moisture absorption (%)	Dielectric constant (1 kHz)	Dielectric strength (V)	Dissipation factor (1 kHz)	Flammability	Service temperature (°C)
Kapton	0.001	23,000	8	10,000	70	3	3.5	7,000	0.003	94V-0	−250 to +250
R/2400 Dacron-epoxy	0.004	5,500	40	50,000	15	1	3.2	3,100	0.015	94V-0	−60 to +150
Nomex	0.002	11,000	49	5,000	10	5	2.0	600	0.007	94V-0	−60 to +120
Mylar	0.001	23,000	15	14,000	100	0.01	3.2	7,000	0.005	burns	−60 to +95

SOURCE: ROGERS CORP.

FLEXIBLE PRINTED-CIRCUIT PROCESSES

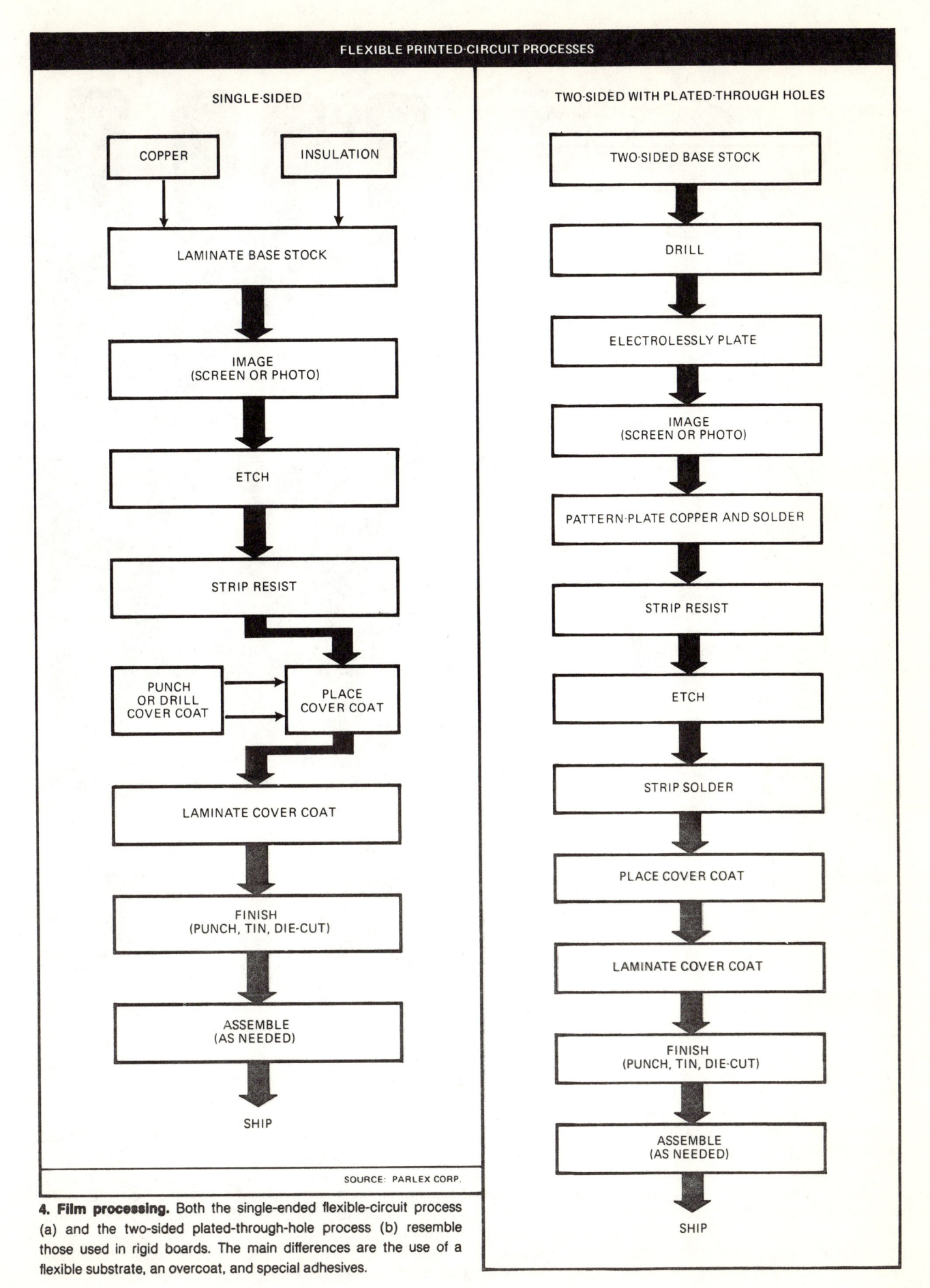

SINGLE-SIDED

```
COPPER      INSULATION
    ↓           ↓
  LAMINATE BASE STOCK
           ↓
       IMAGE
  (SCREEN OR PHOTO)
           ↓
         ETCH
           ↓
      STRIP RESIST
           ↓
  PUNCH           PLACE
  OR DRILL  →  COVER COAT
  COVER COAT →
           ↓
   LAMINATE COVER COAT
           ↓
       FINISH
  (PUNCH, TIN, DIE-CUT)
           ↓
       ASSEMBLE
      (AS NEEDED)
           ↓
         SHIP
```

TWO-SIDED WITH PLATED-THROUGH HOLES

```
   TWO-SIDED BASE STOCK
           ↓
         DRILL
           ↓
   ELECTROLESSLY PLATE
           ↓
        IMAGE
   (SCREEN OR PHOTO)
           ↓
 PATTERN-PLATE COPPER AND SOLDER
           ↓
      STRIP RESIST
           ↓
         ETCH
           ↓
      STRIP SOLDER
           ↓
    PLACE COVER COAT
           ↓
   LAMINATE COVER COAT
           ↓
       FINISH
  (PUNCH, TIN, DIE-CUT)
           ↓
       ASSEMBLE
      (AS NEEDED)
           ↓
         SHIP
```

SOURCE: PARLEX CORP.

4. Film processing. Both the single-ended flexible-circuit process (a) and the two-sided plated-through-hole process (b) resemble those used in rigid boards. The main differences are the use of a flexible substrate, an overcoat, and special adhesives.

119

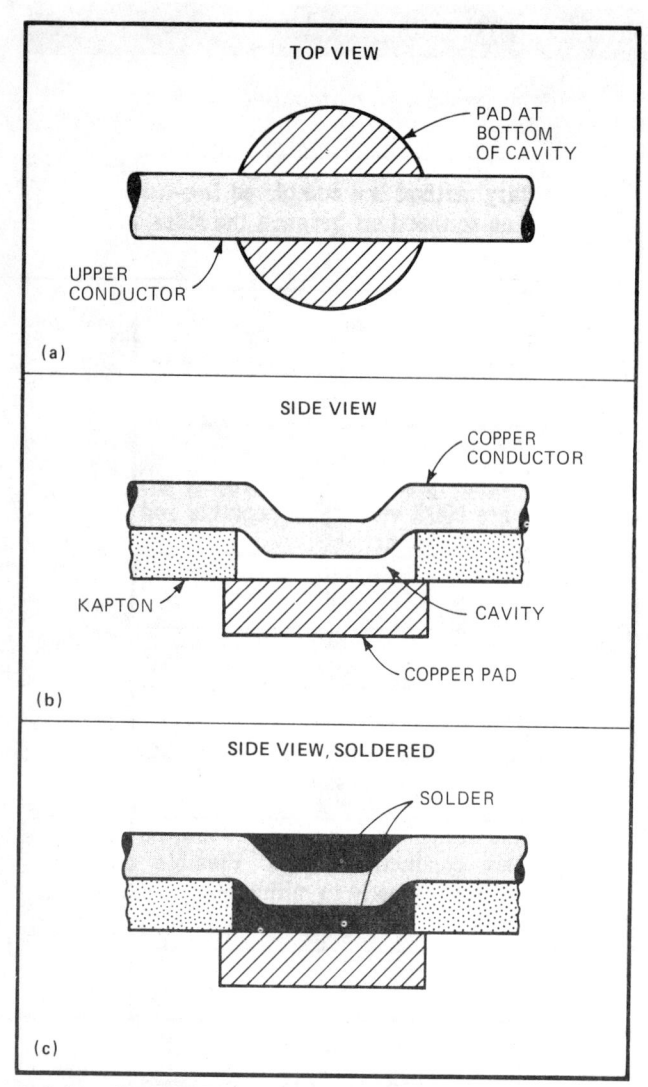

5. Solder-through. Teledyne Electro-Mechanisms uses this method as a low-cost alternative to plated-through holes. A cavity (a and b) is created between an upper conductor and a lower pad. During wave soldering, solder fills the cavity (c) making a through connection.

Corp., Westlake Village, Calif., have additive programs, while Flexible Circuits Inc., Warrington, Pa., combines subtractive and additive plating for some special circuits.

Buckbee Mears has a proprietary additive process for putting copper conductors on Kapton that has been used in missile work. Pactel additively plates copper conductors on thin sheets of polyimide and has supplied flexible circuitry 6 mils thick with 5-mil lines and 10-mil spaces for various military and space projects. It has also produced flexible circuits with 1-mil lines on 3-mil centers and expects to be able to make 0.5-mil lines on 1-mil centers. In addition, Pactel uses its additive process to manufacture strips of film carriers (a film carrier—really a series of repeated flexible circuits—is a copper IC interconnect or spider plated onto an insulating film).

Usually, one of the last steps in making flexible circuitry is to protectively coat the subtractively etched copper conductors (see Fig. 4a). The cover coat, or coverlay, is a clear film that is removed at the points where the circuitry patterns must be tinned. For some

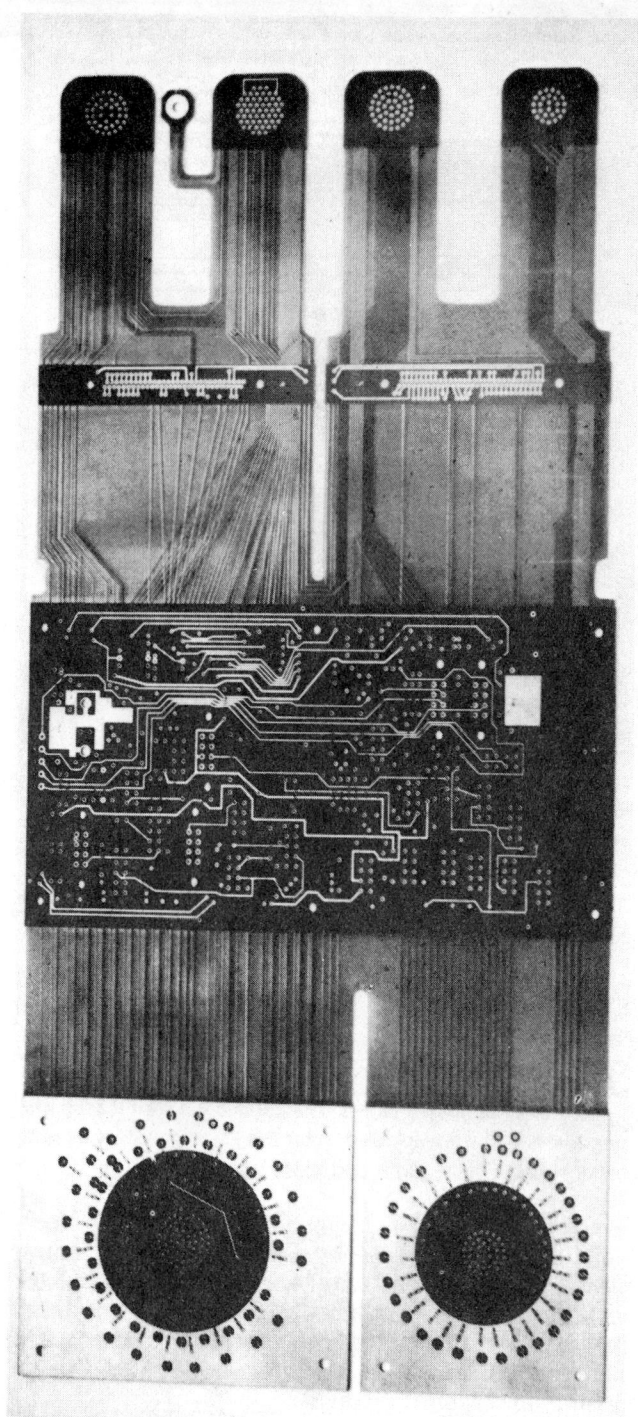

6. Rigid plus flexible. The flexible circuit shown has a dual purpose. It ties the nine rigid multilayer printed circuits together and at the same time serves as two conductive layers of interconnect in each multilayer board. Plated holes extend through all the layers.

time, Flexible Circuits has been additively plating ground planes or traces onto the normally bare cover coat, to act as a radio-frequency shield for the circuitry underneath. Ground is carried up through an additively plated hole extending through all the flexible circuit board's layers.

Yet another method of creating conductors on a flexible substrate is screening a low-temperature-curable

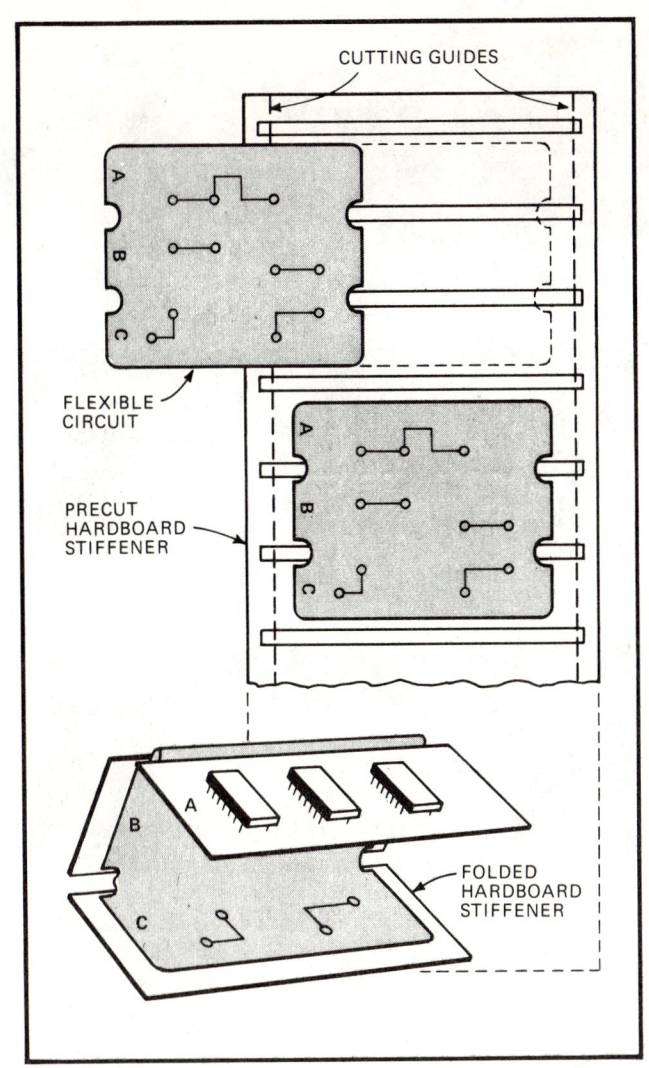

7. Hardboard center. Multiple identical flexible circuits can be mounted on a hardboard carrier. The assembly is then loaded with components and wave-soldered. After this step, the individual units are punched out of the carrier and folded into modules.

conductive ink onto a Mylar film. This low-cost technique is now being used by Chomerics Inc., Woburn, Mass., on Touch-Tone telephone keyboards, calculator keyboards [*Electronics*, July 7, 1977 p. 42], and other applications that combine keyboards, circuit boards, and interconnects, and it is ideal for single-sided flexible circuitry in consumer products.

It is also possible to screen resistors onto the film. However, the problem of soldering discrete components to the conductive ink has yet to be solved.

No matter what the etching or plating technique, flexible circuits can appear in three forms—single-sided, two-sided with plated-through holes, and multilayer.

Bendable boards

Single-sided flexible printed circuits are generally used for the simplest low-cost jobs. The great majority of flexible circuitry made today is two-sided with plated-through holes connecting the patterns on each side. The holes are usually plated through by an additive (electroless) process after the main circuitry has been subtractively etched (Fig. 4b), although two-sided boards have been made with braised eyelets or pins in the holes.

One company, Teledyne Electro-Mechanisms, has come up with a novel process for electrically connecting layers of a two-sided flexible circuit. Starting point of the proprietary method is a completed two-sided flexible circuit with no connections between the sides. Assume a conductor on one side is to be connected to a circular pad directly underneath it on the other side. A cavity is created under the top conductor and over the bottom pad by removing the insulating film in between (Figs. 5a and 5b). During wave soldering, the cavity is filled with solder connecting the upper conductor to the lower pad (Fig. 5c).

Soldering through costs less than electroless plating of holes. Two other pluses for the process are that the connections are 100% visually inspectable and reparable with standard equipment.

Flexible multilayer circuits

The most complex flexible circuits made today are multilayer types. The process for making them is similar to that used for making rigid multilayer boards, in which a sandwich of layers is laminated together in a large press under heat and pressure. Flexible boards have been made with as many as 23 layers, but the use of too many layers results in a loss of flexibility. Most manufacturers agree that the limit for a truly flexible multilayer circuit is five to six conductive layers. Flexible multilayer boards are confined mostly to military work but are now finding their way into computers also, because of increasing interconnection and component densities in the new machines.

These multilayer circuits are not simply flexible copies of rigid multilayer boards. Each layer can be extended separately to serve as a wiring harness to connect elsewhere in the system, and the extensions can go off in many planes. This allows points in the overall system to be connected to specific points in a particular layer.

Flexible circuits can be laminated between the layers of several rigid multilayer boards, forming Parlex Corp.'s rigid-flexible multilayer board system in Fig. 6. This entails making plated holes extending through both the rigid and the flexible boards. The flexible printed circuit serves as a preformed, prewired harness for connecting individual rigid boards and the rest of the system. In addition, the flexible circuit adds two conductive layers to each rigid multilayer board.

Hardboard carrier

This combination in turn has lead to a cost-saving variation that is especially popular in the production of small instrument modules. A two-sided flexible pc board composed of multiples of a particular electronic circuit is laminated to a large hardboard base of the type shown in Fig. 7. The rigid board has predrilled holes to accept the components of each flexible circuit and punched-out areas corresponding to the spaces between the circuits on the flexible pc boards.

At the customer's plant, parts are automatically inserted into the assembly and then wave-soldered. If any repairs are needed, they are made on the assembly.

8. Socketed. This is a flexible-printed-circuit–hardboard combination used in an airborne fuel gage. To make changing ICs easier, sockets are mounted on the hardboard and soldered to the flexible circuit. The flexible board is both a circuit board and a multitermination harness.

Then the borders of the hardboard are sheared away and the circuits are folded up as at the bottom of Fig. 7 or in any other configuration desired.

In general, components are soldered to flexible circuitry. Soldering can be done with hand tools or by wave, dip, or infrared reflow soldering. In film carrier applications, IC chips have been temperature-compression bonded to copper conductors; however, the conductors require special gold plating.

In general, before soldering it is important to know the temperature limitations of the flexible insulating film and adhesive used. It is also important to keep in mind that a component can be replaced many fewer times on a flexible circuit than on a rigid pc board. For instance, at Gull Airborne Instruments Inc., Smithtown, N.Y., a manufacturer of avionic instruments, a component may only be replaced twice on a flexible circuit as compared with a dozen times on a hardboard. Excessive soldering on a flexible circuit can cause layers to come apart or even cause copper conductors to peel off.

Pluggable circuit

Gull gets around the problem of component replacement simply. The instrument assembly in Fig. 8 is a combination of flexible circuitry and hardboard stiffeners. Sockets for heat-sensitive ICs and light-emitting-diode displays mounted on the hardboard allow the devices' removal without resort to desoldering.

As was noted before, a user now has many ways to connect his flexible circuitry to the rest of his system. There are numerous types of connectors, including rectangular, cylindrical, edge, crimp-on, insulation-piercing, and pressure-contact, designed specifically for flexible circuits. Also, many manufacturers wave-solder, weld, or braise pins onto the circuitry to eliminate a male connector. In some applications, the ends of the circuits are tinned and soldered directly to the next interface points, such as a rigid or another flexible pc board.

A DIP connector

A novel approach used at Teledyne Electro-Mechanisms is built around a lead frame that has its outer leads on 0.1-inch centers. Tinned termination leads from a flexible circuit are soldered to the inner leads of the lead frame. Then the outer leads are bent down and placed in a plastic dual in-line package's cavity. The resulting connector, shown in Fig. 9, can be plugged into a rigid printed circuit, a backplane, or a flexible circuit.

Flexible printed circuits found early applications in guided-missile electronics. As more and more circuits

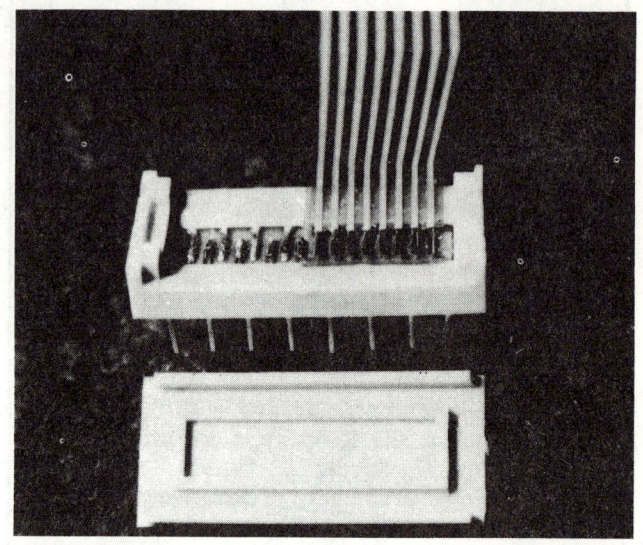

9. DIP connector. In this unit, a flexible circuit's tinned ends are soldered to the inner leads of a leadframe having outer leads on 0.1-in. centers. Placing the assembly in a plastic dual in-line body results in a connector that plugs into other rigid or flexible boards.

were crammed into smaller spaces, the density and number of electronic interconnections grew rapidly. General Dynamic Corp.'s Pomona division, Pomona, Calif., which has worked on missile programs for the Navy since 1964, quickly turned to flexible circuitry.

"Our driving force was, first, circuit density and then reliability," recalls Marvin Abrams, chief of advanced technology at GD Pomona. To meet a density requirement that grew from 775 conductors per square inch in 1964 to about 10,500 by 1972, the GD division started out in 1965 with designs that provided 825 conductors per square inch with 50-mil lines and spaces. These were two-sided flexible printed circuits on a Kapton film. Overall size of the flexible boards varied from 5 to 100 mils thick, 4 to 8 in. wide, and 6 to 24 in. long.

Today, GD Pomona, where all circuit fabrication is in-house, is making a six-conductive-layer flexible pc board up to 34 in. long, with lines and spaces of 25 to 10 mils. Used in the preproduction Standard Missile 2, it permitted reduction of a subsystem on the Standard Missile 2 to one quarter of the size of a comparable subsystem on the Standard Missile 1.

The advantages of flexible circuits were also quickly recognized in avionics—another area where space is a major concern. At Grumman Aerospace Corp., Bethpage, N.Y., they are now a way of life: a recent directive specifies that new equipment must be designed with

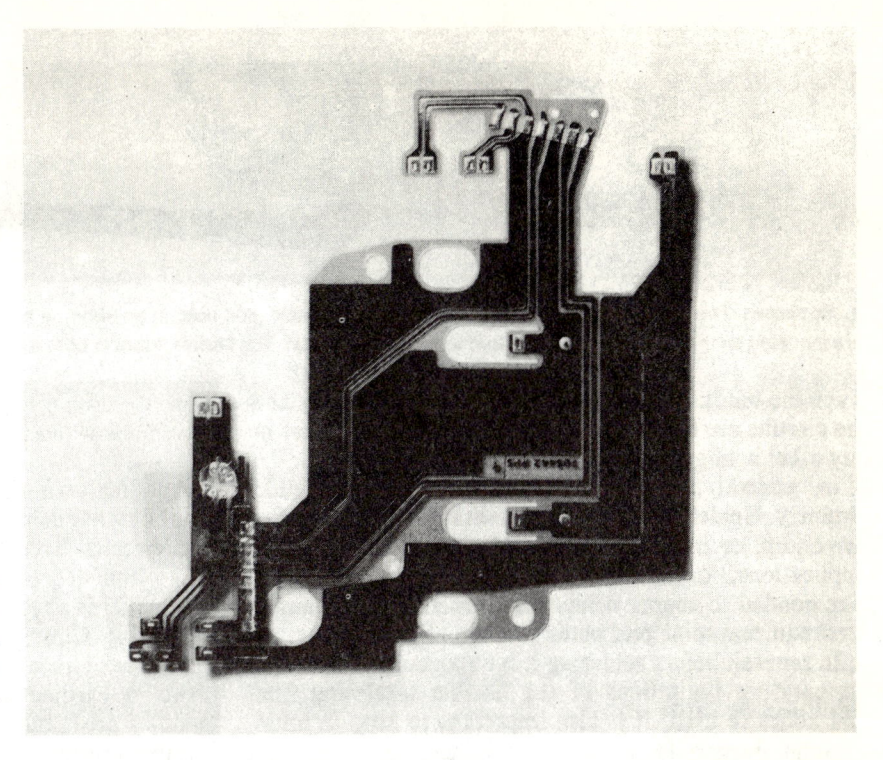

10. Camera circuit. An 8-mil-thick flexible circuit, produced by Sheldahl for the Polaroid SX-70 camera, distributes power to electronics, switches, film roller motors, and the shutter solenoid. The part flexes whenever the camera is opened or closed.

flexible circuits if at all possible, and as further evidence of their importance, the company is preparing its own design manual on the subject.

Michael LaTorre, group head of design engineering at Grumman Aerospace, states, "It is obvious to us that this is a superior method of interconnect." As an example, he cites a case in which a system's assembly time was reduced from 45 hours in the hard-wired version to 2 hours with flexible circuits.

Saving space

Telephonics, a division of Instrument Systems Corp., in Huntington, N.Y., has been using flexible circuitry since 1964 on avionic hardware. Telephonics' mechanical designers got their first exposure to these circuits while packaging the multiplexed entertainment system for the Boeing 747 jumbo jet. As a weight-saving measure, flexible pc boards were used instead of discrete harnesses and card-edge connectors to connect system modules. Single-sided Kapton printed circuit were the main substitutes.

As Telephonics' engineers grew more experienced with flexible circuitry, they went to more sophisticated designs, like the electronic packaging for the headset of the Lockheed 1011's cabin galley intercom. Originally, all the electronic circuitry was packed on two rigid pc boards within the set's small volume.

In 1971, new requirements called for many more electronic components and more wiring to be added to the already cramped unit. It was soon evident that more hardboards could not be added to the headset. Therefore, the designers decided to combine all the old and new electronic components and the harness wiring into a two-sided Kapton flexible circuit. The updated version is an example of an electronic package possible only with flexible circuitry.

Now, the firm's engineers are applying the combined hardboard-carrier — flexible method described earlier to assemble a small modular power controller in which the cut-out boards are folded up and placed in a small cubical package.

Another Long Island avionics firm, Gull Airborne Instruments, uses its flexible-hardboard combination in Fig. 8 to act as both a circuit board and an interconnect for digital fuel metering. The combination has resulted in an overall savings in assembly costs of up to 30% over a straight hardboard package.

Dick Holtz, manager of manufacturing engineering at Gull, has found it cost-effective to apply flexible circuitry to systems that have a production run as small as 25 units and to harnesses with as few as 10 wires. The company converted many units to flexible circuitry two years ago, and the field failure rate has turned out to be less than ¼%.

Consumer products

The largest area of growth for flexible circuitry in the next few years will be consumer electronics. Already, this interconnect method has found its way into cameras, calculators, watches, citizens' band radios, pocket pagers, video games, and microwave ovens. Within the next few years, it will be found in almost every consumer product that contains electronics.

Polaroid Corp. and Eastman Kodak Co., the two leading U.S. camera makers, both have been using flexible circuitry for some time. Polaroid in particular has applied it to various cameras for two basically different reasons, depending on the camera type.

The Cambridge, Mass., firm's SX-70 is designed to fold up into an extremely compact package. To accomplish this, Polaroid's designers selected a foldable pc board of Kapton to house the electronics in the back of the camera and behind the shutter. The board, which folds and unfolds as the camera opens and closes, is a

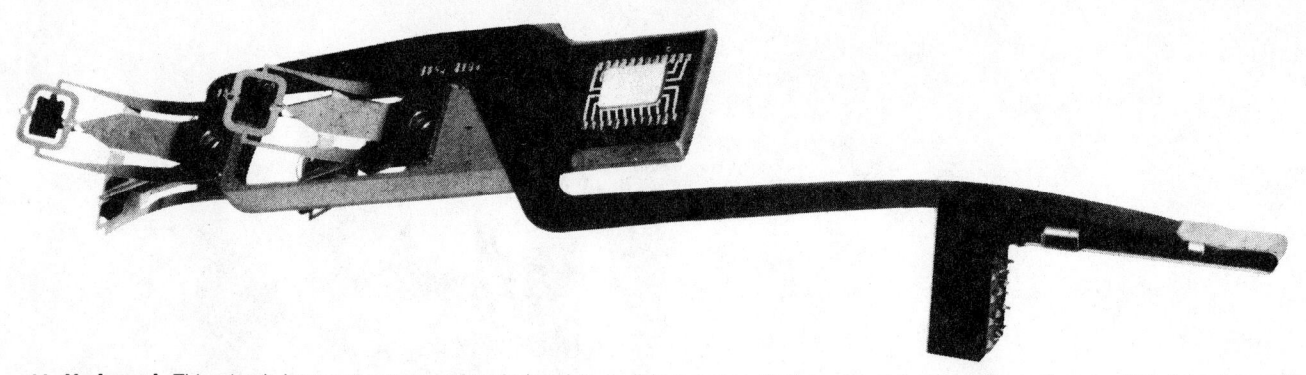

11. U-shaped. This circuit interconnects a magnetic head in a disk memory. During normal head motions, the circuit is folded in a U configuration and flexed in excess of 400 million times. Special treatments and processes are used to make this extremely flexible circuit.

single-sided type that is 8 mils thick (Fig. 10).

On Polaroid's Pronto and One-Step instant cameras, which do not fold, flexible circuitry was chosen for its ability to make multiplanar connections. It was used to accomodate camera wiring that turned at right angles and that had to be connected to many different points at different levels.

Multiple flexures

John Burgarella, director of engineering for product electronics at Polaroid, says that flexible circuitry has proved to be about 25% less expensive than the wiring in early SX-70 designs. He adds that it has also proved extremely trouble-free. The only hitches showed up early in the prototype stage when circuits were sometimes bent too sharply and tore. Modifying the bending radius solved this problem.

In most applications of flexible circuits, the board is bent, shaped or folded only initially. However, in the computer peripheral and rotating memory field, flexible circuitry completely lives up to its name. For instance, a circuit designed by Rogers Corp., Chandler, Ariz., connects signal-processing circuitry to a magnetic head for a disk memory (Fig. 11). Over the course of normal head operation, the circuit may be flexed more than 400 million times. Rogers uses rolled, annealed copper, rather than electro-deposited copper, which work-hardens as it is flexed, resulting in broken conductors; in addition, the company employs special treatments and processes to make the circuit truly flexible.

Flexible heater

Copper is not the only metallic material that can be laminated to a flexible base. Sheet nichrome can be also. Using this principle, Parlex Corp., Methuen, Mass., among others, is making flexible heaters by selectively etching on a nichrome-covered Kapton base. Like its circuit-bearing counterpart, the flexible heater can be folded and made in unusual shapes and patterns. In addition, the heat at various portions of the circuit can be controlled by varying the conductor pattern and thickness. To prevent heat from being applied in selected sectors, the nichrome can be plated over with copper. Circular flexible heaters are now being used to control the temperature of missile gyro packages.

Another area that thrives on the space savings and reliability of such circuitry is medical instruments. Flexible circuits have been part of pacemakers for some time—and acceptance of any technique or part in a pacemaker is a testimonial to its reliability.

Flexible Circuits manufactures a flexible pc board that is used in another highly reliable medical instrument—a cardiac event recorder. Fairly heavy hybrid modules are mounted unsupported on the 12-mil thick Kapton substrate, which is not usually done with most flexible boards, and the 38-in.-long circuit and a miniature tape recorder are stuffed into a 5-by-2½-by-3-in. package that is hung on a patient's belt; the recorder has a mean time between failures of three years. Only flexible circuitry can supply the tight packaging and reliability needed for this instrument.

The future

Flexible circuitry is certain to be one of the growth areas in the wiring field. Most of its manufacturers see a 25% growth rate for the next few years. A large part will occur as it spreads into more consumer products areas. In the automotive field, flexible applications will no longer be limited to the present componentless dashboard circuits; instead, new uses will be in the control circuits the industry is developing for ignition control, fully electronic fuel injection, and pollution control—as true flexible pc boards with components soldered on.

In the technology of flexible circuitry, several trends are starting to appear. One, brought on by pressures for even denser circuits and by the use of unhoused IC chips, is a general shift to 3-to-5-mil lines and spaces from the present 10-mil widths.

According to David Cianciulli, marketing manager for Hughes Aircraft Co.'s Connecting Devices division in Irvine, Calif., "There should be a large increase in the use and fabrication of flexible multilayer boards, since manufacturing methods are now fairly well established." Again, demands for increased packaging density can only be met by the density of the multilayer board and the multiplanar feature of the flexible circuit.

Another aspect of flexible circuitry that will increase is the use of combined rigid-flexible boards. The hardboard carrier scheme is becoming more and more popular, while applications that combine a rigid multilayer or two-sided pc board with a layer of flexible circuitry will also become more accepted. □

Philips says yes to resistless, etchless printed-circuit boards

A two-year-old process for delineating printed-circuit board patterns without using resist may be getting a shot in the arm from Philips' Elcoma division. The Dutch group is considering switching much of its subtractively etched board production to its own version of the American-devised photographic process. Philips intends to license the process, which it calls PD-R.

"The prestige of a giant like Philips is going to do a lot to legitimize the process in the eyes of potential users," asserts John Dennis-Browne, manager of technology sales and licensing for the Kollmorgen Corp.'s Photocircuits division, Glen Cove, N.Y. His unit introduced the resistless process, called Photoforming, in 1975.

Several companies in the United States and abroad have secured licenses but, according to Dennis-Browne, additional development time was required to adapt the procedure to a manufacturing environment. Photocircuits and its licensees are now at the prototype stage of their operations.

The Philips and Photocircuits processes are very similar. Both use ultraviolet light to project an image onto a printed-circuit board with a chemically treated surface. The surface reacts with the light and the image is then developed using photographic techniques.

What results is a metallic pattern that can be built up to circuit-board thickness using electrolesly plated copper. Gone are the screened-on or film resists used in the conventional additive and subtractive processes and the masks through which the pattern is screened on. Gone, too, is the stripping operation in which the unwanted resist is removed chemically. Unlike the subtractive process, there is no etching away of copper in the resistless method.

Savings. All of this adds up to a 30% reduction in cost over Philip's standard subtractive etching of pc boards, according to a company spokesman in Eindhoven. Also, it is also possible to plate conductors and spaces as narrow as 6 mils, compared to 13-mil conductors and 16-mil spaces possible with subtractive etching. Additive-process limits are conductors and spaces about 10 mils wide.

The Elcoma division in Eindhoven is using the PD-R process for prototype telecommunications boards. In its process, Philips first coats a standard FR-4 glass-epoxy board with a layer of titanium oxide. After holes are drilled, the board goes through a swell-and-etch step to ensure a firm surface for the conductor patterns.

This roughened laminate is then activated by the palladium ions in a solution of palladium chloride. Conductor patterns can then be printed onto the board with ultraviolet light. After exposure, the remaining palladium chloride is merely washed off, leaving a pattern of metallic palladium. Copper can then be plated onto this pattern to a thickness of 25 micrometers.

While PD-R works solely with FR-4 laminate, Photocircuits' Photoforming can be used on a variety of glass-epoxy boards and other materials like paper and phenolics. Moreover, Photoforming does not need a catalytic coating like titanium oxide but relies on an adhesive coating to hold the copper. Finally, Photoforming's light-sensitive material is based on non-noble copper salts rather than palladium. □

Handy breadboard systems speed the development of prototypes

by Jerry Lyman, *Packaging & Production Editor*

With five major types to select from, designers are able to putter and probe in the style that suits best

☐ The classic "rats-nest" homemade breadboard—the age-old proving ground where designers checked out their designs—is no more. In response to the growing complexity and density of today's electronic compo- nents, a whole new generation of breadboard systems is speeding circuit development from paper to production.

The functions of the breadboard are the same: to let designers make quick wiring changes repeatedly without damage, insert new component values, and provide easy access for measuring instruments to any point in the circuit.

Ideally, it should approximate the actual circuit environment the system will face, and it should be inexpensive. It should be laid out in one plane with a reasonably good wiring density. Wiring should require no

1. Low profile. This Garry custom IC-socket panel has a 0.5-inch profile because its wire-wrapping pins are mounted on the same side as the DIP sockets. Standard IC panels, with pins and sockets on opposite sides, have 1.2-inch profiles.

special tools. And a breadboard should be easily connected to other breadboards and should be suitable for analog and digital work.

Important characteristics not all users may need include easy conversion to production printed-circuit form, interchangeability with a pc in the production configuration, and good resistance to actual operating temperatures.

There are five main breadboard systems, sold by a variety of firms:
- The IC-socket panel in which rows of wire-wrappable IC sockets or socket pins are mounted in a glass-epoxy board.
- Glass-epoxy circuit boards with hole patterns for accepting dual in-line packages plus plated-on ground and power busses.
- Perforated plastic boards for push-in wiring terminals.
- Solderless plastic breadboard sockets.
- Glass-epoxy panels with rows of DIP sockets, wiring plugs and special wiring jumpers. The table on p. 99 compares the five systems.

The most popular breadboard

Of course, no type has all the primary and secondary characteristics, but the IC-socket panel comes closest. Digital circuits with ICs in 14- and 16-pin DIPs are the basis for much electronic equipment, and this may account for the success of the IC-socket panel, which is the most widely used breadboard system. Also called logic panels, they appeared in 1965 when Augat Inc., Attleboro, Mass., combined three technologies to produce a solderless breadboard: printed circuits, DIP sockets, and Wire-Wrap, a wiring method developed by Gardner-Denver, Grand Haven, Mich., in which a special tool wraps several turns of insulated wire so tightly around a square metal pin that a gas-tight connection is formed.

The panel is a board with rows of holes on 0.1-inch centers, plated-on pc-connector fingers, and etched copper ground and supply buses. The holes can accept special DIP sockets with square pins at the bottom or square, wire-wrappable pin sockets that will accept DIP pins.

Dick Grubb, vice president of marketing at Augat, says the highly reliable wire-wrapped connections and sockets mean the IC socket panel is easily transported and can withstand the environment as well as the final design will. In fact, some manufacturers are staying with the panels in production rather than converting to a pc. If the breadboard and production models share the same package and wiring format, then the breadboard test data can be used to check production units.

Logic panels also have a high IC density, are easily modified, and fairly easily convertible to a pc. They have low noise characteristics because of their ground planes. Usually they are used with transistor-transistor logic, metal oxide semiconductors, and complementary MOS. Special multilayer versions are available for high-speed logic families such as emitter-coupled logic and Schottky TTL. The newest development is panels designed for specific microprocessor chip sets.

Logic panels are designed primarily for the digital circuitry in DIPs. However, all the manufacturers supply blank DIP headers to mount passive and active discrete components.

The panels cost $1 to $1.50 per IC position, so they are expensive compared to other breadboard techniques. They also have high profiles (the distance from the top of the highest component to the bottom of the deepest pin). And high profiles don't permit a one-to-one replacement of a panel and pc board.

"There is a way to get around the high panel profile," says Jim Aaron, vice president of manufacturing for Garry Manufacturing Co., New Brunswick, N.J. "We have supplied special custom panels on which both sockets and 0.450-inch Wire-Wrap pins are mounted on the same side [Fig. 1]. This results in a 0.5-to-0.6-inch profile, allowing the customer to interchange panels and the final board." Augat has similar models. A possible drawback to these low-profile panels is the impossibility of wrapping them with fully automatic equipment.

A typical use of panels is the breadboarding procedure used in designing cathode-ray-tube terminals at Applied Digital Systems Inc., Hauppauge, N.Y. Figure 2 shows the highlights of the conversion from breadboard to final-system pc board.

IC-socket panels are expensive, but they can accommodate anywhere from 30 to 180 14-pin DIPs. However, many circuits and subsystems can fit onto circuit boards, which have room for 20 to 50 14-pin DIPs.

The fastest-growing breadboard

These circuit boards are the fastest-growing system in the breadboard market, according to Floyd Hill, vice president of marketing for Vector Electronic Co., Sylmar, Calif. The other two main suppliers are Vero Elec-

Hardware type	Used by	Circuit suitability Digital	Circuit suitability Analog	Digital logic allowable	Hardware price	Profile	Method of mounting passive components	Wiring method
IC-socket panel	Industry, laboratory	Good	Fair	Schottky TTL, ECL with multilayer panel	High	High	Requires special adapters	Wire-Wrap
Circuit board	Industry, laboratory	Good	Fair	Schottky TTL, ECL	Low	Low if soldered	Direct	Soldering, Wire-Wrap
Perforated board plus push-in pins	Industry, laboratory, hobby	Fair	Good	TTL	Lowest	High	Direct	Soldering, but can accept Wire-Wrap pins
Solderless breadboard socket	Industry, laboratory, hobby, education	Good	Good	TTL	Low	Low	Direct	Push-in, stripped, solid wire into special contacts
Jumper panel	Laboratory, education	Good	Fair	TTL	High	High	Requires special adapters	Specially purchased jumpers into on-panel posts

COMPARING BREADBOARD TECHNIQUES

2. Trying out a CRT terminal. Working from a circuit-board schematic, Applied Digital Data System Inc. wire-wraps an IC-socket panel (above, left). After the panel is tried out on a logic-card tester, it is put into the system breadboard (bottom). After the system is debugged, the panels are converted into double-sided pc boards (above right) for use in production terminals.

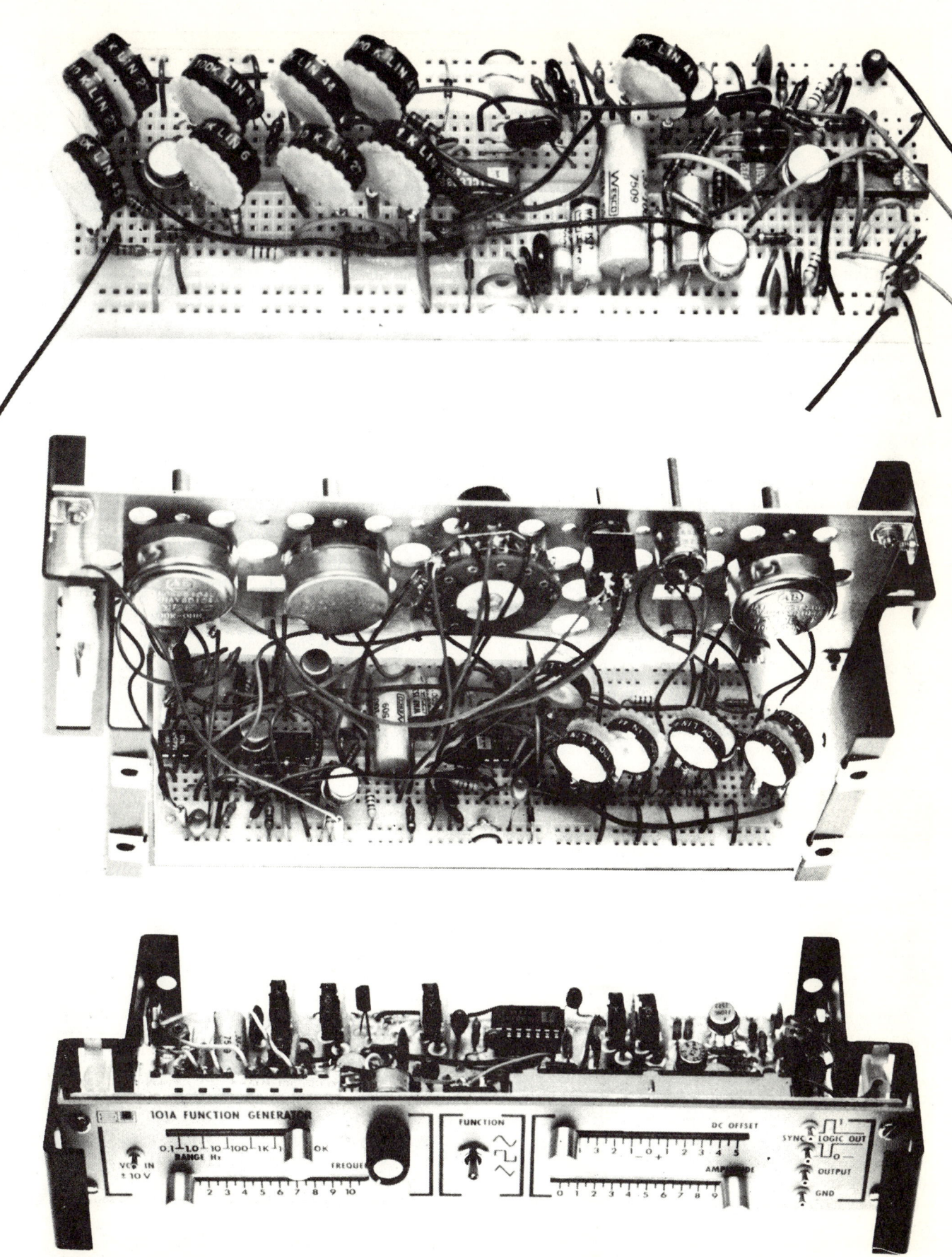

3. Solderless action. Pressing components and stripped interconnecting wires onto an E&L Instruments solderless socket (top) forms a breadboard—minus the controls—of the circuits of a function generator. The breadboard is combined with a universal instrument module to make a fully functional model (middle). The finished product is an E&L function generator (bottom).

tronics Inc., Hauppage, N.Y., and Douglas Electronics Inc., San Leandro, Calif.

All versions have a grid of holes on 0.1-inch centers and plated-on connector fingers. The most heavily used of the many varieties has rows of isolated plated pads for DIPs and copper strips for ground and voltage. Often there are test points and uncommitted pads for passive discrete components. The voltage and ground strips usually are interleaved to cut circuit noise with distributed capacitance.

The most common way to assemble circuit boards is to solder DIP sockets and passive components into place, wire them according to the circuit schematic, and then insert the ICs. An alternative method solders the ICs directly without sockets. A third uses DIP sockets with wire-wrappable pins and additional wrappable pins as needed.

The circuit board has a low profile and is fairly inexpensive—30¢ per IC position. It is suitable for TTL and linear applications, although special versions with a continous ground plane on one side work with ECL and analog designs up to 100 megahertz.

Since circuit boards require soldering, their reliability is only as good as the solderer's. Repairs require unsoldering, and the board's pads are like any pc in their inability to withstand many repairs. Wiring density is good, but it cannot compete with, say, two-level wrapped wire.

Transfer from breadboard to production pc is easy, especially if the dimensions and the connectors of the two devices are the same. Many firms use circuit boards from breadboarding right through to production.

Circuit boards and IC-panel sockets are the only commercial breadboard systems designed for out-of-lab environments, so they are the only ones seen in production equipment. Their sturdiness also makes them ideal for military "brassboards," which are breadboards that can be taken from the lab and tested under field conditions. Industry is beginning to use brassboards in demonstrators or rush field modifications.

The cheapest breadboard

Not all designs call for IC-socket panels or circuit boards. For example many analog systems with bulky discrete components and relays may be better suited to perforated boards—the so-called "perf board."

Pioneered by Vector in 1948, perf boards are made of phenolic or glass-epoxy, or sometimes of copper-clad glass-epoxy for high-frequency work. The original format was 0.093-inch-diameter holes on a 0.265-inch grid. A newer version has 0.042-inch holes on a 0.1-inch grid.

Special metal pins or terminals that accept components with large lead diameters are staked into the boards for all wiring and component mounting. The staking operation usually requires a special tool. The older version is used mainly for analog work with fairly large discrete components, while the newer version can accept DIP components. Soldering is the principal means of interconnecting perf boards, but spring-loaded press-in pins are available.

A 25-square-inch perf board costs 44¢, and 100 pins

CAD vs breadboards

Computer-aided design can bypass the breadboard stage of large digital systems. For instance, at Digital Equipment Corp., Maynard, Mass., only single- or double-sided boards using TTL are tried out on DEC-designed wire-wrappable panels.

"For highly complex pc boards like a multilayer board with high-speed logic, the IC panel cannot duplicate the MLB performance," says Ray Moffa, DEC's engineering manager for components. "Because of this, we go to computer-aided design to produce a prototype board quickly."

The CAD boards are tried out on the bench and then revised as necessary by computer. Certain boards (such as memory boards, which have a regular repeating pattern) are almost always designed by the firm's CAD department, regardless of complexity.

---

cost $1.60, so this is the cheapest of the five breadboard systems. It probably is the slowest construction technique, but it is the most flexible, because any analog circuit can be assembled. It can be used for TTL, but the lack of a ground plane limits some applications.

Perf boards with plated-on connection fingers are interchangeable with similar-sized pc boards. With care, they may be substituted temporarily for the pc boards.

The easily wired breadboard

The three systems described so far need special tools to interconnect wire and components. The solderless breadboard socket, originated in 1967 by A. P. Products Inc., Painesville, Ohio, is wired simply by pushing the stripped ends of 20- to 26-gauge insulated solid wires into grids of spring contacts molded in a plastic body.

They are designed to accept almost any electronic component, from passive to 40-pin DIP large-scale integration. The limiting factor is the diameter of components' leads. Besides A. P. Products, two other companies supply full-sized versions of the breadboard sockets—E&L Instruments Inc., Derby, Conn., and Continental Specialities Corp., New Haven, Conn. Vector offers a version with 8 to 24 positions that plugs into perfboard.

A typical breadboard socket, the E&L SK-10, is shown in Fig. 3a. The entry holes in the acetal copolymer body are on 0.1-inch centers. Running full length are two rows of vertical contacts, which are used for component wiring. On each outside edge are two rows of common busses for voltage and ground distribution. There are eight separate busses, because each of the four rows is split at the middle point of its length.

Breadboard sockets cost about $2 per IC-socket position, so the version pictured, with nine 14-pin DIP sockets, costs around $18.50. However, there is no need for special tools or mounting equipment.

Main advantages of this system are wiring speed without special skills needed, reusability, and ability to accept most components. Since the parts layout is on a 0.1-inch grid, conversion to a pc board with the same grid is easy.

The plastic body more or less limits the breadboard

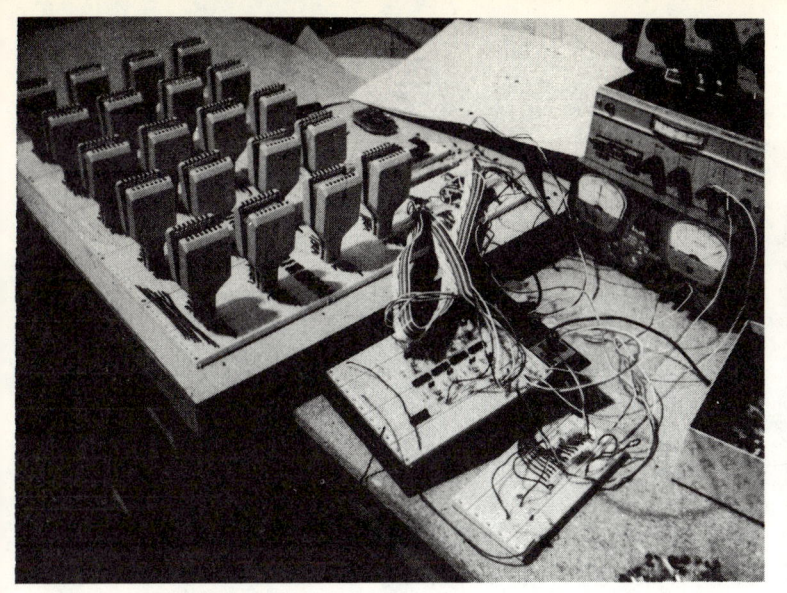

4. Burn-in. Using solderless sockets in burn-in tests allows quick replacement of the units under test. Continental Specialities Corp. tests its logic probes on a setup of its own sockets.

5. Jumpering. Special jumpers (not shown) make quick interconnections to the wiring posts of this Garry jumper panel. Soldering and wire-stripping or wire-wrapping aren't necessary.

socket to the laboratory, so temperature testing must be done with a different breadboard system. The socket is fine for TTL, MOS, and C-MOS, but the 3-picofarad capacitance per contact is too much for ECL or Schottky TTL.

Figure 3 shows a typical application, in design of a function generator. First, a breadboard socket is wired and loaded with components. Next, it is attached to a special panel that holds large potentiometers or other controls. Finally, the breadboard design evolves into the modular instrument.

In other breadboard systems, models vary only in the number and types of DIPs accommodated. But the breadboard socket is available in many forms: packages of multiple sockets on metal ground planes; packages with test equipment, power supplies, signal generators, and switches, and packages for digital or operational-amplifier design exclusively.

To facilitate transfer from breadboard to pc board, solderless sockets may be mounted on epoxy-glass boards with connectors. E&L Instruments even has printed-circuit boards with the interconnection pattern of its breadboard sockets.

The system also is used as a simple connector, as an IC socket in tests, and as a simple burn-in fixture for light-emitting diodes, pilot lights, discrete transistors, and the like. Figure 4 is an example of a breadboard socket used as a burn-in fixture.

The low cost, small size, and ease of wiring of breadboard sockets make them popular with hobbyists and engineers who like to experiment at the desk or at home. The Heath Co., Benton Harbor, Mich. includes a breadboard kit in its catalog of do-it-yourself electronic devices. Other retail outlets carry the socket only.

There are times when an engineer wants to try out a circuit in a hurry and then quickly tear it down. Soldering, wirewrapping, or even stripping a wire will slow him up. So he turns to the jumper panel, the breadboard system designed for fast construction and reuse.

Supplied by Augat, Garry and Cambridge Thermionic Corp., Cambridge, Mass., jumper panels have integral DIPs, flatpacks, or TO-5 sockets. Also on the top are rows of posts connected to each socket pin with printed wiring (Fig. 5). Special jumpers plug into the posts and interconnect the sockets.

Jumper panels are expensive, since the user pays for both board and the jumpers. With the other breadboard systems, inexpensive wires do the interconnections. The jumpers cost 30¢ to 40¢ each, and the boards cost $2 to $2.50 per IC position. However, the entire system is reusable.

A jumper panel has a high profile and a low component density because of the posts on the board top. The system is used in R&D labs and in the classroom, but rarely in industry.

Combining breadboard systems

Breadboards aren't limited to standardized applications of the commercial systems. There are more exotic uses, including combinations of systems, breadboards of more specialized circuits, and homemade breadboards.

Many companies try to settle on one breadboard system for all their designs, but quite a few find that only a combination can work. For instance, American District Telegraph Co., New York, N.Y., tries out large digital systems on IC-socket panels and smaller systems on circuit boards. The cover shows a panel and a circuit board from ADT.

Sometimes ADT's engineers will mount a circuit board on a logic panel to add a small amount of logic quickly to an existing circuit. They often build a power supply module on a perf board in order to power IC-socket panels.

Figure 6 shows another example of combining breadboard systems. The instrumentation division at the U.S. Army's Picatinny Arsenal, Dover, N.J., built the breadboard of a programable calculator with solderless breadboard sockets, a perf board, and a stick-on circuit pattern from Instant Instruments Inc., Haverhill, Mass. (The stick-on is a copper-plated pattern on a 1/16-in.-thick, adhesive-bottomed epoxy-glass board that can be combined with blank boards and soldered

131

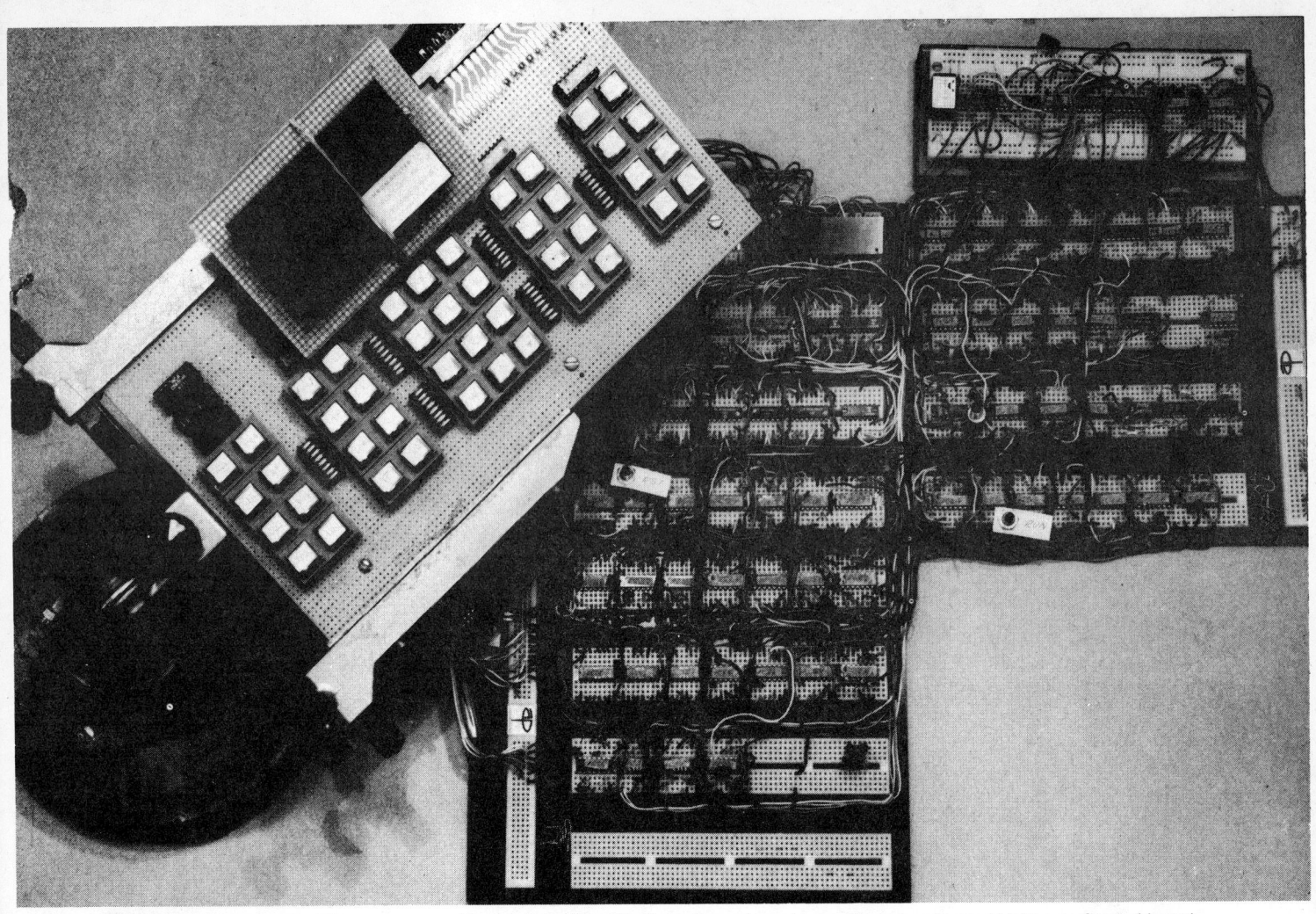

6. A mixture. This breadboard for a programable calculator from Picatinny Arsenal combines different systems. At left, a perforated board with the keyboard and display circuitry is mounted on a holding fixture. Other circuits and the wiring are on the two ranks of solderless breadboard sockets to the right. A stick-on copper wiring pattern serves as connector to the perforated board circuitry.

components and wiring to form a sixth, sometimes-used kind of breadboard.)

The keyboard and the digital readout of the Picatinny design are mounted on the perf board with wire-wrapped connections underneath. Most of the electronic circuitry is on the solderless breadboards. One end of a tape cable is pressed into a socket breadboard. The other is soldered onto a pc board connector that plugs into the stick-on pattern, which is pressed onto the perf board as a connector to its circuitry.

Not always look-alikes

Production models may resemble the breadboards, but they can just as easily differ radically, notably in such applications as development of LSI chips, hybrid circuits, and analog instruments.

Ray Kozak, a design engineer at Teleglobe Pay-TV System Inc., New York, N.Y., is using an E&L solderless breadboard to design complex custom LSI chips. He is developing TTL and analog circuitry in the video frequency range with the sockets. After the circuits have been tried out on the breadboard, they are laid out on an E&L circuit card with the same wiring pattern.

Then the pc board is tested environmentally and used in a demonstration model. Once it is proven out, it goes to an IC manufacturer for conversion to a custom LSI chip used in Teleglobe's production models.

ILC Data Device Corp., Bohemia, N.Y., tries out thick-film circuits on perf boards. Discrete or passive chips are mounted in TO-5 headers, which are then mounted on sockets of a small perf board that simulates the ceramic hybrid substrate and parts layout as nearly as possible. Using a breadboard this way can save the cost of design problems or component failures in the prototype stage.

In analog work, ICs can be in TO packages or DIPs. Discrete semiconductors at all power levels must be accommodated, as must larger components like transformers, potentiometers, relays, and even motors. In many cases, an analog breadboard has no resemblance to the final package, so a board with a connector and provision for many DIPs is not needed.

A case in point is Gull Airborne Instruments Inc., Smithtown, N.Y., whose main product is aviation instruments. These mostly analog instruments are laid out on large perf boards that house discrete passive components, transformers, linear ICs, digital-to-analog and analog-to-digital converters, and the like. Since the boards are big and open, it is easy to make the many measurements necessary to check out a precision instrument design.

After the breadboard passes the tests, Gull's engineers build a prototype, often a small rectangular or cylindrical container with the circuitry of four perf boards

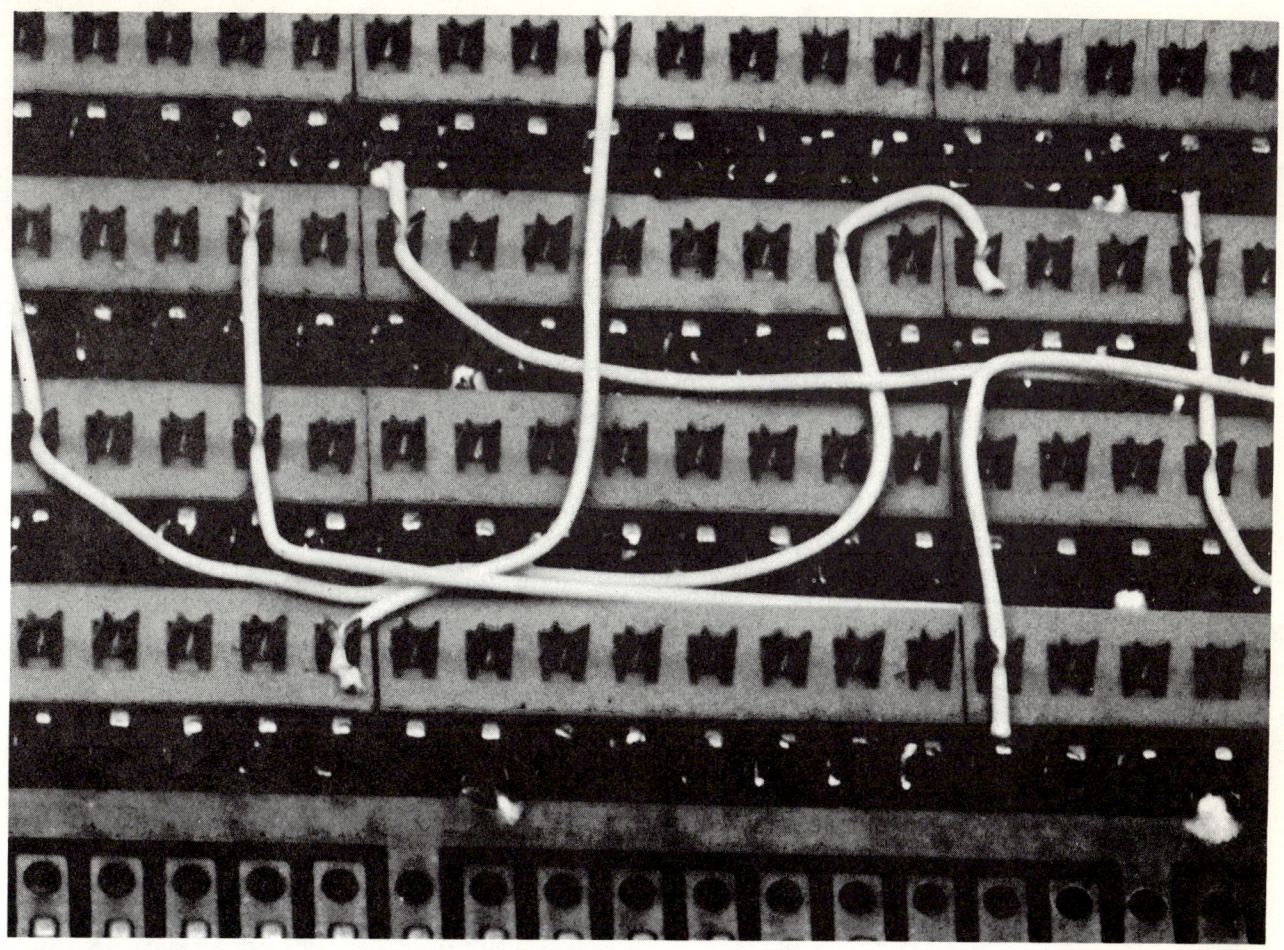

7. Piercing the wires. A new breadboard type from Bell Labs is wired by pressing the wires into rows of special insulation-piercing terminals on the bottom of an epoxy-glass board. DIP components are inserted in rows of pin sockets on the board's top side (not shown).

on one flexible or rigid printed circuit.

At times engineering firms have to develop their own breadboards. For instance, Comtech Labs Inc., a Smithtown, N.Y., firm specializing in satellite communication equipment, tries out low-frequency analog work on perf boards and digital circuitry on IC-socket panels. But analog circuit design in the 1-gigahertz area is tried out on home-built pc boards, because none of the commercial breadboards will work at such high frequencies. The Comtech boards have special printed wiring on one side and a full ground plane on the other.

High-frequency breadboards

At AIL, a Cutler Hammer Inc. division, much of the radar and rf work isn't adaptable to commercial breadboards because of the high frequencies involved. Engineers at AIL in Deer Park, N.Y. try to approximate the final configurations in breadboards. With radar circuitry, special double-sided boards with plated-through holes and most components mounted on standoffs are designed. To simulate rf operational conditions, a rectangle is machined out of an aluminum blank, and a circuit board with components mounted on standoffs is dropped in.

At Bell Laboratories, Holmdel, N.J., the breadboard usually is the prototype, since there is no production there. Charles Von Roesgen, supervisor of optical systems, has tried several breadboard systems in packaging digital prototypes—solderless breadboard sockets, jumper panels, and IC-socket panels.

"With solderless sockets, ICs go in easily, but wire has to be stripped to plug it in," he says. "Larger-size wires often require considerable force and bend rather than engaging the socket contact. Also, when changes are necessary, too many wires often come up.

"Jumper panels can be wired up fast, but they're expensive and have an extremely high profile due to the jumper wires looping above the board." He has found that solderless sockets and jumper panels tend to develop intermittent contacts when stored for any length of time.

For a while, Von Roesgen settled on IC-socket panels, but he discarded them because of their high profile and the difficulty of repairing. "The wrap to be removed always was under two other wraps."

Finally, he invented a new breadboard that is wired by just pressing the wires onto insulation-piercing terminals (Fig. 7). It has a 0.42-inch profile, allowing it to replace a pc board directly in a system. It is a printed-circuit board with socket pins to accept DIPs on one side and the special terminals on the other. Bell Labs doesn't make this breadboard for outside firms, but there is nothing to prevent a firm from seeking a license to produce it commercially. □

Assembling a complex breadboard can be as easy as 1, 2, 3

An updated version of the bus bar helps make
parts layout more orderly and power distribution less noisy

by Bernard J. Carey and Harry Grossman, *Department of Electrical Engineering and Computer Sciences, The University of Connecticut, Storrs*

☐ The breadboard for complex digital circuitry often winds up as a system's stumbling block, simply because a designer has failed to account for parts layout and power distribution. Yet a systematic approach to breadboard building, using an updated version of the bus bar known as the power distribution element, can produce easy-to-assemble wire-wrapped breadboards for microcomputer system modules. Moreover, this approach eliminates much of the high-frequency noise generated by the switching action of the high-speed logic of such modules.

This new prototyping method has been used successfully at the University of Connecticut's microprocessor laboratory to construct central processing units and memory and peripheral-interface modules. It is simple enough for untrained workers to complete complex boards quickly and easily.

Of course, the designer must adhere to the guidelines that follow if semiskilled workers are to be able to build microcomputer boards. Breadboards that were built following these guidelines were tested for noise, and the results demonstrate the noise reduction possible with power distribution elements.

The advantages of using the new prototyping method with these bus bars are several:
- The method imposes an orderliness on board layout that facilitates the layout itself and speeds up the process of creating the breadboard (especially when untrained workers do the job).
- It results in electronic assemblies that have a low noise level on power and ground lines because of the inherent noise suppression of the PDEs.
- It provides a lower-cost technique of fashioning breadboards by substantially reducing the number of decoupling capacitors required to reduce system noise.
- It eases the transition from breadboard to printed-circuit board, because the layout used in the breadboard is reasonably efficient and logically organized from the outset (especially where the bus bars are incorporated into the production pc board design).

Breadboard to pc

This breadboard procedure was designed to exploit the potential of the mechanical and electrical properties of PDEs for realizing a low-noise system. The use of these bars also established an orderly layout.

Students at the University of Connecticut who constructed prototypes with this method were able to conceptualize the PDE as a black box with an input/output relationship that could be used to advantage in constructing new digital modules. Thus they saw the construction process as the positioning and interconnection of a group of functionally related digital elements (logic devices and PDEs).

A close similarity may exist between the layout and signal interconnection of a wire-wrapped prototype and those of the final pc board version of the prototype. This relationship is most evident in the memory matrix of a memory board, but it will also hold true in other boards using random logic, microprocessors, and large-scale integrated devices.

The trend away from random logic to LSI results in a more standard signal interconnection system. That standardization, coupled with the formal organization of the board imposed by PDEs, simplifies layout.

Bus power

Before looking at the actual construction of prototype boards, it is best to examine the power distribution element itself. A PDE (Fig. 1) is composed of two copper conductors laminated into an integral unit with an interleaved thin dielectric. The exterior is generally covered with a plastic barrier material to prevent accidental shorting. One conductor is the power bus and the other is

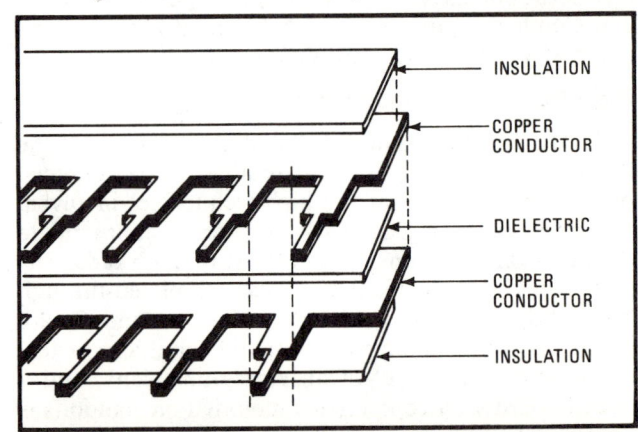

1. Power distributor. The power distribution element (PDE) is composed of a two-conductor bus bar interleaved with a dielectric. This exploded view shows the components of a vertically mounted configuration. Other PDEs are available in horizontal configurations.

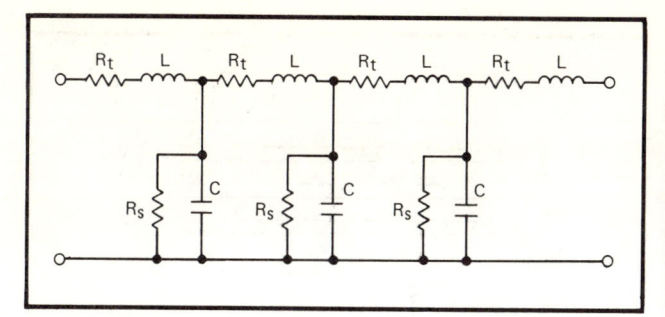

2. Equivalent circuit. The circuit equivalent of a PDE resembles a very lossy transmission line. The buses' low impedance and low inductance make them effective elements for suppressing high-frequency noise on power and ground lines.

a ground return. Pairs of leads from the two conductors are brought out at regularly spaced intervals for insertion into standard pc-board holes. Power and ground for a particular integrated circuit are connected to a pair of leads, while the pair of leads at either end serve as the bus bar's I/O points.

A PDE is equivalent to a very lossy transmission line (Fig. 2) with a low characteristic impedance on the order of several ohms. Because of its low impedance, current transients will produce lower noise voltages than on pc wiring, which has a characteristic impedance as high as 50 ohms. Because a bus bar is quite lossy, it also will attenuate the noise voltage and energy more per unit length than buses printed on the pc board.

A PDE is more efficient at suppressing high-frequency noise than are decoupling capacitors—in spite of the fact

3. Interface breadboard. This 8080 interface card is an illustration of a circuit built with the method described. The thin upright components between the rows of DIPs are PDEs. All system wiring is done on the underside of the board by wire-wrapping.

that commercial bus bars usually have much lower capacitances than typical decoupling capacitors. Above 20 megahertz, the inductance of the leads of most decoupling capacitors renders the capacitance ineffective. While the PDE is limited in capacitance (225 to 1,200 picofarads per square inch), it is an extremely low inductance element at noise frequencies above 20 MHz.

PDEs are available from such firms as Rogers Corp.,

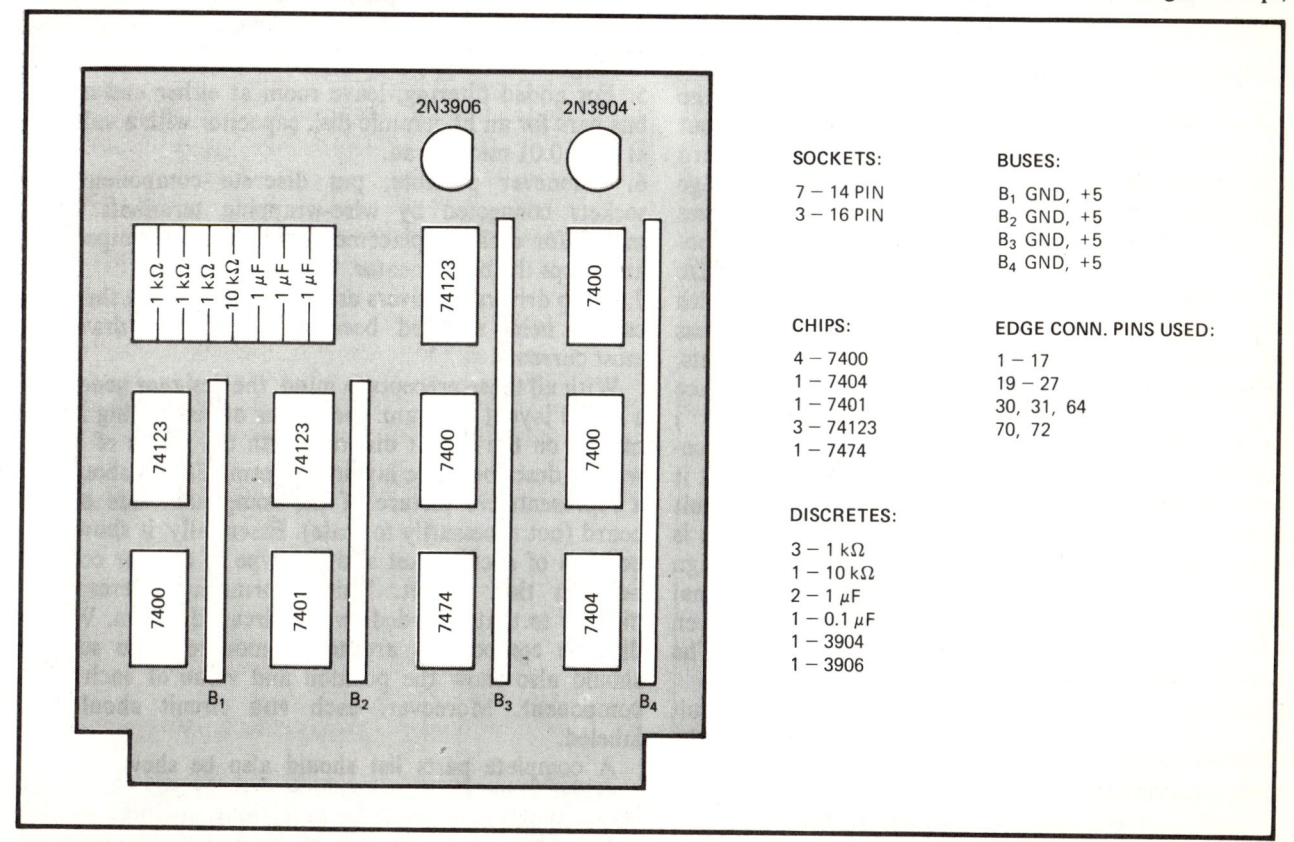

SOCKETS:
7 – 14 PIN
3 – 16 PIN

CHIPS:
4 – 7400
1 – 7404
1 – 7401
3 – 74123
1 – 7474

DISCRETES:
3 – 1 kΩ
1 – 10 kΩ
2 – 1 µF
1 – 0.1 µF
1 – 3904
1 – 3906

BUSES:
B_1 GND, +5
B_2 GND, +5
B_3 GND, +5
B_4 GND, +5

EDGE CONN. PINS USED:
1 – 17
19 – 27
30, 31, 64
70, 72

4. Board layout. In this construction method for breadboards, sockets are arranged in rows or columns, with a PDE for each. Every PDE has its own separate ground and power returns to the edge connector's plated-on ground- and power-connector fingers.

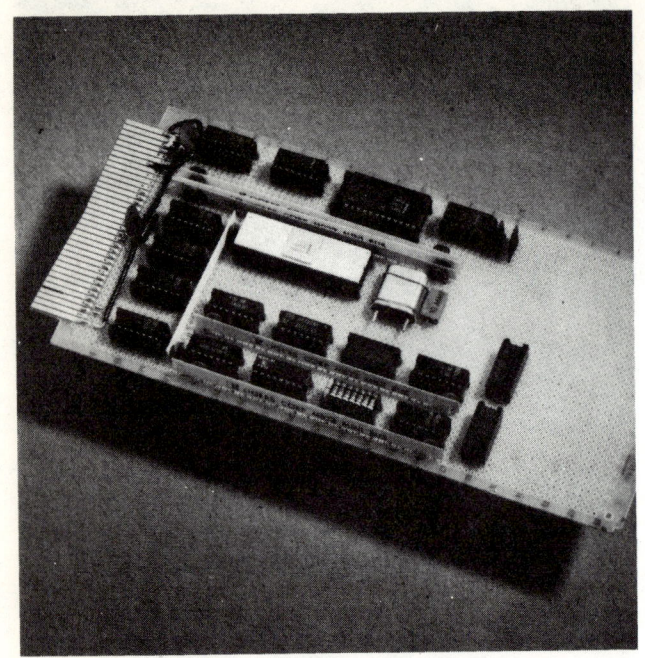

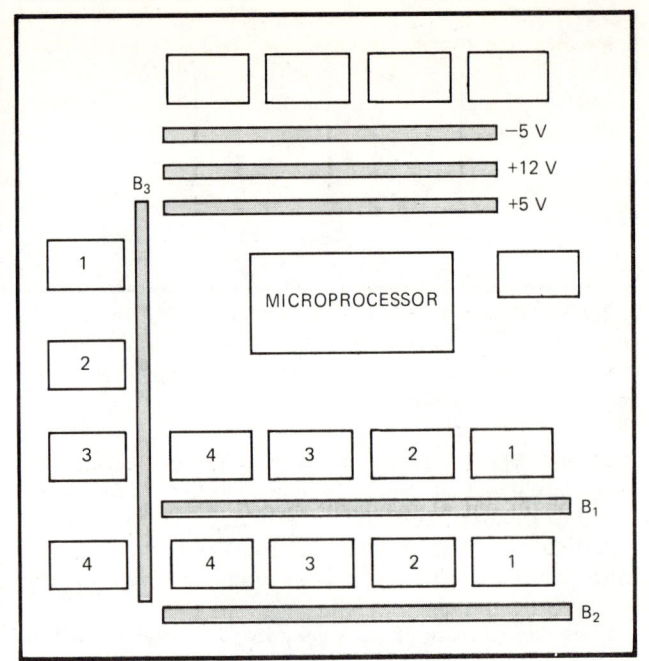

5. Microprocessor. This board (left) has an 8080A CPU plus interface circuitry. It has a mix of TTL and MOS circuitry. Layout (right) must account for three power supplies rather than one as in a memory card, and this prevents a symmetrical layout.

Rogers, Conn.; Eldre Components Inc., Rochester, N. Y., and Bussco Engineering Inc., El Segundo, Calif. The units come in vertical configurations, as in Fig. 1, and horizontal configurations with pins at 90° angles to the edge so that the bar will lie flat on the board. The type in the circuits described in this article is the Rogers M-823, a vertically mounted, two-conductor PDE.

Breadboarding with the PDE

The main component used in the development of the breadboard is a simple glass-epoxy pc board without printed power or ground planes. This inexpensive board has holes predrilled on 0.1-in. centers and 72 edge contacts. It accommodates a mixture of different sizes and families of ICs, in association with discrete components. Standard wire-wrappable sockets ranging in size from 14 to 40 pins are inserted into the board along with standard 9-in.-long PDEs with solder tabs, costing less than $1 each. When combined with discrete components, the result is a breadboard such as the 8080 interface circuit of Fig. 3.

A specific example can best demonstrate the procedures involved in layout of a typical breadboard. But it should be clear that these procedures apply to any circuit design. The goal is to produce a buildable board that is immediately operable (assuming no engineering design errors) in a minimum amount of time, after minimal effort, and at the least possible expense. There are seven specific layout considerations in this procedure. The breadboard builder should:

1. Generate a complete parts list from the circuit diagram so that every part that must appear on the breadboard is known.
2. Note any relationships between IC functions that might make wiring easier. For example, consolidate gates electrically interconnected on the circuit diagram into as few packages as possible, and then locate the packages as close to each other as possible.
3. To get a systematic grounding and power-distribution system, arrange sockets in rows or columns, with a separate PDE for each row or column. Segregate all transistor-transistor-logic and metal-oxide-semiconductor devices by family types into these rows and columns.
4. To prevent ground loops, make power and ground connections to each PDE at one end only and as close to the edge contacts as possible.
5. For added filtering, leave room at either end of the bus bars for an hf ceramic disk capacitor with a value of at least 0.01 microfarad.
6. Whenever possible, put discrete components in sockets connected by wire-wrapping terminals. This makes for easier replacement of defective components and keeps the board neater.
7. Keep drivers/receivers as close as possible to the edge contact being serviced, because these devices draw the most current.

With all these precepts in mind, the designer generates a board layout diagram: the means of correlating information on the circuit diagram with the pieces of hardware it describes. The layout diagram (Fig. 4) should be a representative picture of the component side of the board (not necessarily to scale). Essentially, it shows the position of each socket and the type of chip or component in that socket. This information corresponds directly to that recorded on the circuit diagram. Where discrete components are to be mounted, the sockets should also show the position and value of each such component. Moreover, each PDE circuit should be labeled.

A complete parts list should also be shown on the layout diagram to describe everything mounted on the board. This list should include the type and quantity of chips, sockets, and discrete components needed, all PDEs and the voltages they carry, and a list of all edge

6. Memory. The memory board (below) is a 2-k-by-8-bit static memory using 16 Intel 2102 MOS random-access memories and TTL control logic. The board is divided into families of logic types (right) each having its own PDE for distributing +5 V and ground.

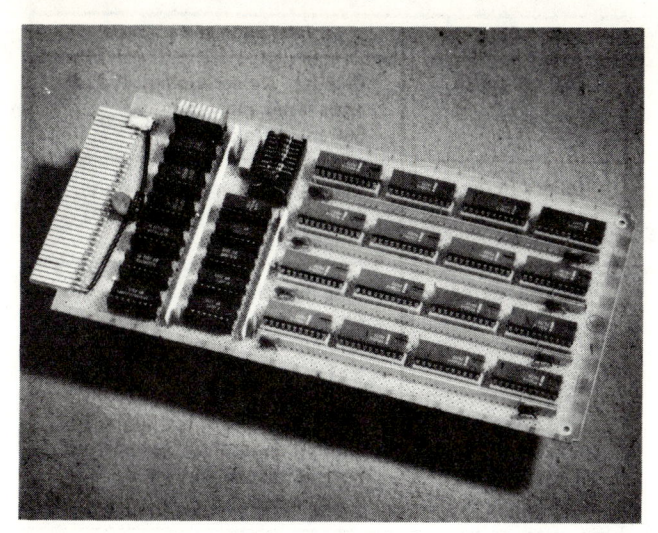

connection pins required. The board layout diagram, together with the circuit schematic, contains all of the information necessary to generate the signal-path wire-wrapping list.

With the board diagram in hand, even an unskilled worker can construct the breadboard. He or she should take the following steps, in order.

First, position all sockets as shown on the layout diagram and secure them to the board with adhesive. Discrete components not in sockets should be attached to the board with a soluble adhesive for easy removal.

Next, solder wire-wrappable pins in place only at those edge contacts that are to be used. Each edge contact carrying the same supply voltage should be positioned and tied together at the edge connector. To bring a good ground to the board, connect three adjacent edge pins and their opposites. All other power supply connections may be made on one edge pin and its opposite connected together.

To cut noise, decouple each power supply voltage brought on to the board edge contacts. In some cases, this is the only decoupling necessary. The recomended decoupling values are 0.1 μF for an hf ceramic-disk capacitor paralleling a tantalum capacitor of 20 to 30 μF.

Then make connections to PDEs. Define one side of the bus bar as ground and the other as a supply voltage. Connect power and ground to the ICs composing the row or column to which the PDE is assigned. Connections are made by soldering a wire to the tabs on the bus bar and by wire-wrapping to appropriate IC sockets. Thirty-gauge wire can be used for wire-wrapped connections. With all the sockets connected to their appropriate bus bars, connect each PDE to its appropriate ground and power supply at the board's input edge contacts.

To avoid couplings between the PDEs, which would increase the noise level, each bus bar should have a separate ground and power connection, with both connected at the same end of and as close to the edge contacts as possible. Minimum wire size for these connections should be 24-gauge multistranded. Of course, the wire gauge should become heavier as the number of devices hung off the PDE increases.

To illustrate the noise-reduction properties of this construction method, a microcomputer system composed of an 8-bit CPU and a 2,048-word-by-8-bit MOS static memory was built. The cards used a mixture of MOS logic, TTL and Schottky-TTL along with various PDE layout configurations. A special test program was designed so that noise would be injected into the power and ground lines of the two-board microcomputer system. The resulting noise signals were monitored at test points on the various bus bars.

Cutting noise

The microprocessor board (Fig. 5) contains an Intel 8080A CPU, a system clock, and interface circuitry between the processor and other modules in the overall system. A variety of power supplies (+5 volts, −12 v, −5 v) and dual in-line packages (14-, 16-, 22-, and 28-pin) are used, together with a heavy proportion of random logic. Of the 14 packages, 12 are small-scale integration and 2 are LSI. As the source of the control, data, and address variables that activate the memory board, the CPU consists of a mix of MOS and TTL.

The 2-k-by-8 memory board (Fig. 6) has an LSI array of 16 Intel 2102 MOS random-access memories and an address-decoding and control section of SSI 7400 series TTL devices. The memory matrix is a 4-by-4 array and is logically organized so that the lower 2-by-4 array is the first 1,024 8-bit words of system memory, while the upper 2-by-4 array is the next 1-k-by-8-bit portion of system memory.

Each column on the board has its own PDE for distributing +5 v and ground, so switching noise is isolated between columns. In addition, the logic organization of the memory matrix is such that noise behavior for each column is essentially the same. The 8-bit word is distributed across the columns so that exactly two devices per

137

TABLE: EXPERIMENTAL TEST DATA

Noise in MOS memory matrix

Instruction	Test point	Decoupling	Noise observed	Explanation and analysis
Read / 18 ms	B_1, IC_1	None	860 mV pk/pk	Case (a). No decoupling. Typical noise levels approximately 860 mV pk/pk.
Halt	B_1, IC_1	None	68 mV pk/pk	Case (a). No decoupling. Only system clock and some support functions operate.
Write / 18 ms	B_3, IC_{15}	0.01 μF	460 mV pk/pk	Case (b). Noise at decoupled supply node of PDE. Note noise is reduced by approximately 2:1 over case (a).
Write / 18 ms	B_3, IC_3	0.01 μF	520 mV pk/pk	Case (b). Noise at end of the PDE without decoupling. As expected, noise is less (460 mV) at decoupled node.
Write / 18 ms	B_4, IC_{16}	0.1 μF	440 mV pk/pk	Case (b). Decoupling values were increased from 0.01 μF to 0.1 μF. It was concluded that 0.01 μF decoupling capacitors were adequate in this breadboard condition.
Write / 18 ms	—	0.01 μF	360 mV pk/pk	Case (c). Decoupling of 0.01 μF at both ends of the PDE reduced noise level to 360 mV pk/pk.

Noise in TTL memory control logic

Instruction	Test point	Decoupling	Noise observed	Explanation and analysis
Write / 18 ms	B_5, IC_2	None	1.2 V pk/pk	Case (a). No decoupling. Typical noise levels were 1.2 V pk/pk.
Write / 18 ms	B_5, IC_1	0.01 μF	440 mV pk/pk	Case (b). Noise at the decoupled supply node of the PDE. Note noise is reduced by about 3:1 over case (a).
Write / 18 ms	B_5, IC_5	0.01 μF	740 mV pk/pk	Case (b). Noise at end of the PDE without decoupling. Again, noise is greater (740 mV) than at the decoupled end of the PDE (440 mV).
Write / 18 ms	B_5, IC_1	0.01 μF	300 mV pk/pk	Case (c). With decoupling at both ends of PDE, noise is typically reduced by about 4:1 over case (a).

Noise in microprocessor

Instruction	Test point	Decoupling	Noise observed	Explanation and analysis
Write /18 ms	—	None	940 mV pk/pk	Case (a). No decoupling.
Write / 18 ms	—	0.01 μF	400 mV pk/pk	Case (c). With decoupling at both ends of the PDE, noise was reduced typically by 2:1 over case (a).

column will be active at any given time. A maximum of two devices will be switching and injecting noise into the power- and ground-distribution system at any one time.

The layout of this board (Fig. 6b) is much more orderly than the microprocessor layout (Fig. 5b). However, both boards follow the layout rules enunciated earlier, with resulting low noise characteristics for both. Timing of noise appearing in the system will be a function of two variables: the program being executed and system clock rate, which specifies the rate at which a program is sequenced.

Microcomputer testing

Two instructions, load accumulator and jump, were used in a test program to cause internally generated switching noise to appear on the system's power and ground lines. The program controls the characteristics of the noise to cause generation of worst-case noise. It switches 11 address variables simultaneously, along with almost all of the devices on the memory board—putting maximum hf noise into both power and ground lines of the PDEs.

Noise tests on the memory board were divided between the transistor-transistor and MOS logic families, each operating under three power and ground noise-distribution conditions: (a) PDEs without decoupling capacitors; (b) PDEs with $0.01\text{-}\mu\text{F}$ decoupling at one end; and (c) PDEs with $0.01\text{-}\mu\text{F}$ decoupling at both ends.

The configuration in (a) provides a basis of comparison for determining the need for and amount of decoupling capacitors. Noise generated in the columns of the matrix and rows of control logic was measured with the CPU exercising the memory through a read/write loop.

With configuration (b), power and ground from the main power supply were individually connected to the bottom of each PDE. Noise flows toward power supply connections where it is dissipated, since this is the lowest-impedance path back to the main supply. For this reason, $0.01\text{-}\mu\text{F}$ decoupling capacitors were added to the bottom (connection end) of each PDE, and noise measurements were taken as in configuration (a).

Noise may be generated at any point along a bus bar and is initiated at any PDE/IC node. Since the noise power is dissipated as the spike propagates from its point of origin along the PDE in either direction, decoupling at both ends—configuration (c)—should provide some improvement over decoupling at one end. So noise measurements were obtained as in other two cases.

Noise tests on the microprocessor board were carried very much as they were on the memory board. The table shows the test data obtained on the memory board (for both the MOS memory matrix and the TTL control logic) and on the microprocessor. The tests were performed on +5-V and ground lines at points referenced in the table by bus and IC numbers.

As the table shows, decoupling both ends of a PDE proved to be the most effective method of noise reduction for both boards. Noise reduction from 2:1 to 4:1 over the configurations without decoupling were obtained. Decoupling only at the supply end gave some improvement, but not as much as with double decoupling.

Finally, a test was performed to determine the effect

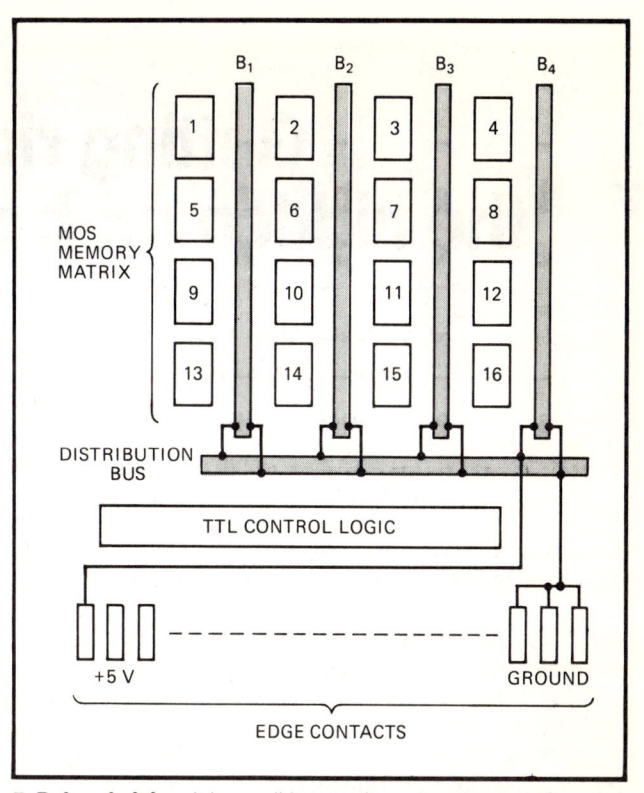

7. Daisy chaining. It is possible to replace all separate PDE ground and power returns with one distribution bus on the memory board of the microcomputer. However, this approach actually increases system noise levels by introducing extra coupling between PDEs.

of daisy chaining: replacing the individual power and ground connections to each bus bar with one pair of wires. A memory board was configured with a PDE used as a distribution plane supplying all columns in the matrix (Fig. 7).

As predicted, noise in the matrix is greater than in the configuration without daisy chaining, because the coupling between the bus bars is closer and the impedance at the junctions of the distribution bus and the PDEs is higher. A comparison of the measured noise voltages with those in the table shows that daisy chaining the power and ground connections increases noise about 33% at the junctions and 60% at the top of the PDEs.

To provide a comparison, a commercial wire-wrappable microcomputer board was tested in the same manner as the microcomputer breadboards. The commercial board's arrangement is such that power and ground are distributed on pc wiring, and a $0.01\text{-}\mu\text{F}$ decoupling capacitor is used with every chip. Signal interconnections are wire wrapped.

Measured noise voltage at power and ground points on this board range from about 350 millivolts peak to peak to 800 mv pk-pk. Of course, there are 25 decoupling capacitors, compared to the 10 when one PDE per column is used on the breadboards. Even with all these added capacitors, the noise characteristics were worse than with the equivalent breadboard. Wire-wrapped boards with PDEs appear to offer clear design and cost advantages over commercial wire-wrapped boards with etched power- and ground-distribution systems and "brute force" decoupling. □

Getting rid of hook: the hidden pc-board capacitance

Those previously unexplained discrepancies in circuitry output can be traced and measured and then designed out

by Wallace Doeling and William Mark, Tektronix Inc., Beaverton, Ore.
and Thomas Tadewald and Paul Reichenbacher, UOP Inc., Norplex Division, LaCrosse, Wis.

☐ Lurking beneath the apparently stable surface of many printed-circuit boards is a mysterious and relatively unknown phenomenon: hook. What this seemingly capricious and thoroughly confusing effect can do to an electronic circuit is a design tragedy: apparently well-designed high-impedance circuitry performs below the specifications that both theory and experience firmly establish as reasonable claims. Even perfectly designed and assembled test instruments read incorrectly.

Hook may be defined as the effect on a signal's voltage caused by a change in pc-board capacitance with frequency. Board capacitance is created between pc-board conductors separated by dielectric material. It can change the response time of a square wave and can bring about erroneous responses at certain frequencies of sine waves.

In an effort to lay bare the mysteries of hook, Tektronix and Norplex undertook investigations of the phenomenon. The results tell a great deal about the causes and the effects of this hidden menace. They also make it imperative that designers no longer ignore the electrical parameters of the laminates onto which their brainchildrens' components and printed wiring go.

To that end, this article will discuss the nature of hook and will look at its causes and methods of measurement. It will show how precise measurements are clearing the way to understanding the variables that bring about hook, and it will suggest what both the users and the makers of the laminates can do.

Variance with frequency

The investigations have discovered that the board capacitance that brings about hook varies inversely with frequency, up to about 10 kilohertz. From 10 to 100 kHz, it appears to be relatively flat.

Apparently, above certain frequencies the molecular orientation and dipole-to-dipole alignment of the polymers used for pc boards do not change rapidly enough to cause variations in board capacitance. Below these

1. Hook. These waveforms are at the outputs of three identical high-impedance networks, each mounted on its own pc board. One board (a) has low hook, a second (b) has moderate hook, and the third (c) has a large amount of hook.

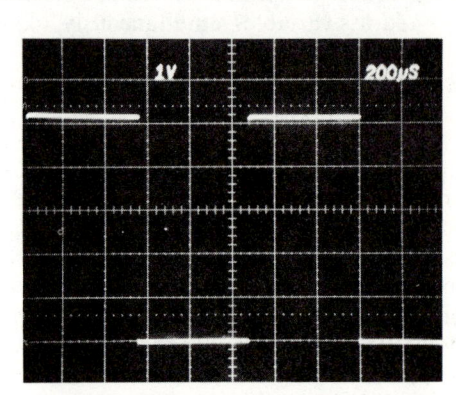

(a) LOW HOOK

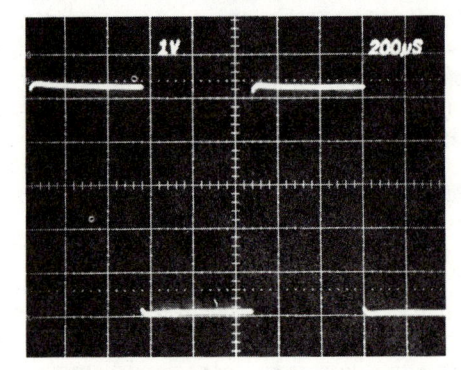

(b) MODERATE HOOK

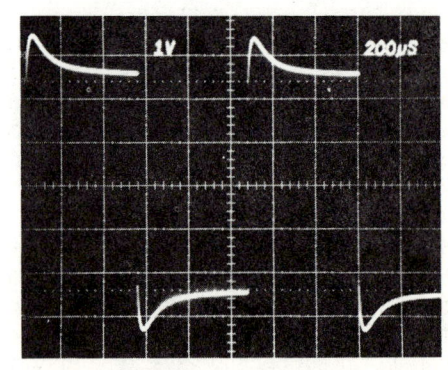

(c) LARGE AMOUNT OF HOOK

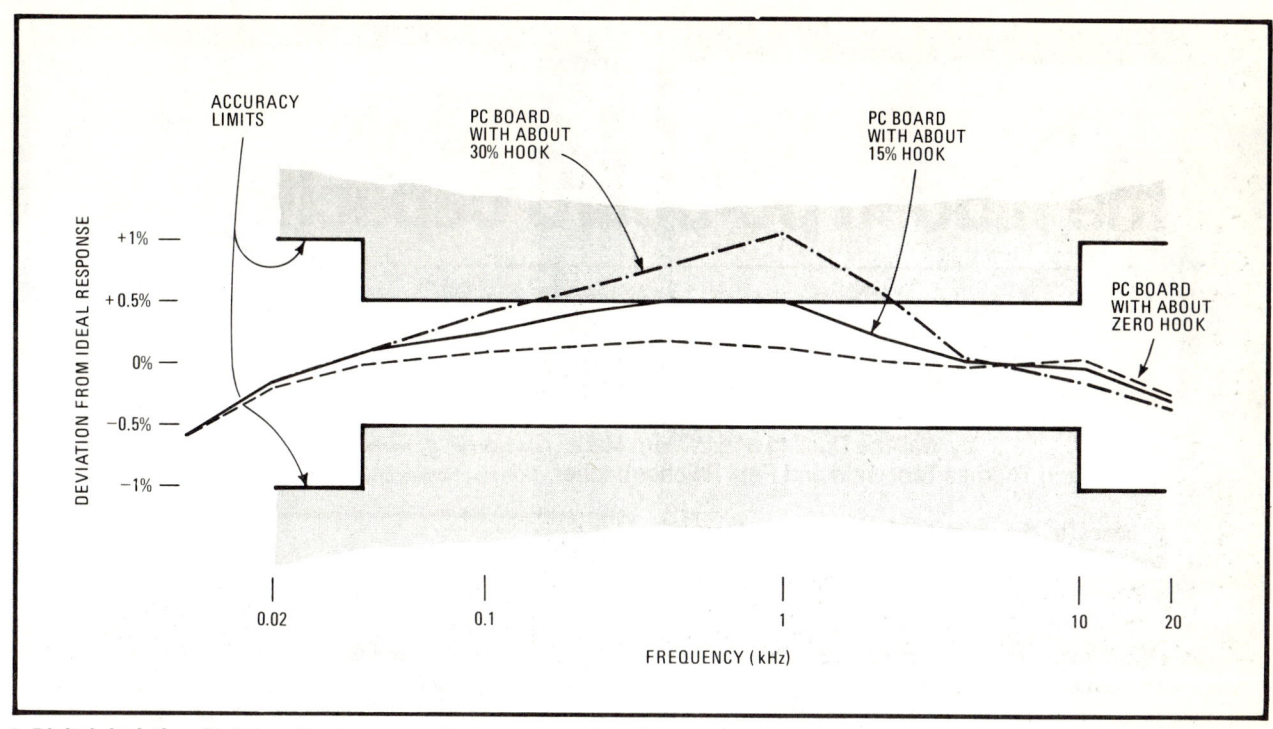

2. Digital deviation. Digital multimeters are particularly vulnerable to large and medium values of hook, which can result in out-of-tolerance deviation measurements. This plot was taken for an ac voltmeter, using 100:1 high-impedance attenuators.

frequencies, the various molecular components orient themselves at differing time constants. Hence, capacitance varies with frequency.

The variable capacitance compounds the circuit designer's difficulties, for example, in the construction of high-impedance attenuators for oscilloscopes and digital multimeters. These attenuators are networks of resistors and capacitors connected to provide a diminished output proportional to the input signal's amplitude for all frequencies within the bandwidth of the instrument. The most frequently used method of interconnecting the components is with a pc board. Hook prevents the attenuator from giving outputs proportional to the inputs for the frequencies of interest.

Hook in scopes and DMMs

Tektronix assigns "hook" to describe a particular type of distortion seen on a waveform displayed on an oscilloscope. Figure 1a shows the proper response of a scope's attenuator to an input square wave. The leading edge should be square, with no overshoot and no undershoot.

The waveforms of Figs. 1b and 1c come from attenuators on printed-circuit boards with a moderate and a large amount of hook, respectively, and leading edges are no longer square. The deviation from a square wave is measured as a percentage of the latter's full amplitude, with an acceptable level of distortion being less than 0.5% to 1.5%, depending on the oscilloscope's performance requirements.

In digital multimeters, hook raises its ugly head in the ac voltmeter section. Attenuators in these meters generally use higher-value resistors than those of an oscilloscope, making the circuits even more susceptible to the high impedance of the board capacitance. The ac voltmeter function of a DMM is most accurate when measuring undistorted sine waves. Its measurement accuracy is usually specified as ±X% of the reading within a given bandwidth. Hook distortion for this function is best viewed by plotting a graph of the deviation of the displayed reading from the true input-signal amplitude.

Effects on accuracy

The effects of various amounts of hook on three DMMs of the same type are shown in Fig. 2. An ideal response curve plotted in this graph would be a straight line following the 0% deviation axis. With two of the DMMs, accuracy limits were exceeded between 700 hertz and 1 kHz. Hook in the pc boards renders these two instruments unsalable.

Hook can affect other circuits as well. In general, it will affect those in which all of the following occur:
- Resistors have high values (between 500 kilohms and 1 megohm).
- Board capacitance is an appreciable portion of the total circuit capacitance.
- Frequencies of interest are below 10 kHz.
- Required accuracy is better than 2% to 5%.

For such circuits, a quick way to gauge the effect of hook is to see if the board capacitance conducts an appreciable part of the signal current at frequencies below the 1- to 10-kHz range. If so, look for improper circuit operation with frequency variation.

Hook is not restricted to pc-board materials. It has been observed in dielectric materials used in some types of switches and capacitors, as well as in the junction capacitance and reverse-biased diodes and in metal-oxide-semiconductor capacitors. The investigations leading to this article, however, have been limited to pc-board capacitance.

Anything that affects board capacitance will affect

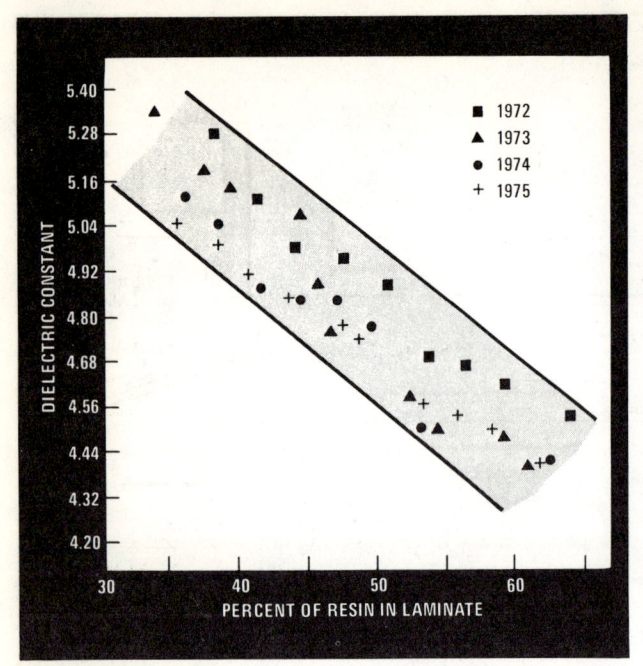

3. Dielectric constant. The dielectric constant decreases as the percentage of resin in the laminate increases. This data is from more than 600 samples of Norplex FR-4 laminates produced from 1972 to 1975, with resin contents varying from 30% to 75%.

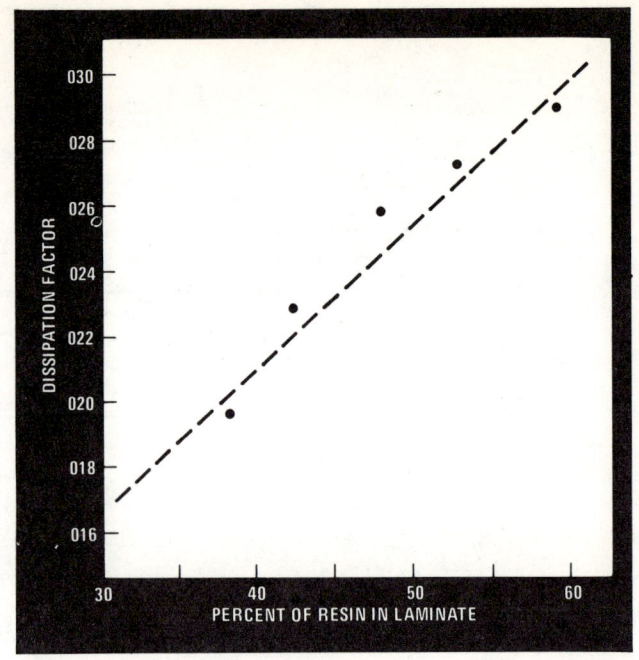

4. Inverse relationship. The dissipation factor of boards made of FR-4 goes up with resin content, unlike the dielectric constant. The variation of both the dielectric constant and the dissipation factor are related to the amount of resin in a laminate.

hook. Thus the board's dielectric constant and its dissipation factor were early objects of the investigation into hook. Inquiry centered on the effects of such factors as resin content upon these two parameters.

Previous work showed that the dielectric constant (ϵ) and the dissipation factor (ϵ^1) of FR-4 (flame-resistant epoxy-glass) laminates to be proportional to resin content; ϵ decreases while ϵ^1 increases with increasing resin content. With F_r = fraction of retained resin in a given laminate, then:

$\epsilon = 6.2 - 3\,F_r$
$\epsilon^1 = 0.0037 + 0.0435\,F_r$

The constants in these equations apply to Norplex FR-4 laminates produced during 1972–75 and having resin contents varying from about 30% to 65%. For humidity conditioning, each laminate (over 600 samples) was submerged in distilled water for 24 hours at 23°C per MIL SPEC 13949E (D-23/23) before testing at 1 megahertz. The regular variations shown in Figs. 3 and 4 are generally consistent with the equations above.

Measuring hook

Because hook is a change in capacitance with a change in frequency, it is necessary to measure board capacitance at various frequencies. An important aspect of the Tektronix/Norplex investigations was to devise an accurate measurement method. Several techniques were studied, with varying results.

The simplest method of measurement is with a capacitance bridge. It can measure low capacitance values, 1 to 50 picofarads, at frequencies ranging as high as between 100 kHz and 1 MHz and as low as 10 Hz. However, the stray capacitance of the bridge can affect the measurement of low-capacitance samples. Another problem: most capacitance bridges are single-frequency devices.

A second measurement method is with a high-impedance attenuator. Such an attenuator (Fig. 5) can be constructed with two capacitances: C_1 being the board capacitance to be tested and C_2 being adjusted to produce an output waveform complying as closely as possible with the input square wave. The deviation, in percentage, of the output waveform from the input square wave indicates the relative amount of hook.

Wide frequency range

Since the square wave is composed of many frequencies, this measurement approach covers a wide range of frequencies at one time. Moreover, this technique is simple and economical, since it requires only an oscilloscope and a square-wave generator.

However, the attenuator method provides no direct reading of the quantity of hook (the change in capacitance), and stray capacitance in parallel may swamp out hook for low values of C_1 (less than 5 to 25 pF). At low frequencies (just where one most wants to measure hook), the 1-megohm resistors become the dominant elements of the circuit: thus the capacitance effect drops out of the picture.

The most exact way to measure hook is with the charge-amplifier circuit (Fig. 6). This is an operational amplifier with extremely low input bias currents—less than 0.1 picoampere is desirable, and 1 pA is maximum. This precision circuit uses capacitors as both the series-input and feedback impedances for setting the amplifier's closed-loop gain.

The output signal of the charge amplifier is a function of three parameters: the input signal; the fixed capacitance, C_f; and the sample's capacitance, C_i.

The relationship between these components is:

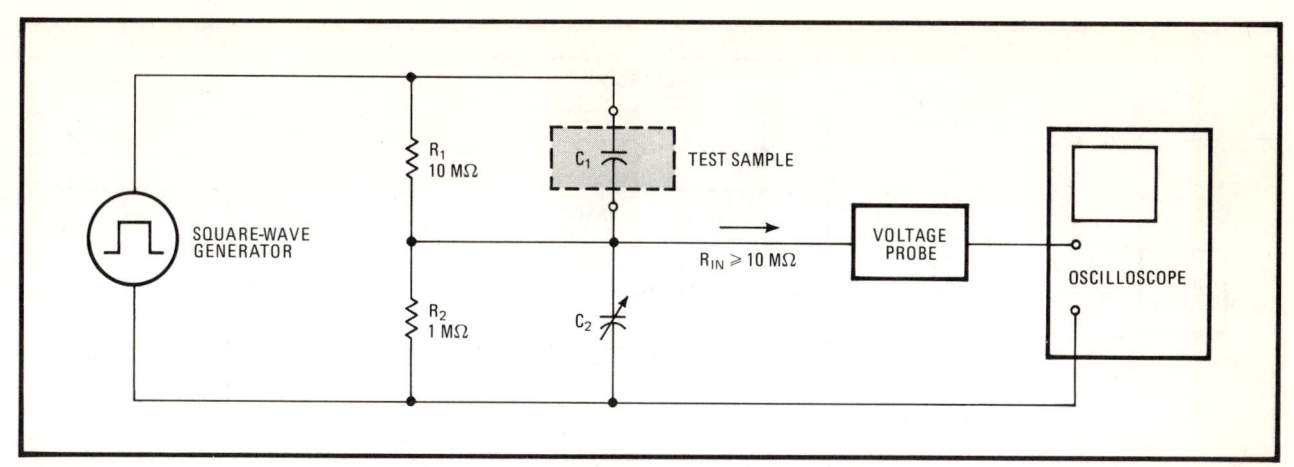

5. Playing hooky. One of the earlier methods of measuring hook uses a high-impedance attenuator with circuit-board capacitance as an element. However, a drawback is that the circuit is incapable of giving a direct reading of circuit-board capacitance.

TABLE 1: LAMINATE HOOK AT 10 Hz					
Type	Resin	Reinforcement	% hook	
^	^	^	Before humidity conditioning	After humidity conditioning
FR-3	FR – epoxy	paper	14	large
CEM-1	FR – epoxy	paper/glass	5	10
FR-4*	FR – epoxy	glass	4	9
G-11	high-temperature epoxy	glass	3	9
Polyimide	polyimide	glass	4	6
N-3	phenolic	nylon	9	14

*Average of 12 FR-4 laminate variations

$$V_{out} = \frac{C_i}{C_f}(V_{in})$$

If C_f is a known-accurate, high-quality capacitor exhibiting no hook, and if the characteristics and accuracy of V_{in} are known, then the characteristics of V_{out} can only be influenced by C_i.

The signal generator used can be either a sine-wave or a square-wave generator. A sine-wave generator allows capacitance measurement at any given frequency within the bandwidth of the system. A square-wave source allows a dynamic capacitance measurement at all frequencies within the spectrum of the square wave at one time; that is, it shows how the hook capacitance changes with time.

Actual measurements

With an acceptable measurement technique established, the evaluation of materials can start. For testing of laminates, two-sided boards with plated-through holes allow hook to be measured along three orthogonal axes. While X, Y, and Z values of hook often show differences for a given laminate, only the effect of laminate type on the average hook value will be presented here.

When measuring a good capacitor (one with no hook), the output signal can be expected to look exactly like the input from the signal generator. NPO ceramic and polypropylene capacitors (units with dielectrics having extremely low dissipation factors) can be inserted as standard C_is in the circuit of Fig. 6 to develop a reference waveform. Any distortion caused by the signal generator, oscilloscope, or charge amp will be seen in the output waveform associated with the standard C_i, allowing for compensation in any future measurements.

Figure 7 shows capacitance vs time measurements taken on a charge-amp circuit at Tektronix on FR-4 circuit-board materials that have low hook, moderate hook, and a large mount of hook. Hook is measured as a percentage given by the distance the waveform's leading edge has dropped divided by its peak-to-peak amplitude. Circuit-board materials measured are the same materials used in the attenuators that have their responses shown in Fig. 1. However in this case, capacitance is directly displayed.

The waveforms display capacitance vertically. The capacitance in Fig. 7a changes very little after the initial step, compared to the larger changes in Fig. 7b and 7c. In fact, the latter capacitances would continue to increase with a waveform of longer period.

Increase in capacitance

A measurement of board capacitance with a variable sinusoidal generator as the signal source of a charge-amplifier measurement circuit is plotted in Fig. 8.

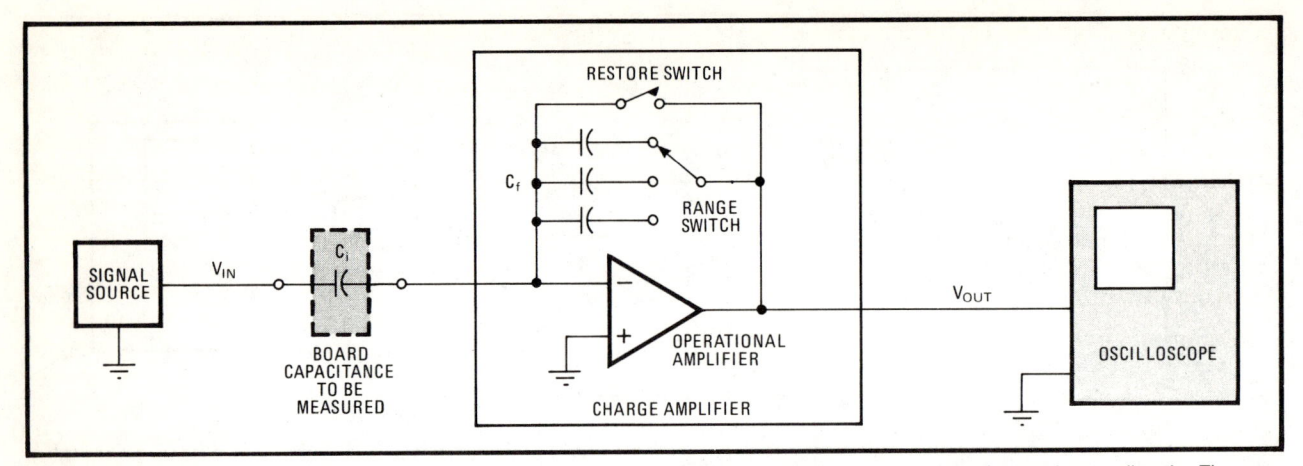

6. Charge amplifier. An operational amplifier with capacitors as feedback elements can measure board capacitance directly. The wave shape of the signal can be either a sine or a square wave. A square-wave drive results in a display of hook capacitance vs time.

Board capacitance was measured at various frequencies from 1 Hz to 100 kHz. The graph shows that, for frequencies below 1 kHz, the change in capacitance rapidly increases.

Capacitance at still lower frequencies was not measured because of equipment limitations. Presumably it would continue to increase for frequencies somewhat below 1 Hz before reaching a limit.

At Norplex, measurements were made on 12 different types of FR-4 laminate, plus five other printed-circuit laminates. Hook was measured with a 10-Hz square wave before and after humidity conditioning. Results are given in Table 1, along with the composition of the various laminates.

Moisture absorption results in greater hook, and, for a given resin system, paper reinforcement is worse than glass fiber (see FR-3 and FR-4 in the table). These observations may be related: paper-based laminates absorb more moisture than those with glass-fiber bases.

Table 1 also shows that other resin systems, such as G-11 and polyimide, can be used to produce low-hook laminates. The N-3 nylon-phenolic laminate was an early attempt to study some less traditional systems.

Other factors

The Tektronix/Norplex investigations established that laminate construction and processing may affect hook as much as the materials used, as Table 2 shows. It offers data on the 12 different FR-4 laminate samples, each made somewhat differently. While hook varies substantially from sample to sample, it is possible to produce an FR-4 laminate with low hook, even after moisture

---

Advice for pc-board users

Current research into hook is beginning to lead to a good economical guarantee of relatively hook-free, FR-4 laminates for printed-circuit boards. But insufficient understanding of the cause of hook still precludes an accurate and reliable formula for the composition of hook-free material.

For instance, hook seems to be a batch- or lot-oriented problem. Years may pass, and several lots of material may be used with hook being low enough not to be objectionable. Then all at once a bad lot of material will be received and cause havoc. All vendors of G-10 or FR-4 laminate material used at Tektronix have supplied both good and bad lots of laminate.

There are, however, several methods that can reduce the effect of hook on circuits.

First, design the circuit and the layout of the printed-circuit board to minimize stray capacitance caused by pc interconnections. It may be necessary to mount all critical components on Teflon standoffs and to do point-to-point wiring to these posts rather than to points directly on the pc board. This keeps the circuit-board capacitance to a minimum, but, of course, drives manufacturing costs up.

Another problem: insufficiently cured laminate material can be noticeably hooky. Baking FR-4 material can be a remedy in some cases.

Another approach is to ask the laminate vendor to agree to a maximum hook specification. A few are now willing to address this requirement. Alternatively, this specification can be in the form of a limit on the change of dielectric constant of the material as measured at 100 kilohertz and at 10 hertz.

The Tektronix approach for FR-4 laminate is a combination of sampling of the incoming material by batch and measuring finished circuit boards for hook before component assembly. A hook test pattern, etched on samples of each batch of material received, checks all three axes. Use of the material in critical applications depends on the results of that test.

For these applications, a 100% check of the finished pc boards before component assembly provides further insurance against scrapping expensive finished boards because they do not meet performance specifications. The cost of a bad board found in an assembled instrument at the calibration stage can be tremendous.

Finally, one important step a laminate user can take is to explain his needs to the vendor. The vendor should realize that today's pc-board requirements go far beyond just supporting and interconnecting components. Today the dielectric properties of a circuit board are decidedly the most important.

absorption (see variation 12, for example).

Again, while high resin content throughout and on the surface of FR-4 laminates generally gave low hook results, the table indicates that other factors also affect laminate hook. It appears that resin-glass interactions are critical.

Hook measurements on laminates also have been made at 1 kHz and 100 kHz. Because of the tentative relation found between resin content and hook, correlations between surface resistance and hook were begun with new test patterns.

Not all variables have been explored in detail. The data was derived from measurements of hook at 10 Hz and at 1 kHz and surface (insulation) resistance at 500 v dc using a special board pattern in four different modes. Tests included four FR-4 types and two other laminates, each measured before and after moisture conditioning.

A note of caution: understanding data from experimental sets where many parameters affect the observed results often requires statistical interpretations to separate important from trivial parameters and to identify interactive parameters. The data bank is not complete enough to permit such rigorous interpretation; however, some trends do seem to be fairly apparent by now.

The data indicates that an increasing moisture content in laminates generally increases hook. The CEM (composite epoxy material) and polyolefin samples tested are not generally recognized laminate materials. The former is a variation of a CEM-3 construction (woven plus nonwoven glass fiber with FR-4 resin), and the polyolefin is Norplex's NZ-932 developmental laminate designed for certain specialized applications.

The results for the various FR-4 types reinforce a conclusion stated earlier. While moisture and material type influence hook, variations within a given laminate show that the process of fabrication also strongly influences hook.

While increased moisture content usually raises the hook of a given laminate, it also increases the material's dielectric constant and decreases resistance, even if no more than traces of ionic materials are present. Therefore, hook would be expected to decrease with increasing laminate resistance.

Test results generally confirm this expectation at each of the frequencies used for these surface hook measure-

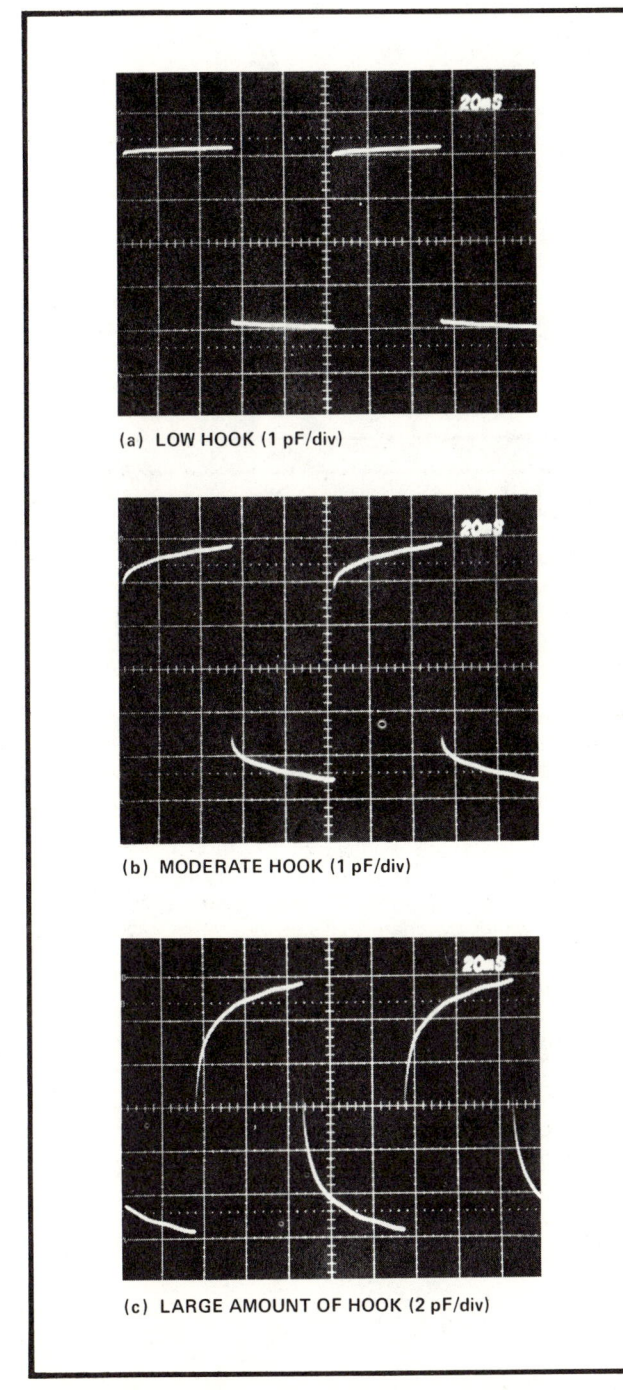

7. Capacitance vs time. Typical capacitance vs time measurements were taken on FR-4 laminates with a charge amplifier. The hook capacitance associated with moderate and large amounts of hook would continue to increase if a longer period were used.

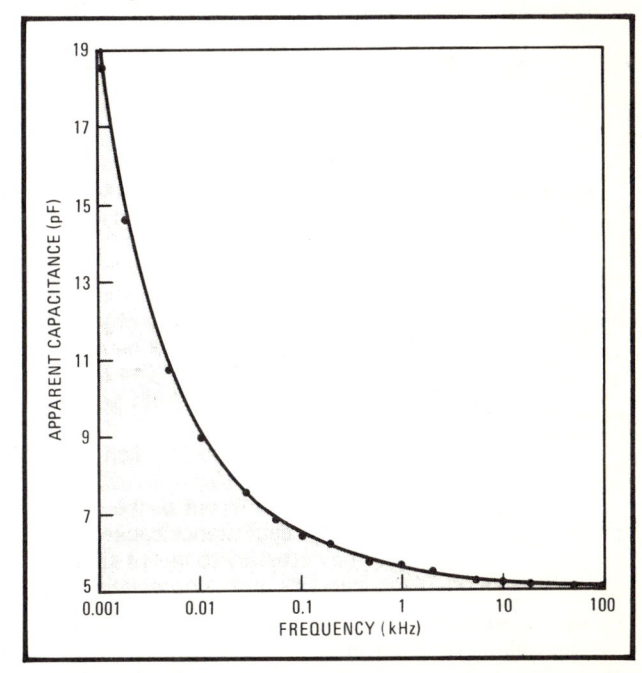

8. Sinusoidal drive. Driving the charge-amp capacitance-measuring circuit with a sine-wave generator produces a curve of this type for a board with considerable hook. Change in board capacitance is particularly rapid in the dc-to-10-kHz range.

TABLE 2: HOOK IN FR-4 VARIATIONS (10 Hz)

FR-4 variations	Resin content	Resin at surface	% hook Before humidity conditioning	% hook After humidity conditioning
1	low	low	5	12
2	high	low	5	6
3	low	low	5	12
4	low	low	4	11
5	low	high	4	11
6	high	high	4	11
7	high	high	4	8
8	high	high	4	7
9	low	high	3	11
10	high	high	3	7
11	high	high	3	7
12	high	low	2	4

What the laminate vendor can do

The present state of the art permits tests of hook in laminate material, which do in fact relate to end-use performance. However, evaluation of the role of laminate manufacturing techniques in the production of hook should be continued and expanded. This work would develop the necessary theory for the consistent production of laminates that exhibit minimal signal aberration over a wide range of operating conditions.

To assure uniform resin composition, incoming materials should be analyzed in a number of ways with an integral role assigned to chromatography, which separates and analyzes mixtures of chemical substances by differential absorption. This technique determines resin molecular distribution, which must be reproducible in order to maintain a variety of constant laminate electrical properties, including low hook.

Additionally, researchers at laminate manufacturers are using chromatographic processes to separate various fractions from the polymers. This activity will allow a better understanding of the hypothesis that molecular polarizability has a dominant effect on laminate dielectric properties such as hook.

The makers of printed-circuit boards primarily use epoxy-glass–supported laminates for most sophisticated applications, such as wide-band, high-speed, low-drift analog circuitry. Earlier, phenolic products chiefly were used, but the increasing need for electrical stability in the electronics industries is beyond the capabilities of the phenolics. There is some concern that as electronics design becomes more refined, the epoxy-glass system will also become inadequate. This is the main reason for continuing to look at other materials in the polymer world.

ments; however, the picture is not entirely clear. Some discrepancies in the results may be partly due to measuring surface resistance at 500 v dc while measuring hook using a 30-v square wave. It is not expected that surface resistance will depend on frequency or amplitude, but this deserves consideration in future inquiries.

More importantly, the laminate test results reinforce the notion that materials and the methods used for laminate construction are important in determining hook. The surface hook for the two experimental laminates, CEM and polyolefin, is especially sensitive to surface resistance changes.

In other words, moisture conditioning of these laminates caused only small changes in surface resistance (relative to most of the FR-4 laminates), while significant changes in surface hook were measured. It is clear that no simple mathematical relationship between laminate hook and resistance has yet been identified. While investigations into hook are well launched, there is a long voyage ahead until all the factors that produce hook are pinned down and a mathematical relationship relating these factors is derived. □

Part 4
Automatic Wiring Technology

Techniques of automatic wiring multiply

Wire-Wrap remains the most popular, but others offer helpful options

by Jerry Lyman, *Packaging and Production Editor*

☐ Automatic wiring originated with the large backplane. This metal or plastic panel, with its thousands of wiring pins on one side and its crowd of printed-circuit-board or cable connectors on the other, turned up first in telephone exchanges and soon spread to computers. But almost immediately, it proved impossible to manually wire one of any size at a cost-effective rate and without excessive errors. Only an automated form of wiring could handle a large backplane.

Automatic wiring today takes many forms, from the well-known Wire-Wrap and Termi-Point through Multiwire, Stitch Wiring, and all the others listed in the table. While still used for backplane wiring, it has also become very popular for wiring panels of integrated-circuit sockets, which have wiring tails that lend themselves to some form of the process.

The techniques in the table, more or less without

SUMMARY OF DISCRETE WIRING TECHNIQUES

Type	Description	Kind of connection	Wire	Applications
Wire-Wrap	Mechanically wraps stripped insulated wire in helical coil about a rectangular post	Redundant gas-tight pressure connection to a copper- or gold-plated rectangular pin typically 0.025 in.^{2} with No. 30 AWG wire; 0.045 in.^{2} posts are used with No. 24 and 26 AWG wire	Teflon, polysulfone-and-polyester-insulated silver-plated copper, No. 24 through 30 AWG (most commonly No. 30 AWG)	Daughterboards, backplanes, and packaging panels
Termi-Point	Connects a solid or stranded wire to a rectangular post by means of a metal clip; 0.031-by-0.062-in. post used with No. 22 through 28 AWG; 0.022-by-0.036-in. post used with No. 28 through 30 AWG	Gas-tight connection at the wire-post interface by means of a spring-loaded clip	T or TE type Teflon-insulated solid or seven-strand wire, No. 22 to 30 AWG	Backplanes and packaging panels
Multiwire	Bonds continuous wire to an adhesive substrate and terminates it to a plated-through hole	Electroless copper-plated connection	Polyamide-insulated, solid No. 34 AWG, 6.3-mil overall diameter	Packaging panels, daughterboards, and backplanes
Stitch Wiring	Bonds continuous wire with insulation displacement	Resistance-weld (diffusion-bond) connection to a stainless-steel surface	FEP and TFE, Teflon-insulated nickel, typically No. 30 AWG solid	Low-profile daughterboards and backplanes
Tiers	Reflow-solders wire through insulation to a solder-coated pad on a printed-wiring board	Solder connection	Low-temperature, heat-strippable copper wire, No. 28 through 36 AWG	Packaging panels and backplanes
Solder-Wrap	Employs a three-stage routing, soldering, and cutting process	Solder connection to component lead	Polyurethane-coated, No. 38 AWG solid copper wire	Packaging panels, daughterboards, and backplanes
Quick Connect	Uses continuous wire with insulation displacement	Gas-tight pressure connection	Typically No. 30 AWG solid copper wire with any nonbonded insulation	Backplanes and packaging panels
U-Contact	Continuous run, insulation displacement between eight contact strips soldered to plated-through holes	Gas-tight, press-fit connection to beryllium-copper contacts with gold-over-nickel plating	Polyvinylidene-fluoride-insulated No. 30 AWG solid wire	Prototype and short-run production

exception, are low-cost, fast, and dense, and facilitate prototyping. Perhaps more important, they all share two big advantages over their main competitor, the multilayer printed-circuit board. First, the computer base used to produce the wiring software can also be made to produce the software needed for automatic continuity testing and for automatic component insertion. Second, changes in a wiring program are easy to put into effect, whereas a modified multilayer pc board can take weeks to deliver.

All wrapped up

Wire-Wrap, the firstcomer in the automatic wiring field, was invented in the 1950s by the Gardner-Denver Co., Grand Haven, Mich. According to Gnostic Concepts Inc. of Menlo Park, Calif., it accounted for over 90% of all automatic wiring in 1976, though the figure could shrink to 70% by 1983. In this process, a special automated (or manual) wiring tool wraps the stripped end of a solid wire five to seven times around a square post forming a gas-tight bond.

Wire-Wrap is the most complete wiring technique at present available. Machines and wiring software, backplanes and IC socket panels, sockets, pins, and test fixtures are all available to convert logic diagrams or wiring lists into finished boards.

The process is heavily used in the telecommunications, instrumentation, computer, military, and industrial fields. It is also popular for prototyping digital circuitry. Its advantages over the multilayer board are high packaging density, easy repairability, a high wiring rate, and low manufacturing costs. Its chief disadvantages are a relatively high profile due to the length of its pins and poor high-frequency performance unless specially designed panels are used.

In 1961, AMP Inc. of Harrisburg, Pa., came out with a competitor to Wire-Wrap. Called Termi-Point, it also employs a gas-tight connection, but one made by sliding a tiny spring-metal clip over the end of a post to press the stripped end of a piece of wire against the post. The spring clip maintains a high-pressure, gas-tight area of contact between the conductor and the post. Up to three connections may be applied to the post either semiautomatically or fully automatically, or even manually.

Termi-Point, which is available only from AMP, is strong in aircraft and avionics because of its three main advantages over Wire-Wrap: it is relatively easy to repair, can handle stranded as well as solid wire, and can accept a twisted pair. But unfortunately, it suffers from the same deficiencies as Wire-Wrap—a high profile and poor high-frequency performance. This situation encouraged the entry of other wiring technologies seeking to supply the missing requirements.

Close to the board

One of the most heavily used of the new low-profile wiring methods is Multiwire. Multiwire was developed during 1970–1971 by the Photocircuits division of Kollmorgen Corp., Glen Cove, N. Y.

The first step in Multiwire is to etch power and ground planes from a copper-clad glass-epoxy board. The next is to coat the board with adhesive. Then a numerically controlled machine lays down a customized pattern of insulated wires in the adhesive, and the board is heat-treated to cure the adhesive. Finally, holes are drilled through wire and board and plated with electroless copper to connect the two.

Unlike Wire-Wrap with its multiple sources, Multiwire is really available only through Photocircuits and one other source, Diva Inc. of Eatontown, N. J. The latter firm multiwires its own disk-storage units but will supply custom boards made on its Multiwire machines. Photocircuits now has three design centers at various loca-

Tooling	Wiring rates (connections per hour)
Manual	50 – 80
Semiautomatic: numerically controlled terminal locating; manual wire terminating	175
Automatic: automatic terminal locating, wire feeding, stripping, and wrapping	900 – 1,000
Automatic: numerically controlled terminal locating; automatic cutting, stripping and terminating	850
Manual: manual termination tool	180
Computerized automatic wire routing, employing conventional printed-wiring technology for power and ground-plane circuit formation	800 from/to wires/panel; runs four panels simultaneously
Semiautomatic: numerically controlled terminal locating; manual wire routing and bonding	450 – 500
Manual systems: manual terminal locating; manual welding	75 – 125
Automatic: numerically controlled routing, soldering, and cutting	375
Semiautomatic: manual routing; automated soldering and cutting	150
Automatic: numerically controlled terminal locating; automated wrapping, soldering, and cutting.	300/panel; runs four panels simultaneously
Manual tooling	100
Semiautomatic: numerically controlled terminal locating; automatic wire feeding; manual terminating	300 – 400
Manual tooling	100

Source: Gnostic Concepts Inc.

tions in the U.S. to produce Multiwire software. In addition, the division sells a complete Multiwire package, including the numerically controlled, four-wiring-head Multiwire machine, as part of its license to use its wiring technique.

Because of its excellent high-frequency performance and extremely low profile, this technique is now being used to wire computers, minicomputers, computer peripherals, and military gear. In most of these applications, Multiwire has displaced multilayer pc boards. Indeed, of all the techniques discussed here, it has shown the fastest growth. However, unlike all the others except Tiers, it is not suitable for breadboarding or even for small production runs. For instance, Stitch Wiring, which originated about the same time as Multiwire, is suitable for both manual prototyping and semiautomatic production.

A stitch in time

Stitch Wiring, as it first appeared commercially in 1970, was a semiautomatic system of point-to-point interconnections in which gold-plated steel pins were pressed into holes in a unclad printed-circuit substrate. Then a programmed wiring head fed and welded Teflon-insulated nickel wire to the matrix of pins.

Such a wiring system proved extremely reliable. For this reason, it was used on electronic equipment for many long-term deep-space missions, despite the fact that the steel pins add a lot of weight to the boards.

Other drawbacks of this form of Stitch Wiring

The world of automatic wiring

Automatic wiring takes many forms, five of which are illustrated here. Shown clockwise from top left are boards wired by means of Tiers, Stitch Wiring, Wire-Wrap, U-Contact (3M Co.'s prototype version of the insulation-piercing technique), and lastly Multiwire. In spite of pressure from its competitors, Wire-Wrap is likely to dominate this field for some time.

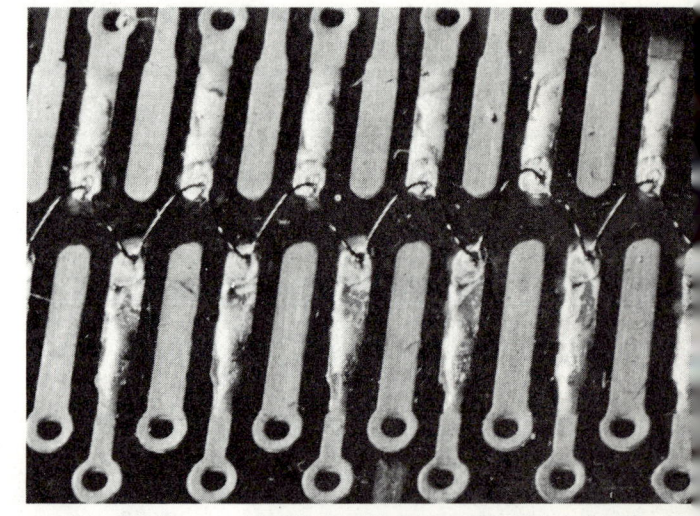

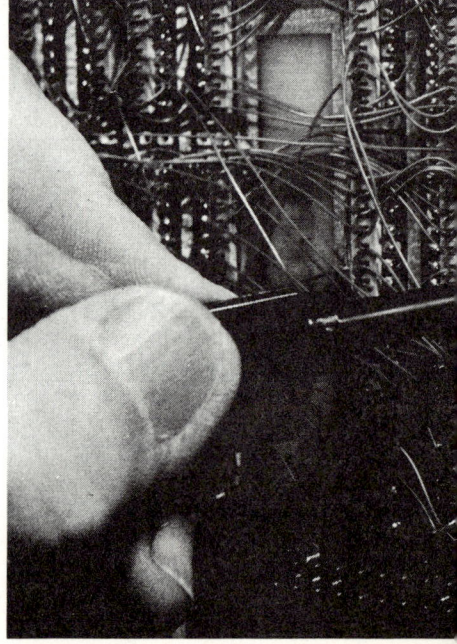

included a high profile—the pins used were 0.200 to 0.450 inch high. In addition, components and wiring had to be attached to opposite sides of the board, adding still more to its overall thickness. Finally, it was not possible to reflow-solder the components, which had instead to be welded to or plugged into the pins.

Within the last few years, Stitch Wiring has adopted a new type of board construction that eliminates the deficiencies of the older type. In the latest version, wire is welded to stainless-steel pads that have been etched from the stainless-steel foil cladding on a copper-plated, glass-epoxy board—sort of a stainless-steel board. These virtually planar boards are much lower in profile and lighter in weight than the earlier welded system and also can have components wave-soldered directly to them, which will cut production costs.

Unlike the basically single-source Multiwire system, the manual and semiautomatic Stitch Wiring machines and wiring services, as well as the special boards required, are available from at least five sources: Augat Inc. of Attleboro, Mass.; Interconnection Technology of Costa Mesa, Calif.; Moore Systems Inc. of Chatsworth, Calif.; the MultiLink division of Odetics in Anaheim, Calif.; and 3G Co. of Gaston, Ore. Augat, a major supplier in the wire-wrappable IC socket panel business, made its recent entry into Stitch Wiring by buying the APAC division of Varian Inc., an Irvine, Calif., firm that was a pioneer in this specialized field.

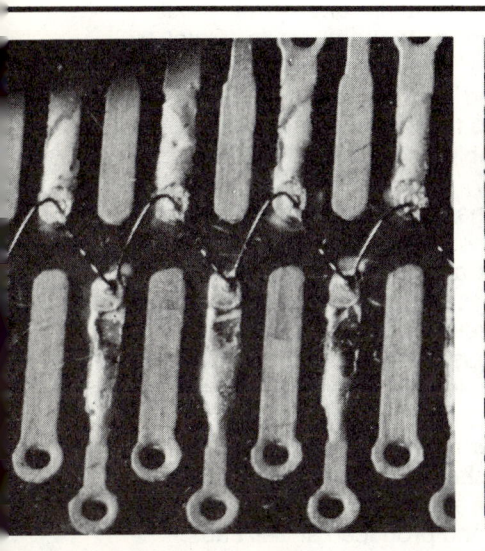

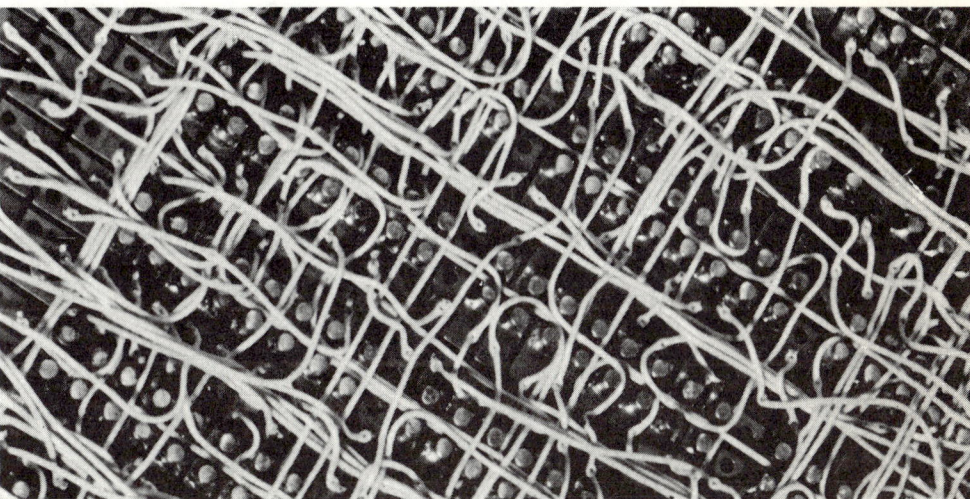

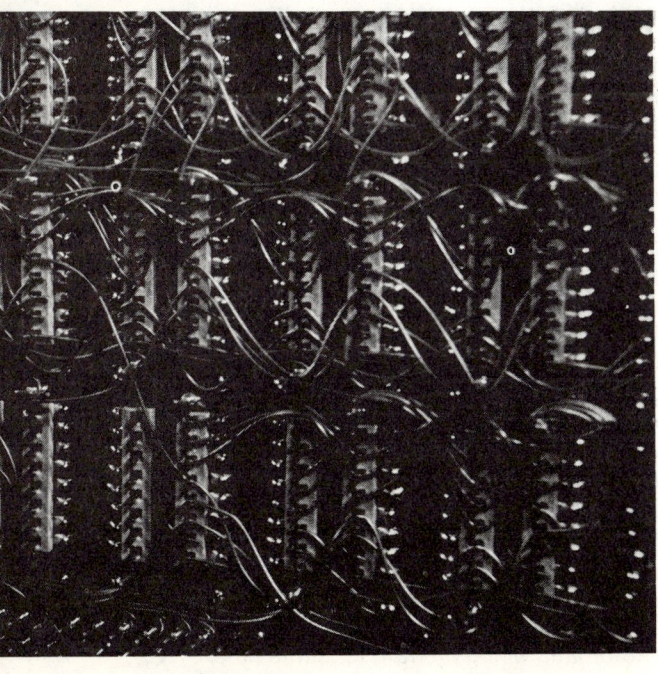

Stitch Wiring is heavily used in military and aerospace applications. For instance, the electronic controls for the space shuttle's rocket motors have been wired by this technique, while Grumman Aerospace Corp. uses it to prototype multilayer boards. In the commercial field, computer companies like Amdahl, Xerox, and Burroughs are in the process of investigating the method. It is interesting to note that the feasibility of the technique for high-speed logic has been demonstrated by Moore Systems, which has supplied a special Stitch-Wired emitter-coupled-logic board operating at 550 megahertz.

Computerized soldering

At about the same time that Stitch Wiring first appeared, two automated wire-routing and -soldering techniques based on the use of conventional pc boards were offered as alternates to both Wire-Wrap and multilayer boards.

Special pc boards with pretinned pads on both sides were required for the automatic systems based on soldered connections produced in the early 1970s by the Infobond division of Inforex Inc., Burlington, Mass., and the Weltek division of Wells Electronics Inc., South Bend, Ind. Numerically controlled wiring heads routed special polyurethane-insulated heat-strippable copper wire to contact the desired pads. After checking, the head moved on to the next wiring position. Then all the wires were reflow-soldered simultaneously to the pads through the insulation. Finally, dual in-line packages were soldered to the other side of the board.

The resulting low-profile board required no special pins or sockets, plating, or adhesive and had extremely dense wiring. But in spite of these advantages and an extremely fast wiring rate, neither Infobond nor Weltek's technique, called Tiers (Through Insulation Electronic Reflow System), really caught on.

Infobond wiring is no longer for sale, but Inforex is still using it extensively in house on its terminals. Weltek is actively pursuing its technique but sells only the semiautomatic- and automatic-wiring machines shown in the table. It neither sells other hardware nor contracts out wiring services.

Tiers is now being used by a large terminal manufacturer for backplane wiring, and several aerospace firms are also in the process of investigating it.

A more successful type of automatic wire routing and soldering called Solder-Wrap was developed by United Wiring and Manufacturing Co., Garland, Texas, in late 1976 [*Electronics*, April 15, 1977, p. 111]. In this system, a special computer-controlled wiring stylus strings a fine insulated wire to the solder tails or leads of sockets or pins in a regularly patterned pc board having wiring guides between the rows of leads. The wires are soldered to the tails by a probe that thermally strips away the insulation at the soldering point. Solder is actually fed to the wrapped joint, which is not done in the older Tiers system. After soldering, the loop of connections is cut at the proper places.

Solder-Wrap produces boards with an extremely low profile and high wiring and packaging densities and, like Infobond, is one of the first systems to make an in-progress check of wiring accuracy. At present, it is employed for wiring circuit cards and backplanes for computers, minicomputers, and peripherals, and its use is growing. A plus for this technique is that it has a manual version with special hand tools for prototyping.

United Wiring and Manufacturing is the sole source of Solder-Wrap machines. But solder-wrappable boards, socket pins, and sockets are available from quite a few sources. As of now, only United Wiring does contract solder wrapping.

Piercing the wiring barrier

Perhaps the newest and certainly one of the most promising automatic wiring techniques is Quick Connect. Developed at Bell Laboratories in Holmdel, N. J., the method is based on the use of insulation-piercing terminals similar to those found in mass termination systems for flat cables. It is easily automated (but can be done manually), is easy to repair, requires only standard pc boards, has a low profile, and needs no wire stripping or soldering.

Basic to Quick Connect are special terminal strips, which have IC socket pins on one side connected internally to insulation-piercing terminals on the other side. Arrays of these strips are mounted on glass-epoxy boards on which ground and power planes have been plated. The boards are wired by pressing insulated wires into the terminals, making a gas-tight termination. Both a special hand tool suitable for breadboarding and a semiautomatic production machine have been designed and produced at Bell Labs.

At present, this wiring system is in use at three branches of Bell Labs on pc boards with conventional or Schottky transistor-transistor logic. Quick Connect is not commercially available, but like many Bell developments will eventually be released to the public.

In a later and similar development in 1977, 3M Co., St. Paul, Minn., came out with a pc breadboard kit based on terminal strips of U-shaped insulation-piercing contacts into which AWG No. 30 Kynar wire was pushed with a special hand tool.

This U-Contact system, unlike the Bell one, requires the dual in-line packages to be soldered in and so far is strictly for breadboarding. However, 3M is developing hardware for plugging DIPs directly into its strips and in conjunction with another firm is looking into automating its wiring process.

Automatic wiring of circuit boards is here to stay. As the faster IC logic comes into service, the lower-profile wiring systems should get a greater play. This could result in a growth for Multi-Wire, Solder-Wrap, Stitch Wiring, Tiers, and possibly the insulation-piercing techniques. Overall, however, Wire-Wrap will remain the dominant automatic wiring method for some time. ☐

High-speed wire-and-solder technique tests connections as it makes them

Automatic method also minimizes production costs and keeps printed-circuit card profiles low for high-density packing

by Bob Whitehead, United Wiring & Manufacturing Co., Garland, Texas

☐ A new and different high-speed automatic wiring technique is beginning to challenge existing methods. It produces boards with an extremely low profile and high wiring and packaging densities, and it is the first to make an in-progress check of wiring. A small group of manufacturers is already using it to cut the cost of wiring circuit cards for computers, minicomputers, peripheral equipment, and digital controllers.

Solder-Wrap (the trademark of the new technique) keeps card profiles low because short solder tails replace the long protruding pins used in other wiring methods. Its end products, though, like theirs, are competitive with two-sided and multilayer printed-circuit boards.

Like the earlier methods, too, it uses stock pc cards with standard conductive patterns for multiple rows of dual in-line packages plus plated ground and power buses (Fig. 1). The three older systems are Wire-Wrap from Gardner-Denver Corp., Grand Haven, Mich., Multiwire from Photocircuits division of Kollmorgen Corp., Glen Cove, N.Y., and stitch-welding, originally developed by Jet Propulsion Laboratories, Pasadena, Calif., and now produced by other companies.

All four automatic systems have several advantages over plating techniques for producing pc boards. The main ones are higher interconnection density, which is frequently comparable to that of multilayer boards; elimination of custom artwork, which is particularly expensive for multilayer pc boards; lower overall costs; shorter turnaround time for design changes, and higher reliability.

Solder-Wrap, in turn, is superior in many respects to the older automatic methods. Most important, it tests its results at every step in the wiring routine; the others cannot check operation until after a board is completely wired. Moreover, the new technique wires boards faster, provides higher wiring and packaging density, costs less, produces better high-speed-logic circuitry, and offers the option of hand-wiring. The manual option, also offered by Wire-Wrap, is useful for breadboarding, low-volume production, and field-engineering changes.

In fact, United Wiring & Manufacturing developed the new system in 1973 as a manual wiring process with

1. All wrapped up. A finished Solder-Wrap board shows No. 38-gauge wire either soldered to socket leads that have been bent over at a 60° angle or wrapped around plastic wire-guide posts.

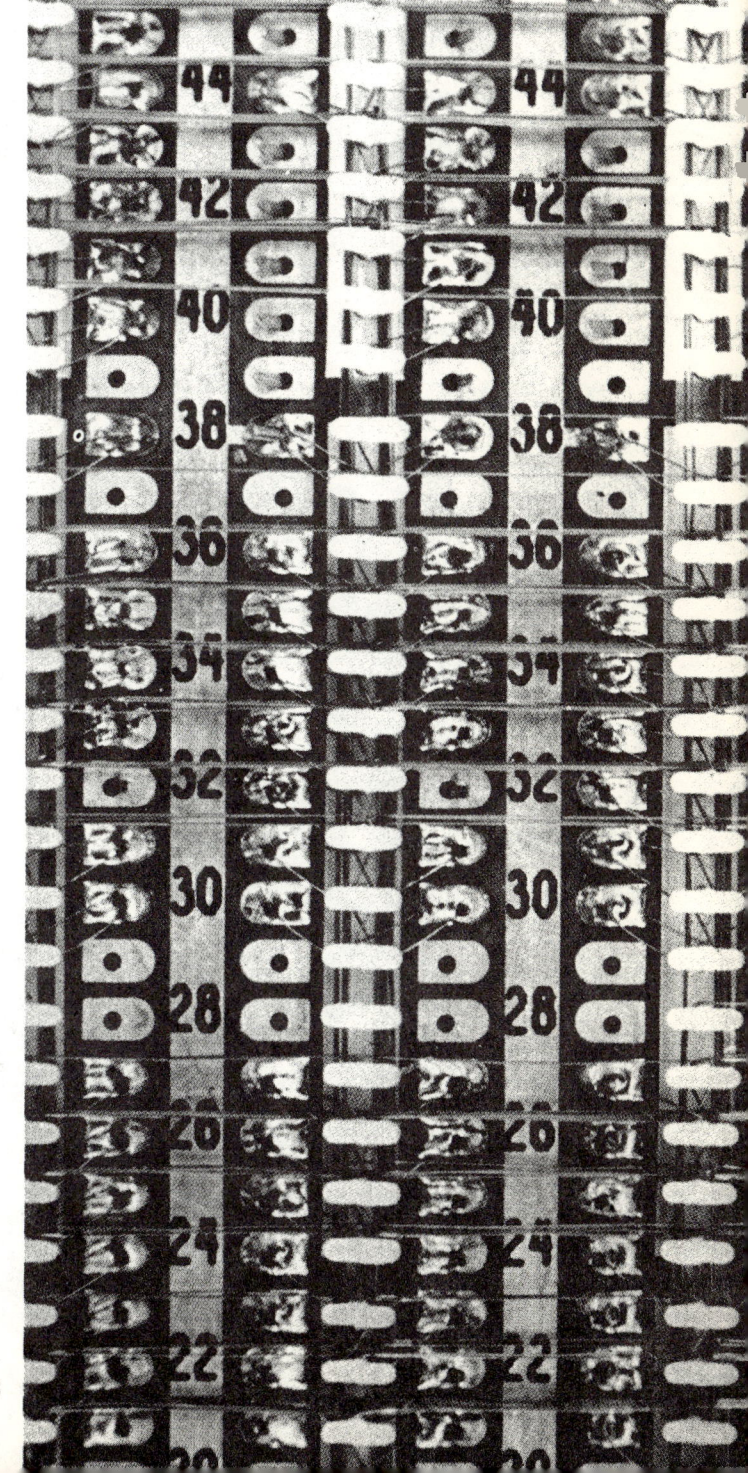

2. Wiring stylus. This special Solder-Wrap tool wraps a continuous wire around solder tails and wiring guides of a special board as directed by a point-to-point computer program.

3. Stripping for action. At speeds as high as 4,800 connections an hour, solder tool thermally strips polyurethane insulation from the wrapped wire and feeds solder to complete the connection.

a view to developing it into an automated system that would overcome most of the disadvantages of the older techniques. By late 1976, the company had developed automatic wiring machines, special hardware, and software for the Solder-Wrap.

Basically, Solder-Wrapping consists of stringing a fine insulated wire to the solder tails or leads of sockets or pins previously inserted into a specially patterned pc board having wiring guides between the rows of leads.

Soldering and stripping

The wires are soldered to the solder tails by a probe that thermally strips away the wire insulation at the soldering point while the sockets or pins are soldered in place. The resulting loop of connections is cut at the proper places.

Machines for the automatic solder-wrapping processes are programmed by paper tapes produced by a computerized data base generated from information from the customer's schematic. Data from this source document is key-punched into three card decks—one each for parameters, device locator, and signals—which the computer checks for errors before they are loaded into memory.

The parameter cards define the physical characteristics of each device used on the board. For instance, a typical IC could be identified as a 7400 with 14 pins—7 on a side—with 100-mil centers. These cards define X-Y coordinates for the locator cards.

The locator coding defines the row/column address of pin No. 1 of each IC located on the board. The signal cards contain logic-element coding for every pin of every IC called out on the schematic.

TABLE 1: SOLDER-WRAP AUTOMATIC MACHINES						
Model	Stringing (connections/hour)	Solder (connections/hour)	Cutting (connections/hour)	Average completed connections/hour	Average wires/hour
100	1,000	1,000	2,500	500	300
200	2,400	2,400	5,000	1,000	666
300	3,600	3,600	7,500	1,600	1,066
400	4,800	4,800	10,000	2,000	1,333

The older computer-controlled systems have fairly high design costs. Except for Multiwire, they do not lend themselves to production of high-speed logic circuitry, especially because wires are not placed near enough to the ground plane. Wire-wrap and stitch-welding produce high pin profiles that limit the number of cards that can be stacked in a given space.

A Wire-Wrap machine inserts a matrix of square metal pins spaced on 100-mil centers on the card. Then, under computer control, a special tool wraps several turns of insulated wire around the pins so tightly that the connection is gas-tight.

In the multiwire system, an automatic wiring head lays down a network of magnet wire, insulated with polyimide, on an adhesive-coated epoxy-glass board. Terminations are formed by drilling through the wire and board, then electroplating the sides of the holes.

Stitch-welding employs a semiautomatic tool to cold-weld Teflon-insulated nickel wire to a board with a matrix of either stainless-steel pins or stainless-steel circuit lands. Like Solder-Wrap, it can provide a low profile by eliminating the pins.

Running the Solder-Wrap routine

With Solder-Wrap, the wiring stylus shown in Fig. 2 can string No. 30, 34, and 38 gauge polyurethane magnet wires. After routing the wires, the machine steps into a solder cycle using the tool shown in Fig. 3. The solder head has three different timing cycles—strip, preheat, and post-heat. During the 400-millisecond strip cycle, the resistance solder head is heated to 800°F to strip the polyurethane from the magnet wire directly under it. After preheating the electrical pins or component leads, the solder-feed mechanism is activated.

The post-heat cycle starts after a predetermined metered amount of solder has been fed to the electrical land. Completing the solder-feed cycle starts the post-heat cycle, which continues as the solder is retracted from the electrical land. The component or socket lead is soldered into the plated-through hole in the land at the same time the polyurethane-coated magnet wire is soldered to the lead.

Last, the tool shown in Fig. 4 begins cutting into separate networks the single continuous wire running throughout the board. Figure 5 illustrates the cuts made in a typical wire network. In this example, a 14-pin device has been wired into a network that ties pins 1, 3, 5, 7, 9, and 13 together. The routing of wires to pins 1 and 13 differs from that for pins 3, 5, 7, and 9. The automatic cutters separate the networks where slashes are shown.

The four automatic Solder-Wrap machines—models 100, 200, 300 and 400—operate under tape control at different rates. Table 1 indicates their stringing, soldering, and cutting speeds. Connection speeds range from 500 to 2,000 per hour and stringing from 1,000 to 4,800

4. Cutting edge. After a network has been strung and soldered, a third tool, controlled by the wiring program, cuts wire along the network at desired points at a rate as high as 10,000 points per hour.

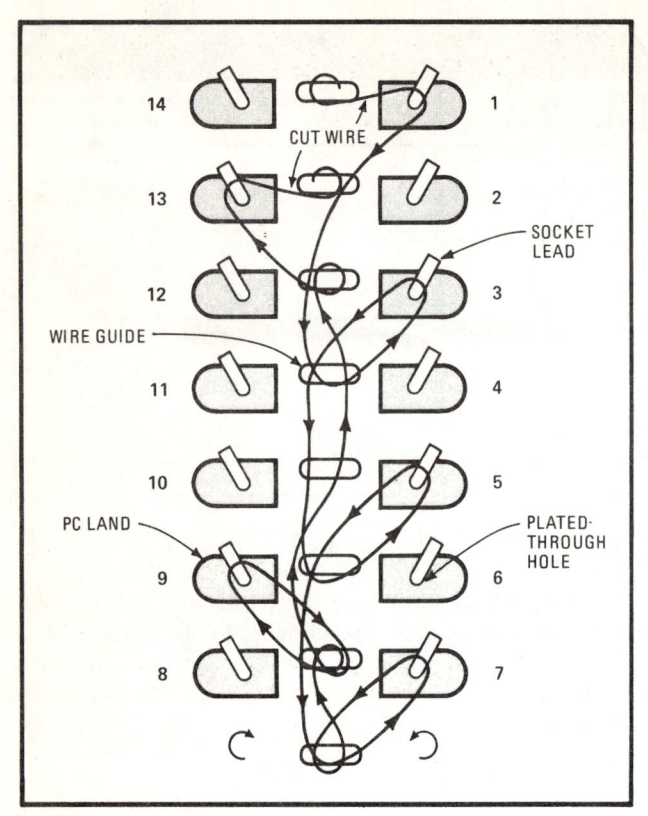

5. Wire routing. Routing pattern is designed to string a 14-pin socket. Wiring guides channel some wires either in a clockwise or counterclockwise direction. This orientation aids in determining whether a lead is going into or coming out of the network.

wires per hour. The model 100, priced at about $35,000, is a new completely automatic single-headed machine aimed at engineering laboratories and small companies. When not being operated in its normal mode, it may be used for semiautomatic wire-wrapping or component insertion.

Models 200 through 400 are designed for medium- to high-volume production. The four wiring heads of the model 400 (Fig. 6) can lay down more than 400,000 wires a month.

Wiring the cards

Like the other three automatic-wiring methods, a Solder-Wrap machine lays wires in the desired configuration on a specially modified and patterned pc card. As do Wire-Wrap and Multiwire cards, the epoxy-glass blanks have plated rows of dual-in-line-package patterns spaced on 0.1-in. centers. Ground and power buses on the component side are connected via plated-through holes to similar conductive patterns on the wiring side. On the component side are mounted IC sockets with solder-tail leads, special socket pins with solder tails, or the components themselves. The solder leads are inserted in the plated-through holes to the wiring side, where they are bent at opposing 60° angles in alternating columns, lowering the wiring profile even further.

Finally, rows of plastic wiring guides are attached between the rows of plated-through holes on the wiring side of the board. These guides, which have oblong posts, are used in the wiring process depicted in Fig 5.

However, the user can adapt manual Solder-Wrap

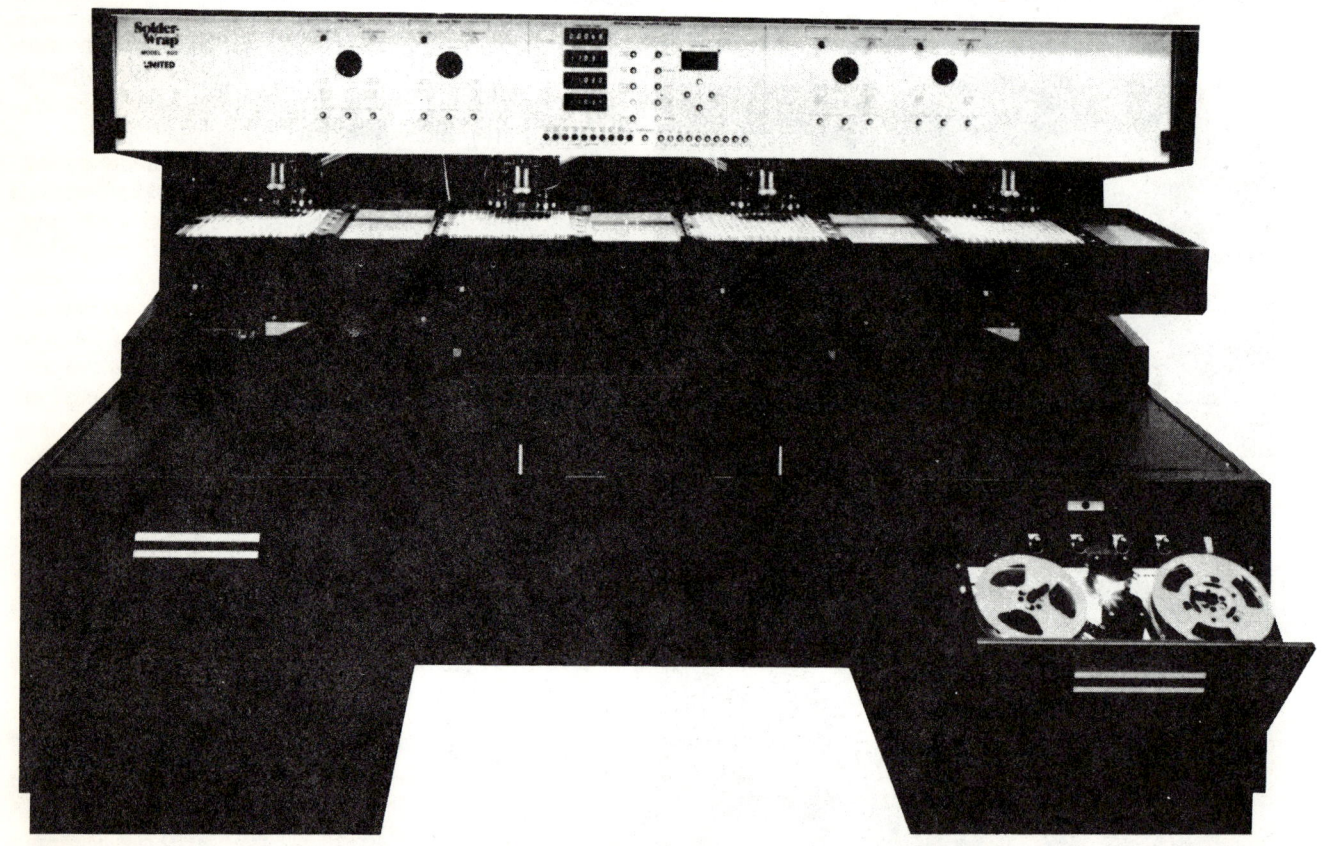

6. Automated wiring. United Wiring and Manufacturing's Solder Wrap model 400 is a fully automatic quadruple-head wiring machine. The wiring pattern is controlled by a paper tape generated from an off-line data base derived from the customer's schematics.

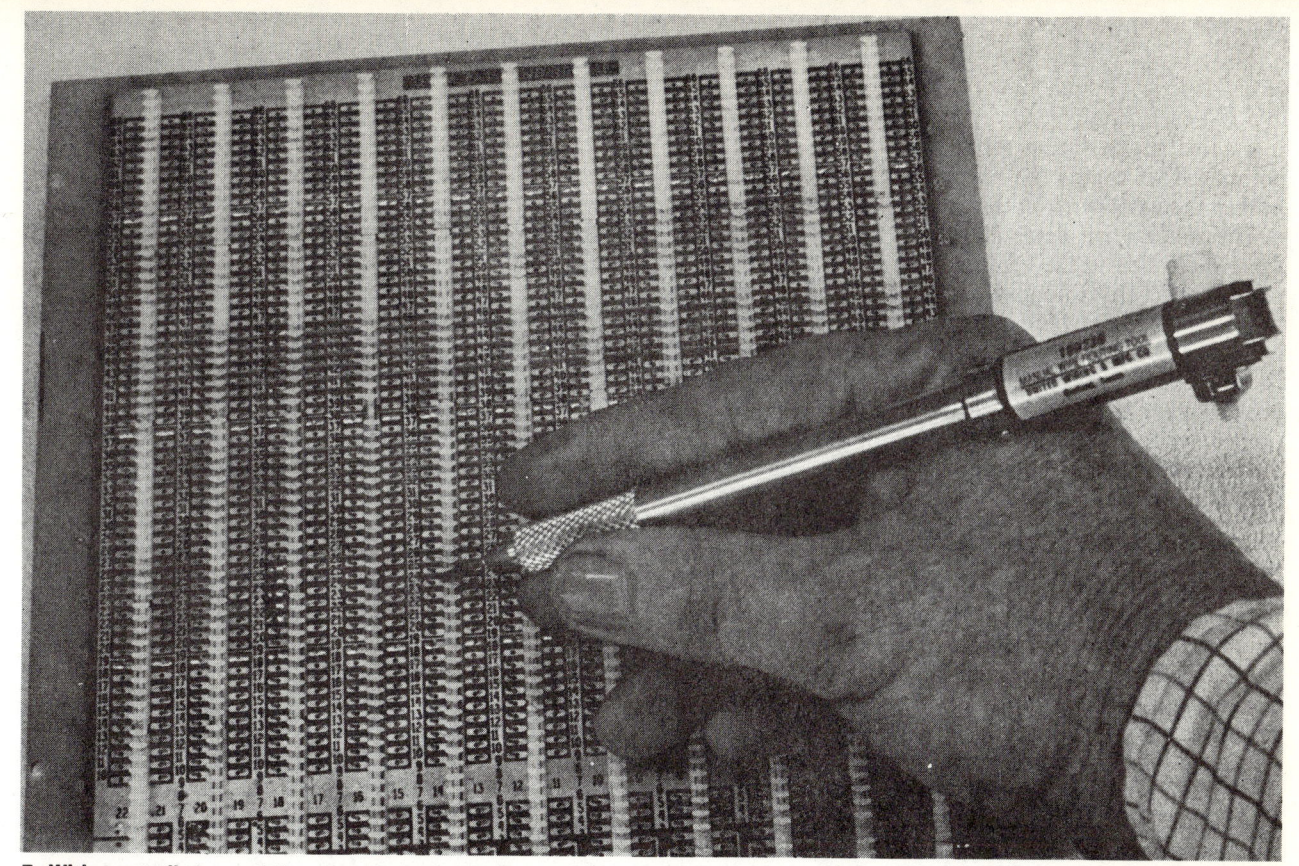

7. Wiring pencil. A pencil-like tool is used to wire manually a Solder-Wrap board at a rate of 200 to 300 wires per hour. Standard hand-soldering and -cutting tools are used to finish the wiring. A special wire cartridge loads the tool with the wire type required.

tools to breadboarding, low-volume production, or field-engineering changes. The manual wiring tool in Fig. 7 can lay down 200 to 300 wires per hour, a rate 3 to 4 times higher than is possible with a comparable hand wire-wrapping tool. In the course of breadboarding, an engineer would load the wiring tool with a special wire cartridge, wire a string, strip away the insulation, and solder the pin connections with a low-wattage soldering iron and then cut the string with diagonal pliers at the proper points.

Mounting circuit cards

Solder-Wrap cards made either with pin-in-board construction or no sockets at all can be spaced on 500-mil centers in a card cage, and the boards with sockets can be spaced on 600-mil centers in the same applications. The only suppliers of boards for its process, United Manufacturing & Wiring offers a library of card designs that includes units for Schottky, transistor-transistor, and emitter-coupled logic. For companies that want to design their own boards, Robinson/Nugent Co., Albany, Ind., has developed both low-profile and high-reliability IC sockets that can be wired with Solder-Wrap equipment. With the high-reliability type, 34 solder-wrapped boards on 0.500-mil centers can be packaged in a 19-in. card cage.

The big advantage of the Solder-Wrap back-panel system is its high cubic density on the wiring side of the backplane. The wiring cost is also lower than previous Wire-Wrap back-panel systems. A Solder-Wrap backplane system would be equivalent to an 8- to 10-layer board.

For applying back-panel wiring, the Milton Ross Co., Southampton, Pa., has designed a special Solder-Wrap edge-board connector that has a solder-tail-lead design on centers of 100 by 300 mils. This connector will be available in several pin configurations for use with present back-panel connector systems. Winchester Electronics division of Litton in Oakville, Conn., is also developing a complete line of Solder-Wrap back-panel edge-board connectors to compete with present available Wire-Wrap back-panel systems.

Automatic Solder-Wrap is the only automatic wiring process available with a complete in-line test of wiring interconnections. All Solder-Wrap machines have four lights on each wiring head that indicate failures of the parameters checked by the self-test circuitry. These lights indicate, respectively, missing pin, broken wires, stripped wires, and solder errors. Since every failure halts a head's tool where an error occurs, a glance at the indicator lights and the position of the tool is enough to identify and locate the error.

Self-testing

While stringing is in process, the wire network being put down is checked for missing pins, strip-wire errors, and broken wires. Missing pins are detected by monitoring for a slack wire. A system to monitor wire-tension errors stops the machine and lights the missing-pin indicator when it detects a slack wire.

The circuit of Fig. 8 checks for strip errors and breaks. Strip errors occur when insulation is missing on the magnet wire, a fault that could short two networks together. In the circuit of Fig. 8, the insulated wire normally tied to V_{cc} passes through the stringing tool, which has its case grounded. If the insulation is broken, terminal A of the circuit will pick up a ground, lighting the strip-error indicator, and the machine stops. Also, if a wire breaks during stringing, V_{cc} is removed from point A. This causes the machine to halt and lights the broken indicator-wire on the appropriate head.

The solder-error-detection circuit of Fig. 9 checks if the ground tied to the heating tool is carried through to point B. If there is no ground, the solder-error indicator lights, and the machine stops soldering. Possible causes of failure that will open up the ground to point B could be a defective solder head, lack of solder, and a missing broken wire. If the solder cycle does not end before a predetermined period set by the operator, a solder-time-out error indicator lights up on the center console, and the machine stops.

Assessing capabilities

Among the advantages Solder-Wrap boasts over the industry leader, Wire-Wrap, are double the packaging volume, lower production costs, self-testing, and lack of cold-flow short circuits (shorts caused when the insulation flows away from wires bearing against pins under pressure). With the newer process, 34 boards can be packaged in a 19-inch rack, which can accept only 13 to 17 Wire-Wrap boards.

A solder-wrapped board costs only 30% to 50% as much as a wire-wrapped board, depending on the type of socket used with the latter. The wire used in the older process is 40 to 120 times more expensive than the magnet wire used in solder-wrapping. The 30-gauge Kynar is $4 per 1,000 feet, and 30-gauge Milene is $12 per 1,000 feet; in contrast, the magnet wire is only 9 cents per 1,000 feet.

For low-to-medium-scale runs, solder-wrapped boards are much cheaper to produce than two-sided boards (Table 2). Since solder-wrapped boards are equivalent to 8- to 10-layer boards, their advantages over two-sided boards are applicable also to multilayer boards. Solder-Wrap costs 72% for front-end tooling, and the new board designs can be turned around in two weeks—only a third as long as the older method requires.

For engineering changes, the newer system does not require artwork, and existing boards can be reworked in the field with manual Solder-Wrap tools. In addition, Solder-Wrap provides twice as much surface density as two-sided boards, and the shortness of leads brings about better operating characteristics in high-speed-logic applications.

The capital investment required for Multiwire is higher than Solder-Wrap requires, and the latter has lower production costs. However, the surface and volume packing densities of the two techniques are comparable. The biggest advantage Solder-Wrap has over Multiwire is the capability to make changes either in production or for prototyping. The only way to make a single board with the Multiwire process is to make it on the automatic equipment.

Solder-Wrap is 30% to 50% more cost-effective than stitch-welding, and packing volume is higher because of the lower profile. What's more, stitch-welding machines are semiautomatic—not fully automatic.

Economizing with Solder-Wrap

It is instructive to calculate the cost of 25 circuit boards to be manually solder-wrapped using a total of 100 16-pin sockets. Typically, the cost for a set of boards without sockets would be $77, cost for the boards with profile sockets would be $112, and cost for boards using P/B (pin-in board) sockets would be $170.

The average cost to manually Solder-Wrap the set of boards would be about 10 cents per wire, or $100 for 1,000 wires. To this must be added the comparative total costs of the system, which would thus be: socketless boards, a total of $177, low-profile sockets, $212, and P/B sockets, $270.

Service costs for automatic wiring of one to four

8. Wiring check. During the wire-stringing cycle, this circuit can detect and indicate either a stripped or broken wire. If terminal A picks up a ground from the stringer, the stripped-wire indicator lights. If the terminal senses loss of V_{cc}, the broken-wire indicator lights.

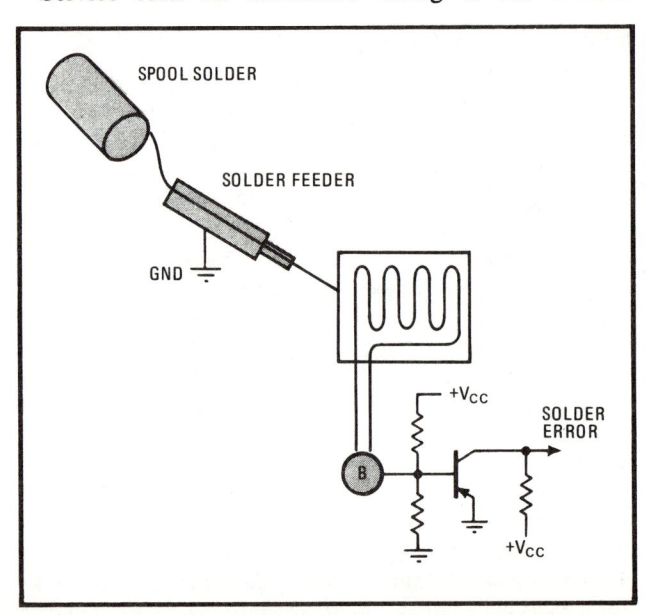

9. Solder error. Malfunctions such as solder-head failure, lack of solder, and missing or broken wires are sensed by this circuit. Any of these errors disconnects terminal A from ground, activating the circuit. This, in turn, lights up an indicator and stops the machine.

10. Fan-Fold. In the Fan-Fold configuration, the three solder-wrapped boards are fastened side by side by sheets of Mylar, and all card-to-card connections are automatically wired together. The flexible sheets and boards can be folded into a compact packet that does not have either backplane or printed-circuit connectors.

TABLE 2: COMPARATIVE SYSTEM COST

Number of systems	Cost per system (Solder-Wrap)	Cost per system (two-sided)
5	$ 1,626	$ 8,131
10	813	4,065
25	325	1,626
50	163	813
100	81	406
200	41	203
500	16	81
1,000	8	40

boards, on the other hand, vary from 4 to 8 cents a wire, depending on the number of boards. For 100,000 to 10 million wires, prices vary from 7 to 3 cents per wire.

To compare the packaging cost of Solder-Wrap with two-sided pc boards, consider a system with a 48-square-inch backplane that holds multiple circuit cards mating with connectors on the backplane. To handle the 900 16-pin ICs, Solder-Wrap requires 12 circuit boards, while the two-sided-pc-board approach requires 23 units. In addition, the Solder-Wrap backplane requires 975 wires, whereas the backplane for the conventional boards requires 1,863 wires.

Nonrecurring engineering costs for both approaches over a range of system quantities are listed in Table 2. The total nonrecurring engineering cost for one system, consisting of software plus engineering changes, for the automatic wiring approach, comes to $8,130 compared with $40,656 for the pc-board method. Recurring costs for this system would be $507 for two-sided boards and $493 for Solder-Wrap. For even 1,000 systems, total Solder-Wrap system costs are lower than they are for two-sided boards, but the biggest savings are at five systems. At the 200 level, system costs for the two-sided packaging start to close in on Solder-Wrap.

Applying Solder-Wrap

United uses its model 400s in its three contract-wiring centers in Dallas and Longview, Texas, and Santa Ana, Calif. The company plans to open more of these centers. United also plans to develop an automatic machine to handle twisted pairs.

A large computer system made by Scientific Machines Corp., Dallas, is completely wired by the new process. The system uses TTL/Schottky logic. SMC has developed software that contains the board's wire list and generates the logic diagram, timing diagram, board artwork, and parts layout, and wire routing for the automatic solder-wrapping system.

Corporation 1171, Dallas, and United Manufacturing have developed a Solder-Wrap Fan-Fold process on a microprocessor system for controlling drug-store inventories. This Fan-Fold package can handle more than 600 16-pin ICs in a package of 6 by 6 by 16 in. This packaging method eliminates the need for backplanes and edge-board connectors.

In the Fan-Fold package, solder-wrapped boards are tied together by rectangular sheets of Mylar. Typically, the Mylar sheet would be fastened from the right side of one board to the left side of the adjacent board as shown in Fig. 10. Then, all adjacent card-to-card connections are automatically wired together, with all these new wires lying across the Mylar sheet. These wires are then sealed into the Mylar with an adhesive. The resulting package of boards with alternating Mylar interconnects can be folded (like a map) into an extremely small space.

Corporation 1171's package houses the system's memory, processor, floppy-disk controller, switching power supply, and interfaces for the printer, cathode-ray tube, and keyboard. All of this circuitry is contained on six Solder-Wrap boards connected by the Mylar sheets with wires embedded in them. The Fan-Fold package has a low production cost, there are no limitations on I/O signals from board to board, and elimination of backplane connectors results in high reliability.

Innovated Systems, Dallas, Texas, is using Solder-Wrap in traffic-control systems; IBM, Austin, Texas, on typewriter-production test equipment; Texas Instruments, Dallas, in communications systems, and National Computer Systems, Houston, in a high-speed computer system. Also making use of the system are Motorola Semiconductor, Phoenix, Ariz., and Western Geophysical, Houston.

Solder-Wrap need not be limited to making circuit boards. Wiring of electrostatic printer heads looks like an attractive possibility. Right now, all terminations of the printer heads are being wired manually. More than 4,000 No. 30 and 37 gauge wires are being connected manually, thereby incurring high labor costs for each head. A Solder-Wrap machine could cut costs by automating the manufacturing process, since it routinely handles wires of this size.

Another special application involves automatically applying solder in tight packages such as digital watches or heart pacemakers. Some companies have tried solder preforms and infrared heating, but leaves' an excess amount of touch-up and repair. A Solder-Wrap machine could solve the problem by being programmed to automatically add solder to selected locations. □

Impulse-bonded wiring is economical alternative to multilayer boards

For a complex system design that remains volatile well into production, impulse bonding creates densely wired boards that, unlike multilayer circuits, are easy and inexpensive to alter

by F.G. Schulz, D.C. French, B.E. Criscenzo, and P.T. Klotz *Bendix Corp., Teterboro, N.J.*

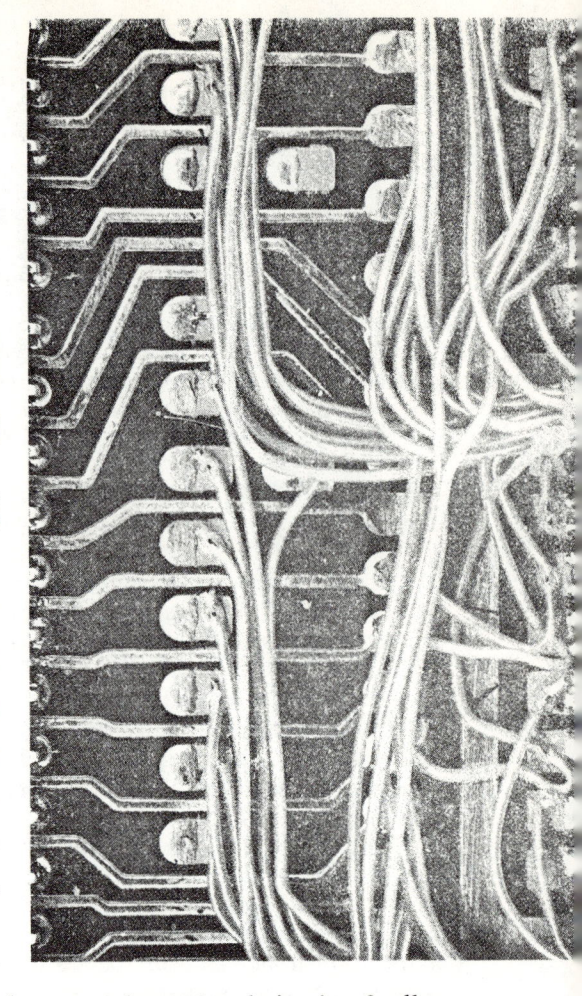

☐ Multilayer printed-circuit boards are often the most cost-effective way of handling high-density wiring in a system in high-volume production. But they become very impractical when circuit changes continue to be made well into the manufacturing phase.

Impulse bonding, however, makes it possible to produce electronic-component assemblies that are as dense as multilayer boards but faster and cheaper to redesign. This wire-bonding process, in combination with automatic component insertion, mechanized flow-soldering, and computerized testing, is economical enough for moderately high production volumes.

Impulse bonding, as described here, is a variation on the pc-board-wiring technique called stitch wiring. It differs in using a specially developed pc board, which accommodates wire interconnections and components on the same side, and in using a brief high-current pulse to diffusion-bond the wiring to the unique pc-board pads. The interconnections are routed and bonded by programed, semiautomatic machines.

This construction takes much less drafting time than an equivalent multilayer board and adds almost unlimited flexibility in making minor or major circuit changes, since neither the artwork nor the drilling nor the photoplating process of the basic board is affected. In addition, the basic board can easily be converted to a multilayer equivalent once the system design has finally stabilized.

Bendix' impulse-bonding facility, consisting of 12 semiautomatic machines, has wired over 50,000 circuit boards for such products as the flight guidance systems of the DC-10, S3-A, and Mercure aircraft, and for automatic test equipment for the F-15. A bond reliability of 0.002 parts per million hours has been attained.

Time studies and time-keeping job cards produced by the impulse-bonding computer program show that the direct recurring factory cost, including both labor and material, is about twice as much for impulse-bonded wiring as for multilayer-board processing—a 4.7-by-6-inch, eight-layer, basic multilayer board costs about $50, and its impulse-bonded equivalent costs about $100. However, the greater economy of the multilayer board is quickly offset if the board has to be redesigned several times or if the initial, nonrecurring cost of the artwork and photoprocessing stages must be absorbed over a short production run.

The multilayer rationale

Many of today's printed-circuit boards require as many connections between devices as an entire blackbox chassis harness did 10 years ago. Such a mass of wiring or conductor runs would make the boards so

1. I-Bond pads. On impulse-bonded circuit boards, bonds are made directly to the rectangular area of the pads, shown magnified here 10 times. Components are mounted and wave-soldered in the plated-through holes beside the pads.

bulky as largely to cancel out the space-saving advantages of using IC packages.

For the most part, the aerospace electronics industry has solved this problem by adopting the multilayer printed-circuit board. However, these boards are not only tricky to manufacture—they also take a long time to design and are costly to move into production, so that they are uneconomical in small lots. Moreover, once the necessary artwork, photomasters, and drill tapes are completed, the interconnection network is hard to change without costly and long-drawn-out rework—not the ideal situation when a design is volatile.

The logic circuits of a commercial flight control system, for instance, may be subject to change well beyond the initial production phase, as flight experience with the system accumulates, as different airlines choose different customized options, and as the aircraft progresses through several levels of FAA certification. In the case of one such system, major configuration changes affected almost every one of its 50 different multilayer boards right at the start of the manufacturing phase. To develop new artwork and get new boards into production would have taken three months per board—an impossible delay when new aircraft were constantly approaching flight service and requiring to be provided with equipment immediately.

The interim solution usually adopted was to cut pads and add jumper wires to a board, even though the initial costs and rework involved in doing this were heavy. Then, as a given multilayer board accumulated more changes and the modification costs and number of jumpers became excessive, the board was redesigned. Three months later, when production shifted to the new multilayer boards, further design changes had already occurred and the whole process immediately started up all over again.

After this experience, which lasted 18 months, the search was on for an alternative to the multilayer board.

The requirements

Above all, this other wiring method had to allow rapid and efficient changes to be made in the wiring of an electronic system until the design was stable. The wired modules had to be interchangeable with multilayer boards in function, size, and weight. High reliability of the wired connections between the components

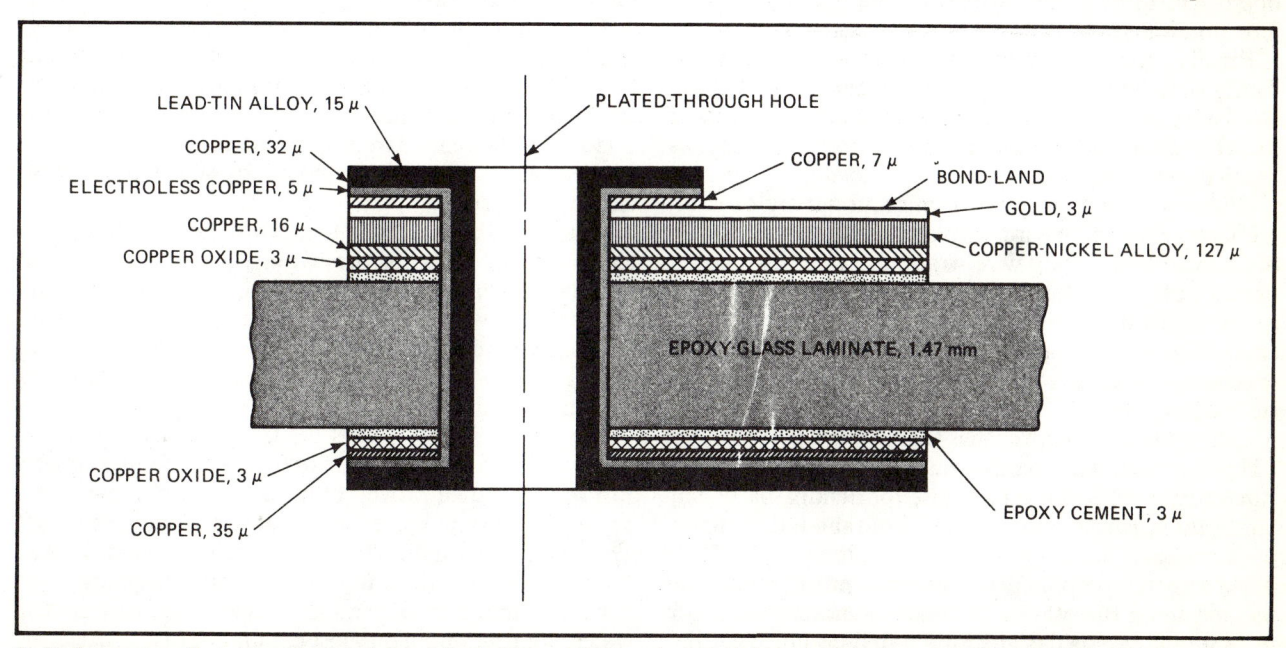

2. Pad cross section. Normal printed-circuit laminating, etching and photoplating processes are used to deposit an I-Bond pad on an epoxy-glass board. A special material is used for the pad—a copper-nickel foil alloy, which is impulse-bonded with a nickel wire.

Three automatic board-wiring techniques

Three commercial systems have come close to meeting Bendix' automated wiring requirements for avionics systems. But in two, the components and wiring end up on opposite sides of the wiring board, increasing bulk, and in two the wiring stage precedes component insertion and soldering. The three systems are:

- **Infobond**, an automated system of point-to-point wiring on the back of a two-sided printed wiring board (the components are on the front or other side). The 38 AWG copper wire used is solder-bonded to terminations by an automatic soldering gun. This process was developed by the Inforex Corp. of Burlington, Mass.
- **Multiwire**, an automated interconnection system in which 33 AWG polymide-insulated magnet wire is first laid down on a adhesive-coated epoxy-glass board. Then terminations are formed by drilling through the wire and board and electroplating the sides of the hole. The resulting tubelets are then used for component insertion and soldering. Components and wiring are on the same side of the board. Multiwire was developed by the Photocircuits division of Kollmorgan, Glen Cove, N.Y.
- **Stitch-wire**, a semiautomatic system of point-to-point interconnections in which gold-plated steel pins are pressed into holes in conventional printed-circuit boards. Teflon-insulated 30 AWG nickel wire is bonded to the pins. Electronic components are then soldered to terminal projections on the opposite side of the board. This process was developed by the Micro-Technology division of Sterling Electronics, Westlake Village, Calif., and by the Accra-Point Arrays Corp., Santa Ana, Calif.

---

mounted on the new boards was essential, and the joint failure rate had to be as good as a soldered wired connection.

The economic success of the new wiring method would depend largely on the ease of making minor modifications, substituting parts, and repairing the finished board. This is necessary for normal rework operations as well as field repairs.

Design costs, too, could be expected to be lower, since the artwork for a single two-sided board would be both much simpler and very much less than for a multilayer board. The artwork for an eight-layer board is very much more complex than that for four two-sided printed-circuit boards since the mechanical layouts, registration, and electrical interconnections of all the layers are interrelated.

Other economic considerations were the compatibility of the new process with existing automatic component-insertion and flow-soldering equipment. Also, the board had to be such that it could be redesigned back into a multilayer board once system design stabilized.

Finally, any alternative to multilayer packaging would be a linear wiring process replacing a cost-effective batch process. It would therefore have to lend itself to automation and computer processing of data bases as much as possible.

With these requirements in mind, a survey of commercially available board-wiring techniques was carried out. Several, including wire wrapping, were discounted because of wiring density or cost considerations. At this stage, a key decision was reached: to make the interconnecting wiring the last step in the manufacturing operation—that is, after all the components have been installed and flow-soldered in place.

Out of the three approaches that seemed possible (see "Three automatic board-wiring techniques," above), stitch wiring offered most promise for Bendix' particular application. However, it suffered from the following set of drawbacks:

- Having the components on one side of the board and the wiring on the other side creates a module too thick to fit into a chassis designed ultimately to hold multilayer boards.
- The steel pins used as device terminals are too heavy for some avionic products—they typically add 3 ounces (85 grams) to a 5-oz (142-gm) module.
- The gold plating on the pins dissolves when a component is unsoldered from them, so that subsequent resoldering is difficult and this hampers repair and rework.
- The cost of the pins and of their insertion, even with automatic equipment, is quite high.
- Manufacturing costs also are relatively high, because neither installation nor mechanized flow-soldering of electronic components is practical.

The impulse-bonding technique

The impulse-bond (I-Bond) wiring method finally adopted uses the same type of bonds and bonding equipment as stitch wiring. The steel pins of stitch wiring are eliminated, however, and instead the impulse bonds are made directly to unique pads on a specially developed board. Wiring and components are on the same side of the board.

The composite I-Bond pad (U.S. patents pending) combines a plated-through hole with a rectangular bond land (Fig. 1). The hole contains the lead of an electronic device, and the interconnecting wire is bonded to the land. The entire pad configuration is deposited on the epoxy-glass board by standard printed-circuit processes.

Crucial to the success of the technique was the careful choice of materials for the wiring, composite pad, and bonding-machine electrode. Altogether 12 metals and alloys were evaluated for the bonding area of the pad, and six copper alloys (as well as several tip-dressing geometries) were evaluated for the upper and lower electrodes of the weld-head.

The use of stainless steel for the pad was considered because it bonds better than other metals to nickel wire. But stainless steel proved hard to laminate and etch on an epoxy-glass board economically and with sufficient uniformity and adhesion. What really ruled it out, though, was the difficulty of soldering anything to it without either special plating or the use of acid fluxes, which are prohibited on most circuit boards.

In the end, a copper-nickel pad foil alloy and a high-

3. An impulse-bonded board. The impulse-bonding process uses a two-sided board with very simple artwork. Each board is designed to accept a wide variety of components including axial-lead devices, discrete semiconductors, DIPs, and crystal-can relays.

4. Top-side wiring. Wiring is impulse-bonded to the board after all components have been wave-soldered. Wiring on the component side allows the board to be tested on the other side by automatic test equipment with "bed of nails" fixtures.

purity nickel wire were found to be the most compatible materials for impulse-bonding. Average peel-strength of the bond exceeds 4.5 pounds (2 kilograms). Pad adhesion, as measured by peel-strength of the alloy from the epoxy-glass board material, exceeds 11 lb per inch (2 kilograms/centimeter) of conductor width. These pad materials are shown in the plating-strata cross section of Fig. 2.

Also shown in Fig. 2 are the intermediate platings needed to insure adhesion to the epoxy-glass board and the materials required to build up the plated-through hole. With the composite pad shown, it is possible to repair impulse-bonded wiring by cutting the impulse-bonded daisy-chained wire and then soldering a copper wire into the adjacent plated-through hole, next to the component lead.

Impulse bonding creates a hybrid resistance-bonded joint by exploiting the molecular diffusion and cohesion that can occur between two surfaces held in intimate contact, particularly when under high temperature and pressure. The term "bond" is used instead of "weld" to emphasize the absence of the localized melting and weld-nugget formation that are characteristic of fusion-welding.

Solid-state bonding

Primary driving force behind the molecular diffusion is the heat generated by the power dissipated when a 300-ampere, 3-millisecond dc pulse is passed through the nickel wire and the gold-plated copper-nickel pad. In general, the heat flux generated at the bond joint is

$$H = I^2RTK$$

where I = current, R = resistance of work (wire, pad, plating interface, etc.), T = time of current flow, and K = a factor representing total conduction, radiation, and convection heat losses.

Actually, each of the above factors is a complex term depending on contact area, pressure, bulk-section geometry and transient heat-transfer properties of the three elements involved—the electrodes, nickel wire, and composite pc-board pad, including its adjacent plated-through hole. The maximum temperature in the immediate bond zone is probably in the range of 1,945 to 2,200 degrees fahrenheit—say 2,100°F. The temperature at the interface between the nickel wire and the copper upper electrode of the weld head may be somewhat higher than 2,100°F, but it is still below the melting point of the wire.

During each bonding cycle, reel-fed wire is passed through a hollow upper electrode, which lowers upon operator command. The rim of the upper electrode (or cathode) bears the wire against the pad surface and applies enough stress to split the Teflon insulation of the wire longitudinally and force the now bared conductor against the cathode rim and the gold surface of the pad. When the force increases to about 9 lb, the nickel conductor pierces through the 3-micrometer-thick gold-plating and sinks approximately 25 μm into the copper-nickel layer of the pad.

When the force applied by the cathode reaches 9.5 lb (4.3 kgm), a permanent magnet in the weld head breaks free and reduces it to 4.5 lb (2 kgm). At this point, the 300-A, 3-ms bond pulse passes from the upper electrode to the lower via the plated-through hole, resistive heating occurs, the gold film in the immediate contact areas melts and is displaced under pressure, and molecular diffusion bonding between the nickel wire and copper-nickel pad takes place.

The board described

The circuit board used in impulse bonding is quite like a conventional two-sided pc board, except that it has very little interconnecting artwork (Fig. 3). The pad patterns on the board are designed to accept a wide variety of electronic components, including axial-lead devices, semiconductor cases, dual in-line packages, and crystal can relays. In general, the interconnecting conductor-runs are limited to power distribution and

163

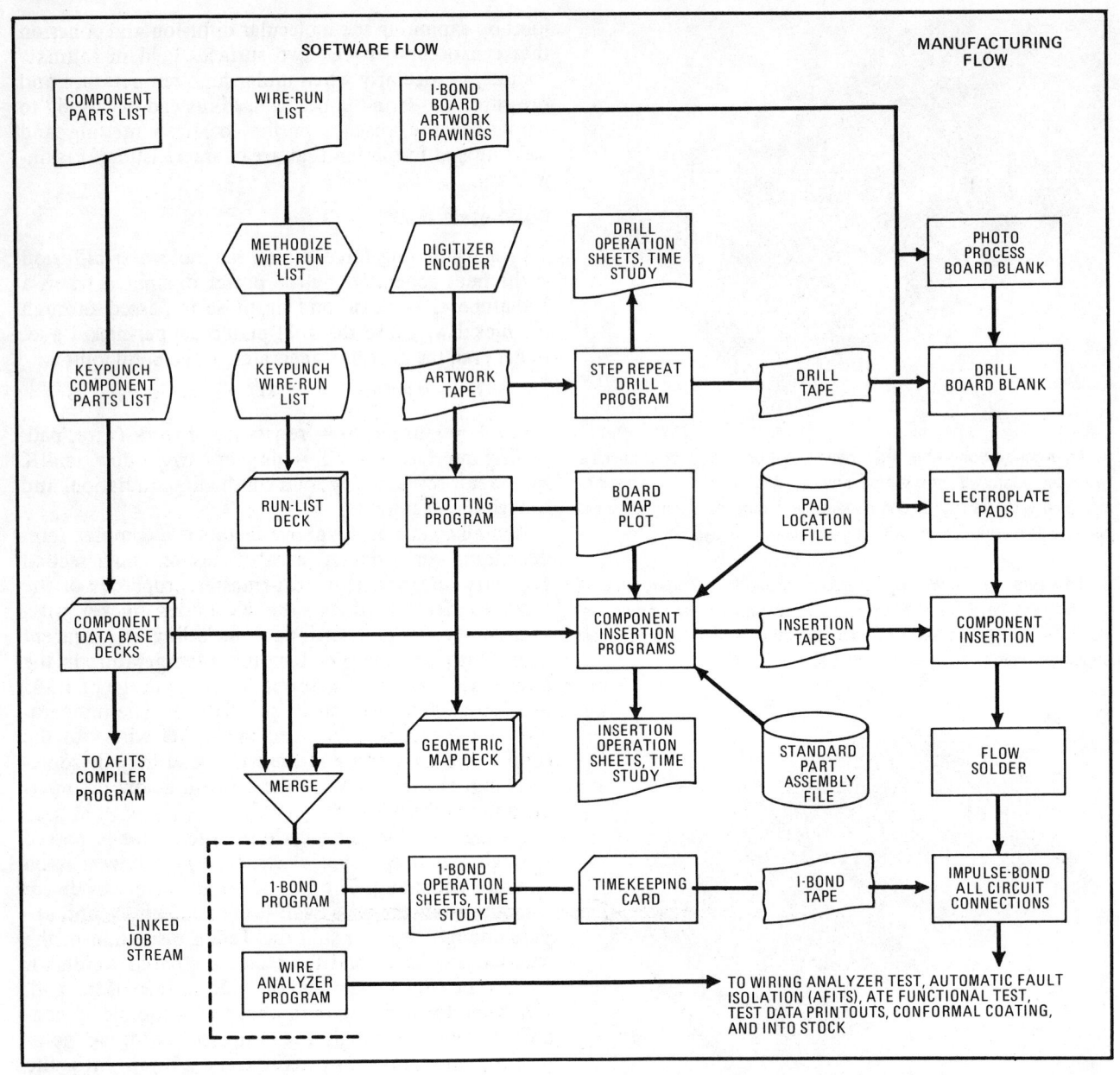

5. Computer-aided manufacturing. Software for board drilling, component insertion, interconnection, testing and time studies are under direct computer control in this process. Colored lines highlight inputs that control the final impulse-bonding step.

ground-plane busses. Layout of this board is quite simple compared with a multilayer equivalent since the designer is primarily concerned with geometric component placement and grouping of related devices.

By keeping the basic board designs sufficiently general, the start-up time, start-up cost and manufacturing inventories for new product designs can be minimized. For example, the total recurring drafting documentation for new electronic-component assemblies that use a standard I-Bond circuit board reduces to: a schematic diagram, parts list, wire run list and an over-all component assembly drawing. Also the commonality of the basic boards would allow both reduced design costs and inventories since one multiple-board could support a variety of assembly modules.

To keep the assembled board from becoming any thicker than a multilayer board, wiring is done on the component side of the board. This also allows the use of conventional automatic and semiautomatic component-insertion equipment, permits the mechanized flow-soldering of all components on the board, and frees the non-component side of the board for use in automatic testing equipment that incorporates the "bed of nails" type of fixtures.

However, component-side wiring does reduce component density by 10% in comparison with a well-designed multilayer board. A completed impulse-bonded module is shown in Fig. 4, before an acrylic conformal coating is applied to seal and ruggedize the board.

In the making

The major steps in the manufacture and test of a typical impulse-bonded assembly of electronic components are outlined in Fig. 5. The process begins with construc-

6. Semiautomatic wiring. The semiautomatic bonding station consists of an X-Y positioning system, weld head, welding power supply, and a digital current monitor and alarm. The display panel indicates route, bond and cut commands to the operator.

7. Test station. Continuity and insulation tests on a completely wired board are done on an automatic wiring analyzer and a special universal contact fixture. If measurements are run continually at a low voltage, no damage results to active components, like semiconductors and integrated circuits, already assembled on the board.

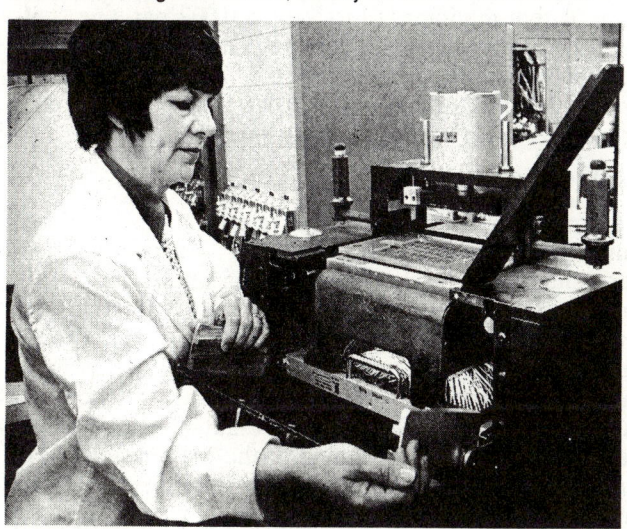

tion of the special I-Bond circuit board from blank copper-clad epoxy-glass material by means of numerically controlled drilling equipment and conventional photoplating processes. Electronic components are mounted either automatically or semiautomatically on the I-Bond circuit board and are secured to the pad matrix by a mechanized flow-soldering process. However, the components are not interconnected until the last step in the assembly process—the installation of the impulse-bonded wiring.

Note that, prior to this step, the component interconnections were not fixed. Changes are easily made in the key-punched run list used in the program that provides tapes for the bonding machines and also the wiring analyzer. After wiring, the assembled module then passes through three levels of tests for wiring, component integrity, and functional operation.

Software and the computer processing of the data bases which define the assembled module are a vital part of the I-Bond manufacturing and test process (Fig. 5). The Fortran IV procedures that generate machine-control tapes for the bonding machines and circuit analyzer are derived from programs used for wire-wrapping equipment. The programs used to generate tapes, time studies and operation sheets for the equipment that inserts components in I-Bond boards are identical with those used for conventional pc boards. Also, the automatic component fault-isolation and functional testing provided in house is supported by general-purpose computer facilities and programs developed for all avionic products.

These procedures help make the impulse-bonded wiring technique economically viable because they reduce much of the documentation cost normally associated with a linear-wiring process. For example, the computer-output of the job stream includes:

■ Operation sheets for the bonding process, including a sequential run list showing route points, bond points and bond-and-cut points, wire-trajectory information, and total node-lengths.
■ The numerical-control tape for the bonding machines which controls machine position, traversing speed, and bond-route-cut indicator displays.
■ The time-study analysis sheet, including details of the I-Bond wiring time-standard for this particular component assembly.
■ A timekeeper's rate card, used to credit each bond-machine operator for work performed.
■ The numerical-control tape for the automatic wiring analyzer machine.
■ Diagnostic trouble-shooting sheets in component terminology, used with the wiring analyzer to isolate specific wiring faults in the assembled board.

Semiautomated wiring

I-Bond circuit boards are wired at the 12 stations shown in Fig. 6. Each semiautomatic bonding station consists of an X-Y positioning system and weld head, manufactured by Accra-Point Arrays Corp., Santa Ana, Calif., a special welding power supply, and a digital-current monitor and alarm system.

The circuit board is mounted on the variable-speed servoed table beneath the fixed weld head. In the weld head is a small display panel that tells the operator whether to route, bond, or cut the wire in front of her. Universal tooling adapters register the board accurately on the table and also provide wire bundling features which aid the operator during wire lay-in.

The display-panel commands, together with table position, and traversing speed, are controlled by the machine's numerical-control unit. The step-by-step advance of the machine is controlled by the operator.

Wire routing paths are not random, but are defined by machine routing commands and the operation-sheet computer printouts (previously discussed) that guide the operator in dressing the mini-harness. The wires are sorted into bundles under the direction of the software program, which groups them in accordance with their functions, pick-up susceptibility and so forth. This assures a repetitive wiring operation and a consistent product.

The welding power supply, manufactured by Hughes Aircraft, Oceanside, Calif., consists of a nickel-cadmium battery bank together with adjustable pulse-timing and amplitude-shaping networks. However, the sensing feedback of the unit was modified to provide constant-current operation instead of constant-voltage operation during pulse discharge.

This modification was necessary because a constant-voltage source allows the total heat-flux in the immediate bond zone to vary with any change in series-resistance. But series-resistance changes radically with only small variations in the distance from the wire-pad contact point to the plated-through hole, which conducts the current to the lower electrode. Variations in the plating strata of the pad and plated-through hole will also alter total series-resistance. Therefore, a constant-current power source was essential to control the temperature in the bond zone.

This temperature is indirectly monitored by the digital current monitor, whose readings are actually proportional to the combined amplitude and duration of every current pulse. Upper and lower limits are set on this value, and an audible alarm sounds when the limits are exceeded.

Process controls and reliability

Like resistance-welding processes in general, the impulse-bonded wiring technique requires careful controls and process monitoring if joint strengths are to be consistant. These controls apply to the bonding-station equipment as well as the nickel wire, the composite pad, and the laminated board. The weld head, positioning system, power supply, and current monitor system at each bonding station are each periodically calibrated to individual equipment specifications and, more importantly, as an over-all system.

Basic to this calibration is the periodic development of a chart for each station that correlates the pulse energy delivered at the electrodes with the peel-strengths of the joints produced on a test board. Other factors in the bond schedule—the conductor-exposure force, bond-pressure force, electrode geometry, pulse duration and waveform—are held fixed by individual calibrations; the cable lengths, power-supply compliance, and weld-head characteristics are predetermined by the particular work-station equipment.

The curve on the chart then is used to determine the optimum input-energy point based on maximum joint-strengths, and the current monitor and alarm system is adjusted to allow only a ±5% deviation from this setting before the alarm goes off.

The current alarm system therefore detects when bond temperature is either too low for adequate diffusion bonding or so high that it may cause pad-burning, zapping or weld-blowout, or localized fusion-welding that would embrittle the joint.

To date, the reliability of impulse-bonded termination has been excellent, as demonstrated by impulse-bonded assemblies in airborne service. For an 18-month flight-service period, with 8.1 billion bond hours accumulated, the joint failure rate is 0.002 parts per million hours (ppmh). This compares very well with the accepted failure-rate predictions of 0.034 ppmh and 0.079 ppmh for the airborne soldered-wire termination and resistance-weld termination, respectively.

Three-level testing

Impulse-bonded component assemblies undergo three levels of testing in which three types of equipment are employed. An automatic wiring analyzer tests the photoplated artwork and the circuit connections installed by the bonding machines. An automatic fault-isolation test system tests the parameters of the individual components using guarded-resistance and capacitance measurements. Finally, automatic test equipment evaluates over-all functioning of the entire assembly to assure its proper operation and to guarantee the interchangeability of assemblies of the same type.

The wiring analyzer consists of a commercial tape-programed circuit analyzer together with a universal contact fixture with a complement of pressure-pin test modules (Fig. 7). It was adapted from a test facility originally developed for circuit testing of bare multilayer boards. Both the analyzer and the fixture are equipped for kelvin (four-wire) measurements of continuity and insulation resistance of up to 1,500 points on the component assemblies.

The test complex has been invaluable in screening out the various manufacturing defects that can occur in the I-Bond wiring process. An interesting feature of this testing is that by running the continuity/anticontinuity measurements at a very low level (less than 0.5 volt), it is possible to ignore the components already assembled on the board. □

Diffusion bonding explained

In molecular diffusion bonding, the molecules of one metal enter (diffuse into) the crystalline lattice structure of another and vice versa, forming a so-called "solid solution." It only happens when the two metals are similar enough in molecular and crystalline structure for their atoms to fit into each other's lattice without recrystallization.

In the bonding process, heat, together with high pressure on the contact face, accelerates the diffusion. But the temperatures in the bond zone must not be so high that they cause localized melting or fusion and recrystallization and the formation of a weld nugget. If this fusion and unannealed recrystallization take place, the bond joint is weakened, being compromised by embrittlement.

Generally, when surface diffusion occurs between a metal and its alloy or between two alloys, the atoms tend to move from a region of high concentration into a region of lower concentration. In the case of impulse bonding, for instance, the nickel atoms from the wire tend to migrate across the contact face to the copper-nickel pad, and copper atoms from the pad tend to diffuse into the nickel lattice of the wire.

Planar stitch welding yields higher board density at lower cost

Low board profile is another benefit of interconnection technique that welds nickel wire to stainless steel pads

by Rene Lemaire and Ray Calvin, Augat Interconnection Products Division, Augat Inc., Attleboro, Mass.

☐ High wiring densities and low profiles are a combination often hard to beat when loading printed-circuit boards with components. When such a wiring method can lower costs over the established high-density low-profile techniques, then it is a strong contender for a myriad of electronics applications.

This advanced interconnection technique is planar stitch-wiring. It is based on welding nickel wire to stainless steel pads on special pc boards. Especially for lower production runs, it already is competitive with multilayer board techniques.

Welded wires

In planar stitch-wiring, point-to-point circuit interconnections are made by electrically welding lengths of Teflon-insulated solid nickel wire to stainless steel pads on the pc boards. On both sides of the board, the weld pads form a grid around the plated-through interconnections into which the components are inserted. Figure 1 shows an example of a finished board.

The interconnections may take the shortest distance between points, or they may be routed indirectly according to the designer's wiring map. Twisted-pair connections may be made; also, weld pads on the components side may serve as interconnects, so long as the wires are channeled around the components' locations.

Like the Multiwire technique, planar stitch-wiring can bond wire continuously.

The insulated wires used in both techniques can be layered, which provides the high density of interconnections. Multiwire achieves its low profile by bonding the wire to an adhesive substrate and terminating it to a plated-through hole. Planar stitch-wiring achieves its low profile with its stainless steel pads—in contrast to the similar stitch-wiring, which welds the wire to stainless steel pins, giving a higher profile and increased production costs due to the pin insertions required.

Another high-density low-profile alternative is the tried and true multilayer board, which may have as many as 30 internal conductive layers for the interconnection tasks. These complex boards generally are restricted to military, aerospace, and high-end computer applications, although a simplified method is beginning to make inroads into commercial uses.

The planar stitch-wired technique came from the APAC division of Varian Associates, aiming at military and aerospace applications. APAC became the Planar division of Augat in October 1977 and has been developing the method for use in commercial applications.

The solid, Teflon-coated nickel wire is 0.010 inch in diameter and has a minimum tensile strength of 45,000 pounds per square inch. Nickel wire and the stainless steel used for the weld pads have the bonding points and corrosion resistance necessary for electric welding, as well as the required differences in resistivity. The nickel provides a strong connection, resistant to shock and vibration, because it bonds without recrystallization or embrittlement.

The weld process

In the welding process (Fig. 2), the insulated wire automatically feeds through a hollow upper electrode and is pressed against the weld pad. The electrode is under pressure and mechanically displaces the Teflon insulation until a metal-to-metal contact is made to the weld pad—connected by the plated-through hole to a similar pad on the other side of the board, where the lower electrode is resting. The wire is slightly flattened

1. Stitch wiring. Combined are views of the wiring and component sides of a planar stitch-wired board. On the wiring side, insulated solid nickel wire is electrically welded to stainless steel pads. Wiring and component pads are connected by plated-through holes.

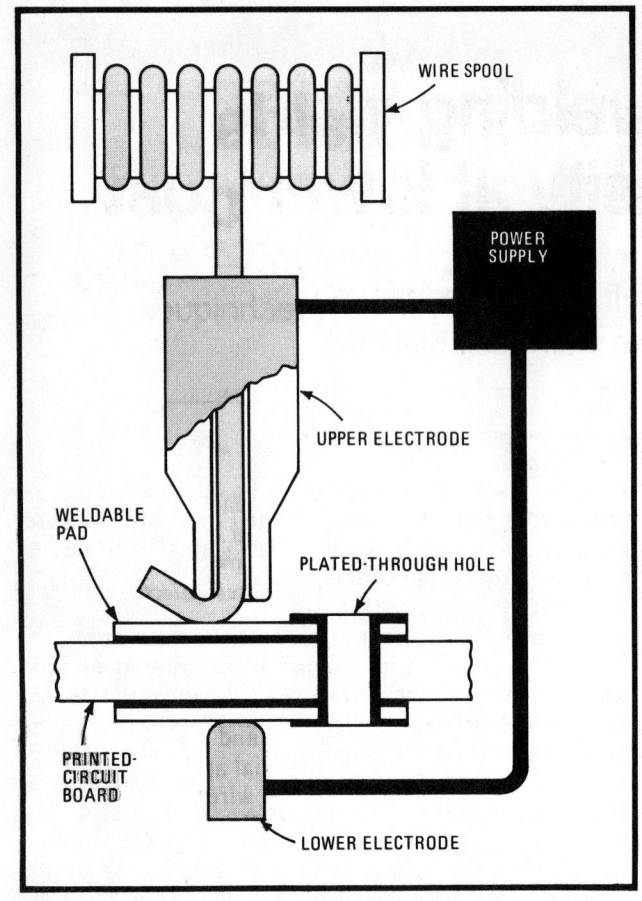

2. Electric bonding. In stitch-wiring, insulated wire is fed through an upper electrode and pressed against a board pad, displacing the insulation. A current pulse is then fed through the two electrodes, bonding the nickel wire to the stainless steel pad.

3. Semiautomatic. In Augat's System 78 planar stitch welder, the desired board pad is automatically positioned under the fixed welding head. The operator then actuates the head and cuts the wire, and the machine moves the board to its next position.

for a increased surface area, and a 2.5-millisecond high-current pulse from a capacitive-discharge power supply produces the bond.

Typically, the bond strength will be 85% of the ultimate tensile strength, varying only ±5%. Average heat generated during the bonding process is low, having no effect on the laminate or solder plating of the board. Underneath the steel pads is copper plating, which acts as a heat sink and minimizes the heating effect and changes in resistance caused by welding.

The welding may be either a manual or a semiautomatic process. In the manual approach, the operator follows a wire list in positioning the board for each weld. In the semiautomatic approach (Fig. 3), a punched-tape program positions the board on its X-Y axes, waiting until the operator has used the weld-activating foot pedal before moving on to the next interconnection. The difference in output is considerable: more than 100 wires an hour for the manual machine; more than 500 wires an hour for the semiautomatic machine.

A steel-clad board

Planar stitch wiring does require a special pc board, which begins with a core of FR-4 flame-retardant epoxy-glass. Laminated to both sides are sheets of 0.003-in. stainless steel, electroplated on their inner side with a thin layer of nickel and with a 0.0015-to-0.0020-in. layer made of copper.

After the stainless steel lamination, holes are drilled through the laminate to form the required pattern. Then the nickel film is plated to the stainless steel, and the epoxy-glass in the holes is chemically sensitized so that it, as well as the board surfaces, can be plated with the copper to a minimum thickness of 0.0015 in.

What follows are the basic steps to form a pc image: coating, exposure, and developing. Next are two etch cycles, one removing all metal layers where they are unnecessary, and the other selectively removing the copper from on top of what are then the weld pads.

During the etching, the plated-through holes and the connecting terminals are protected by an applied mask. After removal of this mask, solder-coating the holes and the terminals completes the manufacture of the basic printed-circuit board.

Myriad board designs

To connect integrated circuits (ICs) or discrete parts in dual in-line packages (DIPs) to a planar stitch-wired board, the patterns for the pads and plated-through holes may be custom designs or standard configurations. The off-the-shelf designs include patterns for 14-, 16-, and 24-lead DIPs, as well as universal layouts. The power and ground planes may be etched directly into the pc board to connect directly to the components, or they may be connected to separate weld pads at each IC position.

Pads formed outboard of the plated-through holes are standard. For even greater density, they may be inboard with the components mounted over them.

Computer-generated design means that the boards may be tailored to the most sophisticated custom circuitry that an engineer may require. Additional voltage and ground planes may be gained with inner layers tied

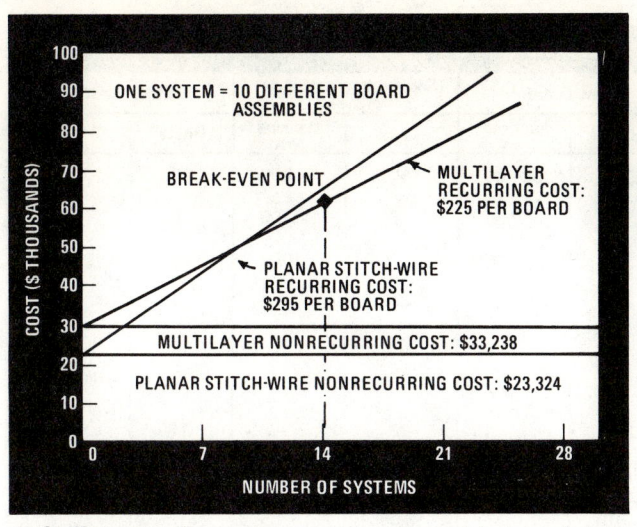

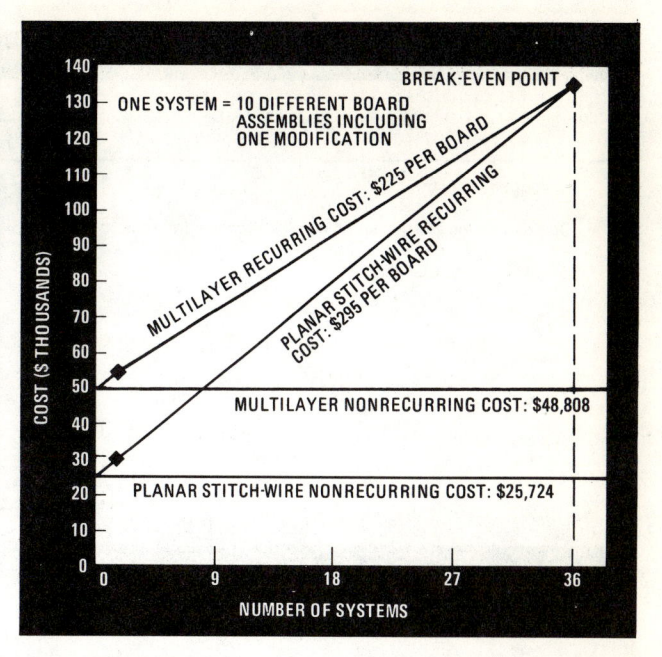

4. Getting even. The break-even analysis to the left shows that, for small quantities, stitch-wired boards are less costly than multilayer boards. One board modification to the same system can radically shift the break-even point upward (right).

through the plated-through holes to weld pads. Large aluminum heat sinks have been designed and bonded to component planes, as well.

A wide range of input/output interconnections are available, including standard nickel-gold edge connectors, two-piece right-angle connectors, and flat cable inter-connectors. Moreover, component mounting is available in a number of options: direct mounting, wave-soldered low-profile dual-in-line-package sockets, or machined socket pins.

A prime advantage of planar stitch-wired boards is the high wiring density—the ability to layer a large number of interconnections in a small amount of space. The density is comparable to that of a 15-layer pc board made by the multilayer method.

Other advantages

Even with this high density, the low profile of planar boards permits packaging in a standard cabinet with as little as 0.5 in. in between. By insuring compatibility with wire-wrapped boards, many systems using wire-wrapping can be expanded by selected board replacement using planar stitch-wiring. Since all wiring in the stitch-wired system is adjacent to the ground plane, the low profile also minimizes any delay of logic signals due to impedance variations, especially at emitter-coupled-logic speeds.

Another advantage of this wiring method is its flexibility, which makes it suitable for prototyping. Since it is easy to apply wires to the interconnection points or to change them, there is no need for breadboards when designing and debugging a new circuit or when expanding an existing one.

The planar boards also work well in full production runs. All connections are readily accessible, so if change is necessary, the wires being replaced are simply cut out and the new connections welded to the stainless steel pads. As many as four bonded wires may be accommodated on each pad. Even if a welding machine is not readily available, there is room for soldered connections at the terminal area of the plated-through holes.

Multilayer boards are the most widely used methods for obtaining high-density low-profile systems. However, they are cost-effective primarily in high-volume production of circuits that are past design change.

Planar stitch-wired boards have the flexibility and capability for being changed in production runs. Also start-up costs and turnaround times can be cut by a third to a half, compared to complex multilayer board designs.

Cutting costs

A cost study recently conducted compared the planar stitch-wiring and multilayer techniques. It found that the initial cost for a standard planar board is about half that for a multilayer board, while a custom stitch-wired board was about 65% of the cost of a multilayer unit. The study was fairly extensive, comparing nonrecurring and recurring costs for a typical small system with and without modifications.

The resulting computations, in the table and Fig. 4, are based upon doing most or all of the nonrecurring work in house, as well as the wiring. If Augat were to do the artwork, engineering, and programming, nonrecurring costs for stitch-wiring should be less than half the amount shown in the table.

For 10 board assemblies per system with no modifications, stitch-wiring has a $9,900 savings in nonrecurring costs. However, it has a higher recurring cost, resulting in a break-even point, for this example, of 14 systems, or 140 boards (left in Fig. 4).

With modifications

If just one modification per board is introduced into the sample system, the crossover point goes much higher. Nonrecurring costs for the first 10 boards are the same as before, but there is an additional $13,000 savings in nonrecurring costs for the stitch-wired board when making the modifications or changes. Also, production of the first system with the change results in another $1,840 savings with stitch-wiring raising the break-even

TYPICAL NONRECURRING COSTS

Multilayer boards	New board Hours	New board Material cost	Second board of same configuration Hours	Second board of same configuration Material cost	Modification Hours	Modification Material cost
Layout and artwork	90	$24.00	60	$24.00	45	$24.00
Board detailing and preparing numerical-coded drill tape	6	2.00	6	2.00	6	2.00
Assembly and parts list	24	1.00	8	1.00	2	1.00
Schematic	40	1.00	40	1.00	4	
Tooling	7					
Setup	6					
Drawing check	32		23		11	
Subtotal	205	$28.00	137	$28.00	68	$27.00
Labor cost at $22.50 an hour	$4,610		$3,080		$1,530	
TOTAL	**$4,638**		**$3,108**		**$1,557**	

Planar stitch-wire boards	Custom board Hours	Custom board Material cost	Standard board Hours	Standard board Material cost	Modification Hours	Modification Material cost
Layout and artwork	38	$ 8.00				
Board detailing and numerical-coded drill tape (computer-aided design)	6	2.00				
Cover detail	4	1.00				
Assembly master	24	1.00				
Assembly drawing and parts list	6	1.00	6	$ 1.00	2	
Schematic	40	1.00	40	1.00	4	
From-to wire list	24		24		1	
Keypunch and computer service	5	30.00	5	30.00	1	$15.00
Tooling	40	20.00				
Setup	5		5			
Drawing check	24		9		2	
Subtotal	216	$64.00	89	$32.00	10	$15.00
Labor cost at $22.50 an hour	$4,860		$2,010		$225	
TOTAL	**$4,924**		**$2,042**		**$240**	

point to 36 systems or 360 boards (right in Fig. 4).

Users of planar stitch-wiring are reporting excellent results in applications that have limited space or require rugged construction to resist the stresses of shock and vibration. For example, the Aerospace and Electronic Systems division of Westinghouse Electric Corp is using planar boards for satellite-launch vehicle control and integrated logistic support. Hughes Aircraft Co. is using them in an electronics package in an experimental military tank, which has a harsh shock and vibration environment. Cubic Corp. is flying a number of planar boards on an electronic package for a drone aircraft.

Commercial applications

Besides these aerospace and military uses, there are many commercial applications, including Itek Corp.'s use in imaging reproduction systems and very dense packaging for sophisticated aircraft instrumentation from the Avionics division of International Telephone and Telegraph Corp. Another aerospace firm used the planar method in a back panel.

To enhance the attractiveness of this wiring method, there is some down-to-earth research needed on equipment development. For instance, a study is being made of the feasibility of a relatively inexpensive field tool facilitating stitch-wiring changes at locations where a welding machine is unavailable. Such a tool will likely be comparable to the hand-held Gardner-Denver wiring gun now used with wire-wrapped boards.

Another development that may appear within the next few months is an economy line of single-sided stitch-wired boards, which promise to substantially cut production costs. It is based on a technique called parallel-gap welding, which will be designed for incorporation into both the microprocessor-controlled semiautomatic welding machine and the manual unit. ☐

Part 5
IC Packages and Connectors

Chip carriers are making inroads

New technique, in which leadless ceramic devices are mounted
on ceramic mother boards, means less weight than DIPs or hybrids

by Jerry Lyman, Packaging & Production Editor

A relatively new packaging technique is quietly muscling in on the market for complex, high-density electronic functions heretofore the preserve of component-laden printed-circuit boards and multichip hybrid integrated circuits. The reason: that new technique—mounting leadless ceramic chip carriers on a ceramic mother carrier—means smaller and lighter though denser systems and better thermal and electrical performance and reparability than does the DIP-pc approach.

The chip carriers, developed by 3M Co. of St. Paul, Minn. are now being produced by 3M and Kyocera International Inc., San Diego, Calif. They are square, cofired ceramic packages with gold bumps on their bottom face. They are made the same way as high-reliability ceramic dual in-line packages but are much smaller.

At first, the packages, which have 16 to 64 input/output counts, are reflow-soldered to some type of interconnect substrate. In 1975, 3M developed simple ceramic multilayer boards for mounting four chip carriers on a substrate the size of a dual in-line package. That ceramic mother board had side-brazed leads that could be inserted into holes on a large pc board. Then came other ceramic-mother-board approaches that interfaced the mother carrier directly with a connector.

The ceramic mother boards are presently available from 3M and Kyocera. But because of delivery problems, some firms have to make their own while buying the chip carriers from those suppliers.

There are four pioneer users of the new technique. Texas Instruments

PACKAGING COMPARISONS				
	Weight (grams)	Area (cm²)	Yield	Cost
Printed wiring board	52	81	high	low
Chip carrier/mother carrier	12	12	high	medium
Hybrid	10	6	low	high

Inc. in Dallas, Honeywell Aerospace in St. Petersburg, Fla., and RCA Corp.'s missile and surface radar operation in Moorestown, N. J., all use it in products for the military. The fourth, Motorola Inc.'s Fort Lauderdale, Fla., operation, apparently is the first to incorporate it in commercial systems.

In papers delivered late last month at the International Symposium on Microelectronics in Baltimore, TI and Motorola pronounced themselves well satisfied with the new technique: not only does it permit packing densities approaching those of multichip hybrids, but it permits them to pretest the chips mounted on the carriers and thus avoid the low hybrid yield that is caused by low chip yields.

In putting together a hypothetical circuit containing 20 digital ICs (see table), Texas Instruments compared three techniques: DIP and pc board, chip carrier and mother carrier, and multichip hybrid. Not surprisingly, TI calculated that the combination of chip carrier with mother carrier would weigh less and occupy less space than the pc board, yet would have the same high yields. As for cost, it ranks second to pc boards.

Says Texas Instruments' Jon Prolop: "With increased automation, the chip-carrier method could approach the cost of the pc board. And, though it will never eliminate the complex multichip hybrid, which is the ultimate in high density and performance, it is indeed an attractive alternative."

At Motorola, says Subash Khadape, "we are taking steps to reduce carrier assembly costs on pocket pagers and two-way radios." Motorola buys its chip carriers in a 330-piece array that is snapped apart along scribed lines. For the future, the firm's manufacturing engineers are considering automatically wire-bonding identical chips into each chip carrier of the array. Then the loaded chip carrier could be split into individual carriers, tested, and reflow-soldered to individual mother boards.

Big reduction. The size and weight advantages of the chip carrier over DIP counterparts are considerable. On average, chip carriers available today are a sixth the size of DIPs. And TI says that it has found it a simple matter to obtain size and weight reductions of 60% in assembled circuits.

Another point made by TI is not widely known. That is the thermal improvement achievable with a chip-carrier–mother-carrier assembly:

173

PACKAGE YIELD AS FUNCTION OF CHIP YIELD

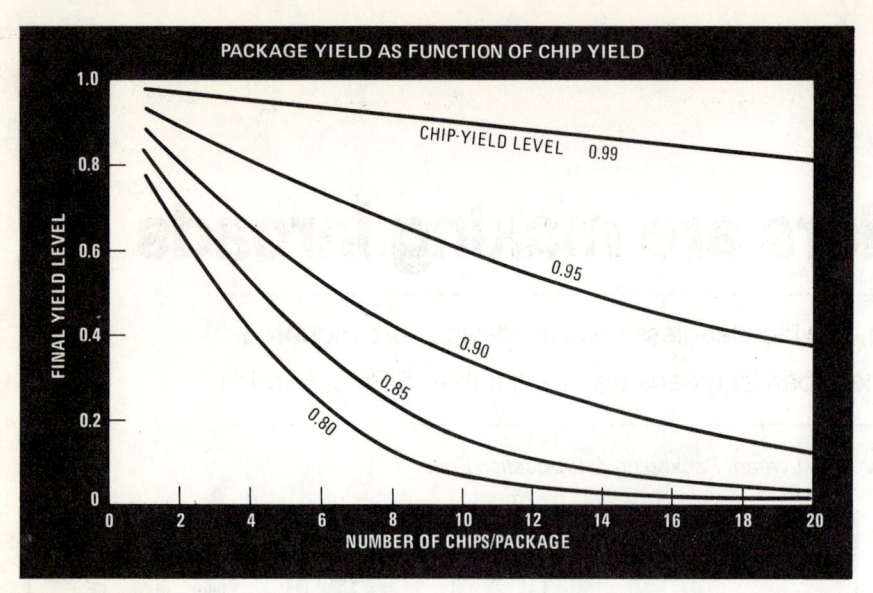

[Graph showing Final Yield Level vs Number of Chips/Package, with curves for Chip-Yield Levels of 0.99, 0.95, 0.90, 0.85, and 0.80]

TI's thermal analysis indicates that one particular unit using conventional pc board assembly had a junction temperature of 125°C, while the functionally equivalent chip carrier assembly registered 99°C. This improvement is due to the superior thermal conductance of ceramic—more than two orders of magnitude above that of G-10 printed-circuit material.

Removable. Finally, the chip carrier is distinctly easier to repair than the pc boards. Extremely large DIPs are very difficult to remove without damaging the board. In a series of tests, TI subjected an array of chip carriers reflow-soldered to a mother carrier to a simulated repair cycle in which a chip carrier was removed and replaced an averge of 15 times before it or its mother carrier was damaged.

Texas Instruments and Motorola have used the technique in a variety of applications ranging from the exotic to the commonplace. For example, designers at TI have applied the new packaging concept extensively to avionic and missile packages using both custom and standard chip carriers and mother carriers. In a charge-coupled-device integrated reformatter package, the Dallas company's designers use five custom chip carriers plus one standard chip carrier on a 10-conductive-layer cofired ceramic substrate with a brazed-on lead frame. One of the chip carriers houses one custom CCD, two custom complementary-metal-oxide-semiconductor drivers, and one transistor. In TI's arrangement, four of the carriers are mounted on a pc board.

Also at TI, eight CCD memory chips are packaged in eight chip carriers and reflow-soldered to a thick-film mother carrier. This scheme allows 524,288 bits of memory to be packed into 7% of the space required for a conventional pc board assembly.

For a missile, TI's designers go to a custom, circular mother carrier to house a microprocessor, read-only and random-access memories, and input/output buffer devices all on chip carriers. In this system, chip-carrier packaging techniques reduce volume by 67% and weight by 75%. Because of these reductions, it was possible to cut the length of the missile itself.

Motorola also has some novel applications of the technique. In one circuit using chip carriers, a purely resistive substrate mounted on a carrier with a transparent cover is laser-trimmed to fine-tune an active filter. In another small chip-carrier—mother-carrier system, Motorola's designers have mixed leadless inverted devices in with chip carriers.

Failure-free. For its part, RCA's Missile and Surface Radar division has been using chip carriers, chiefly in avionics programs, for about four years. John Bauer, manager of design and development, says, "Reliability has been excellent, and no consistent failure mechanism has turned up in either carrier or mother board." The division has an advantage: it can buy many of the chips it needs already packaged and tested from RCA's Solid State division. And both RCA and Honeywell Aerospace are experimenting with a package that combines the film-carrier chip-handling process with the ceramic-chip-carrier packaging process, permitting automatic bonding of the chip on film to the carrier's cavity, which is then sealed.

Honeywell's experimental work comes after about two years of using the chip carriers with its own ceramic mother boards. Calvin Adkins, a senior designer on many of the division's military programs, says: "We went to this technique when our hybrid substrates got larger than 2 by 2 inches. At this point it was difficult to seal the overall hybrid, and the chip carriers, each with its own hermetic seal, solved the problem. No field failures have been reported in systems built around the chip carrier." Honeywell is trying to use only four sizes of carriers: devices requiring 16, 18, 24, and 48 pinouts.

Will the chip carrier win a larger following? The success of the concept, says Motorola's Khadape, "hinges on two vital factors: standardization and automation." Some progress has been made in standardization but much remains to be done in automating the mounting and soldering of chip carriers to mother carriers. □

Special report: film carriers star in high-volume IC production

by Jerry Lyman, *Packaging & Production Editor*

Automated bonding machines can package thousands of film-mounted ICs per hour with great reliability

☐ In a move that's gathering momentum around the world, more and more semiconductor manufacturers are adopting an automated film-carrier approach to assembling integrated-circuit packages. For the manufacturer, film carrier has meant lower production costs and higher profit margins on high-volume products. For the user, it means higher reliability at low prices.

With the technique, thousands of IC packages can be produced per machine per hour. Using manual wire bonding, a single worker can process about 60 chips in the same period of time. Moreover, the wire bonds are weak, having a pull strength of only about 15 grams, and they also cause corrosion about their point of bonding, especially in plastic dual in-line packages.

The automated method derives from the Minimod film-carrier technique that General Electric developed in the 1960s. Its name comes from its dependence on a sprocketed, nonconductive film with a copper surface etched into spidery IC interconnect patterns. Reels of this film or tape are fed along with specially prepared IC chips into a bonding machine. When the tape emerges, it is studded with one chip to each interconnect pattern, ready for bonding by other machines to lead frames.

Alternative names for the process are film-carrier gang bonding, beam-tape automated assembly, and tape automated bonding. The gang bonding refers to the fact that all a chip's terminals are attached simultaneously and not serially to the copper interconnects, variously known as beams or leads.

But under whatever name, the technique lowers the cost of the IC to the user and also enhances its reliability. First, the IC terminal pads are protected against corrosion by special metallic barriers. Second, the gang bonding doubles bond pull strengths to 30 grams and more. Third, the thick copper leads, typically 3 mils wide by 1.4 mils thick, allow the chip to dissipate much more heat than it could through the 1-mil wire bonds.

In addition, the film-carrier method presents an engi-

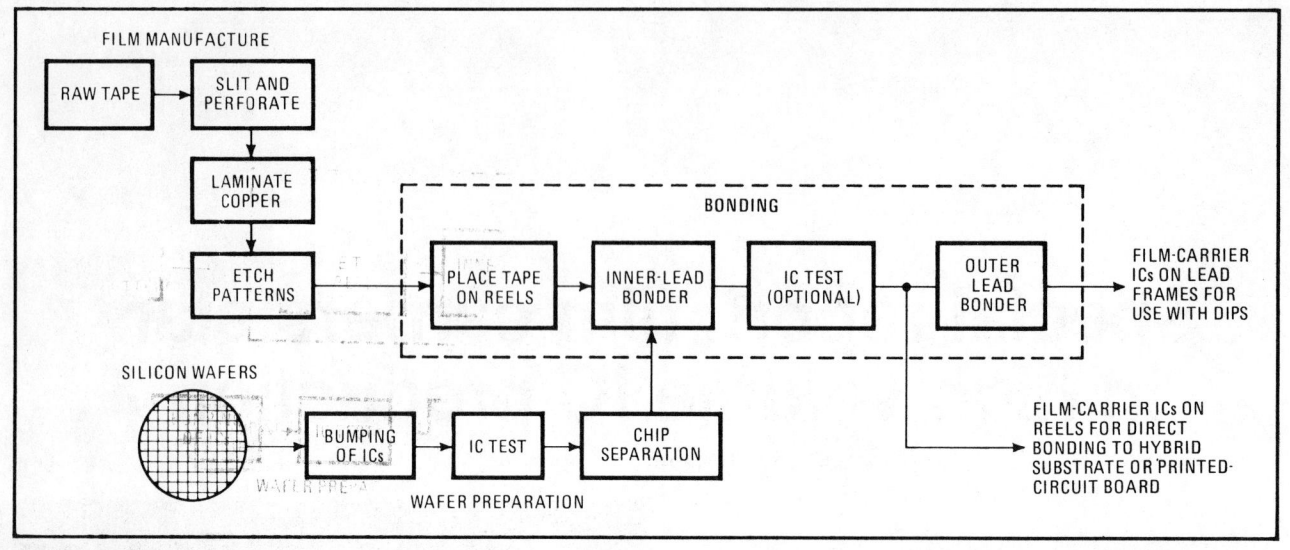

1. ICs on film. Before the highly automated inner and outer lead bonding of ICs to film carriers begins, copper must be laminated to tape and etched into spidery interconnects, and wafers must have gold bumps added to the aluminum terminal pads of each integrated circuit.

neer with a new option. Instead of buying the film-carrier ICs already in DIPs, he can buy them on reels or strips of film to be soldered directly to rigid or flexible pc boards or hybrid substrates. In effect, the new method has produced an IC package—a frame of film—ideal for high-density packaging but with easily attachable leads.

Today, many, but not all, IC manufacturers and users in the U.S., Japan, and Europe have either adopted film-carrier assembly or are experimenting with it. In the U.S., much of the film-carrier output is small-scale transistor-transistor-logic ICs in plastic 14-pin DIPs. International companies are supplying MOS as well as bipolar ICs, both linear and digital, on reels and strips of film. Production rates worldwide range between 1,000 to 2,500 ICs per hour and are expected to rise even further soon.

Figure 1 charts the flow of a typical film-carrier or beam-tape manufacturing process. At the heart of the system is the inner lead bonder, so called because it bonds the chip terminals to the copper leads at their inner ends, round the square hole at the center of the copper interconnect pattern. Conversely, the outer lead bonder later attaches the outer ends of the leads to a metal lead frame for dual in-line packaging. Alternatively, the loaded tape is placed on a reel. Note that the nonconductive nature of the film becomes important when the ICs are to be tested on tape.

Preparing the wafer

Standard IC wafers cannot be used for the film-carrier process without modification. Normally thermocompression or alloy-type bonding is used to join the chip's terminal pads to the copper leads, and to protect the pads from collapsing under the pressure, a metal bump about half a mil high must be built up on each of them. Moreover, these bumps must include a barrier metallurgy system at their interface with the die. Otherwise, the heat will diffuse the bump metal into the die, where it will alloy harmfully with the silicon substrate or aluminum pad.

This barrier system, incidentally, also seals off the aluminum contact pads from corrosion. Thus it is one of the factors making the chips used for film-carrier work inherently more reliable than unmodified chips.

Many types and combinations of metals are used for the barrier between bump and die. They include chromium, copper, titanium, tungsten, nickel, and gold. A section of a typical gold bump is shown in Fig 2. To form the barrier in the bump illustrated, a layer of chromium is placed on top of the existing aluminum pad. Onto the chromium is evaporated a layer of copper 2,500 angstroms thick followed by a 1,000-Å layer of gold. Finally, a gold bump about 4 mils square in area and 0.5 to 1 mil high is plated on.

An alternative way to add bumps is in use at RCA's Solid State division, Somerville, N.J. There, gold bumps are built up over successive layers of silicon nitride passivation, titanium, platinum, and gold.

Keeping chips in their place

After construction of the bumps, the ICs in a wafer are all tested, and any rejects are marked. Next, to allow for individual chip handling, the wafer is stuck to some sort of adhesive base before being sawed apart.

At National Semiconductor Corp. of Santa Clara, Calif., for instance, the wafer is mounted on a flat substrate with a proprietary heat-released adhesive system and then sawed through just to the substrate. Other manufacturers simply attach the wafer to a plastic base by a layer of wax before cutting it apart. In both cases, the heat of the bonding process later on is enough to soften the wax or adhesive and release the chip.

To anyone contemplating going into beam-type assembly, the choice of tape or film must seem quite bewildering. There are three tape constructions (single-layer, two-layer and three-layer), at least eight tape widths (8, super 8, 9, 11, 14, 16, 35, and 70 millimeters) and four film materials (polyimide, polyester, and the as-yet developmental polyethersulfone and polyparabanic acid).

At present, the tape used in commercial applications

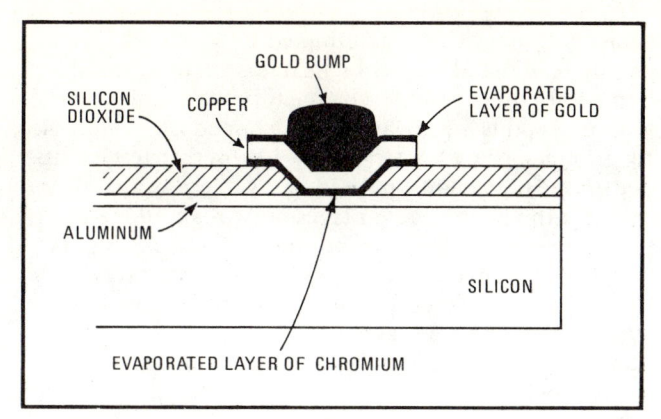

2. Adding height. To resist bonding pressures, chips used in the film-carrier process need special gold bumps built up on their normal aluminum pads. Metallic barrier between bump and pad protects the aluminum from alloying with the gold and also from corrosion.

is three- or two-layer in instruction. Single-layer tape is simply copper foil, and though some experimentation is going on with it, it is still in the laboratory stage (see "The film-less film carrier").

Three-layer tapes were used on the original Minimod, and they are still the most widely used today. This type of tape consists of copper foil 1.4 mils thick that is bonded to polyimide film with a layer or barrier of polyester or epoxy base adhesive. Normally, the polyimide film is bought in large rolls, coated on one side with adhesive, and slit to the correct width. The sprocket holes and all other openings, such as the square holes for the IC chips, are punched out, and the tape is then laminated to the copper foil. Next, the foil is selectively etched into the radiating patterns that form the IC interconnects and, as a last step, the beams are plated with gold, tin, lead, or solder. The tapes come in the so-called cine widths—8, super 8, 16, 35, and 70 millimeters.

A disadvantage of this tape type is the middle layer—the adhesive, which is rated for only 20 to 30 seconds at 200° C and therefore limits the temperature at which the tape can later be processed.

Doing without the glue

The adhesive layer is eliminated in the two-layer tape originated by and still unique to the Electronic Products division of 3M Co., St. Paul, Minn. 3M developed a proprietary method of casting a layer of 0.5-mil-thick polyimide onto rolled, annealed 1-ounce copper foil. Subsequent manufacturing steps involve applying resist and developing an image of the interconnect on both sides of the tape, milling the polyimide chemically, and etching the copper.

Views of both sides of a typical 3M tape are shown in Fig. 3. It comes in 11- and 16-mm widths and will soon also be available in a 14-mm width.

The temperature limitation on the two-layer 3M tape is set only by the thermal characteristic of its cast polyimide—about 400°C. On the other hand, the thinness of the polyimide requires the copper foil to extend the full width of the tape, to strengthen the sprocket holes. This is unnecessary on three-layer tape, which is easier to manufacture in this respect.

3. Adhesiveless film. Top and bottom views of a two-layer tape, manufactured by 3M Co., show layer of copper plus a layer of polyimide film. Temperature-limiting component of this tape is the polyimide film rather than the adhesive used in three-layer tapes.

As must by now be obvious, most beam tapes currently being manufactured use as their base a polyimide film that costs about $30 per pound. The other option is polyester film, which costs only $3 or so per pound. However, a polyimide-based tape can survive 400°C for a short time, whereas polyester is limited to 160°C.

This difference is important if the film-carrier ICs are to end up in DIPs. Carmen Burns, director of National Semiconductor's automated assembly program, chose polyimide "because of the temperature bonding requirements of molding a DIP to a lead frame." Other companies prefer polyimide because it exhibits less thermal shrinkage than polyester.

Nevertheless, polyester has its uses. It can serve as a structural carrier when the IC is removed from the film, say, in a hybrid application.

Two new film materials in development are polyethersulfone (PES) from ICI United States Inc. of Wilmington, Del., and polyparabanic acid (PPA) from Exxon Corp. Each costs about half as much as polyimide and is classified as a Type H (high-temperature) film, though their maximum short-term temperatures are lower than for polyimide—the figures are 220°C for PES and 275°C for PPA.

Of the two, PPA is at the earlier stage of development, and other information on its characteristics is very sketchy. PES, on the other hand, is available in developmental quantities and is currently being evaluated for use as an IC carrier by major IC and film companies. Ac-

The film-less film carrier

The film in present film carriers preserves the configuration of the fragile copper IC interconnect or spider and acts as an insulator. But its presence adds to the final cost of the IC. In addition, many plastics contain metallic ions that could contaminate an adjacent chip. For these reasons, several companies are now experimenting with etching IC patterns on copper foil unsupported by plastic film.

According to Dmitry Grabbe, a staff engineer at AMP Inc., all-copper automated bonding systems may be less than two to three years away. These systems will use 3-mil-thick copper foils etched from both sides and should dissipate heat better than the 1.5-mil-thick copper foil of present film carriers.

There are a couple of drawbacks, though. As Thomas Angelucci, president of International Micro Industries, points out, "An all-copper system will not be possible to test in process—say, at the point where all inner bonds are made and the semifinished ICs are reeled up. Only an insulated tape is capable of this. Also, present-day machines are designed to run with plastic films, not copper foil, so considerable machine redesign may be needed."

In any case, all-copper tapes are a good possibility for low-cost DIPs containing small-scale TTL ICs, when these need not be tested prior to outer lead bonding. However, for LSI chips that require testing, two- and three-layer tapes are a better choice.

---

cording to Charles Miller, manager of new products at ICI United States Inc., it has better dimensional stability than polyimide film, absorbs less moisture, and doesn't tear as easily. But it also is flammable and is attacked by some common solvents. In addition, this new material is potentially suitable for flexible printed circuits and flat cables.

Which tape type or tape width is best? The proper choice of tape width is determined by IC die size, number of leads, and the chip heat dissipation required. Sometimes users go to large tape formats simply to provide test points which are later chopped off. Figure 4, for instance, shows a tape with 14 discrete test points per frame interconnect pattern.

As for the relative advantages of two- and three-layer tapes, both types are currently being used for bipolar, metal-oxide-semiconductor, and complementary-MOS ICs without any special problems. In fact, as Arnold Rose, manager of packaging technology for RCA's Solid State division, observes, "What really counts is the cost-effectiveness, delivery capacity, and reliability of any tape system."

Many large IC manufacturers and users in the U.S. and overseas choose to make their own tapes. They include Texas Instruments, Fairchild Semiconductor, Motorola Semiconductor, and Honeywell Bull.

Those who prefer to buy tape can obtain the three-layer type with copper patterns already etched on it from International Micro Industries, Cherry Hill, N.J., and AMP Inc., Harrisburg, Pa. Three-layer tapes with unetched copper foil and prepunched sprocket and IC holes are available from Rogers Corp., Chandler, Ariz. Only 3M Co. and AMP Inc. supply two-layer tapes.

Inner and outer lead bonding

Automation really enters the beam-tape assembly process in the action of two complex machines, the inner lead and outer lead bonders (Fig. 5). Inner lead bonding mates the IC chip to the patterned film, while outer lead bonders fasten the tape-mounted ICs to a lead frame or substrate ready for packaging.

If the wafers have a high yield of good chips, all these steps can be done at extremely high speeds. The rate can range from 1,000 to 2,500 units per hour in the inner lead bonder.

The type of inner lead bonding used is determined by the metallurgy of the chip bumps and the tape beams. A popular system uses thermocompression bonding of gold to gold—gold bumps to gold-plated copper beams. Here, the heat and pressure between them force the mating parts to adhere.

Typical bond parameters for such gold-to-gold bonding of a 14-lead device, as supplied by Jade Corp. of Huntingdon Valley, Pa., are: a dwell time of 0.25 second, a bond force of 1.25 kilograms (for a 14-pin device), and a thermode or bonding tool temperature of 550°C. In this method, bonding temperature may fall anywhere between 300° and 700°C, and it is held at a constant level.

Another method, eutectic bonding or pulse soldering, is used to bond gold bumps to tin- or lead-plated copper beams. A temperature pulse is put into the thermode, producing 280°C for less than half a second and causing a gold-tin alloy to form between the bump and the plated beam.

The diagram in Fig. 6 illustrates the process of automated bonding. The first step is to align the chip's bumps to a particular beam pattern. This normally requires using a microscope, sometimes in combination with closed-circuit television. Next the bonding ther-

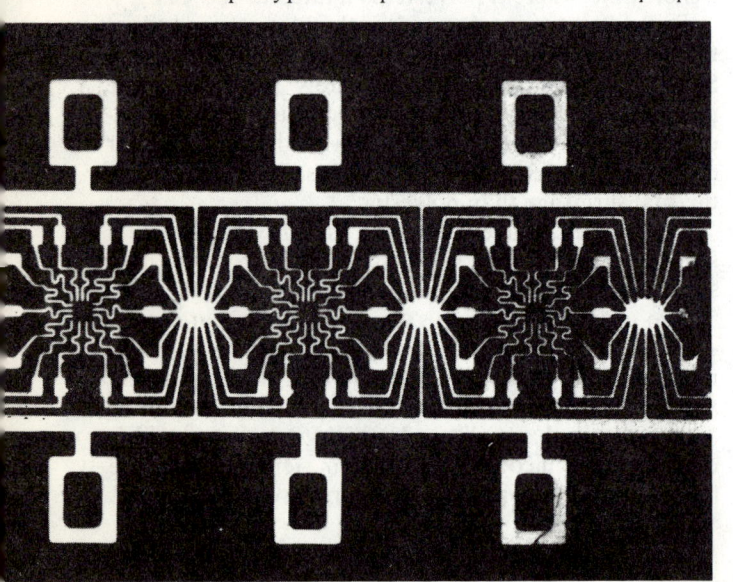

4. Test points on tape. A 16-mm beam tape allows on-tape testing of chips following the inner-lead-bonding step. A test point is attached to each lead, but is removed before final outer lead bonding.

5. Manufacturing. Texas Instruments' large-scale film-carrier automated manufacturing facility has a production rate of 2,000 ICs per hour and, with the recent addition of improved machinery, a potential of 400 ICs per hour. Present output of this line is TTL LSI in DIPs.

mode descends and applies heat and pressure, joining all beams and bumps at the same time and also melting the adhesive or wax holding the chip in place. In the third step, the thermode rises and the array of as-yet unbonded chips is lowered, leaving the newly bonded chip stuck to the copper pattern on the tape. Finally, a new tape pattern and chip move into position for the next operation.

Outer lead bonding connects a film-carrier beam to the outside world via a DIP lead frame, hybrid substrate, or a flexible or rigid printed-circuit board. The ideal automated outer lead bonder should therefore have interchangeable heads capable of thermocompression bonding, reflow soldering, and welding. But in fact, only machines with interchangeable thermocompression-and eutectic-bonding heads are at present commercially available.

Jade Corp. has one of the few commercially available machines that thermocompression-bonds copper beams to a metal lead frame. It operates as shown in Fig. 7. The bonded chip and interconnect are removed from the tape by a punch and raised against the lead frame. A thermode forms all bonds in one stroke. Then the punch is retracted, and finally both the lead frame and tape are indexed.

Once more, many manufacturers and users make their own equipment, both inner and outer lead bonders. An example is the extremely large group of tape bonding machines, shown in Fig. 5, that are in use at a Texas Instruments manufacturing facility.

Only three U.S. machinery manufacturers currently supply commercial bonders suitable for film-carrier work. Industrial Micro Industries, Cherry Hill, N.J., has automatic, semi-automatic, and manual inner lead bonders suitable for thermocompression or eutectic bonding. Jade Corp. has manual and automatic inner and outer lead bonders that use the same two types of bonding. Kulicke and Soffa Industries Inc., Horsham, Pa., has made custom automatic outer lead bonders for welding film-carrier ICs to DIP metal lead frames. Prices of these bonders range from $11,000 to $33,000.

Most U.S. companies, particularly IC manufacturers, jealously hide almost all details of their automated tape-assembly operations. Least reticent are Texas Instruments, National Semiconductor, and RCA, while Fairchild and Motorola Semiconductor Products at least acknowledge that they have film-carrier manufacturing facilities. Bell Laboratory's facility at Allentown, Pa., and Honeywell Information Systems (HIS), Phoenix, Ariz., also maintain film-carrier manufacturing capabilities.

Behind the film curtain

By acquiring the original Minimod from General Electric's Integrated Products department, Syracuse, N.Y., around 1971 or 1972, Texas Instruments has by now built up the most experience in the film-carrier process of any U.S. firm. Since then, the Dallas firm has invested in and modified the process extensively. Today it makes its own three-layer, 35-mm polyimide tapes and has developed its own bonders to turn out mainly TTL ICs of the 14-pin variety. Production rates of 2,000 units per hour have been achieved, and newly installed equipment is expected to increase that figure soon to 4,000 units per hour.

National Semiconductor is perhaps the second largest producer of dual in-line packages containing film-carrier TTL ICs. This firm's beam-tape operation is based on an 11-mm-wide, two-layer polyimide 3M tape, handled by Jade Corp. inner and outer lead bonders with thermocompression heads.

According to Carmen Burns, director of National's automated assembly, the Santa Clara company has been using what it refers to call its "inner lead-bond micro-interconnect system" for about two and half years. Burns has definite views on the terminology of this field.

"It is not accurate to call our system either a beam-tape system or a film-carrier system," he says. "Both designations are inaccurate both when referring to our own work and to what is going on generally. I prefer the term automated micro-interconnection technology, which involves connecting the small geometry of the die to the larger geometry of the punched lead frame or substrate with copper beams."

He explains that "the technology that National developed does not use film as a carrier. Instead the copper is used as the carrier. The film is used to support the copper structure; the small film interconnects under the copper go to individual IC DIP pads to hold them stable during bonding and perform very little function in the

THERMOCOMPRESSION BONDING

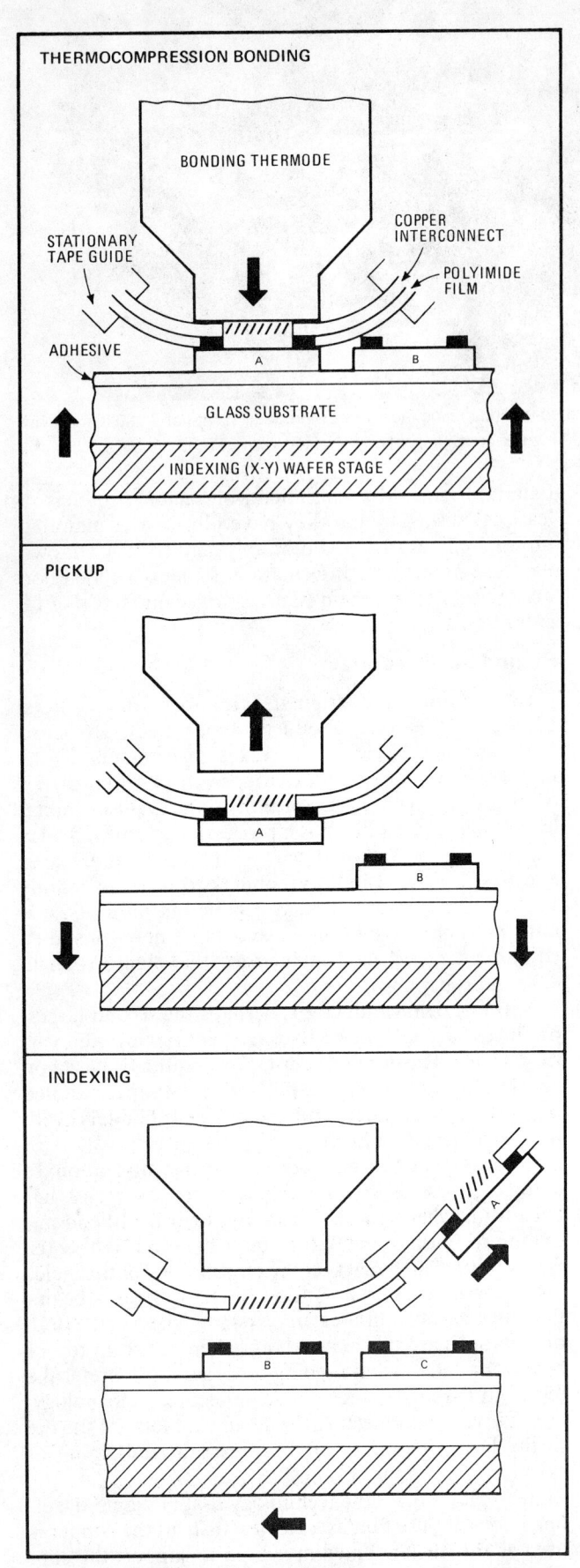

PICKUP

INDEXING

6. Automated inner lead bonding. When a chip is bonded to a tape pattern, the machine aligns chip and tape, a bonding head attaches bumps to beams, and a fresh chip and tape frame are reindexed.

way of mechanical support" of the copper pattern.

For about a year now, RCA's Solid State division has been in pilot-line production of film-carrier ICs packaged in DIPs. In the RCA system, trimetalized chips with gold bumps are thermocompression-bonded to a 3-M 11-mm tape etched with RCA-specified patterns. This tape is fed to an outer lead bonder for thermocompression bonding of the interconnects to lead frames. Jade Corp. bonding machines are used. So far RCA has successfully put bipolar linear ICs and complementary-MOS chips on tape. Most of them end up in 14-pin DIPs, though experimental 32-lead patterns have been achieved on the present 11-mm tape.

Motorola Semiconductor Products Inc. in Phoenix, Ariz., and Fairchild Semiconductor in Mountain View, Calif., are both large producers of film-carrier ICs, too. But outside of the fact that each uses a 16-mm three-layer tape format, very little is known about their film-carrier operation.

Slightly more is known about Bell Laboratories' film-carrier assembly method, which it has been evaluating for about two years. Today, its Allentown, Pa., facility has a developmental operation assembling Bell's own bipolar and MOS ICs onto a two-layer, 11-mm tape and then into 16- or 18-pin plastic DIPs. Both bonding equipment and tape were purchased from makers outside the Bell System.

Outside the U.S.

In Japan, Tokyo Shibaura Electric Co. Ltd. (Toshiba) and Nippon Electric Co. Ltd. (NEC) have both been applying a modified Minimod for about 18 months.

Toshiba uses a three-layer 35-mm polyimide tape and made all its equipment in-house. ICs are inner-bonded with a gold and tin eutectic process. Outer ends of the leads are solder-plated for attachment by the customer. The final package (Fig. 8) is tape with a molded encapsulation over the chip. Note the extra points on the beam tapes, for use in testing or as interconnects.

Toshiba's various film-carrier ICs contain the circuitry for the oscillator, frequency driver and pulse motor driver of an analog electronic watch. At present, production rates are about 500 units per hour. Since demand is small, the finished tape is supplied to customers in strip form rather than reels.

NEC also employs a 35-mm three-layer polyimide format with similar inner lead bonding and solder-plating of outer lead ends. Again the chip is protected with molded resin. The finished ICs, which are bipolar linear circuits designed for cameras, are sold either in strips or as individual devices. Production rate on NEC's in-house-designed equipment is about 400 devices per hour, but is expected to rise to 1,000 in the near future.

In Europe, Siemens AG in Munich, Germany, and Honeywell Bull, Saint-Ouen, France, are engaged in work on film carriers. Philips Gloeilampenfabrieken in Eindhoven, the Netherlands, is also known to be experimenting with the technology.

For nine months Siemens has been supplying customers with sample 24-pin bipolar linear circuits on film, for use in cameras, measuring instruments and hybrid circuits. The German company uses a super-8-mm, three-layer polyimide tape carrier and has already achieved assembly rates of more than 1,000 units per hour. It's now about to transfer the technique into full production. The finished products are sold in reels containing 1000 chips.

A hybrid approach

Of all the companies in Europe, France's Honeywell Bull has had the greatest experience with the film-carrier approach. The French computer firm prefers to call it tape automated bonding (TAB) and uses it specifically to automate the bonding of ICs onto multilayer thick-film hybrids. The procedure is already in use for prototype hybrid modules.

Honeywell Bull manufactures its own 35-mm three-layer tapes, using polyester (Mylar) film as the insulating carrier and tin-plated copper beams as the IC interconnect. François Gallet, manager for advanced development at Honeywell Bull, says "Mylar costs about one tenth as much as Kapton [polyimide], and in our technique the tape does not become a part of the final package and thus doesn't need the stability of Kapton."

Strictly speaking, the film-carrier ICs at Honeywell Bull do not wind up in any kind of package. Instead, after all film has been removed, the gold-bumped mounted chips are soldered onto small alumina substrates measuring 1 by 1 or 2 by 2 inches and containing three thick-film screen-printed layers of interconnects. In essence, the substrate is a multilayer printed-circuit board for unpackaged chips.

To date, the hybrid modules have carried bipolar logic chips for use in a central processor. Now the technique is being used for MOS memory chips and programable read-only memories and will be extended to C-MOS. Chips with as many as 40 to 60 connections have been mounted on substrates.

Honeywell Bull's tape has three zones: an inner zone that is bonded to the IC chip, a center zone that mates with the hybrid substrate, and an outer zone that provides connections for on-film testing of chips.

After inner lead bonding, all ICs on tapes are tested. Then an automatic machine developed by Honeywell Bull gang-bonds all the beams of the central zone of the lead frame to the hybrid's pads, and all unwanted tape is removed.

Back in the U.S., Honeywell Information Systems (HIS) of Phoenix, Ariz., is also doing prototype reliability and testing work with the tape-automated-bonding technology. But its operation differs from the French in some respects.

HIS uses a three-layer 35-mm polyimide rather than polyester, even though these chips, too, are excised from the film and end up in hybrid devices. Bipolar digital chips with gold bumps are gang-bonded to tin-plated copper leads (Fig. 9), and the bonded assemblies are spooled on reels and submitted to 100% electrical test. The tape from these reels then passes through equipment that excises the chip from the film, positions the device at a particular site on a pretested multichip hybrid substrate, and reflow-solders the formed leads to the substrate pads. Eventually, the hybrids end up in a large prototype computer system. A section of a HIS

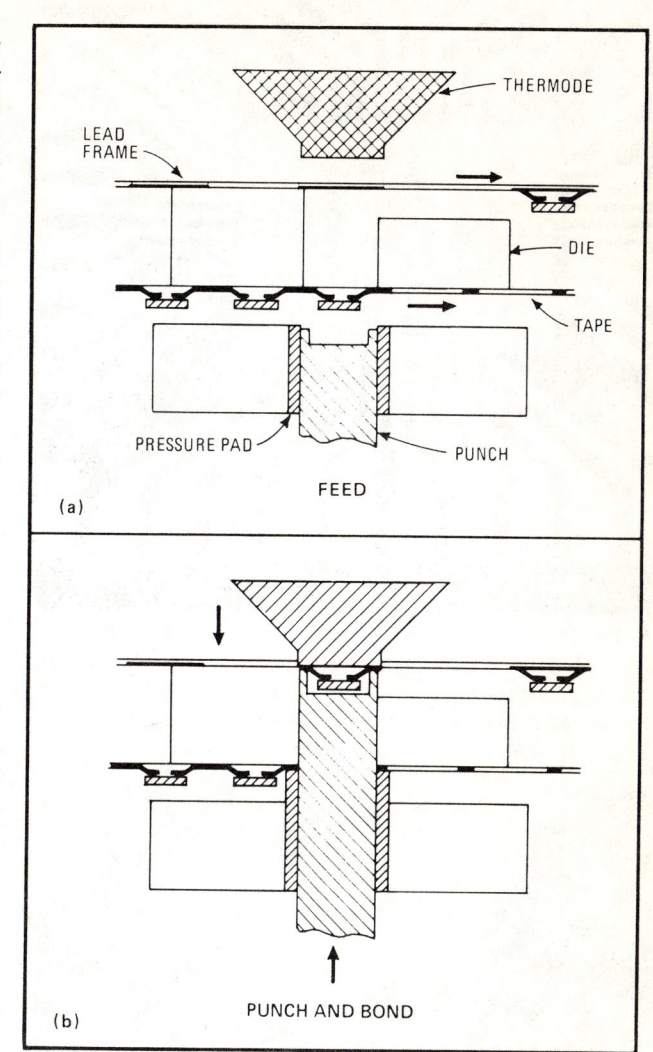

7. Beams to lead frames. Basic elements of outer lead bonding to a metal lead frame consist of: advancing both the tape-bonded chip and the lead frame, punching the device from the tape and bonding it to the lead frame, and then retracting the punch and indexing for the next IC. Welding or thermocompression bonding may be used.

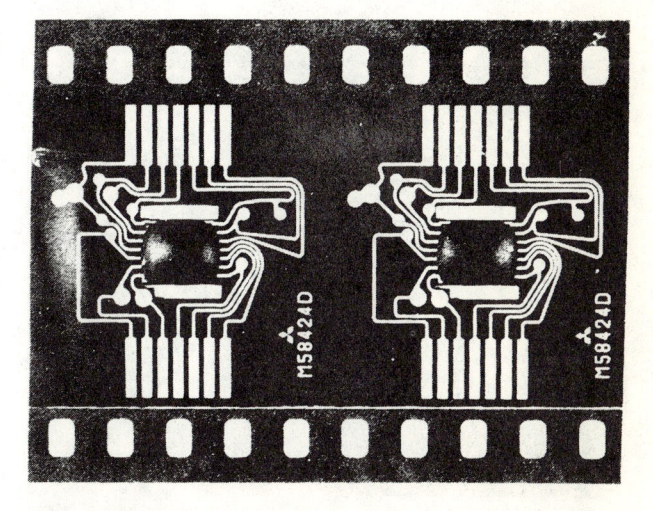

8. Thirty-five mm. Toshiba film-carrier ICs, shown on a 35-mm three-layer polyimide-based tape, are sold to potential customers in film strips rather than in reels. Note the resemblance to a printed-circuit board and the molded encapsulation that protects the chip.

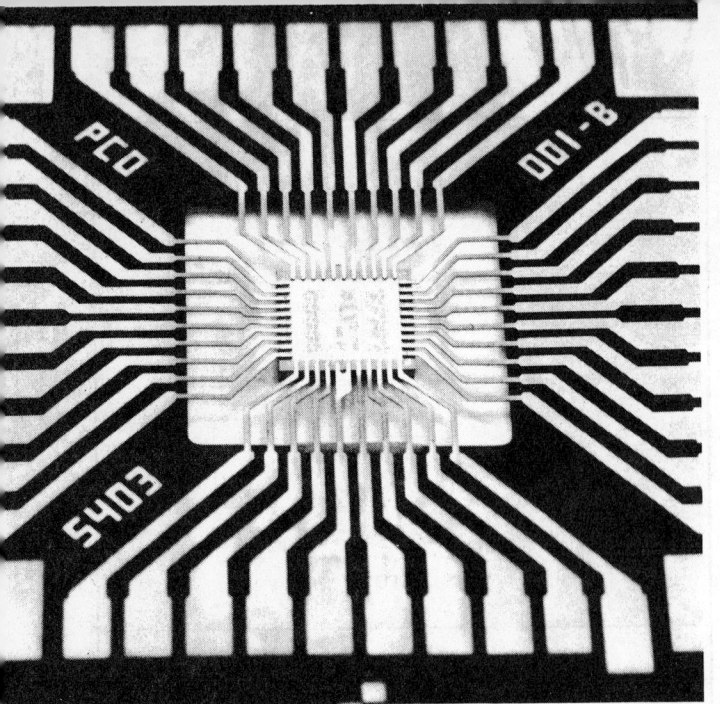

9. LSI on tape. This bipolar digital chip on 35-mm tape was manufactured by the tape-automated-bonding process at Honeywell Information Systems in Phoenix. After inner lead bonding, ICs are spooled onto reels and subjected to 100% electrical test.

10. A hybrid application. At HIS, Phoenix, reels of tape studded with LSI devices are sent through machinery that excises each chip from the film, forms its leads, and positions and solders it on a hybrid substrate. The hybrid is used in a prototype computer.

Phoenix multichip hybrid substrate is shown in Fig. 10.

Neither the Honeywell Bull nor the HIS Phoenix outer lead bonders are commercially available. However, such machines, and inner lead bonders, too, are needed in the potentially large military hybrid market.

The influence of the military

To quote Claire Thornton, director of semiconductor and integrated device development at the Army Electronics Command (ECOM) at Fort Monmouth, N.J.: "Fifty percent of the Army's electronics is going to be hybrid in format for a long time to come. Most of this material is now being made using conventional IC chips attached by wire bonding. The military prefers beam-leaded chips, but these have never been available in the right types and quantity. Now the Army thinks that film-carrier ICs with their substantial copper beams look like a direct reliable alternative to beam leads."

The Army Electronics Command is now evaluating bids from the hybrid industry to develop an over-all industry system for assembling film-carrier ICs into hybrids. The goal is low-cost reliable gang bonding of chips to substrates in small quantities of 100 to 1,000.

When the over-all material system and associated techniques are developed, small hybrid manufacturers will be able to put purchased chips directly on to film-carrier strips rather than reels, which in turn would be gang bonded to thick- and thin-film substrates.

The next few years

The technology of film-carrier systems still has room for improvement. Simplification of the metallurgy of the bumps and beam leads may be expected, possibly quite soon. At least one IC company has already produced an experimental copper-bump-to-copper-lead bond. Others are working on all-copper tapes (see "The film-less film carrier," p. 64).

Extensive environmental testing is also needed to fully establish the reliability of the method. Such test data is often supplied to individual customers by individual firms, but has not yet entered the public domain.

The major IC manufacturers will probably go on putting almost all their film-carrier TTL devices into DIPs. But as the technique spreads, more users will prefer to purchase film-carrier ICs on reels rather than in DIPs. Later, as low-cost manual inner and outer lead bonders become available, even quite small companies will be able to buy standard and custom chips, put them on tape, and package them just as they wish, whether in a plastic or ceramic DIP, a thick- or thin-film hybrid, or a flexible or rigid pc board.

Even now, film-carrier ICs soldered to flexible pc boards are being used in digital watches and cameras. A logical extension to this would be to eliminate the flexible board and to bond all the watch, camera or calculator ICs to specially designed copper interconnects on one or more film carriers. Indeed, several companies are currently working on such a step.

Eventually even light-emitting diode displays and their drives might also be put on film carriers. Such a method would radically cut assembly costs of consumer products like electronic watches and calculators. □

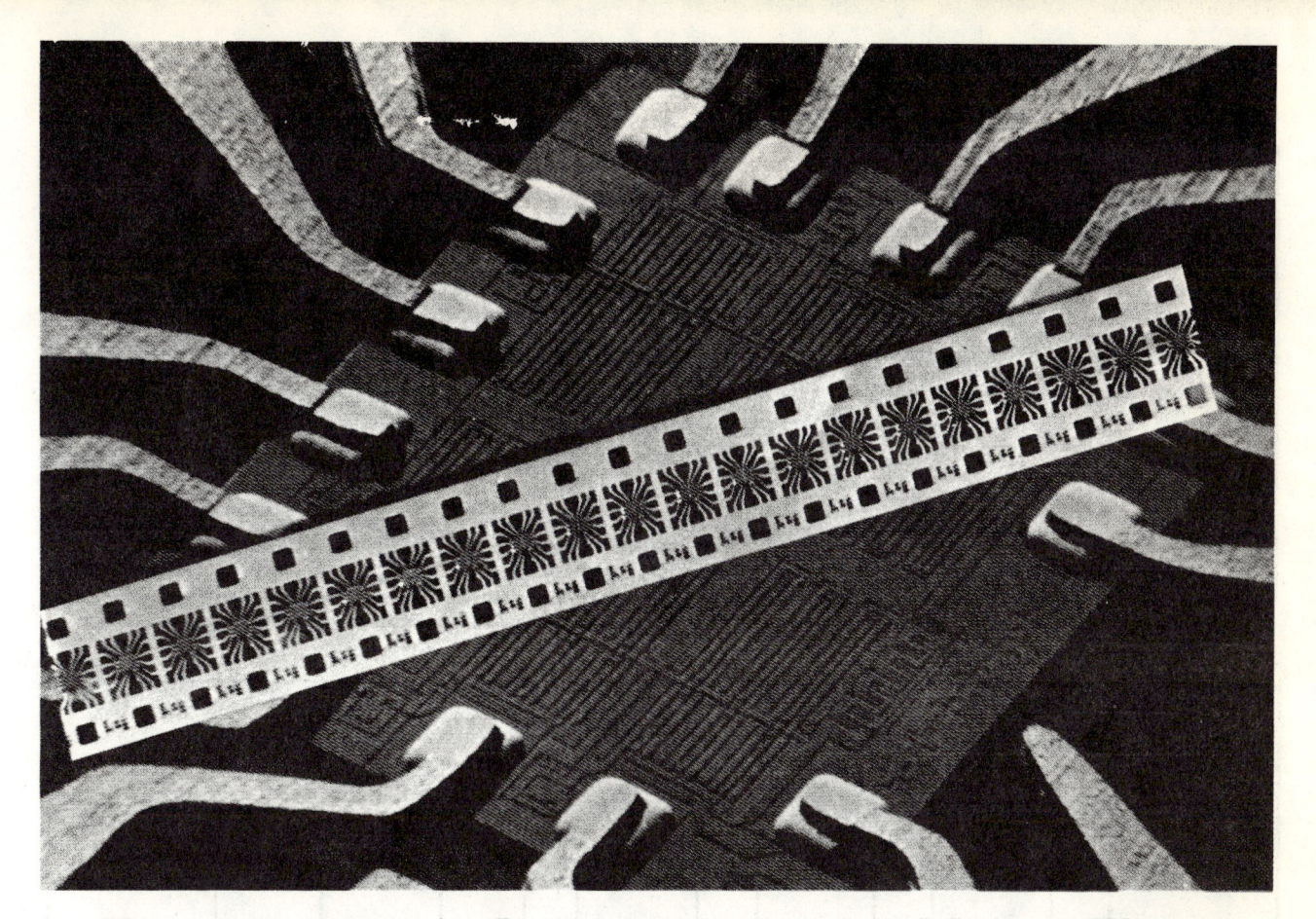

Beam tape plus automated handling cuts IC manufacturing costs

Fully automated finishing operations
make it possible to bring plastic-encased small-scale
integrated-chip assembly back home

by Doug Devitt* and Jim George, Solid State Scientific Inc., Montgomeryville, Pa.

☐ Over the last five years tremendous advances in integrated-circuit manufacturing have occurred. But while wafer fabrication techniques and automated testing operations have leapt ahead, bonding, assembly, and other aspects of finishing have stayed close to the starting line.

An industry that is among the most technically sophisticated has attacked the problem of high finishing costs by shipping devices on a 12,000-mile round trip to have them assembled manually. That was done on the principle of reducing the cost per unit of labor. Now, a more viable approach for plastic-housed small-scale integrated circuits is emerging: reduce the amount of labor by applying a fully automated system built around the beam-tape method of assembly.

Assembly operations contribute significantly to the manufacturing costs of all integrated circuits. In fact, as Fig. 1 illustrates, for devices smaller than 80 mils per side, assembly operations represent the largest single cost of manufacturing. And these devices, the so-called "jelly beans," are the high-volume ICs on which much semiconductor manufacturing is based.

Historically, the experience of the IC industry has been that any manufacturer who is not constantly working to reduce costs risks being undercut by his competitors to the point of bankruptcy. Continually searching for developing nations with low labor rates is not a long-term solution to the assembly-cost problem. In fact, it is a good bet that the cost of offshore assembly will actually rise because of increased airline freight rates, growing inflation in the Third World, and a trend toward higher import duties.

Because of the distance and the turnaround time involved, offshore assembly has two other disadvantages. First of all, a firm must tie up working capital to finance its inventory in the 12,000-mile-long pipeline. Secondly, the system is usually sluggish and responds poorly to changes in market needs.

Under the proper conditions, a better choice for both

---
*Now with Harris Corp., Semiconductor Division, Melbourne, Fla.

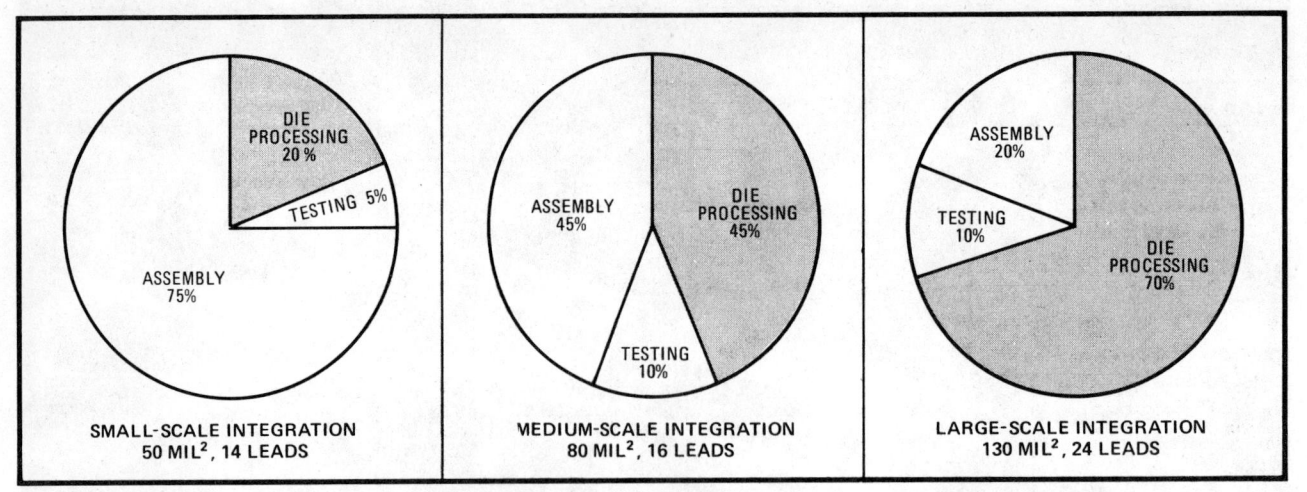

1. Manufacturing costs. As chip area increases, the cost of assembly becomes lower. In small-scale integration, assembly costs dwarf all others. Two methods of cutting these costs are offshore assembly and the use of a fully automated assembly line.

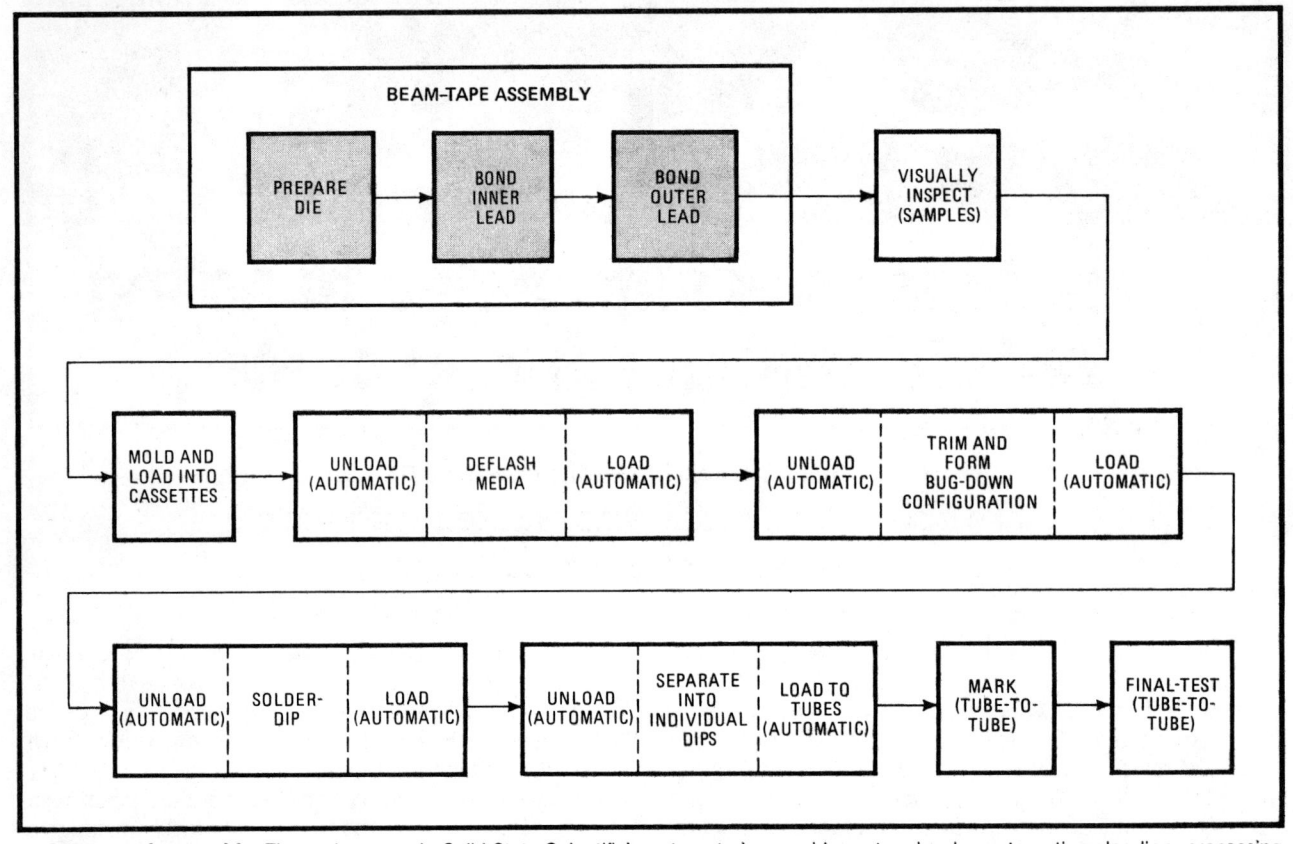

2. Automated assembly. The major steps in Solid State Scientific's automated assembly system involve automatic unloading, processing, and reloading at each step. The individual complementary-MOS chips are never handled by production personnel.

user and manufacturer would be to bring assembly operations back to the United States. What is required is that the manufacturer consider the entire assembly operation, from die attachment to final testing, as an integrated manufacturing system. This system can then be optimized for cost and efficiency.

Recent developments in beam-tape (film-carrier) interconnection systems have encouraged IC firms to re-evaluate their assembly operations. These systems use prefabricated metallic interconnection patterns on a sprocketed insulating film to provide automated, high-speed bonding of the interconnection to both the chip and a lead frame. Integrated into the total assembly operation, such a system allows the IC maker to completely automate the back-end functions. An example of how beam-tape technology has been incorporated into a fully automated assembly line is Solid State Scientific's assembly system for packaged complementary-metal-oxide-semiconductor chips, which is diagrammed in Fig. 2.

In this assembly operation, wafers are given metallic "bumps" by electroplating a copper contact about 25 micrometers high over a chromium barrier (Fig. 3). The barrier prevents undesirable intermetallic compounds

LOOKING AT MAJOR ASSEMBLY TECHNIQUES FOR EPOXY DUAL IN-LINE PACKAGES

	Method of attaching die to lead frame	Remaining assembly steps up to final testing	Geographical location	Period in use
Manual or semiautomated die bonding plus semi-automated finishing	thermocompression by hand or slow machine, 200 – 500 units/hr or thermocompressive gold-ball bonding, 60 – 110 units/hr	molding, deflashing, dam bar removal, tin plating, cutting and forming, tube loading; 500 units/hr (average) per step	low-labor-cost areas of Asia	1965 to present
Automated die bonding plus semiautomated finishing	high-speed thermo-compression, 1,000 – 2,000 units/hr or programmable high-speed sequential wire bonding, 250 – 500 units/hr	same as above, except with newer equipment; slightly faster rate	low-labor-cost areas of Asia	1973 to present
Film-carrier die bonding plus semiautomated finishing	thermocompression using beam tape with inner lead bonding at 1,000 – 2,000 units/hr and outer lead bonding at 2,000 – 4,000 units/hr	same as above	low-labor-cost areas of Asia	1975 to present
Film-carrier die bonding plus fully automated finishing	same as above	same as above, except for automated loading and unloading, and high-speed trimming and forming plus solder dip; 5,000 units/hr	U.S.A.	1978

from forming between the copper contact and the aluminum metalization of the chip pattern.

Each wafer is then electrically probed, and rejected dice are inked. Following probing, wafers are mounted on a substrate, held in place by a thermoplastic material, and then sawed through to the substrate with a diamond saw. Because the wafers are attached to the substrate, separated dice maintain their original orientation after the sawing.

Chips on tape

Actual bonding takes place in two separate operations. In the first step, inner lead bonding, the dice are bonded to an 11-millimeter patterned tape. The tape consists of two layers—a copper interconnection pattern bonded to a sprocketed polyimide carrier for strength. Single-layer pure-copper tapes, which are simpler to process, are just starting to be used by a few firms. However, Solid State Scientific chose two-layer tape for its system because of its commercial availability and satisfactory field performance.

The chip-on-tape resulting from this first step is stored on reels until the second step begins. This procedure, called outer lead bonding, involves fixing the tape's IC microinterconnections onto the copper-based lead frames. After this step, the lead frames are cut into strips of 10 devices and loaded into a magazine for subsequent processing. Devices are then inspected on a sample basis for unsatisfactory bonds and for manufacturing defects that are capable of causing electrical malfunctioning.

Automated procedures

The strip-mounted devices are next routed to a molding facility, where they are epoxy-molded and loaded into a special vertical holder, or cassette. The holder then goes to a deflashing operation. There, the devices are automatically unloaded, excess plastic is removed, and the devices are then automatically reloaded back into the cassette.

The first trimming and forming operation is next. After unloading, the strips are positioned beneath a multistage series of punches. Then, the epoxy between the dam bar and the main body of the device is removed. (A dam bar is a thin metal strip that shorts out all the leads on the lead frame. It prevents plastic from getting on the leads.) The dam bar itself is removed, and finally the devices are formed into the lead-down, or bug-down, configuration. Throughout this operation, the devices remain connected in strips by pins at each end of the package. The strips are then automatically reloaded back into the assembly magazine.

Next, a number of precleaning steps are performed and the devices are dipped in hot solder—all done auto-

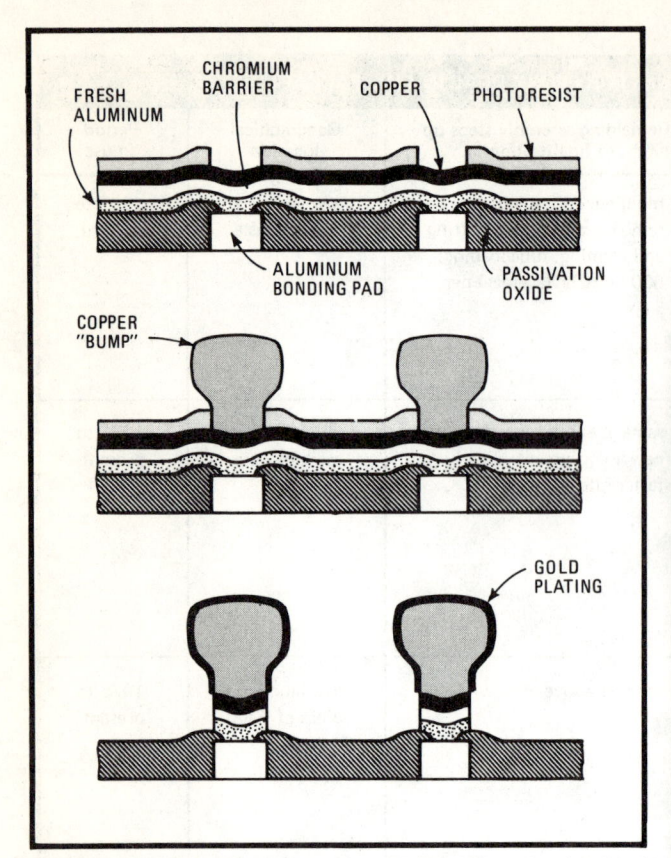

3. Bumped wafers. Bumps are formed by evaporating layers of aluminum, chromium, and copper over a wafer. The copper contact is electroplated up 25 µm; then the resist and unwanted metals are stripped. Lastly, the bump is gold-plated.

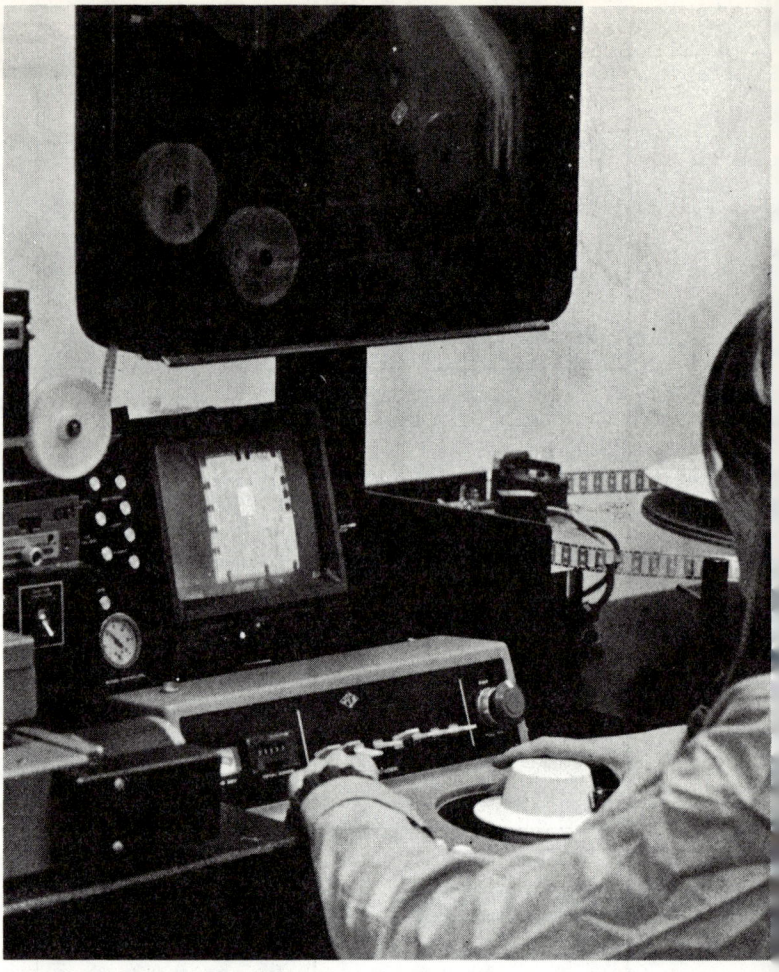

Beaming in. This a view of the semiautomated equipment used at Solid State Scientific for gang-bonding the inner leads of each beam-tape interconnection pattern to the bumped I/O pads of an IC chip. Note that the TV monitor provides excellent visual control.

matically while the devices remain on the strips. Following the final drying step, magazine-loaded strips are transferred to the final trimming operation, in which the devices are separated from the strips and loaded into antistatic tubes. Devices in the tubes are ready either for marking or for final testing, depending on the specific process flow.

A key feature of this assembly system is the automated unloading, processing, and reloading at each step of the sequence. Throughout the entire assembly procedure, individual integrated circuits are never handled by production personnel, thereby eliminating the breakage that will sometimes occur in semiautomated and manual handling.

Automated assembly's impact

The table, which compares the major assembly techniques for dual in-line packages made of epoxy, attempts to put automated systems in perspective. Though not shown, the figure of merit of unit output per operator hour represents the approximate efficiency of the total system from die interconnection up to final testing.

Introduction of high-speed die bonding in 1973 resulted in a very impressive doubling of an operator's production rate from 20 to 40 devices per hour. As beam-tape interconnection systems were introduced, output increased fivefold to an unprecedented 200 devices per operator hour. Integrating beam-tape technology into a completely automated assembly facility leads to a further 50% improvement in output.

For the user, automated assembly systems provide the obvious benefit of reduced costs for those high-volume devices adaptable to automated assembly. Of greater significance, however, is the fact that they equip the IC manufacturer to respond better to the marketplace. On the one hand, during an industry-wide downturn, the response of IC firms is sluggish because much of the work in process is already in the pipeline. The result is rapidly rising inventories, with their concomitant economic penalities.

On the other hand, a period of rising demand causes offshore assembly facilities to run at or close to 100% capacity. Therefore, increasing production during an upturn is extremely difficult and costly, with the result that lead times stretch out and deliveries slip.

Automated assembly can do much to improve this situation by giving the IC manufacturer much better control. In down periods, inventories can be adjusted much more quickly, and the capital that would otherwise be tied up in inventory will be available for other projects. When demand rises, production can be increased quickly by running up to three shifts, seven days a week. □

Growing pin count is forcing LSI package changes

At high pin counts, DIPs waste board space, so designers are trying chip- and film-carriers

by Jerry Lyman, *Packaging & Production Editor*

☐ For all the enthusiasm with which system designers greet each new advance in LSI circuit density or speed, the chilly morning-after thought is always: how does this affect the chip package? Finding the right package for both the system and the chip nowadays raises a long list of questions, involving package performance and ease of handling as well as the basic issue: size.

Only a very large dual in-line package, made of ceramic, not plastic, can supply the more-than-40 pins that the most complex of today's microprocessor or memory chips demand. But since 3-inch-long bodies take up a lot of board space, computer designers in particular are turning to more exotic but more appropriate packages. Particularly popular is the ceramic chip-carrier, and Table 1 lists other challengers to the DIP, including the flat pack, more chip-carriers, and the film-carrier.

Their sheer variety complicates the system designer's task enormously. It is no longer possible to select just by price, as it was when the options were largely limited to plastic versus ceramic versus glass-ceramic DIPs. Shape and size are suddenly critical — how easy

TABLE 1: TYPICAL LSI PACKAGES

LSI package type	Maximum number of leads	Method of attachment to pc board	Removal from pc board	Pc-board area, including leads (in.)	Hermetic seal	Approximate price per lead (¢)	Availability
Cofired ceramic DIP	64	wave solder or socket	difficult	2 x 0.600 (40 lead)	yes	4	readily available
Cerdip	40	wave solder or socket	difficult	2 x 0.600 (40 lead)	yes	2½ to 3	readily available
Plastic DIP	40	wave solder or socket	difficult	2 x 0.600 (40 lead)	no	1	assembled in house
Cofired ceramic chip-carrier	64	socket	simple	0.460 x 0.430 (40 lead)	no	2	readily available
Leaded chip-carrier	64	reflow-solder	simple	0.770 x 0.770 (36 lead)	yes	slightly more than ceramic chip carrier	available on special orders
Minipak	28, 40 in the future	reflow-solder	simple	0.500 x 0.500 (28 lead)	no	0.4	package only available from GI
Flatpack with leads out all four sides	64	reflow-solder	simple	approx the same as leaded chip carrier	yes	1	readily available
Ceramic substrate with clips on 4 sides	up to 156 (proposed)	reflow-solder or socket	simple	0.650 x 0.650 (44 lead)	yes	not available	in design stage at Berg
Plastic pre-molded chip-carrier (AMP)	up to 156 (proposed)	reflow-solder or socket	simple	0.650 x 0.650 (44 lead)	no	not available	will be available in 24-lead type in mid-1977
LID	40	reflow-solder	simple	0.450 x 0.450 (40 lead)	no	1	readily available
Film carrier	40 to 64	reflow-solder or wire-bond	simple	about 0.312 x 0.132 (40 lead)	no	not available (depends on chip, tape)	not generally available in U.S.

Source: Electronics

will it be to redesign a board, once it has been laid out? Reliability levels and power dissipation vary, hermeticity may be lacking, not all functions are available in all package types. In addition, details that the engineer could ignore with the old reliable DIP may now be stumbling blocks—for instance, ease of assembly on and removal from a circuit board, and the method of attachment to a board.

To pass on all the high performance of large-scale integration to the system, the package must be capable of holding a chip maybe 350 mils square, with lead counts of up to 80, and dissipating as much as 5 watts of power. Till now, DIPs have housed about 95% of all LSI. Molded plastic DIPs, however, are comfortable with no more than 40 leads and only at the lower, metal-oxide-semiconductor levels of power dissipation. Cofired heat-sunk ceramic DIPs with 64 leads do better. But the rival ceramic chip-carrier occupies about a third the board space and also degrades chip performance less because its lead resistance is lower.

Not that ceramic chip-carrier is entirely satisfactory. A task force at the Electronic Industries Association recently wrote an unusually comprehensive standard for chip-carriers that solved many of their problems. But even so, chip-carriers share one of the major lacks of the larger, over-24-lead DIP: there is no automatic insertion equipment designed to handle them.

Ease of manufacture and handling is the whole rationale of the film-carrier. Strictly speaking, the film-carrier is not a package at all. It is a series of micro-interconnects—usually etched out of the copper surface of a tape—to which the LSI chips are automatically bonded. Then reels of the tape plus chips can be automatically bonded to a board or substrate or maybe even a plastic DIP.

Since the versatile DIP will no doubt continue to dominate all but the highest levels of the LSI world, a discussion of its several varieties precedes the sections on the various chip carriers and the film carrier.

Dual in-line packages

The dual in-line package may be made of ceramic, a glass and ceramic mixture, or plastic, but the Cadillac of them all is the cofired or multilayer alumina ceramic DIP. Made by 3M Co., Kyocera International Inc., Metalized Ceramics Corp., and NGK Insulator Ltd., it is the most reliable—and the most expensive. Because of its reliability, it is the package in which most new and expensive LSI circuitry makes its bow. But because of its cost, it is usually replaced by other types of DIPs as soon as all the production problems are ironed out and the chip starts being produced in high volume.

The manufacture of the multilayer DIP is an elaborate and painstaking process. The basic element is uncured (green) alumina tape. Screened onto the top of the tape are metalized patterns, which where necessary include vias—square holes filled with the screened-on metal—for connection to other layers. These metalized layers are then laminated together and fired in a furnace, creating a strong, monolithic structure. A lead frame is brazed to the top or side of the package, which can now be shipped to the user along with a metal lid.

Single-cavity cofired DIPs typically consist of three layers—the top one bearing the seal ring, the middle layer bearing interconnections to the lead frame, and the bottom layer for the chip bonding pad. Dual-cavity DIPs could have four or five layers. Multiple-cavity packages, however, are fairly rare. They are used mostly for low-volume custom applications, though they do occasionally turn up in volume production—at Fujitsu Ltd. of Japan, for instance, where three-cavity 16-pin DIPs house current-mode-logic devices.

Still, even the simplest cofired DIP is costly, in time as well as money. Designing a new package requires 10 to 12 weeks and up to $10,000 for tooling costs. Ceramic is fragile, too, making it harder to ship and handle automatically than resilent plastic.

Sometimes, though, only a cofired DIP will do. It is the only DIP type to handle as many as 52 or 64 leads successfully. Not entirely coincidentally, a 64-lead type also happens to be the largest cofired DIP available (Fig. 1). Since this unit is about 3.2 inches long by 0.9 in. wide, it is doubtful whether there will be too many packages made that are larger.

Indeed, it was in an effort to keep them small, despite big lead counts, that some package manufacturers have gone to the slightly more expensive quad in-line package or QUIP. The QUIP has its leads arranged in two staggered rows on 50-mil centers instead of the 100-mil centers of the DIP. Consequently, 42-, 52-, and 64-lead QUIPs are about half the length of comparable DIPs. Rockwell Microelectronics Devices division, Anaheim, Calif., originated the QUIP about eight years ago, and it is now in use at Motorola, Siemens AG, and AEG-Telefunken as well as Rockwell, to package microprocessors, memories, and other dense LSI circuits.

Other advantages of cofired ceramic DIPs are, to quote Harold Ottobrini, president of Metalized Ceramics Corp., Providence, R.I., "superior heat dissipation, true hermeticity, and high package-sealing strength." The ceramic DIPs are superior to plastic DIPs in lead strength, he adds, because the lead frames are brazed on. They are also unlike plastic DIPs, he points out, in giving the user "the ability to test a device in an open package, which saves him money by obviating the final sealing step if the device doesn't work or allows him to repair it."

Finally, in comparison with the glass-ceramic Cerdip, the ceramic DIP can sustain a wider chip cavity, thanks to its solid sidewall. "With a Cerdip," explains Ottobrini, "the sidewall may crack when the leads are bent—a big consideration in very large LSI devices."

The type of ceramic used for the cofired packages is 94% to 96% alumina, which is normally white but can be made black by the application of a coloring material early in the manufacturing process. Black ceramic is now coming into favor for screening out light from photosensitive chips. According to Jim Wade, market manager for ceramic packaging at 3M's Electronic Products division in St. Paul, Minn., "the new trend at many IC companies is to use black ceramic for all ceramic packages since it is the same price as the white."

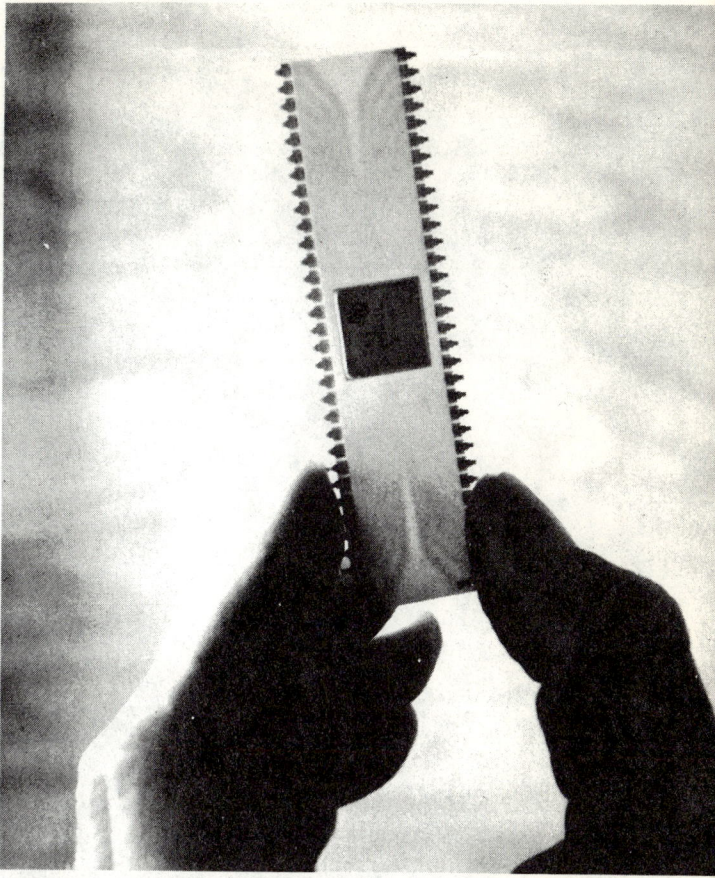

1. Big DIP. The largest commercially available package for LSI chips is this 3.2-by-0.9-inch, 64-lead, cofired ceramic dual in-line package. Packages of this type are only used for complex, expensive LSI circuits like microprocessors and extremely large memories.

Cerdips and window frames

Whether black or white, however, purely ceramic DIPs are losing many IC sockets to a class of packages priced between it and the plastic types. Made of ceramic inmixed with glass, the package comes in either Cerdip or window-frame (Fig. 2) versions.

The Cerdip has a two-piece construction, a cap and base made of pressed ceramic glass. To package a chip, the Cerdip base is put on a heater block till the glass element melts and a lead frame can be embedded in it. Next the chip is attached to a gold pad in the cavity in the base and wire-bonded to the leadframe. Later the cap is sealed to the completed base in a special oven.

The window-frame type is heavily used for memories that are erasable by ultraviolet light (though cofired packages are also available with built-in optical windows). The chip user buys preassembled top and bottom glass-ceramic pieces already fused to a leadframe. A gap or window in the top piece uncovers the die-attach cavity in the base, enabling the user to place the chip through the window into the cavity and bond it in place. Finally, a quartz or sapphire lid is pressed into the opening to act as the medium for ultraviolet light.

Makers of the window-frame glass-ceramic DIP include Coors Porcelain Co., Golden, Colo., and Diacon

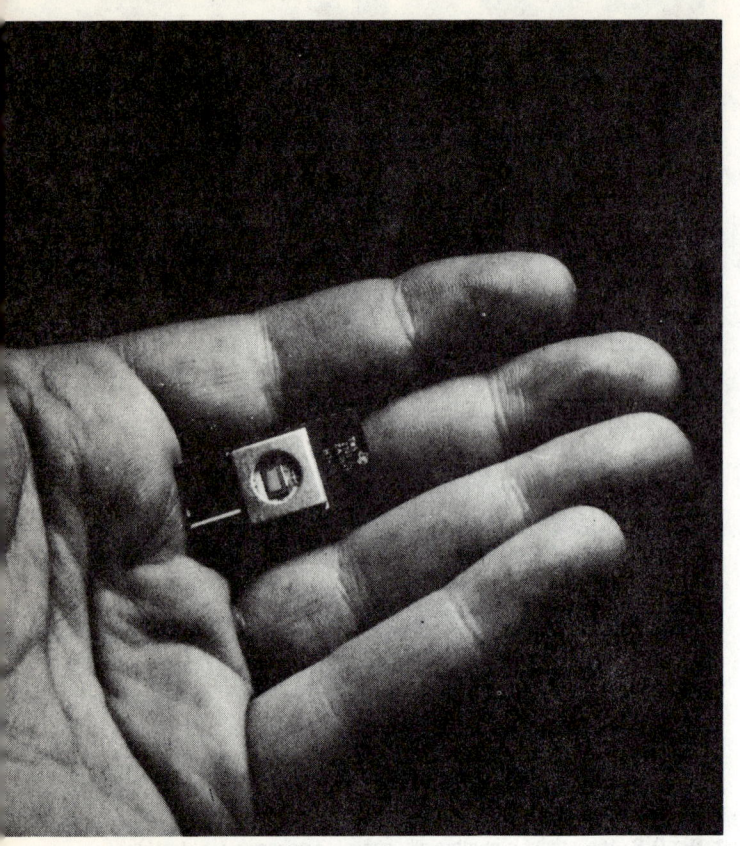

2. Optical window. Special cofired ceramic-glass DIPs, equipped with quartz or sapphire lids transparent to ultraviolet light, are used to house erasable read-only memories. UV light directed through the lid erases a memory's program in preparation for reprogramming.

Inc., San Diego, Calif, and Plessey Frenchtown, N. J. The Diapac, Diacon's version, is about two thirds the price of a Cerdip and is rugged enough to be shipped overseas. However, the Cerdip's assembly is more easily automated.

At present, Cerdips commercially available from Kyocera, Coors, Diacon, and Plessey Chatsworth, Calif., have at most 40 leads, although George Fujimoto, marketing applications engineer of Kyocera International Inc., Sunnyvale, Calif., sees this number eventually rising to 48 to 52. Originally, according to Fujimoto, many integrated-circuit companies were concerned that the temperature required to seal a Cerdip might damage MOS chips. But new glasses that melt at lower temperatures have ended this concern.

The big advantage of the Cerdip is price. An assembled 14-lead cofired ceramic DIP costs about 80 cents, a 14-lead Cerdip costs about 17 cents, and a plastic DIP of the same configuration costs 9 cents.

For hermeticity, of course, the Cerdip is superior to the plastic DIP. It also dissipates power better than the plastic DIP and indeed almost as well as the cofired type.

Unlike the cofired package, though, the Cerdip is not commercially available with more than 40 leads (as noted), it cannot be left open for chip testing, and it lacks the strength of the cofired monolithic structure. But unfortunately like the cofired DIP, it has a tendency to crack during automatic insertion into pc boards.

Nevertheless, at the present time, about 25% to 30% of all ceramic and glass-ceramic types used are Cerdips, and this percentage is growing. The Cerdip is even finding favor in some consumer applications, where it beats out plastic because it can lengthen product life or improve reliability. For example, Atari Corp., the manufacturer of television games, uses mostly plastic DIPs but houses a critical read-only memory in a 24-pin Cerdip.

A plastic world

Other DIP users besides Atari are reluctant to accept the molded plastic variety in all the applications for which it is qualified, despite the reams of test data from IC manufacturers. Exceptions are Data General Corp., Southboro, Mass., a major user of LSI, which packages 75% of its ICs in plastic DIPs, and its largest competitor, Digital Equipment Corp., Maynard, Mass., which uses only plastic packages for standard LSI. At the IC manufacturers themselves, the ratio of plastic to ceramic packages ranges from 50/50 to 95/5.

Almost all plastic packages today are made of Epoxy B or Novalac, which has been in use for this purpose since about 1972. By now hundreds of millions of IC packages using this formulation have been made, and most IC manufacturers are willing to give out their packaging test results (based on MIL-STD-883A).

Representative of these are the results of a Texas Instruments environmental test on a plastic-encased 4,096-bit random-access memory (see Table 2). It shows that most failures occurred in the pressure-cooker test, which is equivalent to putting the ICs in live stream.

This particular test, although the conditions are rarely found in actual operation, accurately indicates that plastic DIPs are most vulnerable to moisture—a fact confirmed by the published test results of other companies. Still, in the real world, plastic DIPs function reliably in most consumer, computer, and industrial applications where high humidity is not encountered.

A lesser limitation is the fact that plastic packages are only available with up to 40 leads. Above 40 leads, molding and leadframe problems limit the production of larger DIPs because of excessive costs. But for up to 40

TABLE 2: TEST OF 4,096-BIT RAM IN PLASTIC DUAL IN-LINE PACKAGE						
Environment	Quantity	\multicolumn{4}{c}{Cumulative failures at hour shown}			
		48	168	500	1000
85°C operating life	214	0	0	0	0
150°C storage life	30	0	0	0	0
85°C/85% relative humidity (biased)	78	—	0	3	—
85°C/85% relative humidity (no bias)	30	—	0	0	—
Moisture resistance (MIL-STD-883, 1004)	38	—	0	0	—
Pressure cooker (15 psig, 121°C)	78	48 / 1	64 / 2	80 / 2	96 / 3

Source: Texas Instruments

The problem of power

Power dissipation, with the new prominence of bipolar large-scale integration and the trend to greater circuit density and higher speeds, is becoming a problem for all package types, not just the dual in-line standards. It has not normally been a problem in packaging metal-oxide-semiconductor chips, which generally dissipate from 0.5 to 1.5 watts at most. Even with present bipolar logic, this figure generally rises only to 3 W. But today, higher-power LSI devices are starting to appear, posing a challenge to package designers.

So far as DIPs are concerned, the typical thermal resistances of their materials are: 25°C/W for cofired ceramic, 26°C/W for Cerdips, 53°C/W for unmodified plastic, and 23°C/W for plastic with a thermal slug.

Ceramic obviously dissipates about twice as much power as a standard plastic DIP, and normally the cofired ceramic DIP is chosen for large chips. But even in a 3-in.-long, 64-lead package, the chip heats up fast enough to limit power to the 2-to-3-W range when the package is to be used in free air.

One option is to increase a DIP's heat dissipation by cementing a finned heat sink to its top. There are also commercial heat sinks that can be clipped to it, to reduce chip junction temperatures or allow the package to be used in higher ambient temperatures.

A case history is to the point. TRW Inc., Redondo Beach, Calif., recently had to decide how to package a new 5-W bipolar product, the commercially available MPY 16 16-bit parallel multiplier. Initially, Jim Buie, senior staff engineer at the TRW Electronic Systems division, used the 64-lead flat pack with a heat-conducting stud cemented onto it directly below the die cavity. The result was excellent in terms of thermal resistance, which measured only 5°C/W between the surface junctions and the heat stud. But the difficulty of mounting or soldering the package plus stud made Buie decide to try a 64-lead DIP.

With the DIP, however, the thermal resistance from the chip junction to the top of the package, even with a heat sink cemented to it, was an unsatisfactory 20°C/W. TRW's solution was simple. It turned the DIP body upside down, by bending the leads in the opposite direction so that now the die cavity faced downward. Then when a finned radiator was cemented to the new top of the DIP, the thermal-conductance path from chip to radiator was much shorter.

The entire 3-by-0.8-in. ceramic surface of the inverted 64-lead DIP is covered with the nine-fin heat sink (shown below). From heat sink to the 300-by-300-mil chip inside, thermal resistance measures only 7.5°C/W. The sink alone exhibits 9°C/W in still air and only 5°C/W in flowing ambient air.

The alternative to heat sinks for power dissipation is to go to a package material with better thermal conductivity, namely a beryllium oxide ceramic. National Beryllia Corp.'s Berlox is more than seven times as thermally conductive as alumina (see chart).

Standard dual in-line packages and flatpacks for LSI also can be made using BeO_2 as substrate. "Cerdips also can be constructed with this material for only about 10% over the price of alumina glass units," says Peter Fleishner, vice president of technology for National Beryllia Corp., Haskell, N.J. As high-speed bipolar technology moves into LSI, Fleishner sees BeO_2-based packages gaining ground since the other possibilities—heat sinks, studs, radiators—become cumbersome above the 1-W dissipation level.

Admittedly, in custom packages the use of heat sinks has led to some exotic designs. The photograph at bottom right shows one of them—an 84-lead ½-in.-square chip-carrier with an integral heat sink. 3M Co. designed it for a large mainframe firm, which needed to house a high-speed emitter-coupled logic circuit that would be capable of dissipating several watts.

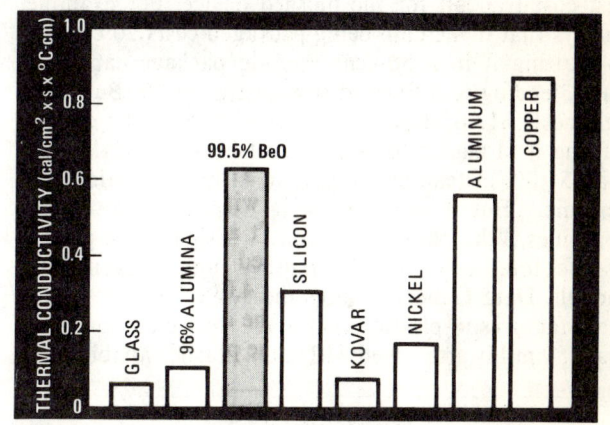

BeO_2. One approach to increasing a package's power dissipation is to try a beryllium oxide ceramic substrate. Beryllium oxide has about six to seven times the thermal conductivity of alumina.

DIP power. A DIP housing a 16-bit parallel multiplier has its chip mounted face down. This gives the shortest thermal conductance path from the chip to the special nine-finned radiator.

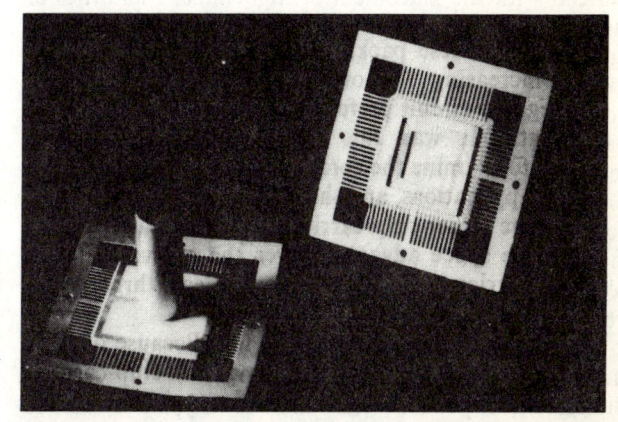

Carrier power. Ceramic carriers can house a high-power chip. This 84-lead, ½-inch-square unit with an integral heat sink was designed by 3M for an ECL chip dissipating several watts.

leads the manufacture of chips in plastic packages is easily automated, and that is one of the prime reasons for their popularity. Adds Carl Carman, Data General's vice president of engineering, "Plastic packages are more uniform in shape than ceramics, again facilitating their automatic insertion into boards. Also the headers of conventional ceramic and Cerdip packages can be put on in such a way that the package won't fit the automatic insertion equipment properly."

The advantages of plastic DIPs should by now be apparent: the lowest cost, the most predictable form factor, and the most resilient (and in this sense the strongest) packaging. Also, "it has the fewest infant-mortality failures of any packaging type," says Robert Beard, director of semiconductor assembly at National Semiconductor Corp., Santa Clara, Calif. "What this means is more systems out the door with less rework."

Minicomputer manufacturer Data General, however, has found plastic packages less reliable than the IC makers predicted, although the company will not go into specifics. Carl Carman says "We can test plastic-packaged parts more because they are cheaper to begin with, which is why we use so many plastic packages."

Carman maintains that the relative costs of the device and the package have to be weighed to come up with the best cost tradeoff for the finished device. For example, he says that if the chip being packaged costs 20 cents or so, putting it in a 80-cent ceramic package can make sense, leading to a finished part cost of $1.00. But if the chip costs $1.00, it pays to cut the finished part cost by putting it in plastic for a total cost of only a little more than $1.00, instead of jumping to $1.80 by putting it in ceramic. "But if you're working with a $10 RAM," he continues, "the package price isn't a significant portion of the total price of the finished unit." Even here, though, Data General houses the 4,096-bit RAM that it manufactures in plastic because the device dissipation is just 250 milliwatts—well within the plastic capability.

Chip-carriers

As LSI chips keep increasing in complexity and size, computer packaging designers in particular have had to contend with the placement of tremendous quantities of the extremely large (40- and 64-lead) dual in-line packages. If used, these packages would require additional boards, increasing the overall system size, and also would degrade circuit performance because of their long lead length. It was this situation that caused the designers to examine the ceramic chip-carrier now in use in hybrid applications, and they soon came to the conclusion that the larger chip-carriers (24 to 64 leads) gave a considerable space advantage over the equivalent DIPs. With the carrier, area reductions of at least three to one were easily achievable.

Package performance improved, too. Because of their radically shorter lead length, the small chip-carriers allowed the upper frequency limit of a typical circuit to be increased by three over a DIP-housed unit. Lead resistance at high pin counts is also much lower in a chip-carrier than in a comparable DIP, which can run into problems on this account at the interface with low-input-resistance transistor-transistor logic. Finally, the ease of removal of the reflow-soldered ceramic was a significant improvement over the DIP.

The concept of the small, leadless, cofired chip-carrier originated at 3M Co., the intention at first being merely to solder them to the ceramic substrates used in thick-film hybrid work. As made by 3M and Kyocera, the chip-carrier is a square multilayer ceramic package, on the bottom of which is a pattern of gold bumps on 40- or 50-mil centers. Also inside, on the bare ceramic, is a gold base pad for chip bonding. A metalized frame makes it possible to put a top lid on for a hermetic seal.

Connections from the lead fingers of the internal IC pad to the external bumps are made through vertical metalized grooves on the side surfaces of the chip. These grooves are fabricated as metalized vias that also provide the perforations at which the carriers are snapped apart. The chip-carriers are then attached to a hybrid substrate by being reflow-soldered to pre-solder-coated lands. The carriers can easily be removed either for replacement or for testing or burning in the chip.

A collection of cofired chip-carrier types is shown in Fig. 3. At present the largest standard chip-carriers are 64-lead types. These units are about one third the length of a 64-lead DIP.

Comparison of the DIP with the ceramic chip-carrier yields some interesting data. A 24-lead cofired ceramic DIP is 1.2 in. long and 0.600 in. wide, has a 0.200-in.-square chip cavity, and costs 80 cents. On the other hand a 24-lead chip-carrier is 0.400 in. on a side, has a 0.235-in. cavity, and costs 40 cents.

Normally the chip-carrier is a three-layer device. But in an effort to lower the cost 3M has made two- and even single-layer chip-carriers. Offsetting their lower cost, however, is the fact that it is harder to seal them and to bond chips to them.

An interesting variation of the chip-carrier is supplied by both 3M and Kyocera. This is the leaded chip-carrier, which is available in 24- and 36-leaded versions from 3M and with up to 64 leads from Kyocera. It resembles a flatpack because its leads extend from all four sides. Compared to the bumped chip-carrier, this leaded chip-carrier has a larger footprint (or pattern on the substrate) but is easier to handle and can be successfully soldered to a pc board or hybrid substrate because of its compliant leads.

As yet, though, most IC manufacturers are still only at the stage of evaluating the ceramic chip-carriers. An exception is Mostek Corp., Carrollton, Texas, which is buying 18-, 24-, and 40-lead chip-carriers and is supplying 4,096-bit RAMs in them. Mostek had previously made the small plastic flatpack shown in Fig. 4 and eventually might make its own plastic chip-carrier.

In-house chip carrier

If Mostek does decide to build its own, it will be following the example of many other IC companies who have designed their own chip-carrier. In the U.S., General Instrument Corp., Hicksville, N.Y., has been supplying consumer-type LSI circuits in its Minipak for about a year. The Minipak (Fig. 5) is a small square glass-epoxy board about a third the size of a DIP. The LSI

3. Carrying chips. The cofired ceramic chip-carriers shown along the bottom of this view come in a variety of forms—note the leaded chip-carriers in the lower left and upper center. Several large and small ceramic mother-boards for the groups of carriers are also shown.

chip is bonded to a plated interconnect pattern that connects to solder bumps on the bottom side of the carrier. A drop of epoxy protects the chip.

This particular carrier can be reflow-soldered to a pc board. The units are available with up to 28 pins, have 50-mil solder-pad spacing, and cost an average of 0.4 cent per lead. By now the company has now supplied hundreds of thousands of calculator, clock and TV-game LSI chips in this package for applications where temperature and humidity are not serious considerations.

In Japan, Nippon Electric Co. has developed the rectangular 56-pin plastic carriers as well as the 52-pin plastic flatpacks. Finally, Fujitsu Ltd. has designed an interesting leaded ceramic chip-carrier (Fig. 6) that can be piggybacked with another unit of the same type.

In Europe, SGS-ATES, the largest Italian IC house, is buying ceramic chip-carriers and developing a plastic chip-carrier with dimensions of 10 by 10 mm. Siemens also purchased ceramic chip-carriers and will eventually supply chips in these and other chip-carriers in the future. Plessey in England is sampling chips in 14-lead carriers, while the large Sescosem division of Thomson CSF in France is just starting to investigate chip-carriers.

Flatpacks and LIDs

One of the earliest IC packages was the flatpack, a flat square package with two rows of ribbon leads emerging from opposite sides. This package found its niche in military and aerospace applications, mainly because of its small size, true hermetic seal, and proven reliability.

Today, newer, more compact flatpacks are surfacing in the commercial world. For instance, Jim Murphy, Diacon's sales manager, says that the increase in size of DIP packages is turning people back to a flatpack type of structure. He believes the new high-power bipolar RAMs will need flatpacks with BeO_2 or alumina substrates.

Flatpacks are available with up to 64 leads on 50-mil centers, and they take up much less space than a DIP. The newer flatpacks, with leads poking out through all four sides, are almost comparable to the chip-carrier in the pc board area they occupy. For instance, the body of a 24-lead flatpack is only 0.375 in. on a side and covers exactly the same board space as a standard bumped ceramic chip-carrier, except for the additional space taken up by its leads. Moreover, price of this glass-

4. Plastic pack. Mostek, like several other IC companies, has designed its own specialized package—a small plastic flatpack with leads emerging from all four sides. The flatpack is used for MOS consumer chips such as digital watch and calculator circuits.

5. Epoxy-glass pack. General Instrument supplies a line of MOS chips in this ½-inch-square, low-cost epoxy-glass carrier called a Minipak. On the bottom are solder bumps that can be reflow-soldered to a pc board. A drop of epoxy protects the chip.

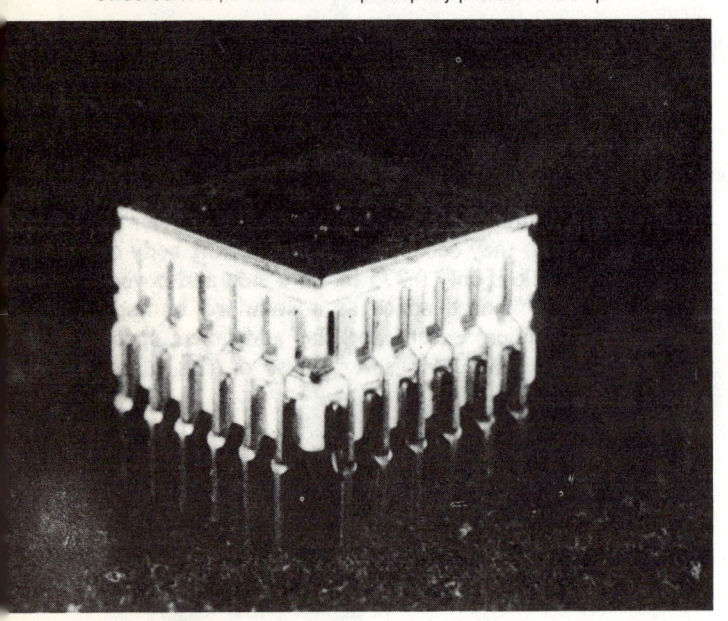

6. Piggyback. Going against the trend to leadless chip carriers is the novel 24-pin ceramic chip carrier used by Fujitsu for memory chips. The 0.5-inch-square unit is designed so that it can be stacked on top of an identical package to increase memory density.

ceramic flatpack is 60 cents as compared to $1.50 for the ceramic chip-carrier.

Two other advantages of the flatpack are testability and low lead resistance. The standard chip-carrier needs a special test socket to accommodate its gold bumps while the flatpack can easily be tested with small alligator chips attached to its ribbon leads. Lead resistance, a possible problem with bipolar chips, is 100 milliohms for a chip-carrier's gold bumps, only 16 to 20 milliohms for a flatpack using a lead frame.

Murphy therefore sees the chip-carrier and the four-sided flatpack finding two separate niches: chip-carriers for MOS work, and flatpacks for the higher-powered, high-speed bipolar logic.

Another ceramic chip-carrier that has been around for some time is the leadless inverted device. Originally it was used only for small-scale-integrated circuitry, but now there are 40- and 44-pad versions for LSI chips.

Built by Plessey Frenchtown of Frenchtown, N.J., the LID has a square ceramic body with elevated terminal pads. The body is built up from three metalized layers, and the terminals are on 35-mil centers. As usual, an IC is attached to the base of the cavity and wire-bonded to the LID's pads. Then the chip and its leads are sealed with a drop of epoxy. Finally, the package is inverted, and its terminal pads are reflow-soldered to a hybrid substrate.

Size for size, a 40-lead LID measures exactly the same as a 40-lead chip carrier—0.450 in. on a side. But lead counts may not go much higher. Also, the LID is not hermetically sealed. Although it is being used on some aerospace projects where the overall system package is sealed, its main use is in commercial hybrid work, in pocket pagers and digital watches, for example, because of its low cost.

Socketed chip-carriers

Still, for high-circuit-density applications, neither LIDs nor flatpacks are as popular as the ceramic chip-carrier, despite several potential difficulties in its large-scale application to pc boards. First, chip-carriers are difficult to solder to glass-epoxy material. Secondly, they are designed for hybrid substrates with 40-mil footprints while pc-board users prefer a 50-mil footprint. Finally, because newer chip-carrier types are arriving on the scene, all people concerned with this device had to be pulled together to generate a standard acceptable to users, IC companies and package manufacturers.

A ceramic chip-carrier that reflow-solders easily to a matching ceramic substrate cannot be directly attached to the glass-epoxy pc board found in most large commercial equipment—the differential temperature coefficient between the pc board and the ceramic is just too great. Then, too, the pc board flexes with temperature, enough possibly to fracture a connection at a reflow-soldered bump of the chip-carrier.

One solution is to mount the ceramic chip-carrier in a socket having compliant (soft, deformable, metallic) leads that could take care of the difference in temperature coefficient of thermal expansion. Another solution, proposed by AMP Inc., Harrisburg, Pa., was to design a plastic chip-carrier with resilient leads that could be connected directly to either ceramic or epoxy-glass multilayer substrates.

The problem of the 50-mil-spaced footprint is covered by a standard produced in early December 1976 by the Electronic Industries Association's Jedec task group JC-11.3.1, consisting of computer mainframe and IC and LSI packaging companies. Chaired by Dan Amey, engineering manager of packaging techniques at Sperry Univac, Bluebell, Pa., the group wrote a comprehensive standard on chip-carriers. The basic idea of the standard is illustrated in Fig. 7.

Four different chip-carriers, each with the same

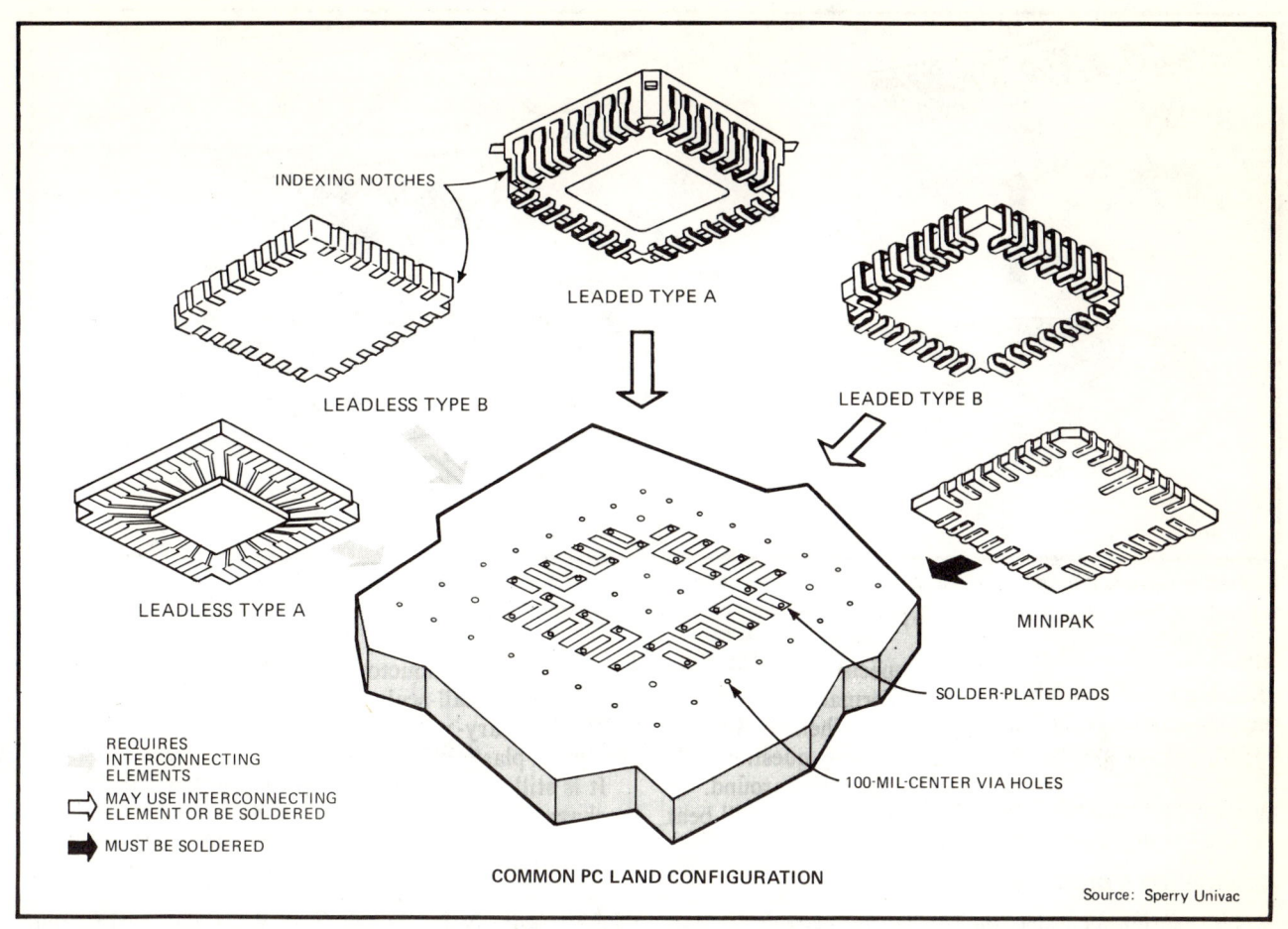

7. Interchangeability. The JC-11.3.1 task group of the EIA has generated an LSI package standard for devices with interconnections on 50-mil centers. The result could be the creation of five different families of chip carriers that would interconnect with a common pc footprint.

dimensions and 50-mil lead spacing, will fit one standard socket, which in turn will fit a standard pc pattern. A fifth carrier with 50-mil lead spacing but with slightly larger overall dimensions will not fit the socket and must be reflow-soldered to the board. Two of the leaded carriers can also be soldered directly to the board. This standard allows a user to freely interchange LSI package types without redesigning his board.

In Fig. 7, leadless type A is a single-layer ceramic carrier presently made by 3M. This unit has to be socketed to go on a pc board. Leadless type B, a multi-layer ceramic chip-carrier must also be socketed.

Leaded type A is AMP Inc.'s design, which is perhaps the most radical. It is the premolded chip-carrier with compliant leads that can be either soldered or plugged into a socket. Leaded type B is a ceramic substrate, suitable for LSI or a multichip hybrid, with compliant metal clips soldered on all four sides. Carriers of this type have been proposed by Berg Electronics, New Cumberland, Pa., along with a socket to accept this carrier. Berg also has come up with a design for a universal carrier socket.

The last unit is GI's Minipak, which was described earlier. It can only be reflow-soldered to a board.

The proposed carrier family would cover devices having from 28 to 156 leads with square packages ranging from 0.450 to 2.05 in. on a side.

The AMP concept deserves closer attention because the manufacturing process is fully automated. It combines a film-carrier approach with a premolded lead frame. Flow of the process is shown in the series of diagrams presented in Fig. 8.

Combining film- and chip-carriers

In the first step, an LSI chip is bonded to a beam tape. Next, the bonded chip on its interconnect is removed from the tape and attached to a leadframe, which has a premolded cavity. After a wash and drying step a bottom cap (which could be a decoupling capacitor) is inserted in the premolded cavity. Then a drop of silicone jelly covers the chip to protect it, and a top cap is forced on. Additional operations trim and form the leads and remove the carrier from the lead frame.

At present, AMP is putting a 24-lead package into production that will meet the new Jedec standard's dimensions. A customer will buy reels of the pre-molded sockets and will then be able to bond and seal LSI chips at his own facility.

Of course chips can be bonded to these carriers by either the film-carrier method (see p. 90) or standard automated wire bonding. AMP has already designed a socket that fits both its own package and the ceramic chip-carrier.

Will the chip-carrier sweep out all other forms of packaging? No, but it should see heavy use from now on particularly in computer and memory applications. Says

(a) BOND CHIP TO SPIDER (b) BOND SPIDER TO LEAD FRAME

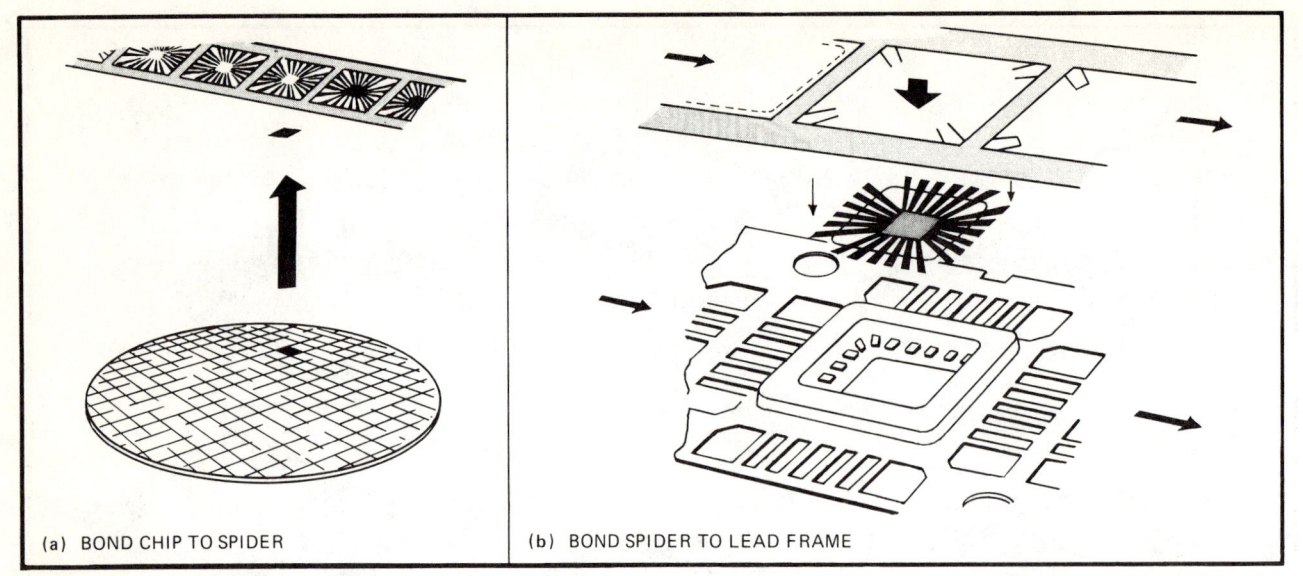

8. Molded carrier. The AMP plastic chip carrier blends the film-carrier and premolded-leadframe approaches. The IC chip is mounted on a film, excised from it, and bonded into the plastic cavity. Then after cleaning, a drop of silicone is placed over the chip and the cavity sealed.

Billy Hargis, senior development specialist for 3M Co., "For high-lead-count and high-performance packaging, the chip-carrier is the best possibility. When you go to a 84- or 128-lead device, a DIP is out of the question. For lower lead counts the DIP will always be around. In memory applications, however, the chip-carrier will beat out the DIP in density, cost and replaceability."

Automation a problem

One continuing objection to the chip-carrier is that there is still no equipment for its automatic placement on a pc board. However, for DIPs with over 24 leads there is no automatic insertion equipment either. A natural progression therefore seems likely from the DIP to either the chip-carrier or the film-carrier.

Film-carriers

All of the previously discussed methods of interfacing LSI chips to the outside world—DIPs, QUIPs, and ceramic and plastic chip-carriers—have a rigid construction that envelops the chip. However, a flexible if skeletal "package" for active chips has been around since the early 1960s—the film-carrier.

In this method a sprocketed nonconductive film has IC interconnect patterns (spiders) etched into its laminated copper surface. Reels of this film or tape are fed along with specially bumped IC chips into an automatic bonder. When the tape emerges, a chip is bonded to each individual pattern. By now all-copper tape is often used, but the basic assembly method remains the same.

The reels of bonded chips can either be fed into other automatic machines, which excise the chip plus its pattern and bond the outer leads of the pattern to a leadframe, or the chips on tape can be used directly on a pc board or hybrid substrate. The whole setup is well suited to automated high-speed handling.

In the U.S., the film-carrier is being used mainly in combination with DIPs. TI, Motorola Inc., RCA Corp., Fairchild Semiconductor, and National Semiconductor Corp. put small-scale transistor-transistor-logic ICs, complementary-MOS chips, and some linear chips on tape and mold plastic DIPs round them.

It is still relatively difficult to purchase a specific chip packaged on a reel or strip of tape in the U.S. But for companies that do want to put their own purchased LSI chips on tapes there are now at least four companies— 3M Co., St. Paul, Minn., International Micro Industries, Cherry Hill, N.J., Pactel Inc., Westlake, Calif., and National Semiconductor's Dynatape division, Santa Clara, Calif.—that will manufacture and supply tapes. For the first time some of the IC companies will "bump" a wafer for a customer (build up the metal I/O pads of each chip in the wafer). So a company with enough resources to buy the special bonding equipment can put its own chips on tape. But this ability is still beyond small and medium-sized companies.

A major U.S. user

One of the few large users of the technique in the U.S. is Honeywell Information Systems in Phoenix, Ariz. HIS is now in full production of a large computer mainframe, the system 66/85 in which the basic packaging module is a ceramic substrate to which are bonded maybe 100 chips taken from various film-carriers.

In the Honeywell system, bipolar LSI chips are automatically bonded to a three-layer tape, tested electrically on the tape, excised from the tape, automatically positioned on a 80-by-80-mm-square substrate, and then reflow-soldered to the substrate. The chips used are mostly Honeywell-designed and -manufactured current-mode logic and have up to 44 leads (only 40 leads are connected to the substrate). The substrate in Fig. 9 holds almost as much circuitry as a standard 12-by-12-in. pc board. A similar method has been in use at Honeywell Bull in France for some time.

To quote Wayne Umbaugh, senior engineer, advanced technology engineering, HIS Phoenix, "The only way we

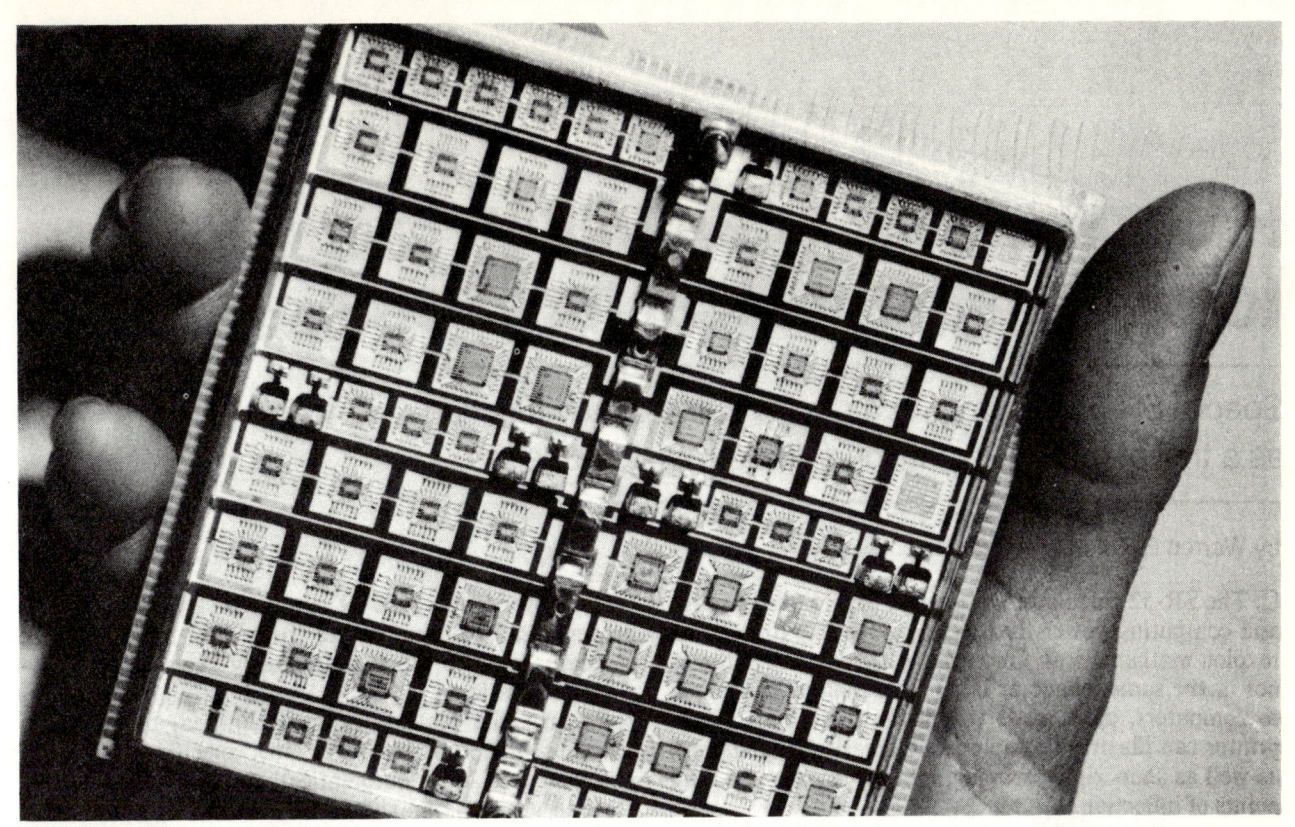

9. Micropackage. Current-mode-logic chips on film-carriers are bonded to a 80-millimeter-square multilayer ceramic substrate that is a module of the Honeywell Model 66/85 computer system. This dense packaging technique is necessary to fully utilize CML's high gate speeds.

could take advantage of the high-speed characteristics of current-mode logic was to go to the dense packaging of the film-carrier/multilayer-hybrid combinations which gave the required high interconnect density."

In its basic bare-bones form, the film-carrier method is more popular overseas. In Japan, both Toshiba and Nippon Electric will supply devices on tape. Many of Toshiba's 35-mm film-carrier devices are actually used in watches. The film-carrier is cut to fit the full size of the watch cavity and also serves as a substrate to mount other components. Other devices of this type are finding their way into Japanese cameras and calculators (especially the thin types). NEC furnishes LSI chips on 35-mm tape for a joint Toshiba/NEC computer, which is built around film-carrier ICs reflow-soldered to square multilayer ceramic substrates.

In Germany, Siemens is supplying MSI chips on Super-8 film and is investigating the possibility of supplying LSI chips on tape. Again, most of this output is going for consumer applications. Gerndt Oswald, marketing manager for Siemens AG's components division, believes the low cost of film-carrier guarantees it a good future, but adds, "One serious requirement has yet to be fulfilled—that suppliers agree on a standard carrier and standard tooling." There have been some efforts along this line in the U.S., but what is really needed is an effort on a par with that just done for chip-carriers by the EIA.

Whither next?

LSI chips on film will obviously have their largest impact on the hybrid market. How quickly and fully this comes about will depend on the availability of bumped chips and low-cost bonders.

Chuck Spence, Mostek's manager of assembly development, says his company has been evaluating the method. "Current technology dictates putting bumps over the entire wafer," he comments. "This is fine for TTL or C-MOS with high probe yields, but for higher-technology devices with low probe yields, the cost of bumping the entire wafer and then using only a small portion of it outweighs the cost savings involved in the beam-tape system. If a process can be perfected that bumps the tape, it would be a cost-effective method for lower-yielding IC technologies, since then standard chips could readily be applied to the bumped 'tape.' "

Actually Pactel Corp., Westlake, Calif., has such tape but cannot yet supply it in the quantities an IC manufacturer would need. Several other IC and packaging companies are researching this problem.

Robert Beard of National Semiconductor believes the film-carrier rather than chip-carrier has a chance of becoming the dominant technique for handling LSI "because it is similar to techniques already used in the mainstream of industry." He says, "Since the chip can be mounted directly on the pc board and protected with a cover, what the technique does is to eliminate the need for packaging. The equipment that mounts the chip onto the board is the only new element. But it's not replacing anything; rather it's eliminating the need for several steps and kinds of equipment."

The AMP approach, of course, combines both the film-carrier and chip-carrier technique (p. 89). More marriages of the two techniques may yet result, for a number of reasons. For instance, devices on single-layer copper tapes cannot be tested or burned in on the nonisolated tape. But placing the excised chip and beams in an inexpensive chip-carrier could solve this problem. □

Packaging technology responds to the demand for higher densities

In the search for ways to cram more VLSI and LSI onto a substrate, several forms of chip-carrier are emerging as leaders

by Jerry Lyman, *Packaging & Production Editor*

☐ The art of squeezing in the most chips per square inch—commonly called high-density packaging—is spreading rapidly from the military and aerospace domain into many other areas of electronics. Designers of all kinds of electronic equipment, particularly large-computer mainframes, are feeling that the real-estate crunch has only just begun, now that large-scale and very-large-scale integrated-circuit technology is spinning off 24-, 48-, 64-pin, and even larger packages. Familiar high-density schemes, such as cramming dual in-line packages onto a single two-sided printed-circuit board, are already inadequate for many applications: the DIPs required by the larger and more complex ICs take up too much space and their internal line resistance and capacitance limit circuit performance.

To meet demands for higher density while maintaining performance and reliability, manufacturers of electronic equipment are now looking hard at these alternatives to the ubiquitous DIP-on-board method: ceramic chip-carriers on multilayer pc boards or multilayer ceramic substrates; film chip-carriers on multilayer ceramic substrates; and bare chips wire-bonded to multilayer thick-film hybrids. Table 1 lists the planar high-density packaging techniques now available to the designer with their relative chip densities and maximum substrate dimensions.

In selecting one of these methods, an engineer must consider: what component density is needed; how many components are to be packaged; what type of digital logic is to be used; environmental and thermal considerations; and cost. At first glance, the conventional bare-chip hybrid or the tape-mounted hybrid automatically bonded to multilayer ceramic substrates might seem the best choices with their highest component-per-square-

1. Polyimide substrate. One way to overcome the size limitation of ceramic substrates is illustrated below. The unit is a Pactel six-layer substrate laminated to an aluminum heat sink. The conductive pad pattern is for either ceramic or film carriers. Substrates as large as 16 by 14 inches may be fabricated using this method.

2. Chip and wire. The Raytheon unit shown here has four conductive layers and carries 29 low-power Schottky chips in a 2.2-by-1.4-inch pluggable package.

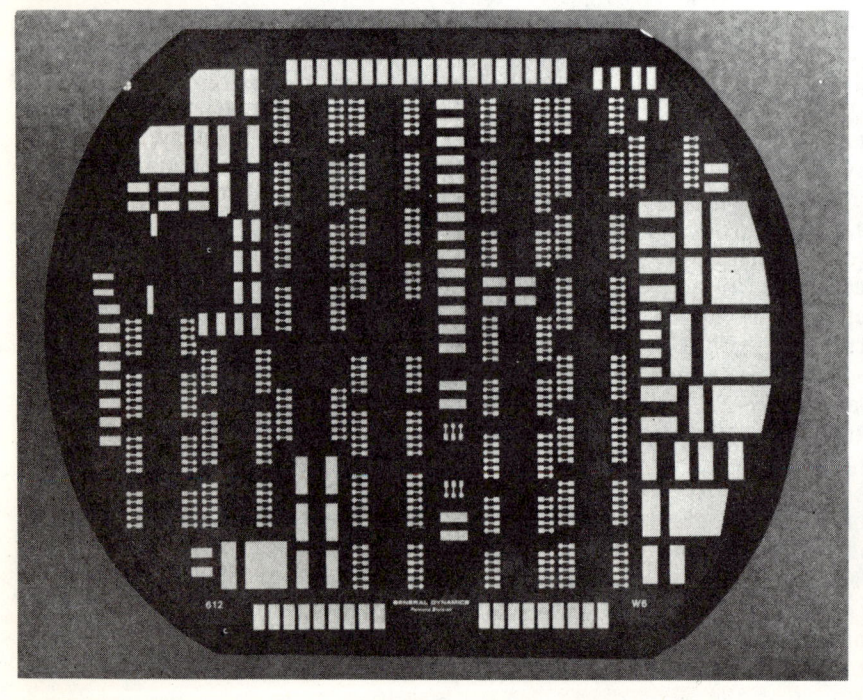

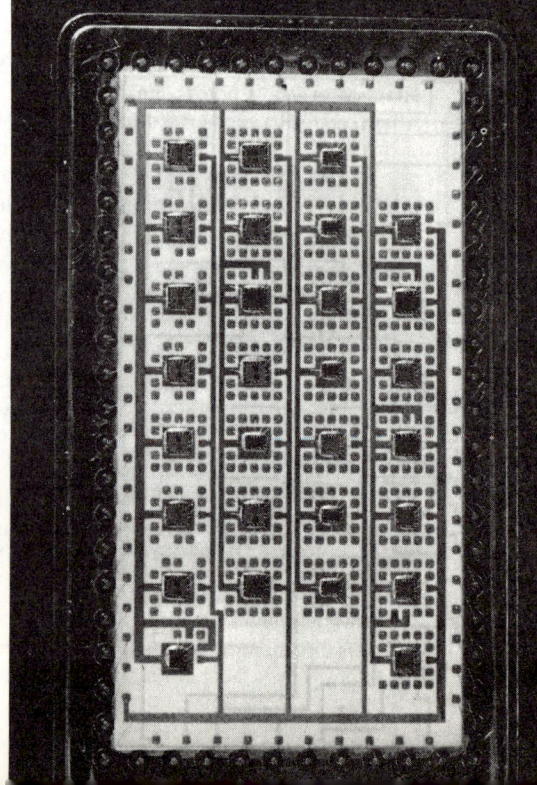

inch density. But the tradeoffs explored in this article suggest that chip-carriers on multilayer pc boards can be a better choice and that DIPs on multilayer boards are still a viable alternative in some applications. First, then, a look at the advantages the older techniques still offer: DIPs and flatpacks on pc boards and conventional wire-bonded hybrids.

The DIP goes on

The familiar dual in-line package is both the most readily available and lowest-cost package on the scene today. This is one reason why the most frequently used high-density packaging method is DIPs mounted on multilayer pc boards. The DIPs may either be wave-soldered to the boards or plugged into sockets.

Computer-aided design programs are available to lay out the interconnections for multilayer boards, and the manufacturing of these boards is now a mature technology. Multilayer boards made of either epoxy glass or polyimide have one outstanding advantage over ceramic substrates: a pc board as large as 22 by 16 inches can be made, while at present most independent and in-house hybrid manufacturers will not take on a ceramic substrate larger than 2 inches on a side. These dimensions limit the total number of chips per substrate to around 100, but computer manufacturers routinely pack 200 to 300 DIPs on a single multilayer board.

In the future, however, the ceramic-coated steel substrates produced by Alpha Metals Inc., Newark, N. J., and Erie Ceramic Arts Co., Erie, Pa. may overcome the area limitations of the all-aluminum substrate. Alpha, for instance, is tooling up to supply ceramic-coated steel substrates or boards as large as 12 by 8 inches. At the bare-chip packaging density specified in Table 1, this could mean a large porcelainized substrate with screened and fired-on conductors capable of carrying hundreds of chips.

Another future possibility for large-scale hybrids is Pactel's use of thin layers of polyimide with additive metal patterns plated on. The layers are laminated to an aluminum heat sink [*Electronics*, July 22, 1976, p. 101]. This method, developed by the Newbury Park, Calif. firm, allows substrates as large as 16 by 14 inches. The unit shown in Fig. 1, a 2-by-3-inch substrate, is a six-layer type designed to accept either ceramic or film carriers.

Still, says Jeff Waxweiler of Algorex Corp., a Syosset, N. Y., computer-based design service, "DIPs on pc boards give the highest performance at the lowest cost." With this type of construction—layers consisting of interconnects, power and ground planes—it is possible to control the characteristic impedance of the circuit board to an extremely close tolerance, according to Waxweiler. This type of control is still not possible with the multilayered ceramic substrate, so the designer can only try to keep his interconnections as short as possible.

Perhaps a better solution for high-speed logic is the fine-line printed-circuit board. In this construction, pc traces and spaces are 5 to 7 mils wide rather than the 10 to 20 mils of a standard pc board. This line reduction yields an interconnect density equivalent to that of an eight-to-ten-layer board on just one two-sided pc board Since it is simpler to control the dimensions and tolerances of the laminate materials used in fine-line boards, the method is well suited to emitter-coupled-logic technology, which re-

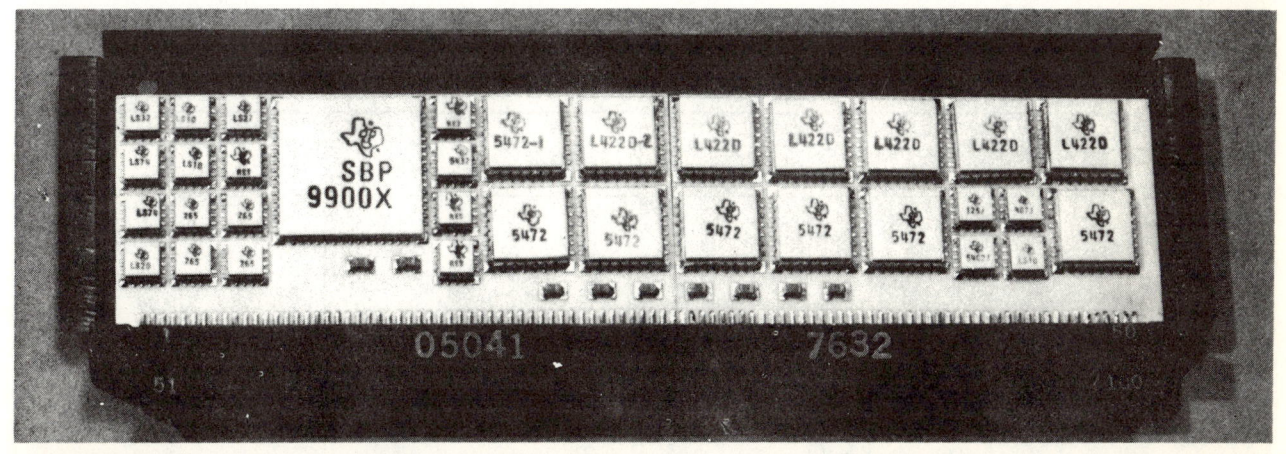

3. SEM. Many of the Navy's standard electronic modules (SEMs) consist of small thick-film substrates mounted to either ceramic or epoxy-glass motherboards. ILC Data Device Corporation uses the latter construction method in the SEM shown above.

4. Motherboard. Texas Instruments combines chip-carriers with a multilayer ceramic motherboard in this unit. The circuitry consists of a complete militarized microcomputer with a PROM, RAM and TI's integrated-injection-logic SPB 9900 microprocessor.

quires a tightly controlled characteristic impedance.

Many large mainframe firms like Sperry Univac, Blue Bell, Pa., and IBM Corp., Endicott, N. Y., combine fine-line and multilayer techniques to create extremely large, dense, and fast circuit structures. For instance, Sperry Univac is producing a 10-layer ECL processor with a 50-ohm characteristic impedance by using fine lines on all the interconnection layers. ECL ICs in 48-pin packages are mounted on the large board, which has four working layers, two pad (outer) layers, and four ground and power planes.

Flatpacks on multilayer boards

Another established technique and the next step up in the high-density hierarchy is the use of hermetically sealed, metal flatpacks on multilayer boards. A flatpack is a small, square package with two rows of ribbon leads emerging from opposite sides (newer versions have leads extending from all four sides). These leads are reflow-soldered to a pc board, rather than wave-soldered, leaving the board's plated-through holes free to connect the various layers. The combination of flatpacks and multilayer boards has always been extremely popular in military and aerospace electronics applications, primarily because it doubles chip density over DIP designs. The metal-packaged flatpack also transfers heat better than a ceramic DIP.

The main disadvantage of the flatpack is simply that not all IC types are available in this package. Another disadvantage is more subtle: as the component density of any circuit increases beyond eight chips per square inch, it becomes almost impossible to put conductors on the outer layer or layers. At the low digital-logic speeds of transistor-transistor logic or metal-oxide-semiconductor devices, this is not a problem, but for ECL applications the substrate now needs a strip-line–like configuration. A structure of this type is difficult to manufacture to tolerances tight enough to control impedance.

Chip and wire

This illustrates a packaging fact of life—the higher the IC speed, the lower the packaging density possible. In many space-limited applications, however, component density takes precedence over performance, cost, and total number of components packaged. Here, the wire-bonded, bare-chip multilayer hybrid is the undisputed leader, as Table 1 shows.

The ultimate system for component density, the chip-and-wire multilayer-ceramic hybrid, saves space by using the smallest package available—the IC chip itself. Multilayering eliminates interconnects from the board surface, allowing the designer to pack wire-bonded chips into the space saved. A typical large digital hybrid is shown in Fig. 2. Substrates can be as large as 2 by 2 in. and can have as many as eight conductive layers, although most companies prefer only four or five. As Table 1 shows, component density for chip-and-wire hybrids can vary from 15 to 25 chips per square inch. The nearest competing technique, the chip-carrier, has only half the density.

But despite its superior packaging density, the multilayer chip-and-wire hybrid has some disadvantages. It is well known that as the number of chips per hybrid goes up, the individual chip yield must approach 100% to get a decent packaging yield. This means that if a company does not probe or pretest 100% of its chips, final hybrid yield will be low, resulting in excessive repair. But the cost of 100% testing at the chip or wafer level is extremely high, so many hybrid manufacturers are looking at two alternatives: subsectioning large hybrids and mounting chips on a testable film carrier or ceramic chip-carrier for pretesting.

Cutting up

Subsectioning can be approached in two ways. Circuit Technology Corp. of Farmingdale, N. Y., makes two 1-by-2-in. substrates for a hybrid design, rather than one large 2-by-2-in. substrate. It then tests each smaller substrate, wire-bonds interconnections between the two units, and hermetically seals the combined hybrid with the two substrates butted against each other. This subsectioned 2-by-2-in. device has proven to have a better yield than the single 2-by-2-in. substrate.

Another approach to hybrid subsectioning is the Navy's standard electronic module, or SEM, which is part of a Navy program to create a library of digital and analog function modules. In this method, ceramic substrates with standard case sizes and printed-circuit

TABLE 1: RATING PLANAR HIGH-DENSITY PACKAGING TECHNIQUES					
Method	IC package	Circuit substrate	Density (IC chips/in.²)	Maximum size of substrate (in.)
1	dual in-line package	multilayer pc board or fine-line two sided pc board	2 – 2.5 (14 – 16 pin)	22 x 16
2	flatpack	multilayer pc board	3 – 4.5 (14 – 16 pin)	22 x 16
3a	chip-carrier	multilayer pc board	8	22 x 16
3b	chip-carrier	alumina substrate	8	≈ 2 x 2
4a	bare chip	multilayer alumina substrate	15 – 25	≈ 2 x 2
4b	chip on tape	multilayer alumina substrate	≈ 15 – 25	≈ 2 x 2

SOURCE: ALGOREX CORP.

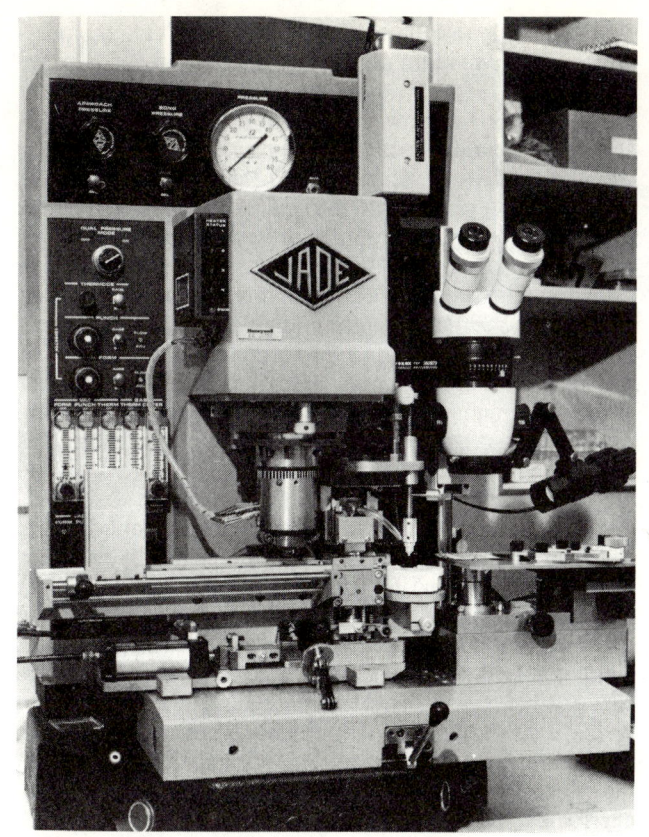

5. Chip on tape. The Jade 1810 system is specifically designed for hybrid work. It excises chips from a tape carrier, forms the outer leads of the chip's spider, die-bonds the chip to a substrate, and mass-bonds the chip's outer leads to a thick-film substrate.

6. Large-scale hybrid. This 2-by-2-inch multilayer substrate houses an array of current-mode logic chips on film carriers. CII Honeywell Bull mounts up to nine of these substrates on an 11-layer motherboard. Note the cooling fin assembly on the bottom of the substrate.

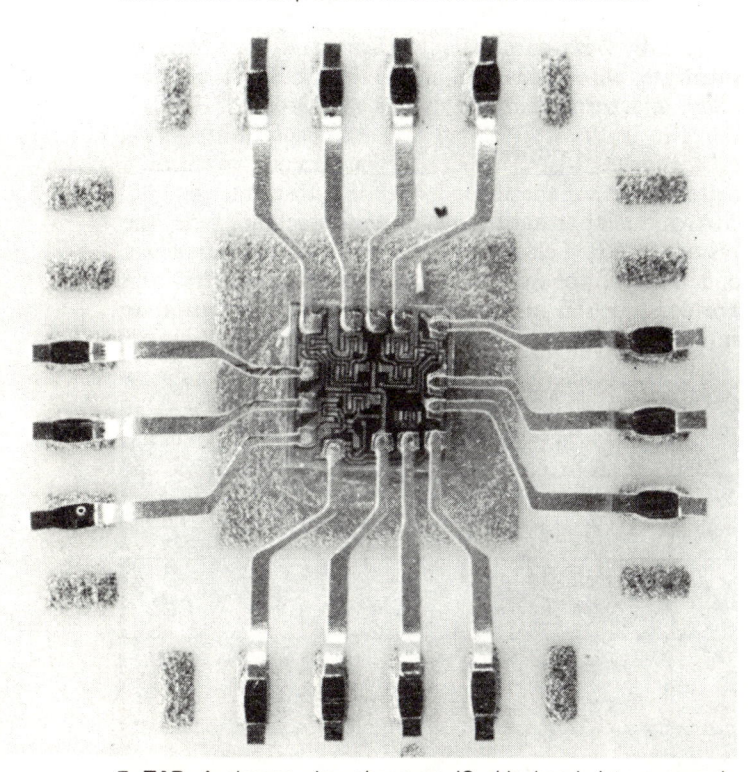

7. TAB. A closeup view shows an IC chip bonded to a ceramic thick-film substrate by the tape-automated-bonding method (TAB). The chip and its spider were excised from a three-layer tape and thermocompression-bonded to the substrate's gold pattern.

connectors or standardized rectangular multilayer pc boards are used as interconnects for smaller thick-film hybrids in sealed packages. A typical SEM, manufactured by ILC Data Device Corporation of Bohemia, N.Y., is shown in Fig. 3.

Each SEM submodule may be pretested and removed if necessary. The SEM sacrifices packaging density for testability and yield. The yield is higher than that of a large-scale single-substrate hybrid, while density approaches that of the leadless ceramic chip-carrier.

Carrying chips

The ceramic chip-carrier is a small, leadless, square package with a gold-plated cavity. Lead metalization is connected internally to gold solder pads on the bottom face of the unit. The alumina package is designed specifically for reflow-soldering to an alumina substrate. The combination of a multilayered-ceramic (componentless) motherboard and leadless ceramic chip-carriers is listed as method 3b on Table 1. Not only does this method allow pretesting and/or burning in of the chips in the carriers, but it also protects the chips from damage while handling. These advantages contribute to a higher yield than the straight bare-chip hybrid method.

Many companies have gone to the ceramic motherboard/chip-carrier technique. These include Circuit Technology; RCA Corp., Moorestown, N.J.; Martin Marietta Corp., Orlando, Fla.; Honeywell Avionics, St. Petersburg, Fla.; and Texas Instruments Inc., Dallas.

Table 2, furnished by Jon S. Prokop and Dale Williams of TI's microelectronic center, compares several packaging techniques applied to a hypothetical case of a digital circuit with 20 digital chips. The methods include conventional DIPs soldered to a pc board, chip-carriers soldered to a ceramic motherboard, and a multichip hybrid. The data shows that the chip-carrier and

The data establishes that the chip-carrier and motherboard combination is more cost-effective than the DIP and pc board approach. But note that the limitation

201

of ceramic substrate size could disqualify this technique for some applications.

One way to get around this limitation is to attach a leadframe to the motherboard. Then many motherboards with chip-carriers can be mounted on a pc board, giving both the advantages of many chips per square inch and many packages per board. This type of assembly has been fabricated at both TI and RCA Moorestown. One such configuration, a SEM using ceramic chip-carriers built by TI to operate from −65° to 125°C, is shown in Fig. 4. The unit is a complete microprocessor module with a programmable read-only memory, random-access memories and TI's integrated-injection-logic SBP 9900 microprocessor.

According to Prokop and Williams, constructing SEMs in the chip-carrier/ceramic motherboard fashion resulted in a 60% volume and weight reduction over conventional pc-board assemblies. In addition, the superior thermal characteristics of the all-metal-and-ceramic construction of these SEMs resulted in a lower temperature rise than in the equivalent assembly.

Chips on tape

In general, the chip-carrier and motherboard design will never replace the densely packed multichip multilayer hybrid, but it is an approach that leads to high yield (because it allows pre-testing and burning in), ease of repair, and lower cost. A technique with these advantages and with a packaging density approaching that of the bare chip method is also available—chips mass-bonded to tape or film carriers.

Since the 1960s, a highly automated method has been used for mass-bonding IC chips to an insulated, sprocketed tape. In this system, known as the film-carrier or tape-automated bonding method, the IC interconnects (spiders) are etched into a copper surface that is laminated on a sprocketed, nonconductive film. Reels of this film are fed simultaneously with specially bumped chips (see "Bumped chips versus bumped tapes") into an automatic inner-lead bonder. The final product is a reel of tape with one chip bonded to each frame.

The reels of chips with spiders are fed to another automatic machine, the outer-lead bonder, for bonding to leadframes that eventually end up in plastic DIPs. This type of manufacturing has been going on for some time in the U.S. at TI, National Semiconductor, Fairchild, and Solid State Scientific. It is mainly used for what the IC manufacturers call jellybeans—TTL and C-MOS small-scale ICs with 14 and 16 pins.

However, the military has long recognized that the chips on tape could be excised out of the tape along with small copper-beam interconnects to form a flexible chip-carrier and that this carrier's packaging area would approach that of a bare chip.

At one time, this flexible film carrier for hybrids was held back by a lack of tape suppliers, a lack of machinery for excising and bonding chips on tape to a hybrid, and difficulty in obtaining bumped chips [*Electronics*, Oct. 27, 1977, p. 139]. Now there are at least seven companies willing to supply standard two- and three-layer tape. They include the 3M Co., St. Paul, Minn.; International Micro Industries, Cherry Hill, N. J.; National Semiconductor's Dynatape division, Santa Clara, Calif.; Fortin Laminating Corp., San Fernando, Calif.; and Pactel Inc.

Machines tape-bond hybrids

Jade Corp., Huntingdon Valley, Pa., and International Micro Industries have come up with machines for excising and bonding chips on tape to thick-film substrates—Jade's 1810 is shown in Fig. 5. Getting bumped wafers, however, is still a problem; it generally requires an in-house bumping facility. Nevertheless, film carriers are

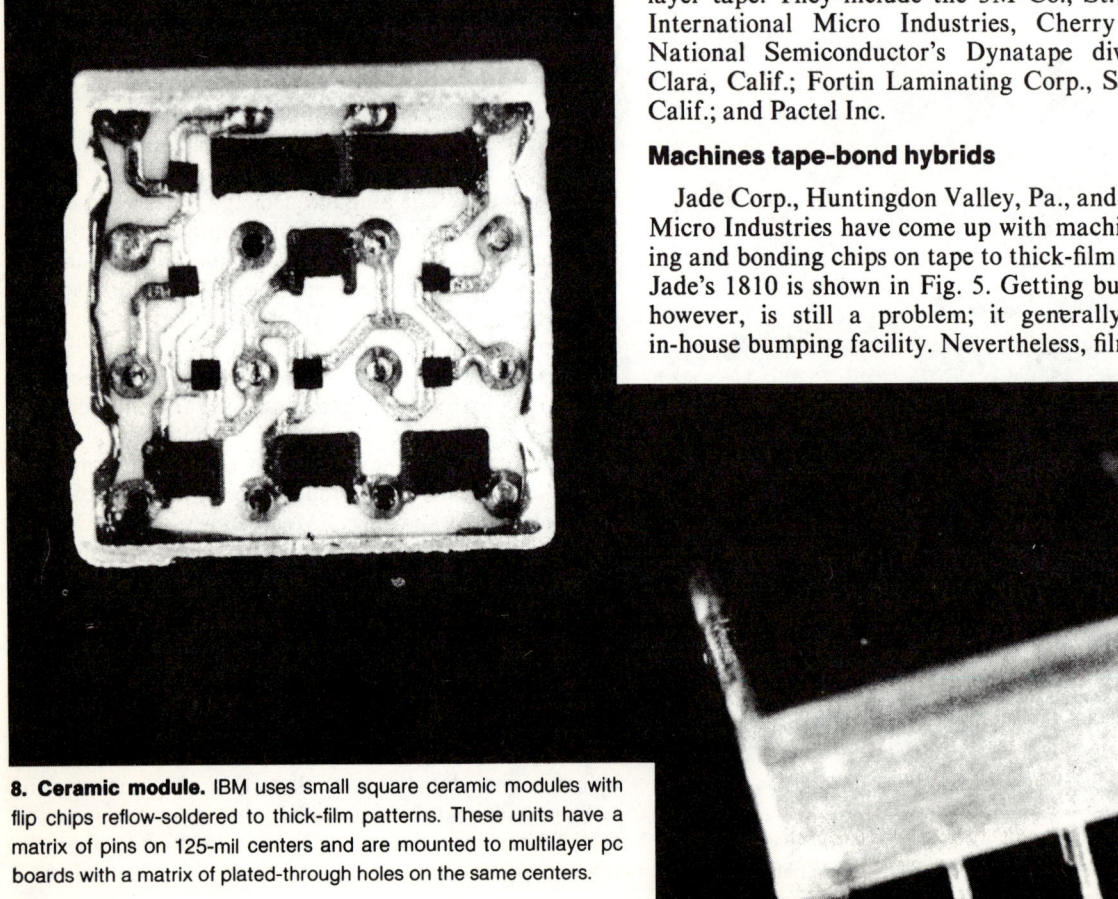

8. Ceramic module. IBM uses small square ceramic modules with flip chips reflow-soldered to thick-film patterns. These units have a matrix of pins on 125-mil centers and are mounted to multilayer pc boards with a matrix of plated-through holes on the same centers.

available now and suitable to a fully automated approach consisting of automatic bonding of chip to tape; testing and burn-in on tape; and excising and bonding to the conductors of a thick-film substrate. On the other hand, there is no denying that this method requires an expensive initial capital investment.

CII-Honeywell-Bull, the major French computer maker, is now installing machines to produce film-carrier packages at its factory in Angers next year. The French firm calls its process TAB—for tape-automated bonding—and will use it mainly for current-mode-logic modules in computer mainframes. There will be some TAB packaging for fast bipolar memory too, but all n-MOS chips will use standard DIPs.

In the Honeywell-Bull process, up to 36 chips-on-film can be mounted on a 2-in.² piece of alumina. This substrate has four conductive layers for connections—one for logic, one for the supply voltage, one for ground, and one for the pads of the chips. A cooling-fin assembly is mounted on the bottom side of the substrate, as shown in Fig. 6.

Up to nine substrates are mounted on an 11-layer motherboard that measures 10.2 by 11.2 in. The motherboard has three logic connection layers, three ground planes, one layer for clock distribution, one for reference voltage, two for supply voltages, and one for the pads of the substrates. Center-to-center spacing between the motherboards is one inch. To cool the boards, a turbine forces air over the fins of the substrates.

The Honeywell-Bull system, with its excised chips-on-carriers reflow-soldered to the substrate, bears some resemblance to a hybrid module that the Phoenix-based Honeywell Information Systems used in the production of its now defunct System 66/85 and to another module in a joint Toshiba/NEC computer [*Electronics*, March 17, 1977, p. 90].

At Honeywell Avionics' hybrid facility in St. Petersburg, Fla., TAB is being used for military hybrids produced in smaller quantities. Honeywell Avionics uses its own 35-mm three-layer tapes and adds bumps to purchased wafers. A Jade inner-lead bonder is used to place the chips on strips rather than reels of tape. The tested and burned-in chips are excised, then the die is epoxy-bonded and the lead spider is thermocompression-bonded to a four-layer substrate by a Jade 4810 machine. A typical film chip-carrier bonded at Honeywell Avionics is shown in Fig. 7.

Honeywell Avionics has so much confidence in the process that it lays out all new hybrids in both chip-and-wire and TAB formats. These digital and analog hybrids, which are mainly on 1-by-2-in. substrates, use both Schottky-TTL and MOS ICs.

Rudolph Oswald, head of the hybrid microelectronics

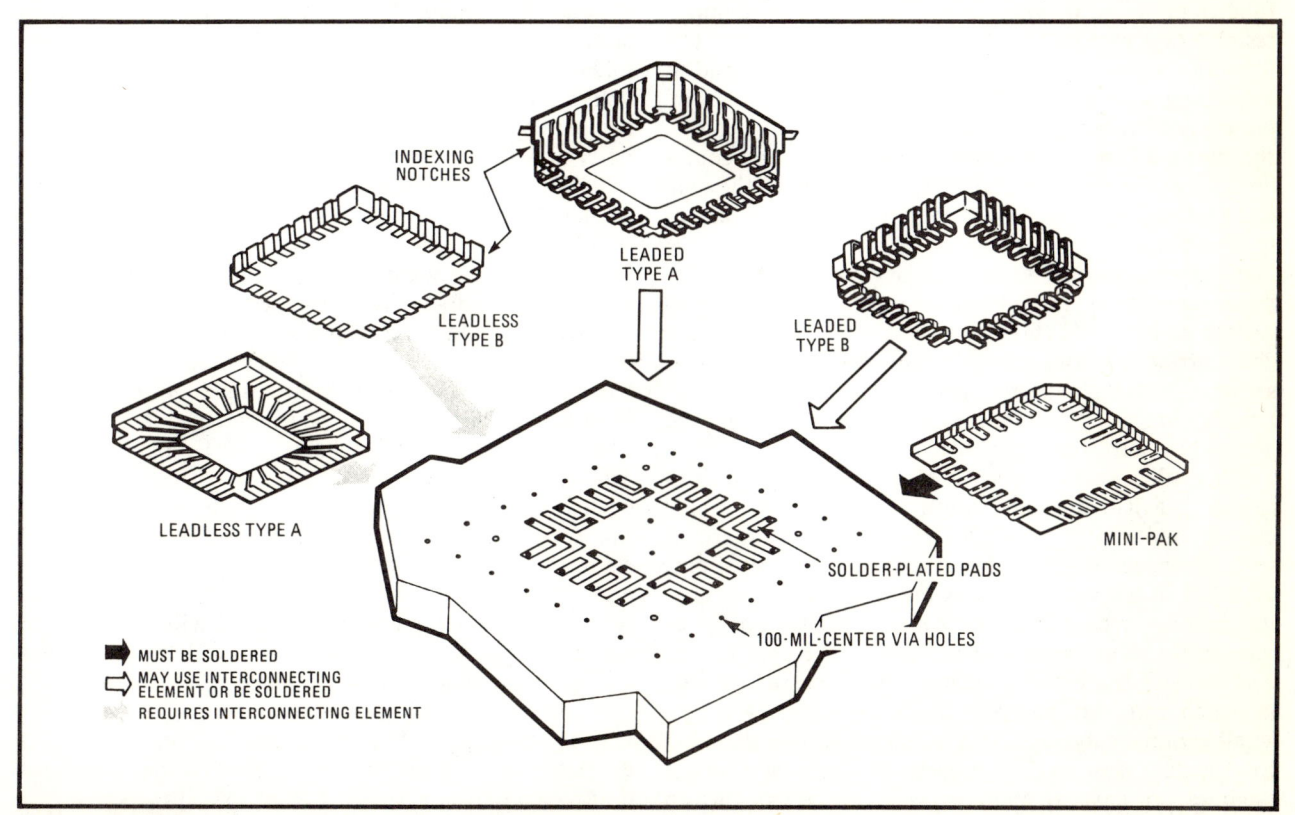

TABLE 2: PACKAGING COMPARISON FOR A 20-CHIP DESIGN				
Method	Weight (gm)	Area (cm²)	Yield	Cost
Printed-circuit board	52	81	high	low
Chip-carrier/ mother carrier	12	12	high	medium
Hybrid	10	6	low	high

SOURCE: TEXAS INSTRUMENTS INC.

9. Chip-carrier family. In order to further the use of chip-carriers on pc boards, the Jedec JC-11 committee created a standard for this family of five different carriers. Each can be attached to a standard pattern of pads on 50-mil centers.

avionics division of Honeywell, says: "We have found the yield with TAB much higher than with a comparable chip-and-wire hybrid. In my opinion, most people in the hybrid field are afraid of TAB because it means acquiring a good technical base covering bumping chips, tape design, and bonding, as well as a familiarity with multilayer hybrid technology."

Oswald sees TAB as the next big hybrid technology. He is concerned that the U.S. will be bypassed by both the Japanese and Europeans in hybrids because, he says, more work is being done on TAB overseas and a great quantity of TAB equipment is going offshore.

Alan Kiezer, director of engineering at Jade, has indirectly confirmed Oswald's remarks. He notes that unlike the U.S., where only military and computer hybrids have been fabricated with the chips-on-tape method, European and Japanese firms are actively engaged in using the same method for high-volume production items like television sets, calculators, cameras and telecommunication equipment.

Bumped tape may simplify production

One reason for the slow acceptance of this method in the U.S. may be the need for bumped wafers, an added expense in wafer processing. A start has now been made in eliminating the need for bumped wafers—a bugaboo to the small and medium-sized hybrid producer interested in acquiring a TAB capability. Dynatape, Fortin, 3M, and Koltron Corp. of Sunnyvale, Calif., are all working on testable bumped tapes, which would eliminate the need for bumping wafers.

General Dynamics, Pomona, Calif., has a Navy-funded program to look at the feasibility of bumped testable tape as a means of increasing hybrid yields. The firm has made its own bumped testable tapes and is evaluating samples from outside vendors on 1-by-1-in. multilayer hybrid substrates. Like Honeywell Avionics, the Pomona firm is working with Jade automatic equipment to excise the chip and spider, epoxy-bond the chip and thermocompression-bond the lead spider to a ceramic substrate.

Packaging engineers for the large computer mainframe companies have been observing the film chip-carrier activity in hybrids for some time. The ceramic chip-carrier, covering roughly half the area of a DIP, seems a better choice to these engineers as a smaller replacement IC package for the large multilayer pc boards that most computers use.

Since 1965, IBM has been using its own ceramic chip-carrier, a half-inch-square substrate with either a single chip or multiple chips bonded to a thick-film interconnect. As shown in Fig. 8, this carrier has a matrix of pins on 0.125-in. centers. The pins are soldered into a fine-line multilayer board that has a matrix of pads and holes also on 0.125-in. centers. The basis of the chip-carrier was and still is a flip-chip technology, pioneered by IBM, in which chips are bumped for reflow soldering to the small ceramic substrate. This construction was the basis for the IBM System 370 models. Newer IBM ceramic modules can have as many as 72 input/output pins on 0.1 inch centers compared to 20 on the 1970 version. In both the old and new IBM computers, dual substrates are often stacked one atop another to increase memory chip density.

The latest IBM computers mix a 1-in.² version of IBM's ceramic carrier with standard ICs in DIPs. Both types are soldered to mounting holes in the same type of multilayer board used in the System 370.

In 1976, interest in the use of chip-carriers on pc boards moved the Joint Electron Device Engineering Council task group JC11.3.1, consisting of mainframe, IC and LSI package firms, to create a standard for a family of small square devices that could be reflow-soldered to a standard pad pattern on a pc board [*Electronics*, March 17, 1977, p. 88]. The standard called out two leadless ceramic types that required sockets, a plastic type with compliant leads, a ceramic substrate with solderable clip-on edge connectors, and a leadless chip-carrier on an epoxy-glass substrate.

All of these units were to fit on a standard square pc footprint on 50-mil centers. Figure 9 illustrates the various types of chip-carriers designed for reflow soldering to a pc board. At present, leadless types A and B shown in the diagram are commercially available. A socket for these types is being produced in sample lots at AMP Inc., Harrisburg, Pa., and in the developmental stage at Berg Electronic division of Du Pont, New Cumberland, Pa., The plastic package with compliant leads has been developed at AMP but is still being evaluated. Leaded type B, a square substrate with soldered-on edge clips on four sides, is to be available soon. The edge clips are in pilot production at Berg. Mini-Pak, the last package, is used by General Instrument Corp. Hicksville, N.Y., for consumer-type MOS chips, but is not available separately.

Dan Amey, chairman of the Jedec JC 11 committee, is engineering manager of packaging techniques at Sperry Univac, Blue Bell, Pa. He is now in the midst of a development program to make the transition from large multilayer boards with ECL ICs in DIPs to the same type of boards and circuitry with chip-carriers.

In one of Amey's projects, Sperry Univac supplied

Bumped chips vs bumped tapes

In the film carrier process special metalization layers are evaporated over an IC chip's aluminum interconnect pads. These built-up pads have come to be known as bumps. Their purpose is to protect the IC's I/O pads from the heat and pressure of mass bonding.

In order to eliminate the need for bumped chips and to allow the use of standard chips on tape, bumped tapes are now being developed. These tapes will have the special metalized bumps needed for mass bonding on the inner leads of each frame's copper microinterconnects or spiders.

Yet another version of the bumped chip is the IBM's flip chip. This semiconductor device is designed for face-down reflow soldering to a thick-film substrate.

In the flip chip process, an initial layer of glass is put down over the IC's surface. Via holes are then etched through to the IC's aluminum I/O pads. Additional metalization layers are plated on to the exposed pads, and then solder bumps, typically 6 mils in diameter and 4 mils high, are built up.

10. Chip-carriers socketed. This is an example of a 68-pin leadless ceramic chip-carrier mounted in a new AMP plastic socket. The socket is reflow-soldered to the printed-circuit board, and a finned heat sink maintains the proper junction temperature for the IC.

AMP with an existing dual in-line package memory array and its peripheral circuit on an 11-by-14.6 inch six-layer pc board. AMP was directed to re-layout the board so that the 84 22-pin DIPs in the memory array could be replaced with the leaded plastic AMP package. Peripheral circuits were to remain in DIPs. With the leaded chip-carrier the area taken up by the memory array was reduced by two thirds, demonstrating the packaging efficiency of the chip-carrier over the DIP. A similar size reduction could be accomplished for the entire unit if all ICs were packaged in the premolded leaded chip-carrier. Despite its promising potential, the AMP unit must still prove that it can handle the environmental conditions encountered in computer applications.

Amey is also evaluating a ceramic leadless carrier in an AMP socket, a 68-pin heat-sink version as shown in Fig. 10, and a leadless carrier with Berg edge clips soldered on 50-mil centers.

In one application, Sperry Univac has put both small multichip hybrids and ceramic chip-carriers on one multilayer pc card. The hybrid, an assembly of ECL chips on a 1.35-in.^{2} three-conductive-layer ceramic substrate, has edge clips that are reflow-soldered to the substrate and then later to the pc board. The ceramic chip-carrier houses an ECL LSI chip and also uses an edge-clip mount.

Amey points out that this mix of hybrids and chip-carriers, both using edge clips as leads, is based on existing technology. Berg, for instance, has been making edge clips on 100-mil centers since 1971. The 50-mil-center edge clips for the ceramic chip-carrier present no new problems, according to Jim Diliplane of Berg.

Using the new space-saving carrier with leads on 50-mil centers will not mean the abandonment of the present-day 100-mil hole spacing on pc boards. Amey points out that with 50-mil hole spacing, pc traces cannot be run between adjacent holes. The optimum layout for chip-carriers, says Amey, is to have all holes on 100-mil centers and all chip-carrier patterns on 50-mil centers. With this layout and 5-to-7-mil pc linewidths, at least two lines can be run between adjacent holes. It should be kept in mind that all the Jedec versions of the chip-carrier are reflow-soldered to a board and that the plated-through holes serve as vias between conductive layers, rather than as mounting holes for component leads.

Going against the tide

The Jedec spec on chip-carriers was based on the assumption that the leadless ceramic chip-carriers available in 1976 (made by 3M and Kyocera International, Inc.) could not be soldered to standard pc substrates. It was thought that the differential coefficient between the carrier's ceramic and the pc substrate was too large and that the board and carrier would separate after thermal cycling. However, there are mavericks like John E. Fennimore, a member of the professional staff in charge of mechanical microelectronic design at Martin Marietta's Aerospace division in Orlando. Fennimore has been reflow-soldering standard leadless ceramic chip-carriers to standard pc substrates for about two years. For exam-

11. Carriers on board. Martin Marietta's technical staff has successfully reflow-soldered ceramic chip-carriers to standard pc laminates. The unit pictured is a 5-by-5-inch polyimide board carrying 82 leadless ceramic chip-carriers reflow-soldered to conductive pads.

ple, Fig. 11 shows a 5-by-5-in. eight-layer polyimide board carrying 82 reflow-soldered leadless chip-carriers that was fabricated at Martin Marietta.

In a theoretical solder-stress analysis, Martin Marietta's research staff found it feasible to fabricate small pc boards with leadless components within typical temperature limitations. Actual measurements on samples from seven sources of fiberglass, epoxy glass, and polyimide showed that the thermal expansions of these materials were within the temperature limitations of the thermal-stress analysis.

Test boards were thermal-cycled to MIL-STD-883 with no failures. Since the tests, Fennimore has built boards using this technique for both cannon-launched electronics and helicopter avionics.

In a parallel development, Capt. Roger Settle Jr., manager of the Air Force Avionics Laboratory Hybrid Printed Wiring Board facility, ran a complete series of tests on samples containing leadless ceramic chip-carriers reflow-soldered to test patterns on alumina, epoxy-glass, triazine, and polyimide substrates. Each board was subjected to repetitive thermal cycling ($-55°$ to $125C°$) until a failure occurred. The results showed that leadless ceramic chip-carriers can be attached to epoxy-glass, polyimide and triazine pc boards and that such a process is quite adequate for benign to moderately severe environments if proper process control is observed. The amount of solder, type of solder, and reflow-solder process were found to be particularly critical. For severe environments, Settle's report recommended the use of ceramic, polyimides or triazine substrates.

In spite of Settle's findings, the Air Force has now awarded a contract to Texas Instruments to develop high-density, low-cost microelectronics packaging, with the work aimed at creating a leaded ceramic chip-carrier. Despite the apparent successful soldering of leadless chip-carriers to pc substrates, the Air Force evidently still prefers a leaded carrier for reliability.

Which carrier?

The Jedec family of chip-carriers only seems to be the tip of the iceberg of carrier variations. According to Allan Keizen of Jade, the next few years will see a proliferation of chip-carriers. In the long run, the film carrier with its advantage of full automation should win out. For the immediate future, it appears that some form of leaded chip-carrier will dominate the field of high-density packaging.

The next few years should see tremendous arrays of leaded chip-carriers on large multilayer boards. The TAB or film carrier method will probably penetrate thick-film hybrid work farther, but only if a viable testable bumped tape becomes available or if extremely large production runs are called for. For extremely dense packaging, the chip-and-wire hybrid will probably still lead the pack.

Finally, it pays to keep this guideline furnished by Jeff Waxweiler of Algorex in mind: "In choosing a packaging method, it is probably best to pick the lowest-density method possible, since the highest sophistication gives the highest cost and lowest yield." □

64-pin QUIP keeps microprocessor chips cool and accessible

Three-part design has low thermal resistance; readily removable chip-carrier has exposed contacts to ease testing

by William Lattin and Terry Mathiasen
Intel Corp. Aloha, Ore.
and Steven Grovender
Minnesota Mining and Manufacturing Co., St. Paul, Minn.

□ Now that a single silicon chip may contain up to 100,000 transistors and have well over 60 input/output pads, both semiconductor and system manufacturers have been forced to reconsider how to package such large-scale integrated circuits. The need for a change is particularly pressing in the case of microprocessors, which are now appearing as complex, 16-bit units.

The 64-pin quad in-line packaging system developed jointly by Intel Corp. and 3M Co. meets all the requirements of these increasingly powerful microprocessors. Compared with the 40-pin ceramic dual in-line package that has been standard for most commercial microprocessor applications, the QUIP is much smaller yet has a larger chip cavity—400 mils square. Its thermal resistance (35°C per watt) as well as its pin-to-pin capacitance (5 picofarads) and lead resistance (0.5 ohm) are so low as not to limit chip density, speed, or ability to interface with transistor-transistor logic.

As for manufacturability and testability, the QUIP is sturdier than the 64-pin DIP and capable of being probed while actually mounted on the printed-circuit board. And that board can be the low-cost, two-sided kind because the QUIP's pins are on 100-mil centers.

Structurally, as shown in Fig. 1, the QUIP is a three-part system, consisting of a leadless ceramic chip-carrier, a leaded socket, and a metal clip. This metal clip holds the carrier in place face down in a 64-pin socket and dissipates heat generated by the LSI device.

The QUIP is shorter than the 64-pin DIP, 1⅝ inches versus 3⅛ inches. This results in shorter internal metal-

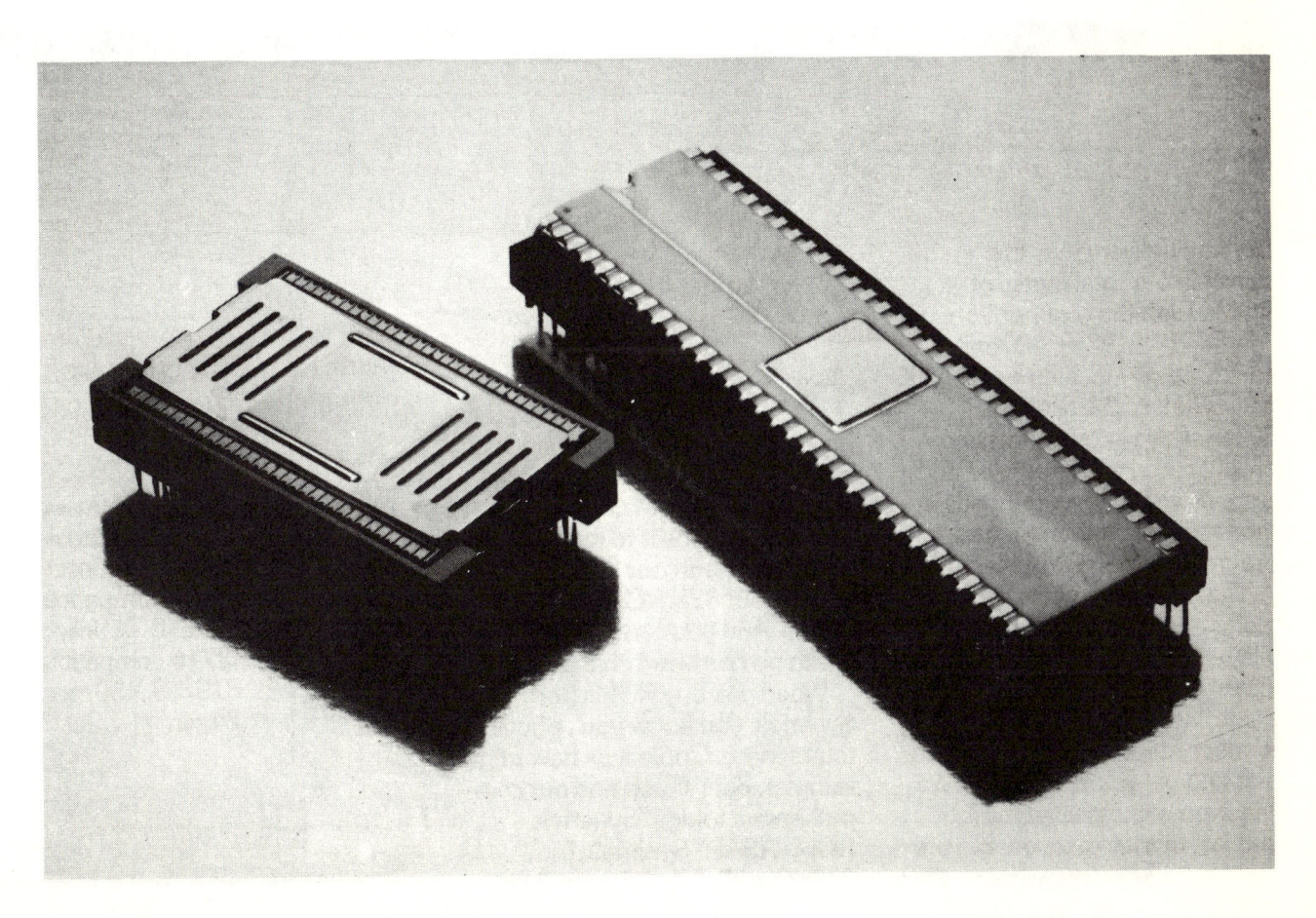

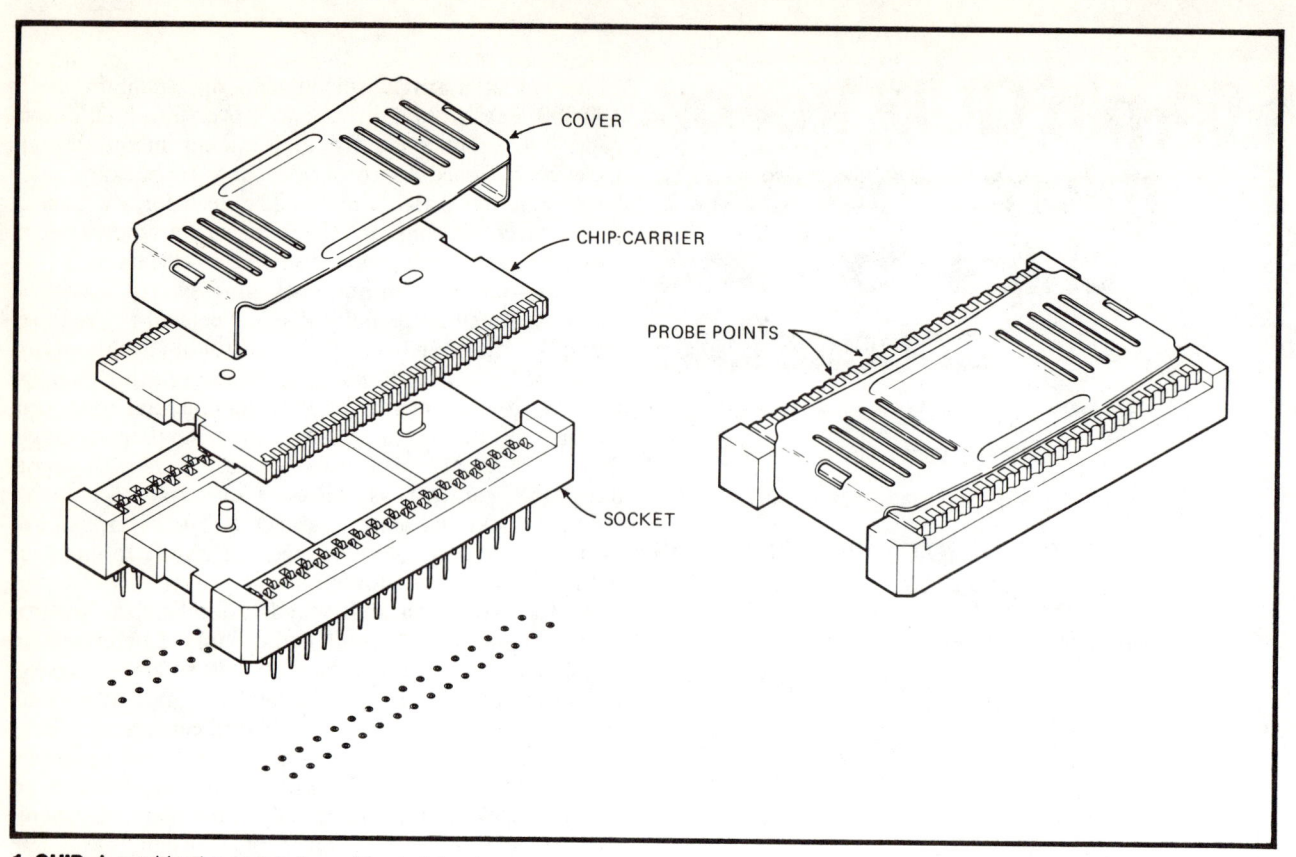

1. QUIP. A combination metal clip and heat sink, a leadless rectangular ceramic chip-carrier, and a zero-insertion-force socket are the three components of the Intel/3M quad in-line package. The carrier I/O pads are on 50-mil centers, the socket pins on 100-mil centers.

	THREE LSI PACKAGING SYSTEMS COMPARED						
	64-lead dual in-line			64-lead quad in-line		
System	Chip-carrier 3M #ST 80232TA	Socket RN #ICN649S5G	Combination	Chip-carrier 3M #ST 88364BC	Socket 3M #3534000	Combination
Board area			3.5 in.²			2.1 in.²
Approximate height above board			0.300 in.			0.350 in.
Circuit-board pad spacing, center to center			0.100 in.			0.100 in.
Method of attachment to board			solder			solder
Package removal from socket			difficult			simple
Nominal thermal resistance in still air (θ_{JA})			35°C/W			35°C/W
Nominal resistance of longest lead			1.1 Ω			0.3 Ω
Nominal capacitance of longest lead at 1 MHz			7 pF			3 pF
Die cavity size	0.325 x 0.325 in.			0.400 x 0.400 in.		
Gold thickness	60 μin.	30 μin.		60 μin.	30 μin.	
Approximate cost per lead in 10,000-unit quantity	8.1 ¢	3.8 ¢	11.9 ¢	5.2 ¢	5.3 ¢	10.5 ¢
Availability	special order	readily available		readily available	readily available	
Burn-in socket information	RN #SB-25-HT burn-in strips, not zero-insertion-force			3M #3.62-0000 zero-insertion-force socket, available		

2. Chip down. This sectional view illustrates how the spring-loaded contacts of the zero-insertion-force socket wipe against the carrier's pads. The chip-carrier is mounted face down in order to make a shorter thermal conductive path between chip and metal heat sink.

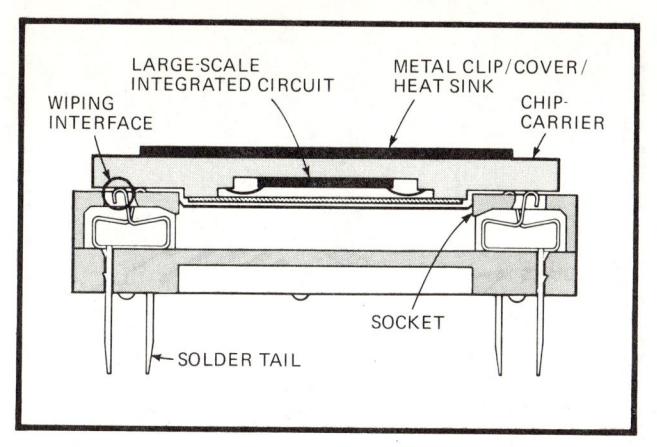

ized conductors on the chip-carrier surface.

The 64 metalized contact pads do not terminate on the socket-interface (bottom) surface of the chip-carrier but extend up the side to the carrier's top surface, where they can be probed during operation of the device.

The chip-carrier contact pads make a gas-tight connection with the spring-loaded contacts of the socket by making a wiping motion against the gold-plated chip-carrier pads. This occurs during carrier-to-socket assembly via pressure supplied by the retaining clip (Fig. 2). Both the carrier contact pads and the 64 socket contacts are situated in two rows of 32 on 50-mil centers, but they emerge as four staggered rows of 16 pins on 100-mil centers to facilitate use with standard pc boards.

None of these design details is arbitrary. All solve specific packaging problems, perhaps the most important of which is the need for low thermal resistance (θ_{AJ}).

Thermal and electrical considerations

Semiconductor designers are restricted by a package's thermal resistance because it boosts junction temperature but must not do so beyond a maximum consistent with chip reliability requirements. Junction temperature is determined by the ambient temperature plus the product of the chip's power dissipation and the package's thermal resistance. But the denser the chip circuitry, the more power it needs to dissipate, and the more heat its package needs to dissipate.

DIPs require heat from the chip to flow through the base of the chip cavity and up round its sides before it can leave the package. In contrast, the QUIP inverts the chip-carrier so that the LSI chip is mounted directly adjacent to the metal retaining clip. This allows heat from the chip to flow straight from the carrier into the metal clip. The chip can easily dissipate 2 W without exceeding an operating junction temperature of 170°C. The nominal thermal resistance of carrier and socket is, as mentioned, 35°C/w.

The reduced lead resistance and lower pin-to-pin capacitance of the 64-pin QUIP, as compared with the 64-pin DIP, are due to its shorter length: the metalized traces on the chip-carrier traverse a shorter distance before connecting with the socket contact.

Too high a lead resistance would affect the 0 level of transistor-transistor logic. When a chip has to drive TTL interface circuitry, the interface will see any voltage drop between the chip and the pc board. The standard TTL 0 level is defined as 400 millivolts; so for convenient interfacing the lead resistance of the package must be less than 0.5 ohm—the 64-pin QUIP's specification.

Lead-to-lead capacitance can degrade the performance of metal-oxide-semiconductor microcomputers when driving large off-chip capacitance loads. To minimize this problem, the QUIP's pin-to-pin capacitance is less than 5 pF.

Besides improving electrical performance, the shortness of the leadless chip-carrier increases its mechanical strength. This plus the lack of leads to be damaged ensures higher yields from the packaging process.

Another physical characteristic of the carrier—its large chip cavity—means it can cope with the larger microprocessor chips that are becoming common. Measuring 25 mils deep as well as 400 mils square, it accepts the thicker die made from 4-inch wafers.

Being leadless, the chip-carrier makes a zero-insertion-force socket possible. In the past, chip packages with high pin counts suffered from the force required to insert them into their sockets. Only a few insertions could be made before the board, socket, or carrier was damaged. In the 64-pin QUIP, however, the socket has spring-loaded pins, and these simply push hard against the carrier's pads when under pressure from the metal clip. Levering the clip off with, say, a screwdriver is enough to release the carrier and allow it to be replaced with another (Fig. 3).

68-lead Jedec standard square		
Chip-carrier #SR 88568AA	Socket AMP #P771506	Combination
		1.4 in.²
		0.500 in.
		0.050 in.
	screw	
		simple
		50° C/W
		0.1 Ω
		2 pF
350 x 0.350 in.		
µin.		
4¢	≈3¢ in prod. quant.	≈7.4¢
ecial order	prototype stage	
	unknown	

For ease of manufacture the system is completely polarized. The carrier cannot be inserted in the socket without proper orientation, having been mechanically keyed to prevent this. The socket is also keyed by the staggered pins, so that it cannot be reversed when being inserted in the pc board.

The staggering of the socket's leads on 100-mil centers optimizes layout even on low-cost two-sided boards. The carrier's two rows of 32 leads on 50-mil centers are converted through the socket into four rows of 16 leads on 100-mil centers. Thus copper traces of standard width pass easily between the pins, yet the board has a package density of one using 50-mil centers.

Easy to test

Finally, for test purposes, the chip-carrier is provided with a special set of probe contacts on the top side, to supplement the ones on the bottom that make contact with the socket. These probe contacts are left exposed by the metal clip, so that the engineer can make electrical measurements on the chip while it is operating in its socket—an impossibility with other leadless packages.

From a semiconductor manufacturer's point of view, the 64-pin QUIP has two outstanding advantages. First, the carrier has no leads that can be bent or broken during the difficult handling and testing stages of manufacture—and incidentally, this leadlessness also eliminates all the manufacturing steps associated with lead frames. Second, the carrier is mechanically very strong, being about half the length of a 64-pin DIP yet made of the same high-quality ceramic.

Being both leadless and inherently strong, it lends itself to automatic gold-ball bonding, which increases productivity by a factor of three over manual ultrasonic aluminum-wire bonding. Also, the alloy seal between the carrier and the cavity lid provides hermeticity and takes place at a much lower temperature than the glass-seal interface used with such package as Cerdips.

Chip users benefit, too

From a chip user's point of view, it is best to compare the QUIP with both the DIP and the leadless ceramic chip-carrier (see table). Evidently, its major competition will come from the square leadless ceramic chip-carrier known as the Jedec type, which occupies an even smaller board area and has slightly better lead resistance and capacitance specifications.

However, the Jedec carrier was originally designed for reflow-soldering to ceramic substrates, so that it is hard to solder to a pc board having a very different thermal coefficient of expansion. So a special socket is made for it by AMP Inc. A screw and nut hold this socket's contacts against the pc board's pads, both being on 50-mil centers. This contact density, plus the carrier's smallness, makes the Jedec package excellent for large-high-density multilayer boards. But it requires specially designed boards and as yet no automatic handling equipment is available for it.

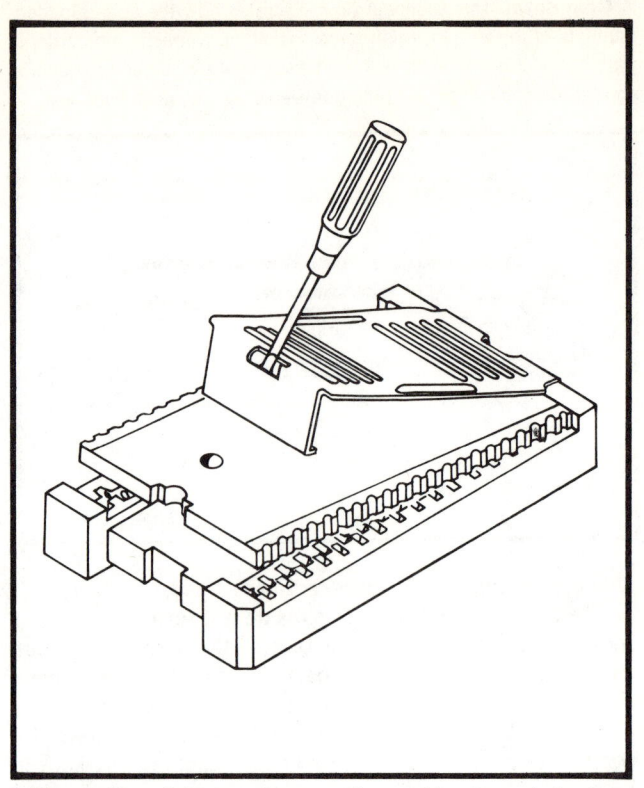

3. Dismantling. Only an ordinary small screwdriver is needed to lever off the retaining clip of the QUIP. The leadless carrier can then be slipped off two keyed posts molded into the socket. A microprocessor emulator or tester may be plugged into the empty socket.

The QUIP, on the other hand, can be used for either complex multilayer boards or low-cost two-sided pc boards. Moreover, since it has an integral socket with wave-solderable leads, it is directly applicable to a two-sided board and can use much of the computer-aided-design board-layout and automatic-insertion equipment developed for applying DIPs to such boards. In addition, the QUIP will use 40% less board space in two-sided pc board applications.

It appears the microprocessor will be the ubiquitous computing element of the future, and the new QUIP has been designed specifically for microprocessor applications. It will let relatively unsophisticated users construct microcomputer systems on low-cost pc boards for such applications as home computers and appliance controls.

For instance, releasing the retaining clip is enough to remove the leadless chip-carrier. Then a cable and connector from a microprocessor emulator or tester may be plugged into the socket for either control or diagnostic purposes. For debugging and maintenance, there is access to all the pins of the LSI microprocessor from the top of the pc board via the probing pads.

The QUIP is already available from one of its co-developers, 3M Co. The other co-developer, Intel Corp., plans to market a device packaged in a 64-pin QUIP some time in 1979. Also available from 3M are high-temperature burn-in sockets that will allow the 64-pin leadless carriers to be placed on burn-in boards. To be developed soon are cables and connectors for mating an empty QUIP socket to an emulator or tester. □

Zero-insertion-force connector ousts conventional backplane

Structural connector opens up new packaging configurations like side entry and tandem stacking

by James Taylor, AMP Inc., Harrisburg, Pa.

☐ Selecting a cost-effective and reliable package for microprocessor systems has become as vital as choosing the right microprocessor. With the change in circuit topologies brought about by the bus-structured microprocessors and the arrival of the zero-insertion-force (ZIF) connector, engineers are turning to some novel packaging schemes that have been able to dispense with the conventional backplane.

Composed of a motherboard with an array of printed-circuit connectors to receive the daughterboards, the backplane is still the predominant circuit-packaging structure today and is common to both digital and telecommunication systems. But traditional backplanes are increasingly unsuitable for several reasons:

■ Microprocessor architecture, based on 40-pin-or-over integrated circuits, is driving the number of contacts required by the daughterboard to 100 or more.
■ Backplane cooling problems have been aggravated by dense concentrations of large-scale integrated circuits that increase the amount of power that has to be dissipated per backplane.
■ Gold contact systems, so prevalent in backplane architecture, have become prohibitively expensive.

The most significant component in every backplane system is the interconnection of backplane and daughterboards. This function has been performed most frequently by what are known as cantilever-beam, card-edge connectors, which have significant limitations.

Big-board problems

One drawback is that when the printed-circuit-board span approaches 17 inches along the board and connector interface, warping is likely. If a long board bows, it causes the contacts to deflect abnormally and they may take on a permanent set. Then, if the warped board is replaced by one that is not bowed, or is bowed in the opposite direction, some of the abnormally deflected contacts may fail to contact the fingers on the card.

For example, when the number of contacts required at the interface between daughterboard and motherboard approaches 50, the performance of a conventional card-edge connector becomes marginal because of the physical length of the converter and the large insertion and withdrawal force required. This force is a linear function of the number of contacts. If each contact-pair (pin and socket) requires 1 pound of force to mate or unmate, then an interface that contains more than 50 contacts becomes generally impractical.

One solution seems to lie in reducing the contact force. But, ironically, designers want to raise contact force so that they can replace gold-to-gold contact interfaces with far less costly non-noble tin and tin-alloy systems.

Thus the challenge is to lower the insertion/withdrawal force while raising the contact force. Fortunately, there is an answer to this dilemma in the recently introduced zero-insertion-force (ZIF) connector.

The term "zero insertion force" identifies a family of connectors in which the force of insertion or removal of the mating part is essentially zero.

Stacking. Configuring daughterboards with stacking ZIF connectors eliminates the backplane and card cage entirely. This packaging architecture lends itself to large printed-circuit boards. Interfacing with flexible cable connectors is possible in this method.

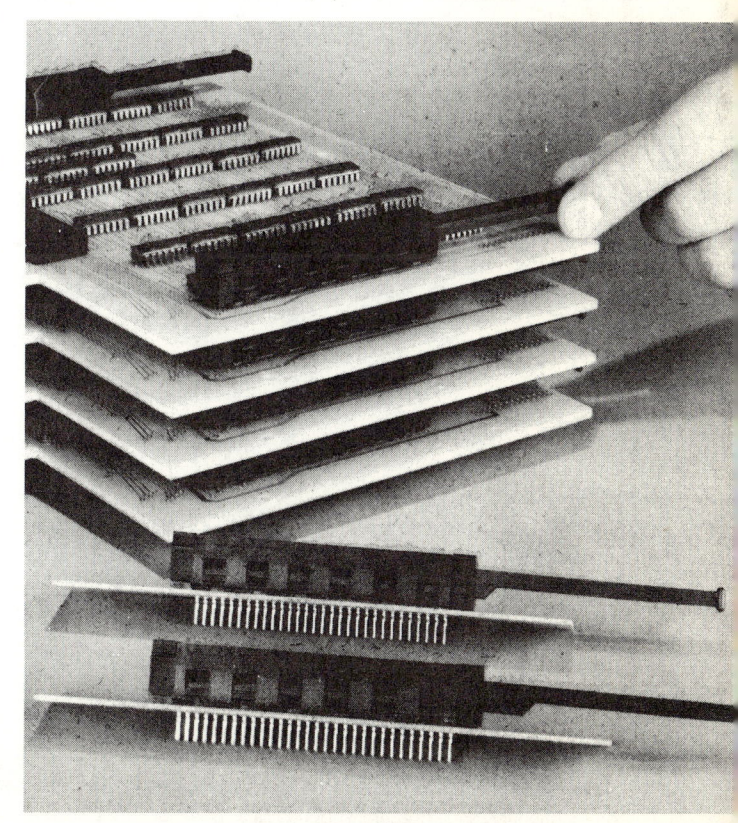

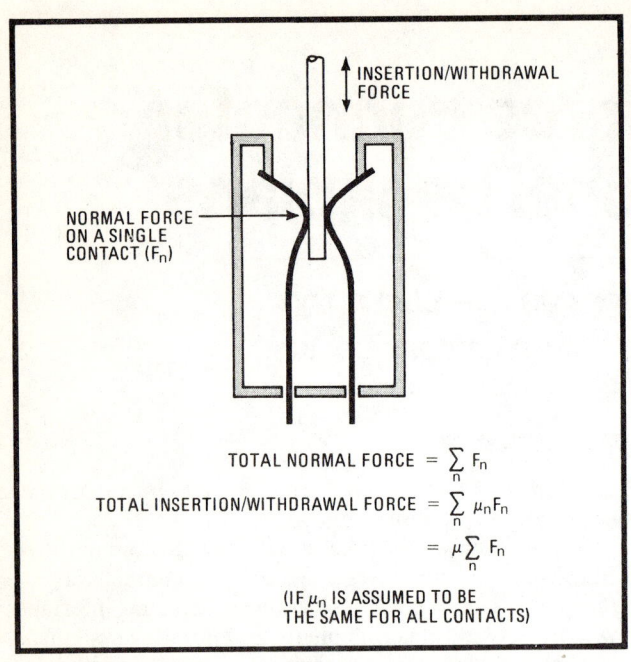

1. Sliding friction. In a card-edge connector, insertion or withdrawal force is due to the friction encountered as the circuit board enters or leaves the connector. The insertion-withdrawal axis is at right angles to the axis of force at each contact interface.

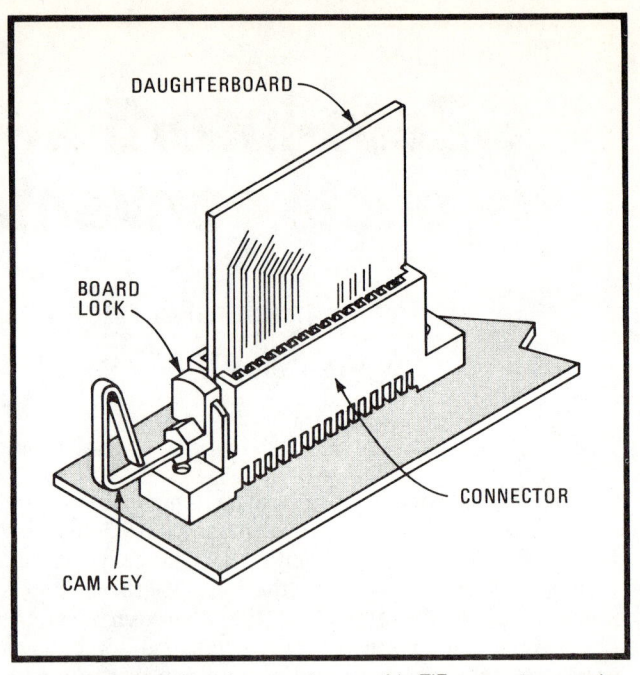

$$\text{TOTAL NORMAL FORCE} = \sum_n F_n$$
$$\text{TOTAL INSERTION/WITHDRAWAL FORCE} = \sum_n \mu_n F_n$$
$$= \mu \sum_n F_n$$

(IF μ_n IS ASSUMED TO BE THE SAME FOR ALL CONTACTS)

3. Key-operated. Rotating cam key on this ZIF connector counterclockwise opens contacts and board lock at near end. This enables easy withdrawal of the daughterboard. The key can be made removable to prevent tampering by unauthorized personnel.

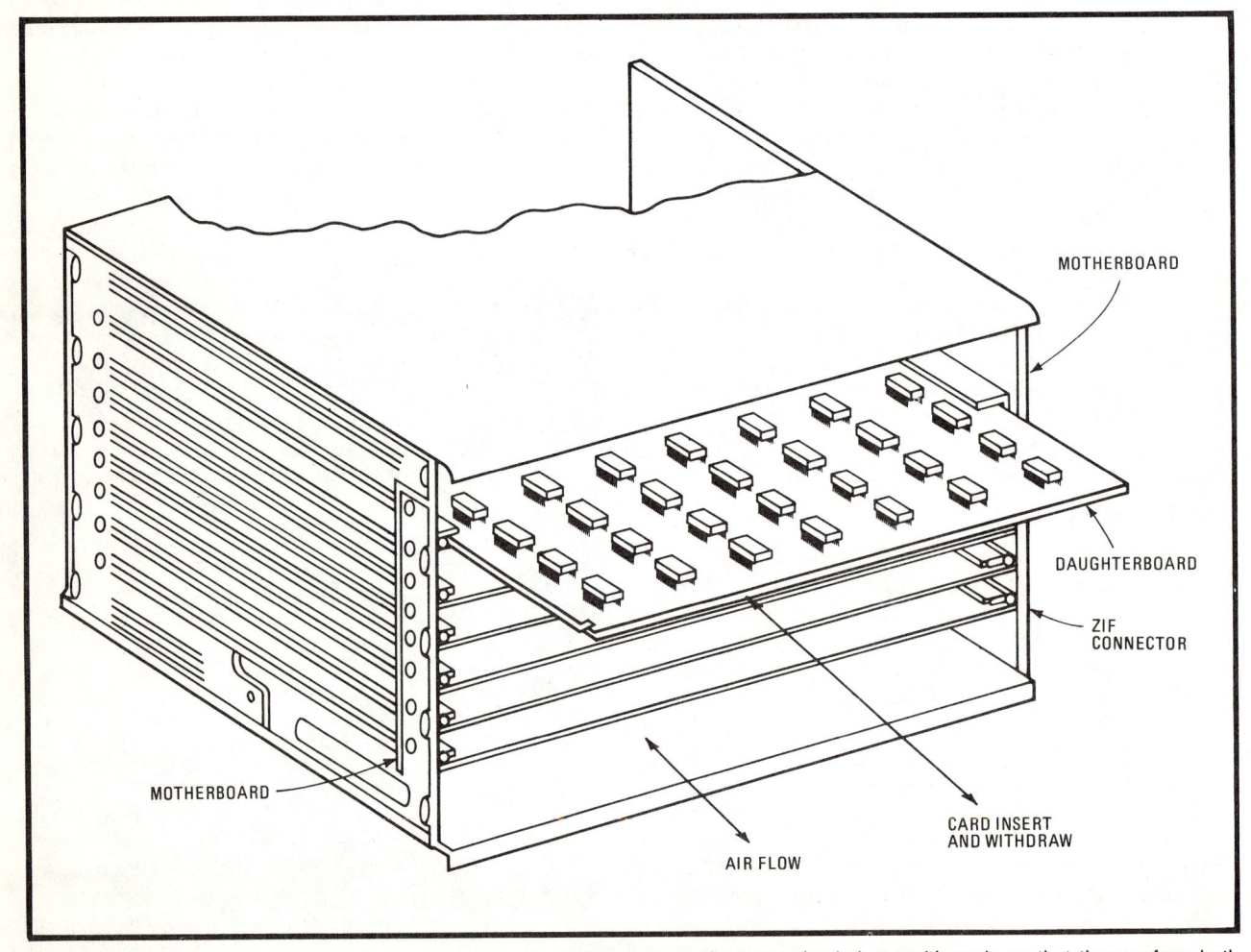

2. Side entry. By using zero-insertion-force connectors, the designer can configure two backplanes with each, so that they perform both interconnection and structural support roles. This type of packaging is well-suited to bus-oriented circuitry.

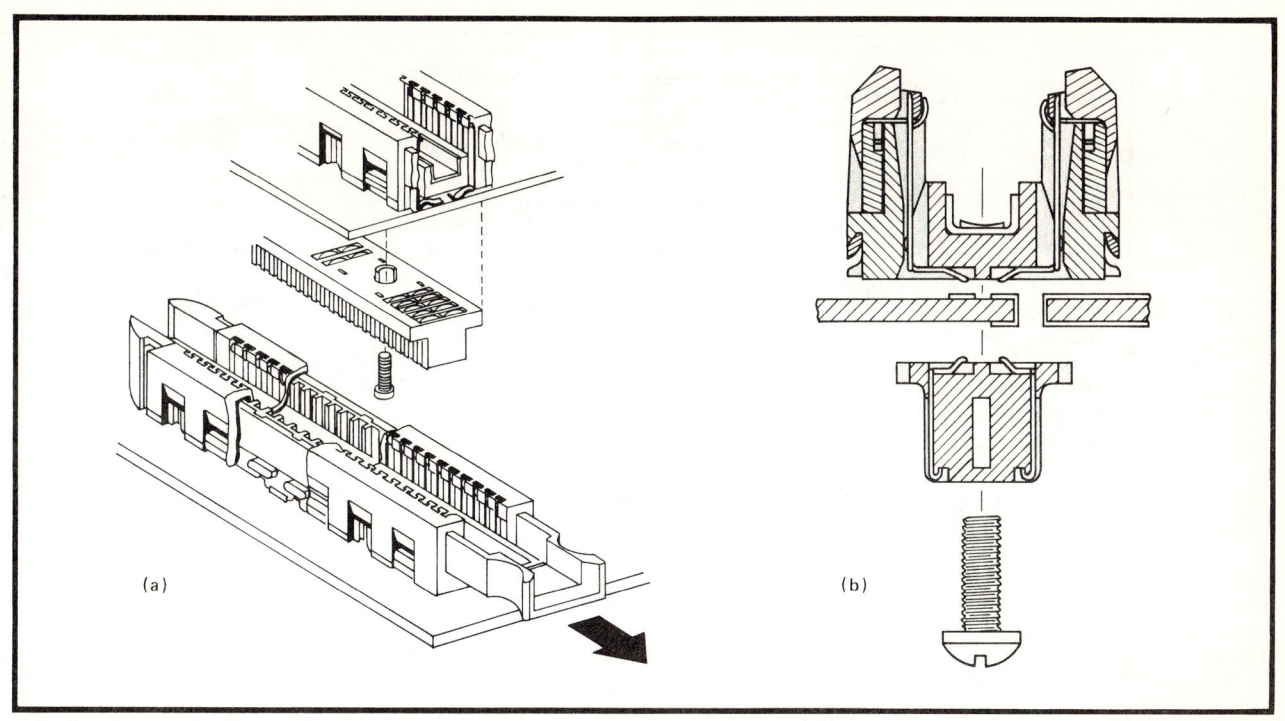

4. Surface mount. Another version of tandem stacking uses surface-mounting ZIF connectors that require no soldering, shown in (a). In (b), protruding tabs contact the pc board. Male and female portions are bolted together, forcing the tabs against the pc traces.

The most common mating part is a printed-circuit board, which usually plugs into a card-edge connector.

When the board is inserted, each contact deflects and develops a force at the contact interface that depends on the amount of deflection. The force at each contact interface must be sufficient to ensure a reliable electrical interconnection.

What results from insertion or removal of a daughterboard is not the sum of these forces ($\sum_n F_n$ in Fig. 1), but the sum of the frictional forces that develop from the normal forces ($\mu \sum_n F_n$). The reason is simply that the insertion/withdrawal axis is at right angles to the axis of force developed at each contact interface.

This insertion/withdrawal force serves a second vital role. It acts to retain the board, keeping it securely seated in the connector.

The ZIF two-step

With a ZIF connector, insertion becomes a two-step process. First, the circuit board is mated, and since no force is required, it is seated effortlessly. Second, the engagement is performed mechanically. This may take the form of pushing a lever or rotating a shaft to engage the contacts. Removal is performed in the reverse order.

A major advantage of ZIF is that there is no limit to the number of contacts because insertion force is zero regardless of how many there are. These connectors can be finger-actuated (engaged or disengaged) for up to 120 contact pairs, or lever or bell-crank operated for up to 280 contact pairs. Equally important, ZIF connectors allow the designer to try out desirable packaging alternatives to the backplane configuration.

One such alternative is side-entry packaging, which eliminates the card guides common with traditional backplanes. With card guides, as with card-edge connectors, insertion and withdrawal forces develop because the daughterboard must fit snugly in the guides to stay securely in place.

The alternative configuration is shown in Fig. 2. Two ZIF connectors support each daughterboard, performing both structural and interface roles, one along each side. Moving the daughterboard interface from the rear to the sides of the board doubles the interconnection capacity. A natural circuit isolation also occurs because of the dual daughterboard structure. For example, signals and power buses can be distributed on alternate sides, or a designer might run certain signals and power along one side and confine other sensitive signals to the opposite.

A second advantage of this approach is that a clear air flow is created from front to rear. Hence an air plenum can easily be affixed to the rear to move the air in a laminar motion across the board surfaces.

ZIF connectors used in side-entry packaging can very easily provide 175 dual positions (350 contacts) with contacts on 0.100-inch centers. The connector housing is sufficiently narrow to allow adjoining circuit boards on centers as close as 0.6 in.

Key rotation

One manner of actuating a ZIF connector is shown in Fig. 3. Rotating the key causes all contacts to move approximately 0.030 in. along axes perpendicular to the daughterboard. The objective is to open each contact pair beyond the maximum board thickness, thereby ensuring that no force is encountered and no contact abrasion occurs during insertion/withdrawal cycles.

Variations of the cam-actuated connector, shown in Fig. 3, are available with a variety of actuating levers, rods, or a hex-shaped cam key. The cam key, which is

Getting acquainted with ZIF

The basic distinction between the traditional cantilever-beam, card-edge connector and the zero-insertion-force connector is shown in (a). With the cantilever-beam card-edge connector, the printed-circuit board itself deflects the beam contacts upon entrance. The insertion force it encounters upon entering is a result of the combined frictional and normal insertion force developed by the deflected contacts.

With the ZIF connector, the board is inserted into the connector housing, and because the contacts are held open (deflected) by an actuating device, it encounters no resistance and sets itself in the bottom of the connector body. Once the board is seated, the actuator (in this case, a rotary cam) is rotated 90°, allowing the contacts to engage the mating surfaces on the circuit-board fingers.

ZIF connectors can be configured in many ways. The objective is always to eliminate insertion/withdrawal force and at the same time to establish a sufficient normal force at each contact once the mating printed-circuit card or connector is put in place.

Shown in Fig. 3 is a rotary, cam-actuated ZIF connector that may be operated by a key or a conventional nut driver. It is available with up to 65 dual positions (130 contacts) on 0.100-by-0.200-inch, 0.125-by-0.250-in., or 0.156-by-0.200-in. centers. Both sequential and nonsequential versions are manufactured.

A second type of ZIF connector, termed board-actuated, is shown in (b) and (c). In both cases, the board actuates the contact engagement. Not shown in either figure are latches or other devices to keep the board firmly mated with the contacts.

The sliding-ramp connector shown in (d) is yet another ZIF design, with a sliding bar configured as a sawtoothlike series of inclined ramps to engage and deflect the

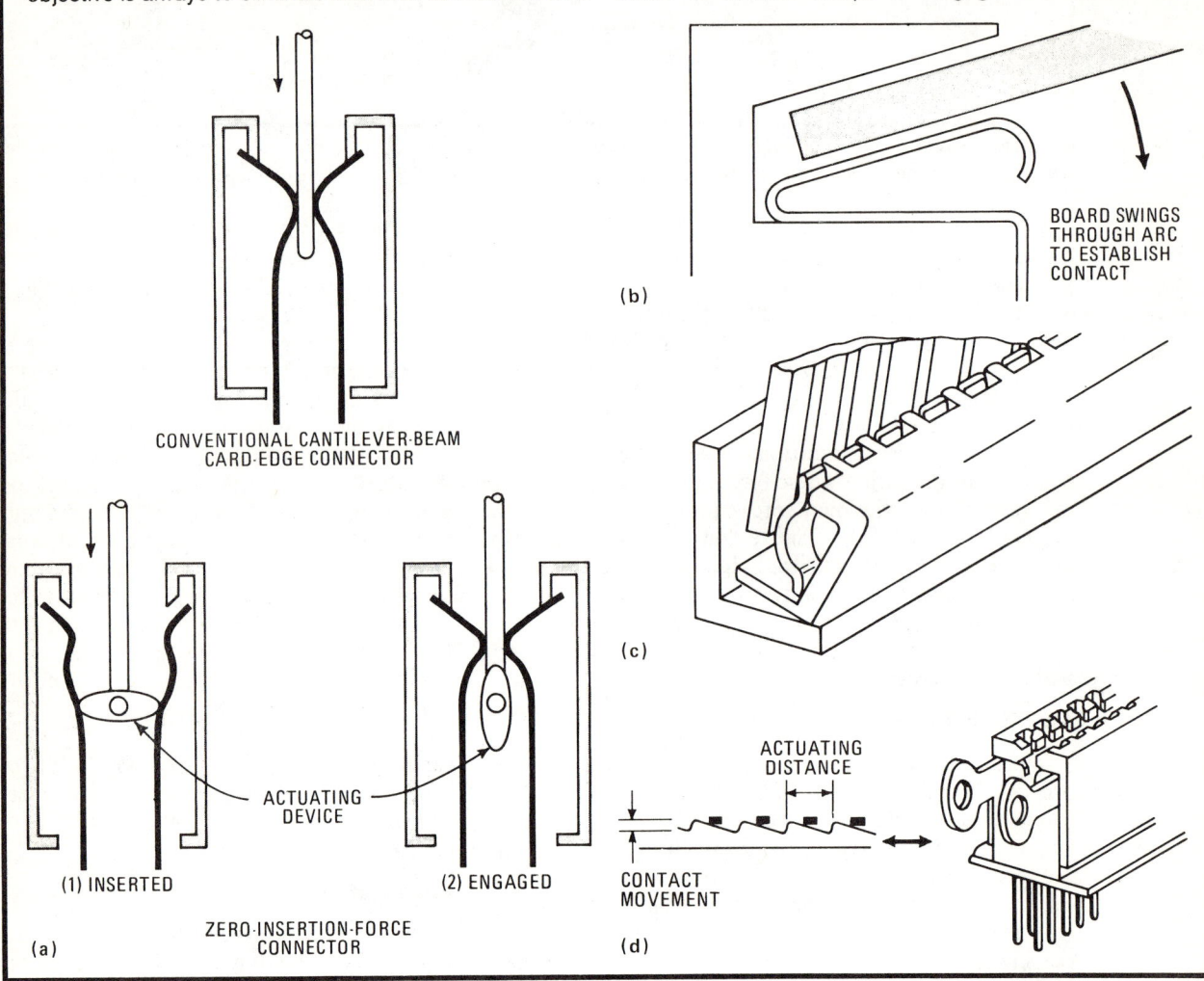

shown in Fig. 3, can be made removable, in order to prevent unauthorized personnel from removing printed-circuit boards.

A further advantage of side-entry architecture is that it enables a packaging designer to think in terms of sequentially operated ZIF connectors. For instance, sequential versions of the rotary, cam-action ZIF connector, shown in Fig. 3, execute a "2-4-remainder" contact sequence upon closing. This means that when the cam handle is actuated, first two, then four, and then the remaining contacts mate with the fingers on the board. Upon opening, the sequence is reversed. An appropriate circuit assignment would connect ground to the first two contacts, power to the next four, and signals and miscel-

214

contacts. With contacts on 0.100-in. centers, though, the ramps cannot exceed that length. Thus a steep pitch is needed to develop momentum. This creates several design problems that discourage its use.

However, the interposer technique shown in (e) is a refinement of the sliding ramp technique and is satisfactory for gang-deflecting the contacts in a ZIF connector. Using this technique, a 175-position (350-contact) connector is entirely practical, even with contacts on 0.100-in. centers. The cam can generate a 0.030-in. contact movement connector. The role of the interposer is to lift the contacts off the printed-circuit board. Thus the contacts are stressed maximally when the connector is in its open position—a small portion of a connector's useful life. This arrangement contrasts with a card-edge connector where the contact stress is at the maximum with the board securely seated in the connector.

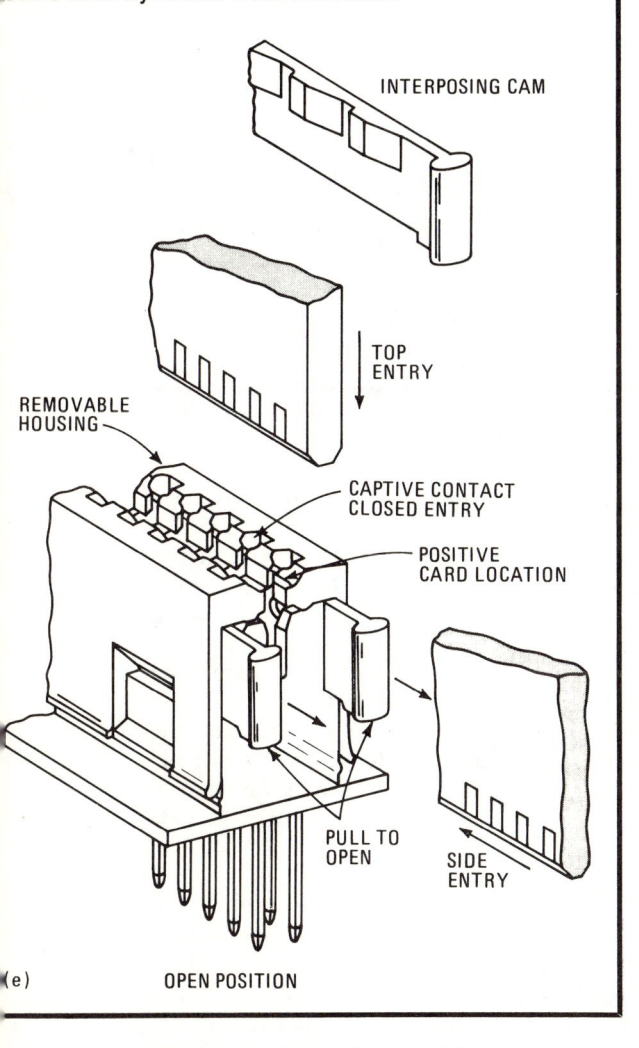

laneous circuits to the remaining contacts.

Though power-sequencing is attractive in many applications, it is particularly valuable in digital and communications systems where it is essential to remove or insert a daughterboard without draining power from the entire system. In these systems, the connectors serve as sequencing switches.

Another advantage of a two-sided backplane geometry is that signal and power routing on the daughterboard are eased. Paths become shorter and fewer dog-legs are necessary in the daughterboard circuit pattern, as the designer has added interfaces along the two opposite edges of the board.

Another alternative to the backplane is illustrated in the photograph on page 211. The outstanding feature of this novel design is the complete absence of any card-cage structure. In this packaging architecture, boards are stacked with a stacking ZIF connector, which mounts on 0.025-in. square posts on a standard 0.100-by-0.100-in. matrix. The posts extend 0.288 in. beyond a 0.0625-in. printed-circuit board. Organization of the buses, if so desired, is developed transverse to the boards through the mating connectors.

Tandem stacking

Stacking-type connectors of the type illustrated have contacts with square, 0.025-in. posts that have been press-fitted into a printed circuit board, like many card-edge connectors and other wrapped-wire post assemblies. The pins are, in addition, completely compatible with a variety of other connectors, including the popular flat-cable connector types.

With tandem stacking, the connectors can be located arbitrarily along the perimeter of the mating boards. In fact, the ZIF connectors need not be confined to the board perimeter so long as actuating levers are accessible for assembly and disassembly.

A further advantage is the ability to add boards when more options are needed, so long as the housing, or envelope, for the overall assembly is ample. The designer simply confines each option to one or more of the stacking circuit boards.

By spanning the connector near the center of the board, tandem stacking lends itself to the configuration of relatively large boards. In this case, the additional stacking connectors are located so as to stiffen an otherwise flimsy structure.

In addition to the ZIF stacking connector, there is a surface-mounted ZIF intercard connector, shown in Fig. 4. In this configuration the male and female portions are bolted to a printed-circuit board so that the spring-loaded, tin-alloy–plated ears of each connector are compressed against corresponding traces on the board. A normal force of 260 grams minimum at each contact will provide a highly reliable interface with tin-lead–plated pads. Stainless-steel support members are contained in each mating connector. By connecting them to a ground plane on the circuit board, a 90-ohm nominal-impedance connector system can be created.

Maintainability is another attractive feature of the tandem-stacking geometry. Once a defect is traced to a single board, the stack can be disassembled and reassembled with the suspected board at the top, giving a technician access to all components on its surface.

The system is bus-interruptible in two ways: first, a connector can be omitted from a series of boards; second, as in the case of the connector shown in Fig. 4b, the 0.25-in. posts can be cut off where desired.

Circuit-board connectors are seldom designed to carry

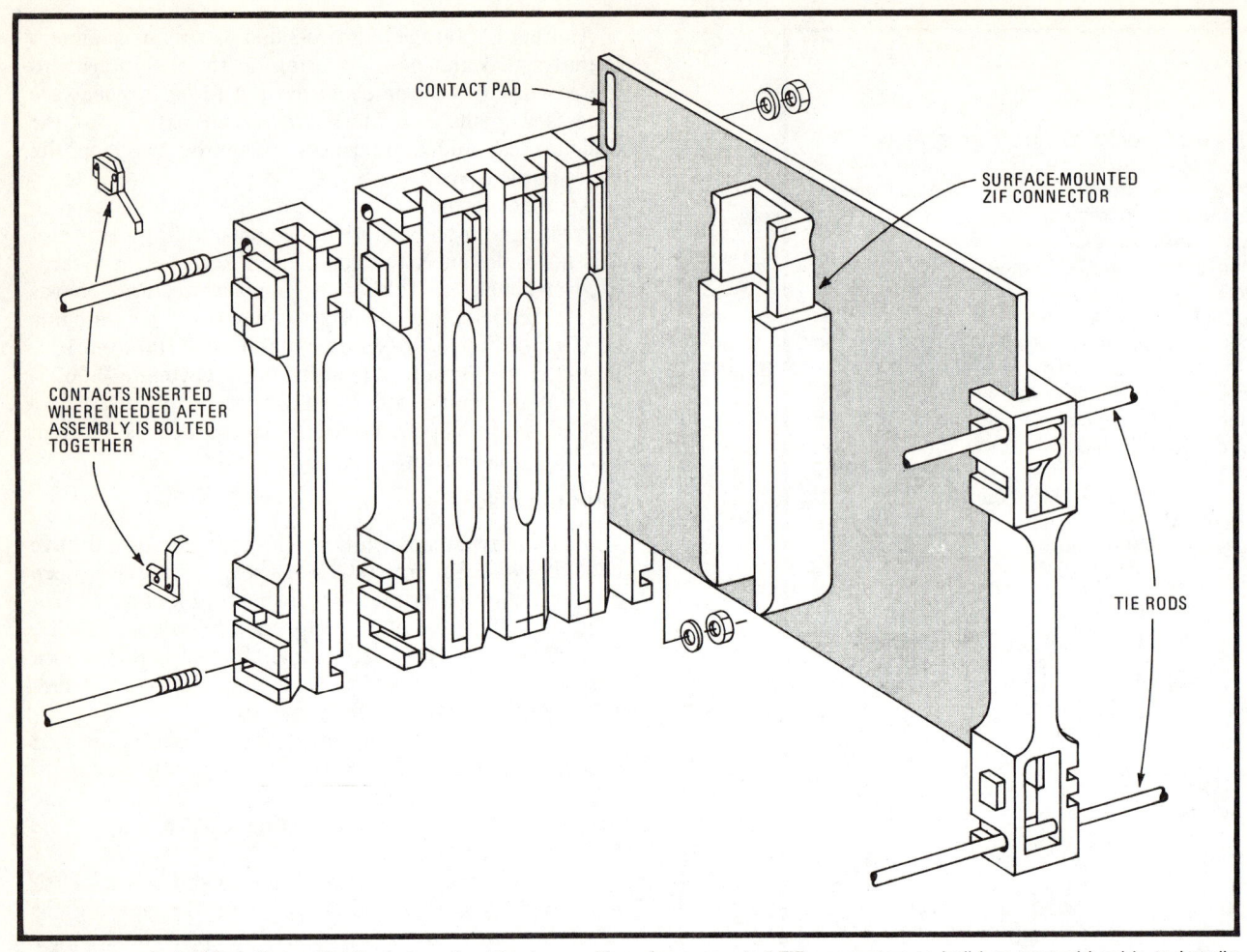

5. High-current stacking. Power-distribution card guides team with surface-mounted ZIF connectors to build an assembly able to handle both low-level signals and supply currents as high as 50 amperes. The tie rods distribute high currents with a negligible voltage drop.

more than 3 amperes. However, in many circumstances a much larger current is needed to drive a number of power-consuming circuit boards. One way to increase a connector's current-carrying capacity is illustrated in Fig. 5, where card guides (rods) at each end of the circuit board also serve as high-current connectors. The tie rods are each capable of distributing 50 amperes with negligible voltage drop.

Dividing signal and power

The ZIF connector in the middle of the board enables circuits to be interconnected on a card-to-card basis and provides a natural division between signal and power. Contacts can be inserted after assembly of the card guides to provide an electrical interface between the contact pad shown in Fig. 5 and the tie rods. By selecting the positions and lengths of the contact pads, power sequencing can be designed to ensure that a prescribed connecting and disconnecting sequence occurs at each card insertion or withdrawal.

Particularly attractive in the architecture shown in Fig. 5 are the short paths followed by signal and power traces to reach components on the board. The fact that signal traces from any component on each board are routed toward the center surface-mounted ZIF connector, whereas power traces fan out toward the periphery of the board, eases the designer's layout problems by a significant amount.

A crucial consideration in many products with digital circuitry is how to interconnect subsystems successfully via buses. Bus architecture has separate lines for data, addressing, and control, and enables a number of modules to communicate. In the traditional backplane there is virtually no isolation of the data signals. These must interface along the only surface available—the connector interface between daughterboard and motherboard. All the ZIF packaging schemes previously discussed are suitable for data buses, since they provide isolated alternatives to backplane routing.

Packing and bus architecture

In the side-entry geometry of Fig. 2, the bus may be routed along one side while power is distributed along the other. Tandem stacking offers the designer of bus architecture even more flexibility. Here the bus connector can be located virtually anywhere on the surface of the board. Also, in power distribution (Fig. 5), ZIF connectors can be anywhere on the board.

A noteworthy feature of both tandem-stacking and power-distribution techniques is the unusually direct transverse path of the bus. It must be routed to the board's circuitry from the intercard connectors. □

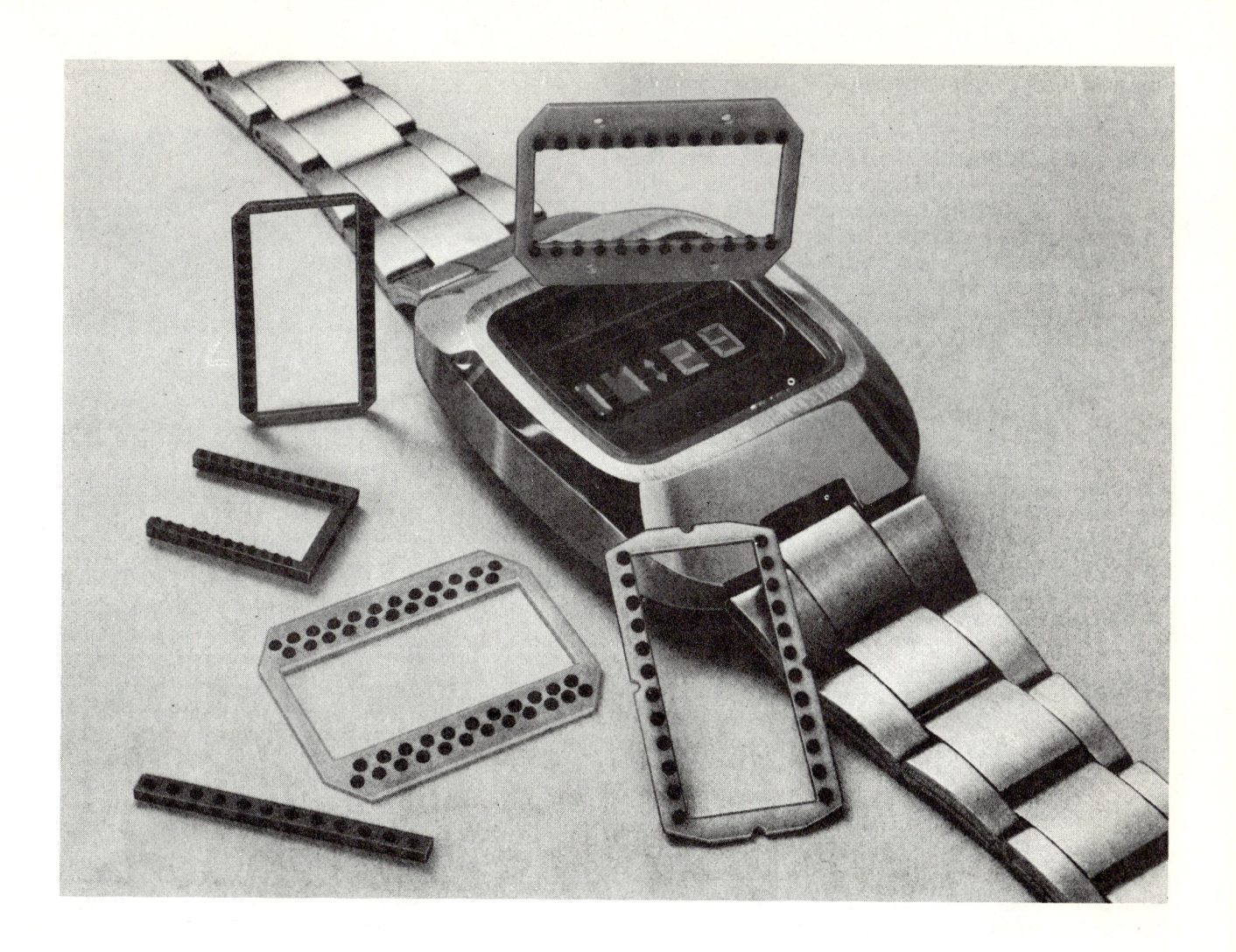

Conductive elastomers make small, flexible contacts

A wide variety of products—including liquid-crystal-display watches—can benefit from interconnections made with elastomers containing carbon or metal particles

by Leonard S. Buchoff, *Technical Wire Products Inc., Cranford, N.J.*

☐ As solid-state devices grow ever smaller, the demand for smaller connectors with closer contact spacings grows larger. And when microelectronic circuitry has to be mated with displays in such increasingly popular applications as hand-held instruments, automobile dashboards, pocket calculators, and electronic watches, the need for connections with greater immunity to shock and vibration also grows.

Both sets of requirements are met by conductive elastomeric connectors, which have contacts formed from an elastomer such as silicone rubber that is made conductive by the incorporation of carbon or metal particles. They cushion against shock and vibration, provide a gas-tight seal that excludes corrosive atmospherics, and are inexpensive in large quantities. The size, shape, conductivity and spacing of the contacts can be varied enormously, and they can be reproduced with great precision, yet still accommodate dimensional variations on the connected assemblies.

Such connectors are replacing other kinds of pressure contacts in applications such as zero-insertion-force, mother-daughter board, and liquid-crystal-display connectors. Elastomeric buttons can be vulcanized onto almost any metal lead or wire, including printed-circuit boards, flat cable, and leadless IC packages (Fig. 1).

The most widely used elastomer is silicone rubber because of its resistance to most environments, including ozone, oxygen, ultraviolet light, water, and temperature extremes—its useful temperature range is typically −80°C to +200°C. In addition, the material is easy to mold and exhibits low compression set.

Its highest strength and resistance to set is obtained when carbon is used as the conductive filler. Carbon filler is adequate if an electrical resistance of 100 to 5,000 ohms per connection is acceptable. Silver, silver-coated copper, or silver-coated glass spheres provide resistances per connection in the 0.1-to-10-ohm range, and nickel or other metals also can be used as conductive fillers. Other factors controlling contact resistance are fabrication technique and shape.

In fact, the way to produce the optimum combination of electrical resistance with mechanical force and deflection is to vary button shape—particularly height. Normally, buttons are designed to be compressed from 20% to 50% of their original height to allow for tolerance build-up on other components in the system. However, care must be taken not to make buttons too tall, or they may short out adjacent buttons or components when compressed or themselves flop over rather than compress.

Buttons through holes

In some applications, the circuitry on one side of a pc board or hybrid package has to make connection with elastomer buttons on the opposite side. Conductive rubber can be molded directly through the substrate, but only if the carrier has a thickness tolerance of ±0.001 inch. Otherwise, flashing of the elastomer will occur around the transfer mold if the board is too thin, while too thick a board will suffer physical damage.

To circumvent the need to grind the board to the required dimension, conductive silicone rubber bumps can be molded to a solderable pin, which is then inserted through the pc board (Fig. 2). The silicone rubber provides a pressure contact on one side of the board, and the pin can be soldered on the other side.

Conductive elastomeric technology can also be used to mount components on a printed-circuit board if, for example, the board cannot be soldered. Figure 3 shows a printed-circuit board on which the circuit and resistors were silk-screened by using epoxy inks. Since soldering to the screened inks would not be feasible, attachment points for discrete components are provided instead by conductive elastomer buttons with holes molded through the center. In addition, since no soldering is involved, a thermoplastic board, which would be damaged by heat, can be used to provide the necessary ±0.001-in. tolerance.

The LCD connection

Liquid-crystal displays offer a major market for conductive elastomeric connectors. Metal contacts are unsatisfactory because they may have burrs and sharp corners that can damage the tin oxide or indium oxide

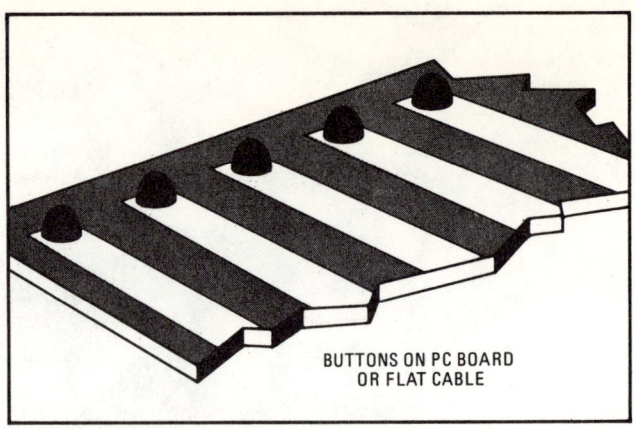

1. Rubber bumps. Conductive elastomeric connector contacts can be vulcanized onto almost any metal lead or wire.

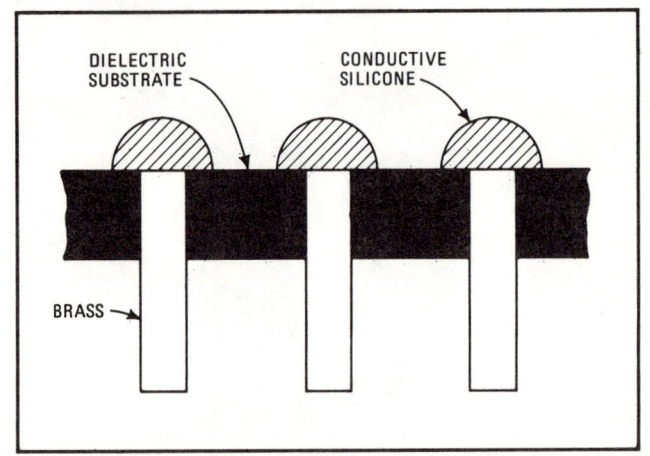

2. Through holes. Brass pins can be used to connect circuitry on one side of a board to elastomeric contacts on the other side.

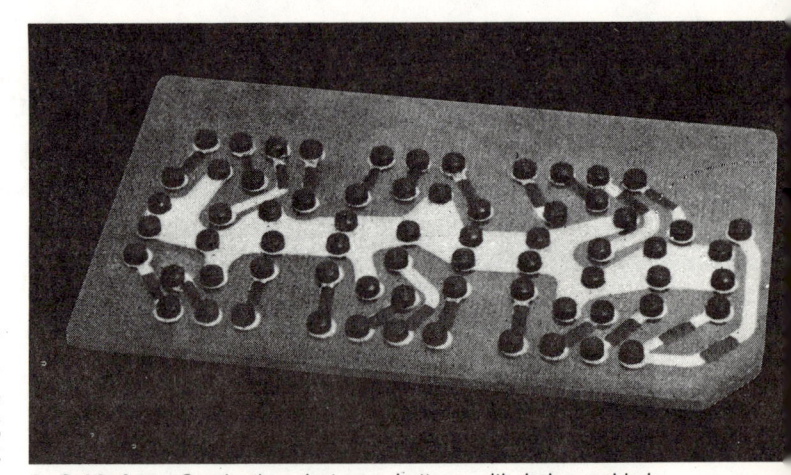

3. Solderless. Conductive elastomer buttons with holes molded through the center can be used for attaching discrete components where soldering isn't feasible.

conductive pads on the glass LCD package. They may also lose contact by being displaced after either shock or vibration.

The LCDs in most digital watches, for instance, are interfaced to the drive circuitry through elastomeric connectors. The requirements are stringent: connectors must be small, have close spacing—0.025 to 0.85 in.—between contacts, be replaceable in the field by inexperienced personnel, cost little to assemble, and accommodate wide variations in the configuration of watch movements.

Usually, connection is made between a series of printed-circuit-board pads or hybrid-circuit conductive pads and the conductive leads on two edges of the 3½-digit LCD (Fig. 4). The connector is positioned over the pc board or hybrid ceramic surface. The LCD is then placed over the connector and clamped in place by a special clip or the watch bezel.

Typically, the connector is a contact carrier frame made of glass-filled nylon and containing a pattern of holes that corresponds to the pads on the drive circuitry and the LCD. Conductive rubber is molded through these holes to form buttons that extend on each side of the carrier frame. In most watch applications, the conductive buttons are carbon-filled silicone with typical resistances of 200 to 5,000 ohms between the pads and the LCD. These values are very small when compared with the 200 megohms across an LCD segment. With carbon- or silver-filled silicone buttons, a deflection as small as 0.001 in. on a 0.015-in.-high button produces a reliable enough connection for LCD use.

The carrier frame can serve other functions besides holding or positioning the connective buttons. Compression stops, a means of clamping the substrate and LCD together as a subunit, and positioners for the LCD can be integrally molded into the frame. In some cases, the carrier acts as a holder for the entire electronic circuit, including the quartz crystal, battery, and chip.

Some variations

The form and material of the carrier frame may be varied. Buttons can be molded into strips which are laid on each side of the LCD. Alternatively, if a minimum connector height is required, buttons can be molded through holes in a 0.002-in.-thick Kapton or Mylar sheet or strip to a total height of 0.008 in. This connector can then be positioned over the corresponding contacts of either the LCD or the pc board and be temporarily cemented in place as a subassembly.

An alternative to the button-in-frame connector scheme used in most watch designs is shown in Fig. 5. In this package, made by U.S. Electronic Services Corp., Sharon Hill, Pa., for Optel Inc., Princeton, N.J., metalized paths run from the chip-mounting pads on the underside of the connector to the top. On these exposed wire ends (shown at left in Fig. 5) are molded conductive rubber buttons (shown at right) that provide contact pads to the LCD. Careful preparation of the connec-

4. Watch the connection. The first major market for conductive elastomeric connectors is in LCD electronic watches.

5. Alternate. Elastomeric contacts can be molded directly on the watch-component carrier for space and assembly economies.

219

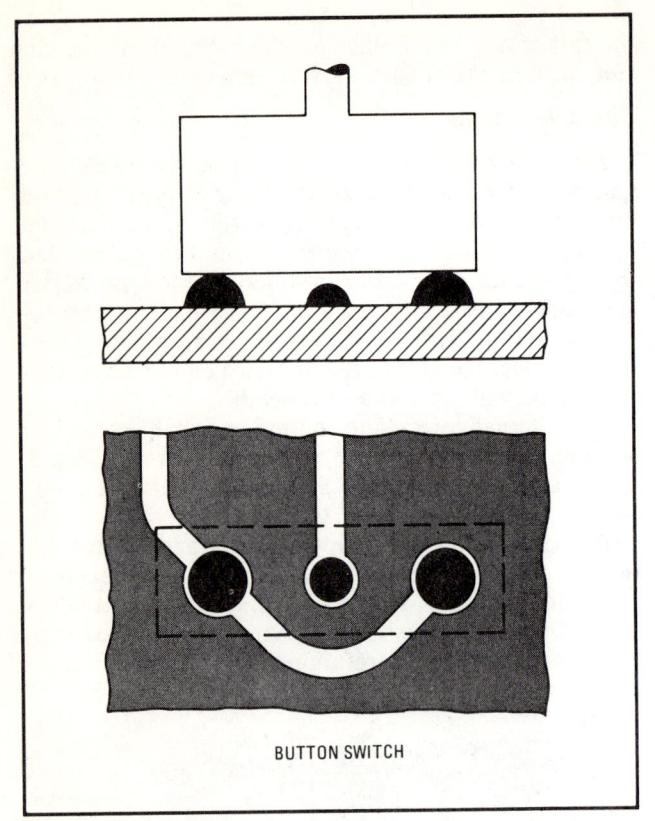

6. A switch. Conductive rubber bumps can serve as switch contacts and spring returns for the activator.

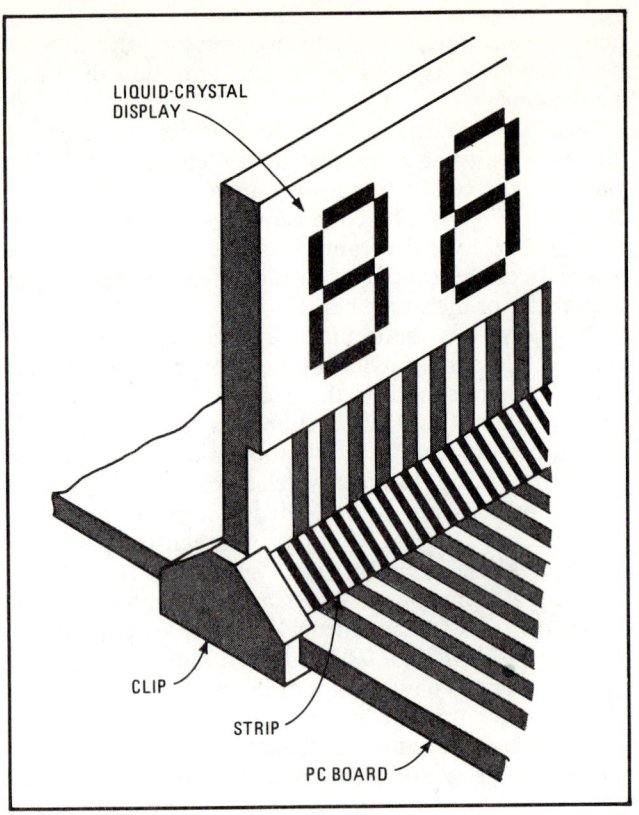

7. Striped interface. Alternate layers of conductive and nonconductive silicone rubber can be used to form a connector.

tor surface has produced excellent adhesion of the conductive silicone.

Connection to the liquid-crystal display is not the only area where conductive elastomers can be used in the electronic watch. Also under development or in production are: connections between the various circuit layers, connections to leadless IC chips, battery connections, grounding to the case, connection to the discrete resistors and capacitors, and connecting paths.

A rubber switch

The conductive-rubber bump switch shown in Fig. 6 is being designed into some electronic watches. Bumps are molded onto an interrupted path on a pc board. The higher bumps act as a spring return for the actuator, holding it in place. When the actuator is depressed, its conductive surface connects the lower bump to the other bumps, completing the circuit. As many as four different button heights can be produced in one group, and the entire array can be positioned in a circle as small as 0.15 in. in diameter. If the bumps are of various heights, continuing pressure on the actuator will create a sequence of discrete contacts, while bumps of different configurations will allow for quite distinctive tactile responses.

The advantage of having conductive rubber doing a number of jobs in a particular circuit is that the buttons can be produced in a single molding operation, greatly reducing the cost per connection. With proper design, only a minimum of space need be taken up by connectors and circuitry.

So far, digital watches have been the focus of discussion. But elastomeric connectors can also be used with larger LCD units and pc boards. For this purpose, a connector concept about to be trademarked "Zebra" is useful. It consists of alternate layers of conductive and nonconductive silicone rubber (Fig 7). These are typically 10 mils wide each and carbon-filled in the case of the conductive layers.

For most of the present units, this spacing is acceptable— that is, where the center-to-center distance between pads is 0.040 in. or more and pad width is at least 0.015 in. But prototype connector strips have been supplied with layers as thin as 0.002 in. and silver-filled conductive paths. In use, there is always at least one conductive path joining every pair of matching contact pads, and there is always at least one insulating area between each set of pads.

Interfacing printed-circuit boards

To interface a printed-circuit board with a large LCD or a daughter board perpendicular to the pc board, the Zebra connector is supplied in triangular shape. It can be put across two butted pc boards to connect the corresponding paths or can be clamped between two pc boards to establish a removable connection. Other shapes can be cut to meet individual requirements.

As for other applications, conductive elastomers provide rapid, reliable connections in test fixtures and burn-in fixtures. They are inexpensive and readily adaptable to complex, multipoint contacts. The goal of continuing development programs is to produce conductive elastomer connectors for use with IC chips having conductive pads on 0.010-in. centers. □

Part 6
Environmental Factors Affecting Interconnections and Packages

THERMAL DESIGN

The thermal demands of electronic design

The dense packaging, fast switching of ICs today create so much heat that designers must bone up on old remedies, find out about new ones

by Stephen E. Grossman, *Packaging & Production Editor*

☐ Heat—the inescapable byproduct of every device and circuit—is looming larger and larger as an electronic design problem. With large-scale integration packing ever more active devices on chips, and with circuit boards carrying IC packages in the hundreds, the pressure is on circuit designers to mind their thermal manners as never before. Still another dimension to the problem is added by each advance in device performance: higher switching speeds in digital ICs, for example, or higher output powers in analog circuits.

The prudent designer will regard heat as his implacable enemy. Poor thermal design at the least reduces lifetime, undermines reliability, and degrades system performance. At worst, it can cause catastrophic failure, unsafe products, and costly, even fatal damage.

Thermal management, then, is too important to be left only to specialists. It should be the concern of all electronic engineers right from the circuit concept stage, through the selection of components, materials, into layout and final packaging. In the end, a heat specialist may be required, but his services may be costly if preliminary attention hasn't been paid to heat fundamentals.

David Hegarty is well acquainted with their importance, for as chief applications engineer at Wakefield Engineering Inc., Wakefield, Mass., a company which specializes in thermal management, he often receives pleas for help which arrive too late. Says Hegarty: "Frequently, the circuit designer lays out the whole board—then he comes to us for help in coping with the heat. Too often it is just too late for an efficient and economical design. So the thermal solution may end up being a kluge."

At the other end of the scale are firms like IBM, which, recognizing the critical nature of thermal management to its business, involves teams of heat transfer specialists and thermal designers in the development process. Richard Chu, who manages IBM's thermal development

1. Taking the heat off. Liquid cooling is an increasingly popular way to remove heat from semiconductor devices with high dissipation levels. This cold plate, made by Wakefield Engineering, can dissipate over 600 watts. Note that the coolant lines pass directly under the devices.

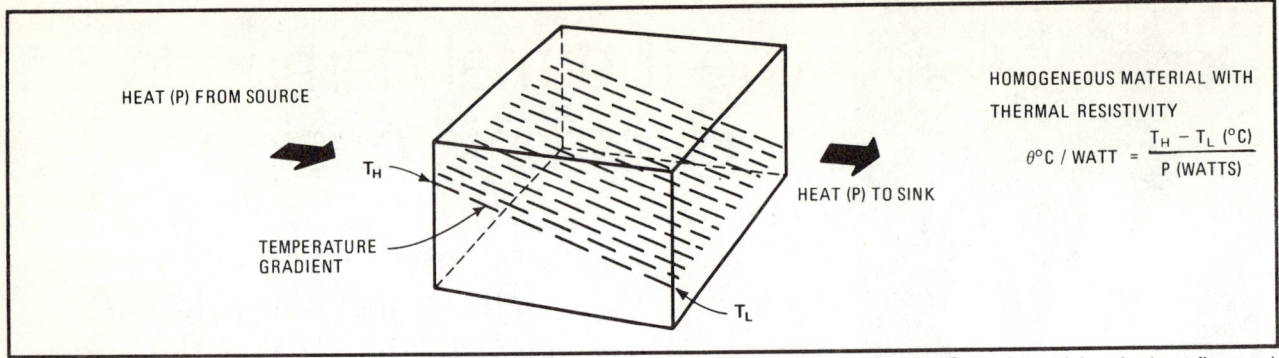

2. Thermal Ohm's law. Components with arbitrary geometries can be assigned a thermal resistance (θ) by determining the heat flow and the temperature drop parallel to the heat flow path and calculating the quotient of temperature to power in degrees centigrade per watt. Thermal resistance plays an important role in solving many thermal problems in electronics.

$$\theta °C / WATT = \frac{T_H - T_L \, (°C)}{P \, (WATTS)}$$

group at the Poughkeepsie, N.Y., development laboratories, describes his company's attitude: "Here at IBM we look at heat transfer from the very beginning of a design. We also make a point of considering the impact of thermal design on both product performance and reliability. It is not simply a matter of saying 'the lower the temperature, the better.'"

So it behooves the electronic designer to get some feel for the terminology and the units of measurement of this important domain, and to keep up to date with its emerging techniques (Fig. 1). Such knowledge will enable him to include more thermal management in his

How heat hits semiconductors

Because the electrical and mechanical properties of electronic devices are temperature-dependent, it comes as no surprise that both excessive heat and erratic, uncontrolled temperature excursions accelerate semiconductor device failure.

There are two prominent failure mechanisms: thermal mismatch, and hot spots.

Thermal mismatch causes a stress known as shear to develop along a bonded interface between two dissimilar materials when the junction is subjected to temperature variation. It occurs as the result of unequal coefficients of thermal expansion in the two mated materials. As an example, if silicon is bonded to a ceramic substrate at a high temperature, the silicon will shrink less than the ceramic when they cool, and the shear stress that develops may rupture the bond. This accounts for the popularity of epoxy adhesive to bond large silicon dice to substrates because bonding temperatures are low—typically 200°C. (Eutectic die-attach temperatures rise at least twice that high.)

Current crowding is a phenomenon that develops in bipolar devices causing localized device heating. It leads to hot spots that raise temperatures to destructive levels and cause premature device failure. Such crowding across the emitter-base junction occurs because an unavoidable voltage gradient develops across the base, adjacent to the emitter.

One technique that device designers have employed to combat this effect is to configure the base-to-emitter geometry as interdigitated fingers. This maximizes the parameter-to-surface area ratio and counteracts the crowding effect.

A second technique is to ballast the emitter by increasing the resistivity of the paths between the emitter contact metalization and the emitter-base junction as shown. It is accomplished by raising the bulk resistivity of the emitter region. A self-regulating effect results because the bulk resistance response to rising current is to develop a reverse bias which prevents hot spots from developing.

Yet another technique is to enlarge the silicon pellet so that the interface area is larger and provides a broader path over which the heat can flow to the sink. But enlarging the interface of the pellet with the heat sink increases the likelihood of a bond failure because of the thermal match problem. Copper is a good sink material because it has a low thermal resistance, but it is a poor thermal match to silicon. Its coefficient of thermal expansion is about 17.5×10^{-6} in./in./°C, much larger than that of silicon at 2.3×10^{-6} in./in./°C. Some manufacturers turn to molybdenum, because its coefficient matches silicon's well (3×10^{-6} in./in./°C). Then they clad it with copper to reduce the thermal resistivity of the interface with silicon.

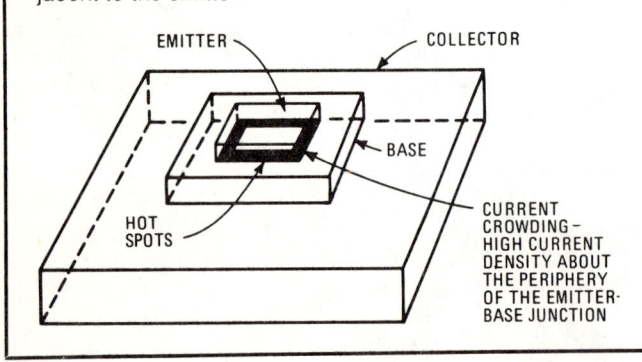

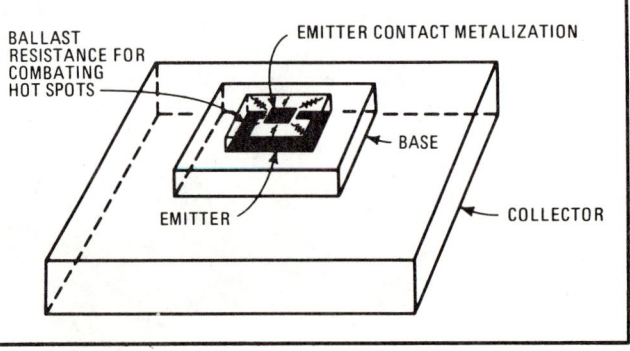

224

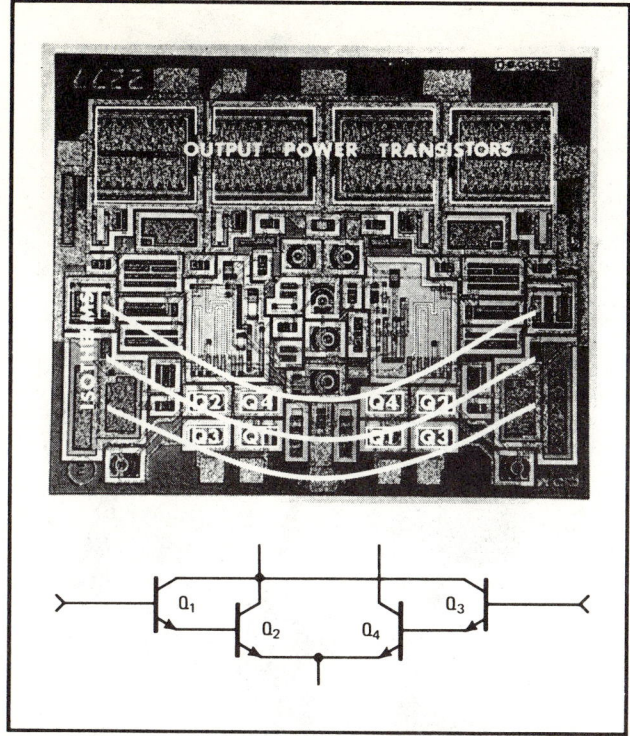

3. Hot lines and cold. Isotherms, shown on this dual power IC chip, are lines that connect points having the same temperature. By judicious placement of the driver devices (Q1-Q4) relative to the isotherms, power gain remains virtually independent of temperature.

future design plans, and will also provide him with the vernacular to communicate with heat transfer specialists should the need arise.

Heat always follows a thermal path that is downhill, so to speak, traveling from the relatively high temperature of the source to the cooler temperature of the sink.

For simplicity, it can be assumed that this thermal path is homogeneous with a uniform cross section (Fig. 2). Under these conditions, the thermal path can be thought of as a thermal resistor, which has the dimensions of degrees per watt instead of ohms. One degree per watt also represents a thermal gradient through which 1 watt of heat may be transferred through the material, and thus also connotes a potential gradient which serves the same role as a circuit voltage. That is, temperature gradient is a potential that determines the power flow, in watts, through a given component.

The concept of thermal resistance then conveniently reduces the analysis of thermal problems to a simple relationship resembling Ohm's law. Thus, in order to raise the heat transfer capability from one point to another, without changing the temperatures of the source and sink, it's only necessary to lower the thermal resistance between the two points as far as possible.

The problem can become complicated, however, when the thermal resistance path is not homogeneous. Then the over-all gradient becomes a function of the composite thermal paths.

The levels

It is commonplace in packaging to speak of separate packaging levels: the device, the surface which supports the device—the printed-circuit board or the backplane—and the enclosure. This division also lends itself to analysis of the thermal aspects of packaging.

IC devices come into consideration because device designers are packaging more and more active circuits into the device package. Though power levels for the commonly-used families of logic gates are in the 10- to 100-milliwatt range, the problem is simply that putting thousands of these gates into a small IC package translates into extraordinarily high power densities never before encountered in electronic packaging. Since silicon cannot withstand operating temperatures much above 150°C, the device designer cannot allow the junction temperatures to rise above the device limit. So he must lower the thermal resistance of the path out of the package. This means either altering the geometry, or selecting packaging materials with lower thermal resistances, or doing both.

Geometry is also the key to assuring stable power-IC performance independent of temperature, for it enables device designers to take into account the fact that gain is dependent on temperature. Figure 3 illustrates the device geometry of a Sprague parallel power amplifier chip. The output power transistors are identified near the top of the chip. Isotherms, which are lines connecting points having the same temperature, have been drawn through the region at the bottom of the chip, where the Darlington-connected driver pairs are located. Note that transistor Q_4 lies on the isotherm nearest to the power transistors. This transistor is the hottest. Transistor Q_3 is a degree or so cooler, while paired-transistors Q_1 and Q_2 are at about the same temperature. However, in both cases the transistor pairs are equidistant from the center isotherm. Consequently, the gain of the Q_1-Q_2 pair is about equal to the gain of the Q_3-Q_4 pair, and over-all gain of each amplifier section remains essentially independent of temperature variation.

The next level of thermal concern is the printed-circuit boards and the mechanically wired backplanes. Here the predominant mode of cooling is convective air.

In free convection the boards are positioned in a vertical attitude so that warm air rises along the board surfaces carrying the heat away. Free convection is economical, because no air-moving equipment is required, and it is also reliable, which explains its popularity where low maintenance costs are crucial—in telephone companies, for example.

But the higher density of modern electronics needs a faster rate of air flow than 0.5 linear feet per second—which is about the highest that can be expected in free convection. Consequently, forced air cooling has become commonplace, and today fans and blowers are performing two principal functions: cooling localized hot spots such as high-power transistors, and flushing hot air from an equipment enclosure, so that the ambient temperature is held below a specified design limit.

But even forced air cooling is reaching its limit for localized cooling as digital designers drive devices at higher switching speeds and raise the power they dissipate. As a result, liquid cooling is being used more and more.

Bill Allen, a manager of large computer circuit design, at Burroughs Corp., Paoli, Pa., points out some of the arguments in its favor: "Temperature variation

4. Breezes to order. This fan weighs only an ounce, fits in a 1-inch cube, and is well suited for microelectronic applications. Rotron, the manufacturer, rates life expectancy in excess of 20,000 hours. Device delivers up to 8 cubic feet per minute.

within a mainframe enclosure plays havoc with both noise margin and propagation delay. But by switching to liquid cooling such variations are easily minimized. Liquid also enables the size of the power supplies, a major source of heat in most large systems, to be reduced. Also, air-moving systems often create a lot of acoustic noise, and liquid cooling rids the enclosure of blowers and fans."

Cold plates are heat-exchanging devices which transfer heat from an electronic device to a moving liquid. The unit shown in Fig. 1 dissipates a lot of power—600 w developed by 21 TO-3 devices and two TO-66 devices. Thermal resistivity from device case to coolant is about 0.3°C/w per transistor, and the coolant flow rate is 1½ gallons per minute.

Cold plates can be operated as either an open-loop system, which would be a simple tap-to-drain arrangement with flow provided by the local water supply, or a closed-loop system. The closed-loop system, though more costly, enables the coolant to remain free of contamination. Frequently, a deionization system is employed which removes impurities from the coolant and also prevents foreign materials from blocking the coolant lines.

Closed-loop systems normally employ a heat exchanger to cool the fluid. The heat exchanger is a device which functions very like an automobile coolant system. A pump maintains the coolant flow. The hot coolant from the cold plate is passed through a radiator which is cooled by a forced air flow. The coolant is then returned to the cold plate.

Tom Coe, president of Wakefield Engineering points out: "Computer designers are rediscovering the heat exchanger, a familiar friend to engineers who have worked in induction heating and high power rf."

Heat pipes

Still waiting in the wings is the heat pipe, which has yet to take hold as a major participant in commercial heat transfer applications even though the technique

5. Heat pipes. These units are designed to cool discrete devices which mount on the tabs at bottom and transport heat to the finned radiator at top. Manufactured by Jermyn, they are among the first off-the-shelf products to employ the heat-pipe principle. The advantage of heat pipes is that they provide thermal resistivities several orders of magnitude lower than the best thermal conductors.

dates back to a 1942 patent. Heat at one end of a sealed pipe causes liquid inside that end of the pipe to vaporize and travel to the opposite, cooler end, where it condenses. The liquid makes the return trip to the hot end by being drawn by capillary action along a wick lining the pipe.

Thermal experts are enthusiastic about heat pipes because these devices function with a very small temperature drop—several hundred times less than the best thermal conductors. Negligible temperature drop permits a designer to position a sink in a remote location and then use the heat pipe to transport the heat from a pad to a finned radiator (Fig. 5). The pads on these assemblies are drilled by the user and the semiconductor devices are secured in place. □

THERMAL DESIGN

Plastic power ICs need skillful thermal design

Minimizing power dissipation and lowering thermal resistance with batwing lead frames and heat sinks enables 5- to 10-watt linear ICs to operate reliably

by Sumner B. Marshall and Raymond F. Dewey, Sprague Electric Co., Mass.

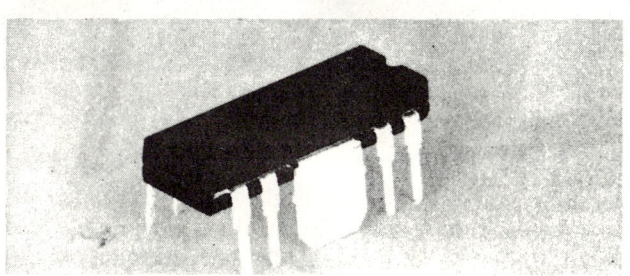

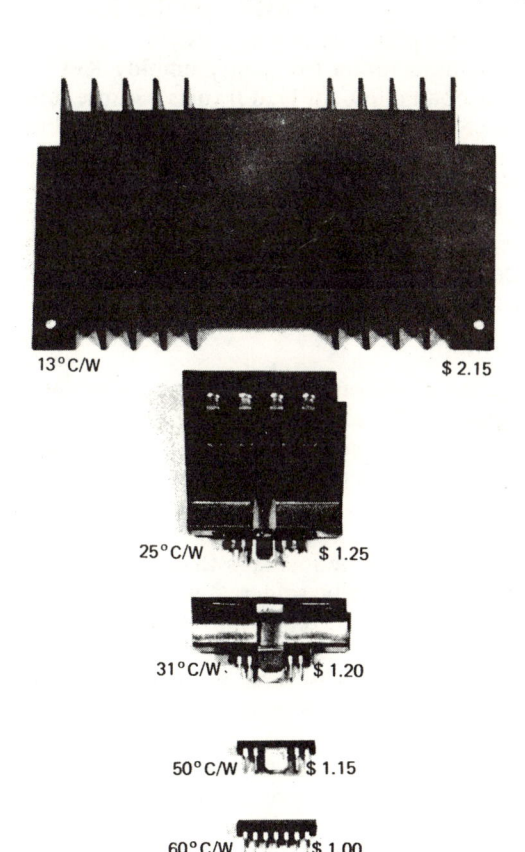

☐ Linear power ICs in plastic packages can now deliver 5 to 10 watts of continuous power and are beginning to supplant the more expensive metal-cased transistor. However, their thermal requirements are much trickier to handle than those of the discrete power device, for there is no simple way to heat-sink them and assure them of a path of low thermal resistance. ICs are soldered to a printed wiring board or, what's worse from a thermal viewpoint, plugged into plastic sockets.

Thermal design can be an important contributor to performance. Merely adding a heat sink costing less than half a dollar to a plastic-packaged IC can raise the rated power output by a factor of eight. But from the reliability standpoint, thermal design is really crucial because failure to heed its requirements will assure the onset of a whole raft of thermally-induced ailments—increased leakage currents, material decomposition, drift, and premature device failure. It falls to the designer, then, first to select a device that satisfies not only the electrical and mechanical requirements but the thermal criteria as well, and then to protect the device from damaging temperatures.

To perform these tasks, he must understand the thermal paths within the plastic-packaged device, and familiarize himself with the roles of heat sinks (Fig. 1) and forced air cooling. With this background he is well equipped to optimize the thermal operating environment of the plastic-packaged power IC.

Start of the trouble

The power IC must be held below its maximum safe operating temperature. This is typically 150°C for silicon.

A second constraint is the temperature of the operating environment of the packaged device, usually termed the ambient. If the plastic package were the only consideration, the ambient could be allowed to rise to perhaps 70° or 85°C. But usually the maximum operating temperatures of nearby components will restrict the ambient to an upper limit of 50°C.

Another constraint is that a minimum temperature differential must be maintained between the ambient and the chip,to ensure that the required power can be dissipated. These relationships are apparent in the thermal equation:

$$P\theta_{J-A} = T_J - T_A$$

where P is the rate of heat flow in watts

1. Sink the heat. Batwing power IC (top) lets heat sink be directly attached to lead-frame tabs, greatly increasing thermal dissipation capability. Larger sink geometries (middle) lower the thermal resistivity and thus increase the power capability. Such arrangements contrast sharply with the high thermal resistance (60° C/W) of the conventional 14-lead device with no heat sink (bottom).

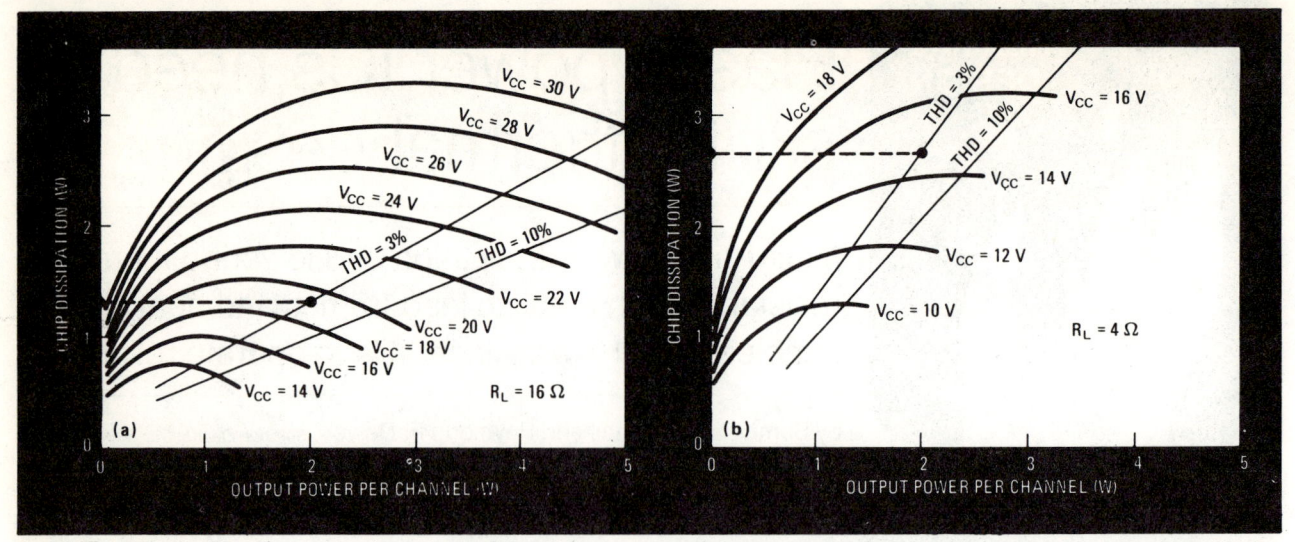

2. Minimize dissipation. The curves shown here apply to a Sprague type ULN-2277 dual power IC. If the amplifier is operated into 16-ohm loads (a), rather than 4-ohm loads (b), chip dissipation is minimized. This also minimizes device temperature and enhances reliability. Dissipation values indicated are for 2 watts of delivered power and 3% total harmonic distortion (THD).

θ_{J-A} is thermal resistance from junction to ambient in °C per watt

T_J is the junction temperature in °C

T_A is the ambient temperature in °C

The design task then is to first minimize the dissipated power and the thermal resistances from the junction of the chip to the ambient, and then to take whatever additional steps are required to lower the ambient.

A good place to begin is to select both the circuit components and the dc supply voltage for minimal power dissipation. The chip power dissipation for various load impedances and supply voltages can be obtained from manufacturers' specifications.

A typical example is the Sprague Type ULN-2277 dual 2-watt audio amplifier IC. As shown in Fig. 2, the power dissipation is determined by the output power required, the maximum acceptable total harmonic distortion (THD), and the dc supply voltage, V_{CC}. If a power output of 2 w is required with a 3% maximum total harmonic distortion, then chip power dissipation is about 2.7 w at a V_{CC} of 15 v with a load impedance of 4 ohms. However, power dissipation is only 1.4 w at V_{CC} = 19 v with a 16-ohm load. In general, the highest load impedance for a given output power is the most desirable, within the output voltage capability of the device.

Once the circuit has been optimized for minimal power dissipation, attention can be turned to the matter of thermal resistance.

The path out

Heat removal from plastic-packaged ICs is difficult. Unlike discrete components, which often have studs and thus fairly low junction-to-ambient thermal resistances, ICs are usually soldered into printed-circuit boards or plugged into plastic sockets, and chip-to-ambient thermal resistance without a heat sink are relatively high.

There are two paths from the chip to the ambient. One is the path from the junction through the plastic case (denoted by the upward pointing arrow in Fig. 3) and has a thermal resistance of between 50°C/w and 100°C/w. The second path (indicated by the downward pointing arrow in Fig. 3) is the sum of the thermal resistances of the silicon chip, the die bond, and the lead frame, and has a much lower thermal resistance. So the designer's best course is to make sure that the thermal resistance of this path is as low as possible.

Device manufacturers frequently employ Kovar, an iron-nickel-cobalt alloy, for lead frames because it has a coefficient of thermal expansion which is quite close to silicon. In this way they minimize the mechanical stress that develops between the lead frame and the chip when the device is subjected to temperature variations. However, Kovar's thermal resistance is about 30 times that of copper, so in high-power circuits a copper or copper-alloy lead frame is preferable.

Batwings help

Manufacturers are further enhancing the thermal path by altering the conventional 14- and 16-lead designs. One such design is the "batwing" IC package (shown with and without heat sinks attached in Fig. 1). It is becoming an industry standard. Size is the same as a conventional 14-lead IC package, but the central lead-frame sections are formed as tabs that measure ¼ inch square. These tabs may be soldered, welded or bolted to a heat sink or inserted directly into some sockets. This geometry achieves a worst-case thermal resistance of about 11°C/w, junction to case, whereas an IC with a conventional 14-pin copper lead frame exhibits a thermal resistance of 19°C/w.

Sometimes even a package that boasts as good a thermal design as the batwing configuration will require a heat sink. The manufacturer's data on a given device will enable the designer to make this decision. The data, which sometimes is presented in the form of curves, takes into account the maximum chip temperature that can be tolerated and the thermal resistance of the IC package.

Actual thermal performance in any design, however, also depends on many other factors, like interference in the air flow by nearby components, heat radiated or convected by other components, atmospheric pressures, and humidity. So, in selecting the thermal resistance of a heat sink, it is wise to allow a generous safety factor.

Heat sinks for plastic ICs can be procured from a number of vendors in a variety of styles. A few are shown in Fig. 1.

As an alternative, sinks can be fabricated from copper sheet. This material is quite effective in reducing thermal resistance from the case to ambient, and dimensions for a range of thermal resistances are given in Fig. 4. These values are for square sinks, 0.015 in. thick, with a dull or painted finish. They are intended to be mounted vertically on either side of the lead frame as shown in the figure. They should be soldered directly to the lead frame. Soldering adds on an interface thermal resistance of the order of 0.3°C/w.

Although unfinished copper is an effective heat sink, it lacks the eye appeal of finished metal. For this reason, and also to prevent corrosion of the raw metal part, heat sinks are often painted or anodized.

The most common finish is black anodizing. Besides being economical and attractive to look at, it enhances thermal performance by as much as 25% because the dull black is the most efficient surface for thermal radiation. On the other hand, an anodized surface is a poor thermal conductor, so those surfaces which mate with the IC must be free of anodization to ensure optimum thermal conductivity.

Iridite and chromic acid are other popular metal finishes because they offer low electrical and thermal resistivities. But like anodization, they also result in poor thermal radiators.

Convective cooling

Convection as well as radiation is at work removing heat from the sink. If the power dissipated is low, then the air is essentially stagnant, and the effective thermal resistance of the sink-to-air interface will be quite high. However, as the power dissipation rises, the air adjacent to the sink heats up and begins to rise. This induced air flow is known as a natural convection, and it sweeps the heated air clear of the heat sink, effectively lowering the sink-to-air thermal resistance.

Forced air cooling can improve heat sink performance by as much as 100%. A rule of thumb is that semiconductor failure rate is halved for each 10°C reduction in junction operating temperature. Even where space is at a premium, the cost of a small, compact fan or blowers can often be justified. Often an air-moving device, intended primarily to flush air from an enclosure, can be so located that it will force a high air flow directly across a plastic power IC. □

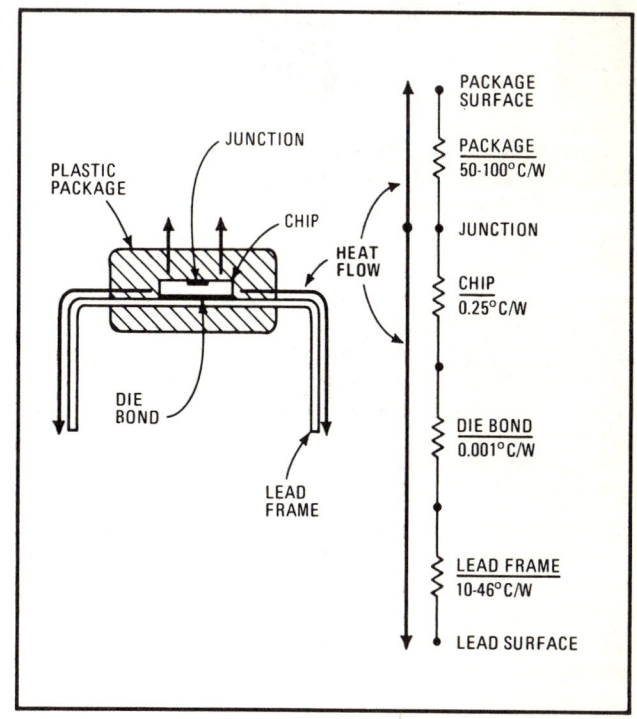

3. Two paths. Thermal resistivity is lower—on the order of 10°–46° C/W—through the lead frame than through the package—over 50° C/W. Device designers can optimize the thermal resistance through the leads by selecting a material with low thermal resistivity.

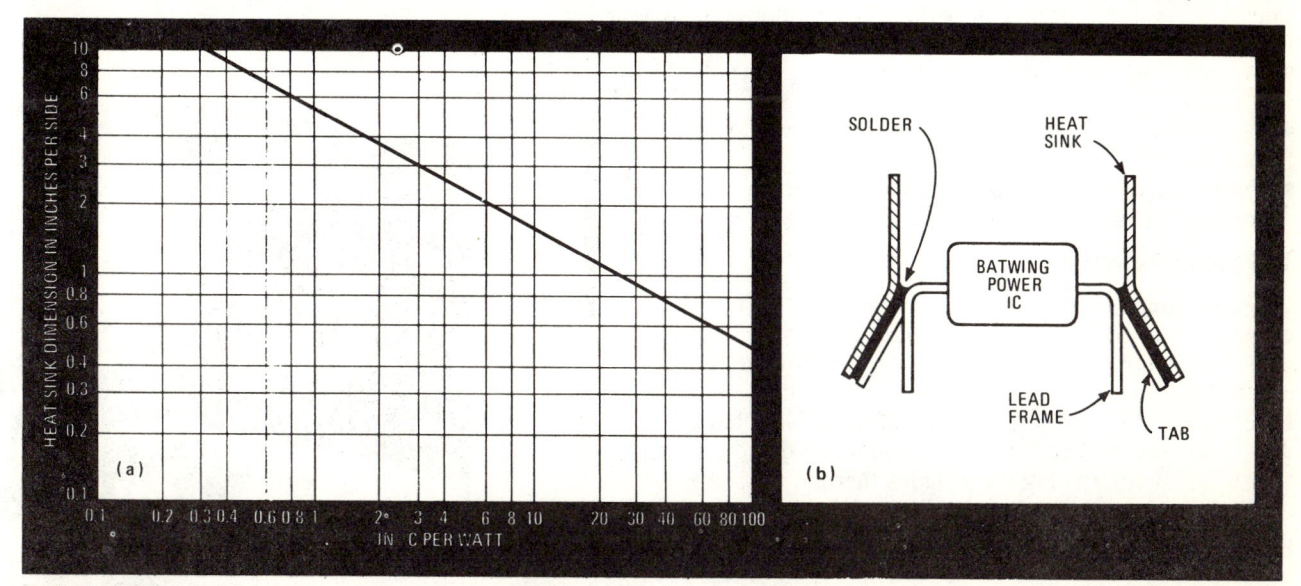

4. Design a sink. Once the required case-to-ambient thermal resistance is determined, the dimensions for pairs of square, 0.015-inch-thick copper plates can be selected (a). Plates are mounted to lead-frame tabs (b). Soldering holds interface thermal resistance to about 0.3° C/W. A dull black finish will enhance the radiating properties of the sinks.

THERMAL DESIGN

Analysis can take the heat off power semiconductors

Designing for thermal requirements with good electrical isolation ensures stability and reliable performance of these devices

by Forest B. Golden, *General Electric Co., Auburn, N. Y.*

☐ Cost-conscious design engineers, well aware that a 1% component-failure rate during the warranty period can easily add 50 cents to a product's cost, are eager to reduce device temperatures. To put it simply, the cooler a device operates, the more reliable it is. Generally, if an engineer can lower its temperature by 20%, he will reduce the device's failure rate by a factor of three. This concept applies to such discrete devices as the power transistor, diode power rectifier, and thyristor family, which includes the silicon-controlled rectifier and the triac.

Thus, keeping semiconductors cool, while often thought of as a necessary evil and a chore to the design engineer, is really an intimate part of the design process. It is not enough for the designer merely to select a device that meets circuit requirements. He must follow up with a systematic thermal analysis to enable him to select a heat sink and mounting technique that will hold device temperature low and thereby assure reliable operation for a long time.

These factors should be considered in making the thermal analysis:
- Determine power dissipation and temperature limits so that the case-to-ambient thermal resistance can be calculated and the proper heat-transfer techniques selected.
- Evaluate electrical-isolation requirements of the device package.
- Ensure that the thermal resistance of the device interface is minimal.
- Verify the calculations by careful temperature measurements under operating conditions.

The thermal circuit

Semiconductor-design engineers learned early to translate heat-transfer units to understandable circuit units—watts for heat and degrees per watt for thermal resistance. Thermal resistance—the inverse of thermal conductance—is the key to selection of the cooling hardware, such as a heat sink, for a semiconductor device.

A thermal circuit (Fig. 1) includes thermal capacitor elements to account for thermal storage. However, since power dissipation of most device designs can be calculated as rms values, thermal capacitance can usually be ignored.

The thermal equation resembles Ohm's law because heat in watts is analogous to electrical current, and temperature difference is comparable to potential difference:

$$\theta_{JA} = (T_J - T_A)/P_D$$

where θ_{JA} is the thermal resistance of the junction in degrees per watt, T_J is the temperature of the semiconductor junction, T_A is the temperature of the ambient, and P_D is the power dissipation in watts.

To calculate the all-important thermal resistance—from junction to ambient—the designer must determine three thermal parameters: the maximum allowable

Kilowatt cooler. This cooling package is typical of some of the highly efficient thermal packages recently developed to cool power-semiconductor rectifiers and SCRs. The enclosure, developed by Thermalloy, is 13 inches high, 14 in. wide and 16 in. deep. It can dissipate approximately 3.5 kilowatts. Devices shown are three-phase full-wave rectifiers that can handle 1,000 amperes. Each of the six modules provides a thermal resistance of 0.07°C/watt from sink to air. Assembly requires 300 cubic feet per minute of air at a pressure drop of 0.9 in. of water to achieve this low thermal resistance.

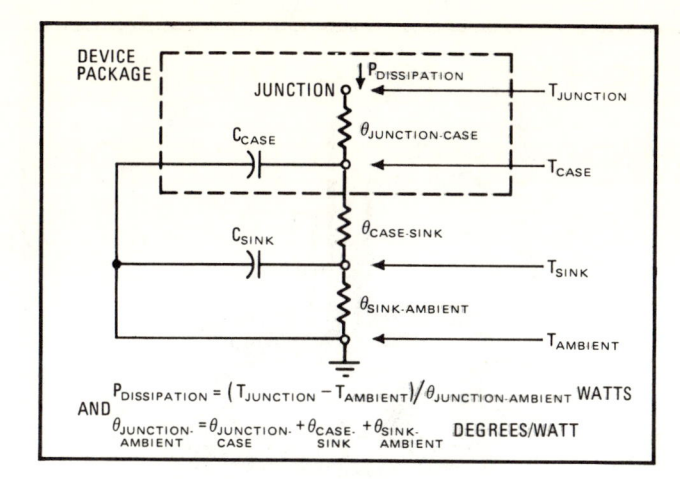

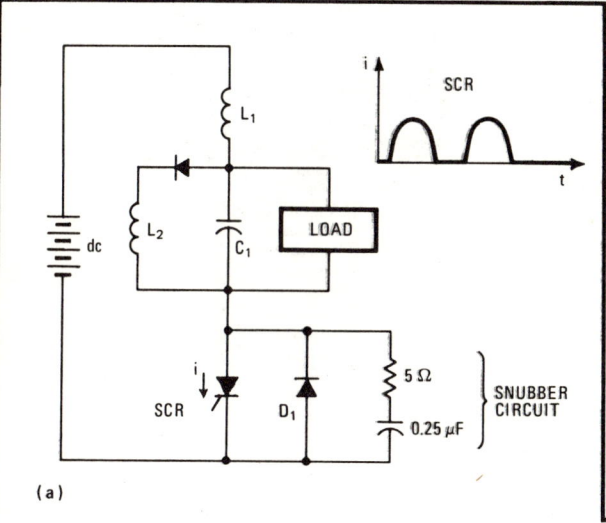

1. Thermal circuit. Circuit analogs enable the designer to analyze the thermal paths. Resistors represent the thermal resistivities of the various components. Capacitors account for storage capabilities, but are usually omitted in steady-state calculations.

$$P_{DISSIPATION} = (T_{JUNCTION} - T_{AMBIENT})/\theta_{JUNCTION-AMBIENT} \text{ WATTS}$$

AND

$$\theta_{JUNCTION-AMBIENT} = \theta_{JUNCTION-CASE} + \theta_{CASE-SINK} + \theta_{SINK-AMBIENT} \text{ DEGREES/WATT}$$

temperature of the semiconductor junction, the maximum ambient temperature, and the maximum power that the device will dissipate.

Applying thermal parameters

Determining the power dissipation of a thyristor or a semiconductor rectifier is not difficult because worst-case values are specified by manufacturers' data sheets. Application of this data is illustrated by calculation of dissipation for a high-frequency inverter circuit to be used in pulse operation. The term inverter applies to a circuit that converts dc to a periodic waveform. The inverter circuit (Fig. 2a) develops a near-sinusoidal current through the SCR. A typical rating curve for such an SCR (Fig. 2b) provides the maximum watt-second loss per pulse for any sinusoidal pulse the SCR can handle.

The worst-case anode-power dissipation is determined by multiplying the repetition rate by the watt-second loss per pulse. This value will be accurate if the values of the gate drive and snubber circuit are comparable to those specified in the data sheets. (A snubber circuit is a low-pass-filter network that prevents spurious thyristor triggering.)

2. Pulse power. The inverter circuit (a) delivers a near-sinusoidal-shape pulse current through the SCR. Curves in (b) provide the watt-second power loss per pulse. Multiplying this value by the pulse-repetition rate yields the power dissipation.

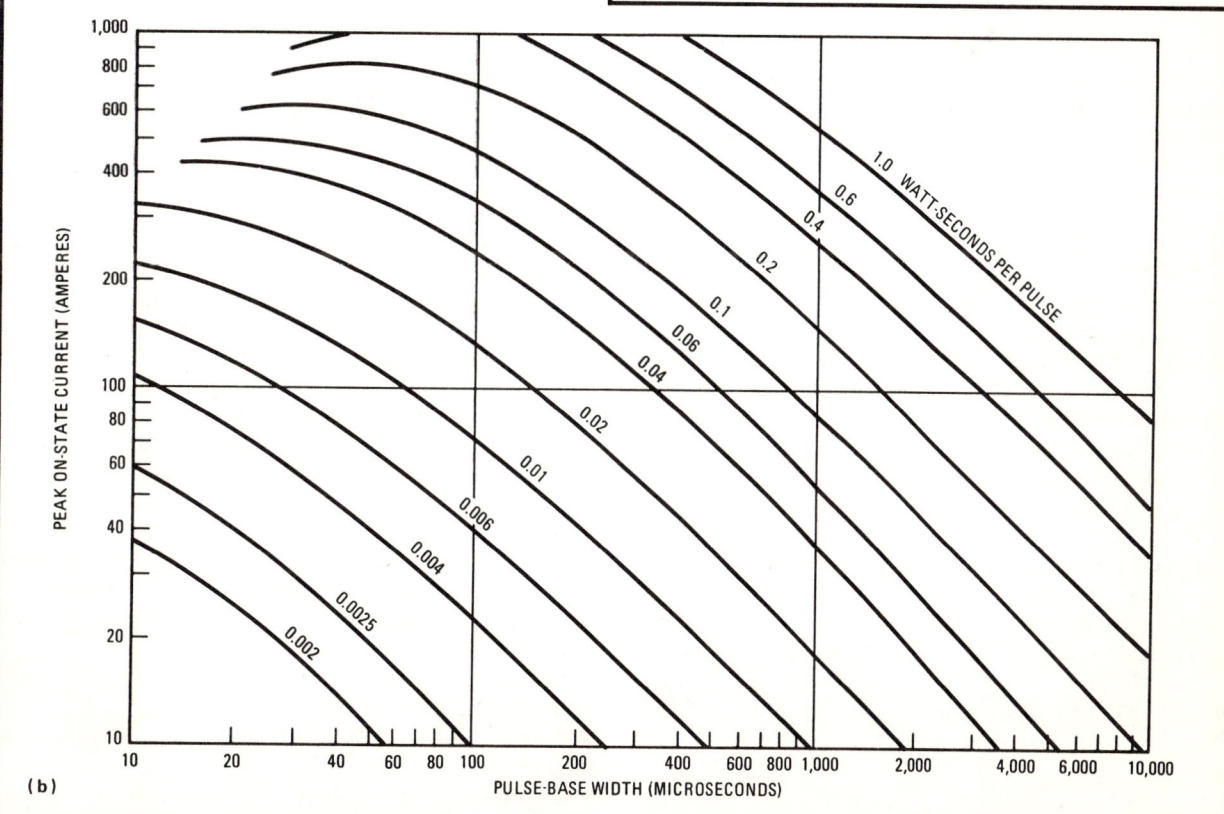

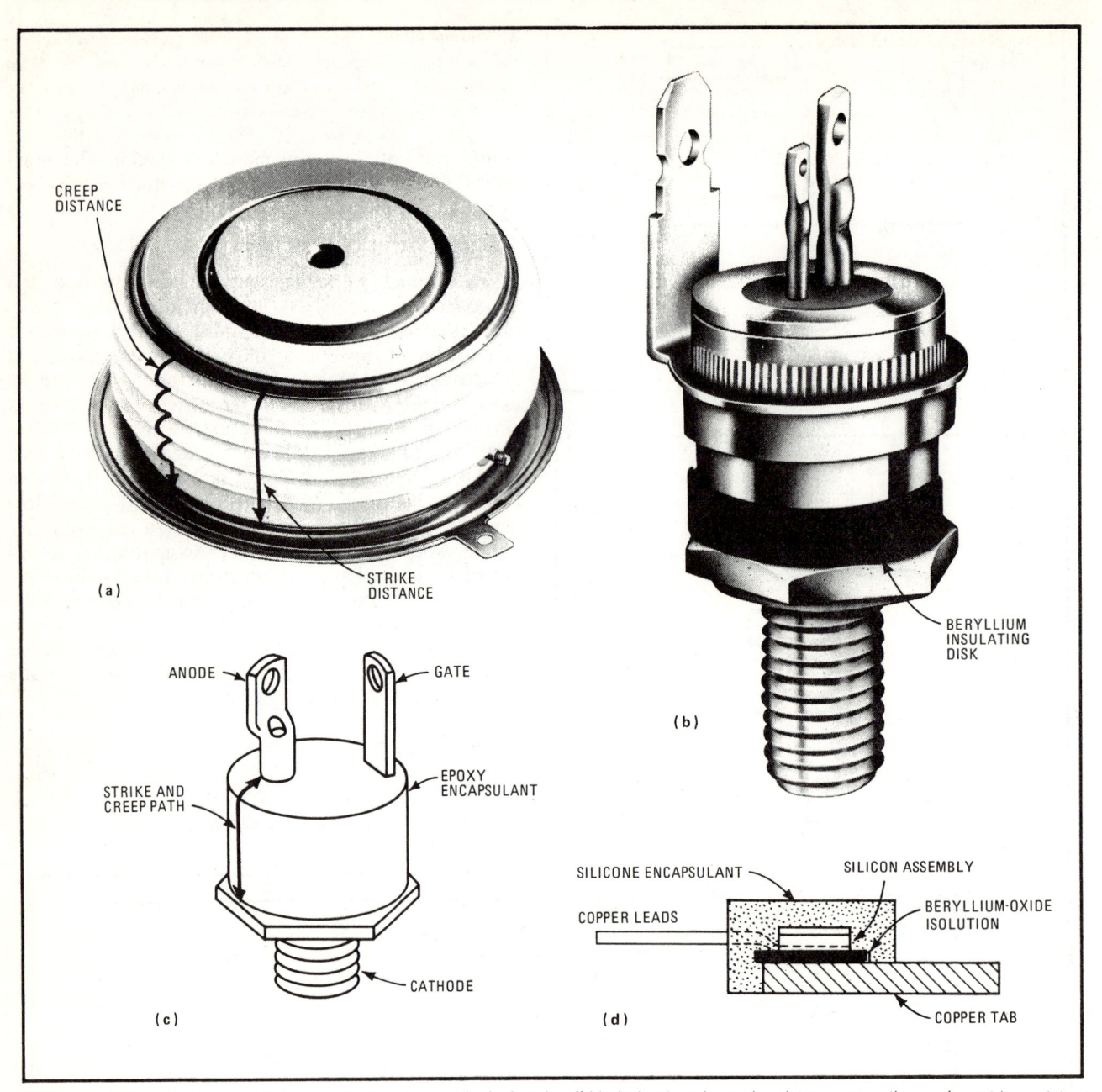

3. Power-device geometries. Electrical isolation is the principal trade-off hindering low-thermal-resistance mounting and must be weighed against the thermal resistance values in manufacturers' data sheets. The designers of the pressure-mounted disk in (a) assure adequate creep and strike distances by ribbing the surface. Thickness is 26 mm. Device can operate at 480 volts rms, and current-handling ability exceeds 500 amperes. Such devices as the stud-mounted unit shown in (b) pose an electrical-isolation problem because thickening the ceramic or increasing the diameter of the ceramic increases the thermal resistance on one hand and develops an interference with the hex-shaped base on the other. Adding an encapsulant to a stud-mounted device (c) greatly enlarges creep and strike distances required to meet NEMA and U/L requirements. By putting a thin beryllium-oxide insulator within the TO-220 package (d), electrical isolation can be obtained at minimal thermal expense. Designer can then attach the copper tab to a mounting surface with good thermal properties, thus doing away with the added cost of purchasing and mounting an independent heat sink.

However, determining dissipation data for switching transistors is considerably more difficult because the required data is seldom available. As a result, the designer has no choice but to determine the wave shapes and amplitudes of the collector-to-emitter voltages and currents, develop graphic plots of the product, and then integrate over an operating cycle to arrive at the rms power dissipated.

To ensure an accurate evaluation, these guidelines should be followed:

- Assume worst-case parameters—that is, figure on maximum voltage and current, lowest and slowest drive conditions, stiffest snubber networks, and highest frequency of operation.
- Include all power components, notably base power and blocking power.
- Test a number of devices to calculate a worst-case figure.
- To avoid measurement error caused by inductive shunting in high-frequency drive circuits, use a coaxial-

Thermal runaway

It is commonly believed that if the duty cycle of a semiconductor device is low, a heat sink is not required. But this dictum simply isn't true. Because semiconductor blocking characteristics are highly temperature-dependent, thermal runaway can develop. Blocking current is the current that flows during the off portion of the duty cycle, and, despite the fact that it may be no more than microamperes, it can contribute to a runaway condition. Runaway describes the positive-feedback chain reaction in which temperature rise causes a temperature-dependent current to increase. This, in turn, further heats the device to cause a still larger current. If unchecked, the current increases until the device destroys itself.

As an example, the peak off-state current (I_{DRM}) of a thyristor doubles for every 10°C rise in junction temperature. This doubling in current means a doubling of blocking-power loss (P_B). Such an increase under adverse temperature conditions can lead to a runaway that culminates in catastrophic failure.

Stability is not solely a function of blocking current; it is also highly dependent on the thermal resistivity of the entire junction-to-ambient path. Thus, a low-resistance heat sink can play a crucial role in assuring stable operation.

The plots below depict both stable and unstable device operation. In each case, the thermal resistivity is superimposed on the power-dissipation curve for the device. In (a), a stable thermal situation is depicted. It is stable because the dissipation capability of the over-all thermal path (P_E)—including the sink and interface—exceeds the power developed (P_D), which is the sum of the on-state conducting losses (P_O) and the off-state blocking losses (P_B). If the ambient temperature were to rise, the (P_E) line would shift slightly to the right, but, since this line would still lie above and to the left of the P_D curve, operation would remain stable. The reason is that the dissipation capability (P_E) still exceeds the power dissipated (P_D). Note that, for stable operation, the blocking losses (P_B) are assumed to be only a small percentage of the total losses (P_O and P_B). However, in such low-duty-cycle operation as a crowbar circuit, the exact opposite may be true. That is, the on-state losses may be negligible, compared to the blocking losses.

Now, suppose a designer attempts to mount a semiconductor on a bracket that has a high thermal resistance and he makes no attempt to enhance the thermal circuit with a heat sink. This situation is depicted graphically in the unstable plot. Because of the high thermal resistance, the power-dissipation curve has a much lower slope (α_2). The designer might assume that the thermal requirement is ample because the operating temperature is lower. However, a small upward shift in the ambient temperature will shift the entire (P_E) line to the right and beyond the (P_D) plot. If this happens, the device dissipation exceeds the dissipation capability of the system, and runaway develops; that is, temperature and current rise until catastrophic failure occurs.

The prudent design engineer will want to adhere closely to the guidance contained in the data sheets for each device regarding maximum thermal case-to-ambient resistance (θ_{C-A}), for which maximum blocking ratings apply. If such data is not specified, the designer should consult the device manufacturer.

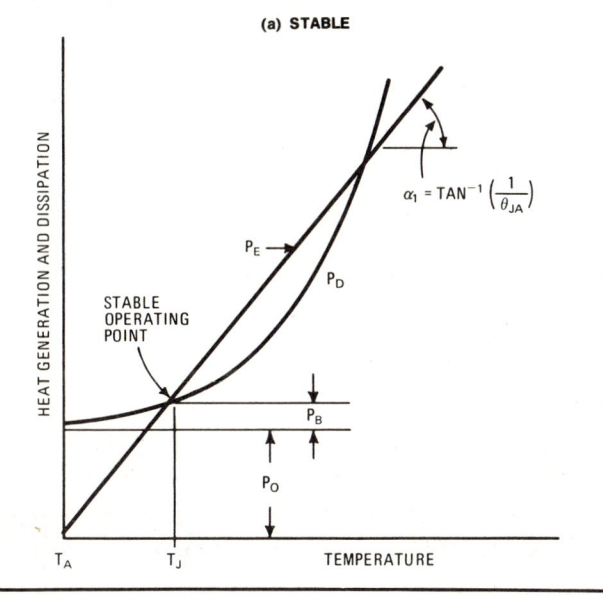

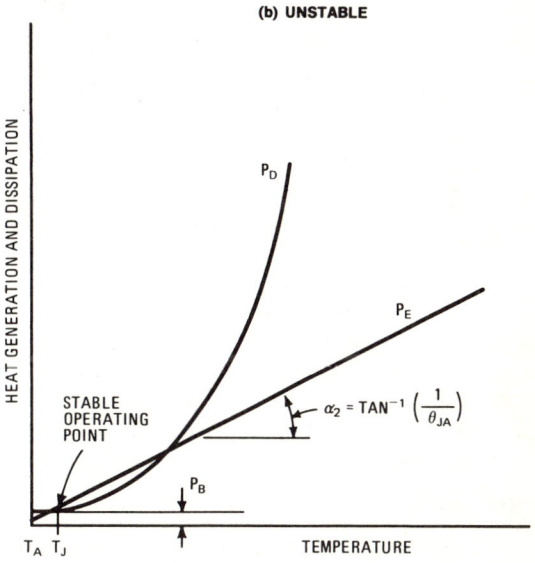

(a) STABLE

$\alpha_1 = TAN^{-1}\left(\frac{1}{\theta_{JA}}\right)$

(b) UNSTABLE

$\alpha_2 = TAN^{-1}\left(\frac{1}{\theta_{JA}}\right)$

current shunt or a quality current-measuring device that has adequate bandwidth.

Worst-case junction and ambient temperatures should be found. The maximum junction temperature, shown on the device's data sheet, should not be exceeded, except where such other parameters as transistor collector-to-base voltage or thyristor anode-to-cathode voltage can be lowered, thereby raising the limit of the junction temperature.

This method of derating makes possible low-voltage SCRs having a junction temperature of 150°C. The normal maximum is 125°C. To assure optimum reliability, design engineers will usually derate junction temperature according to a derating curve furnished by the manufacturers.

The ambient temperature may be specified in advance, or it may have to be determined. When equipment is enclosed in an unventilated cabinet, the ambient encountered by the semiconductor's heat sink may be considerably higher than the ambient outside of the

4. Big sink. Thyristors rated at 480 amperes rms are attached to large heat-spreader blocks, which are attached to the cabinet by an epoxy adhesive. The epoxy provides the required electrical isolation; yet, because of the large heat-transfer surface area, it does not materially hinder the thermal path to the equipment structure.

enclosure, so there is no substitute for careful temperature measurement inside the enclosure. Although papers have been published on the subject of determining ambient temperature inside of enclosures, these design guidelines produce only first-order approximations, rather than final design values.

Evaluating isolation requirements

After power and temperature values have been determined, the designer must examine carefully the system's electrical-isolation requirements. Generally, the shorter the thermal path, the better. However, the requirement for adequate electrical isolation is diametrically opposed—isolation requires a longer path.

The electrical path must be designed to prevent breakdown that might be caused by arc-over or current leakage along the surface between terminals. Sufficient air gap—known as strike distance—must be provided to prevent arcing, and a minimum surface-resistance path, called creep, must be provided between terminals by the intervening insulator surface.

Adequate electrical isolation for a large thyristor with a 500-A rms rating and up with a silicon-pellet diameter from 33 to 40 mm (Fig. 3a) is provided by designing it as a pressure-mounted disk package with a 26-mm height to prevent strike and ribbing the surface as shown to prevent creep. As indicated in Table 1, a package 26 mm thick has more than adequate creep and strike isolation for line voltages as high as 480 V rms.

A tougher problem is posed by stud-mounted semiconductors in the range of 1 A to 50 A (Fig. 3b). Problems develop if the dimensions are altered to improve the strike and creep distances because thickening the beryllium-oxide disk increases the thermal-path length, thereby raising the thermal resistance. And if the diameter of the disk is increased, it interferes with the hexagonal base.

One way to lengthen the creep and isolation distances is to add an epoxy encapsulant (Fig. 3c). The epoxy develops adequate creep and strike distances from both anode to cathode and anode to stud to permit operation at 230 V, which is adequate to qualify the device for approval by Underwriters' Laboratories.

However, electrical isolation inside the TO-220 package shown in Fig. 3 (d) is provided internally by inserting a beryllium-oxide layer between the chip and the copper tab. Thanks to low thermal resistivity, the beryllium-oxide provides a good thermal path for the heat to travel from the chip to the copper tab. Although the chip-insulator interface has a small area that tends to degrade the thermal path, that tendency is offset by making the insulator relatively thin. This is possible because electrical strike and creep requirements are greatly reduced by the protection of the semiconductor package.

The result is roughly equivalent to the externally isolated package in respect to both electrical isolation and thermal dissipation. Because the copper tab is electrically isolated, the device can be mounted directly on an enclosure wall that can serve as the heat sink. Thus, the additional cost of a discrete heat sink can be eliminated, along with the additional labor required for its assembly.

Spreading heat through a sink is the only effective means that has been developed thus far to cope with the thermal requirements of high-current SCRs. Avtek Corp., Burlingame, Calif., has used this method to advantage (Fig. 4). The large 480-A rms thyristors are mounted on thick heat-spreader blocks that are at-

TABLE 1. STRIKE AND CREEP STANDARDS

Voltage (rms)		Distance (mm)	
		U.L.[1]	NEMA[2]
130	Strike	5.08	3.18
	Creep	6.35	6.35
240	Strike	7.62	6.35
	Creep	10.16	9.53
480	Strike	10.16	9.53
	Creep	12.7	12.7

Notes: 1. Proposed U.L. standard for appliance and consumer equipment — May 1973
2. Standards for semiconductor converters — Nov. 16, 1972

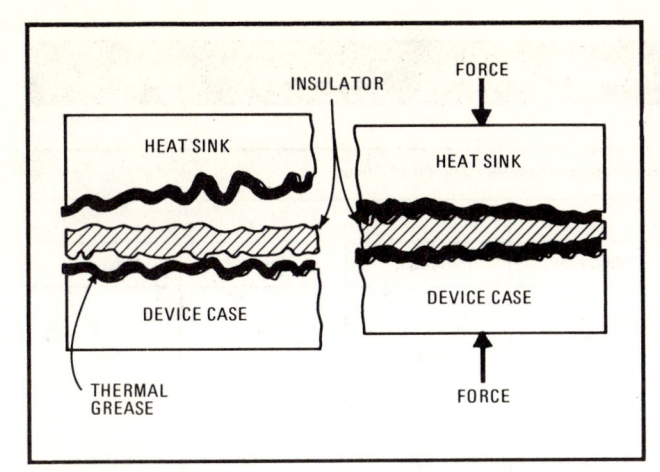

5. Mountainous. Exaggerated profile shows the thermal interfaces between a semiconductor device, an insulator, and the sink before and after engagement. Thermal grease helps to offset poor thermal path between materials that are microscopically rough.

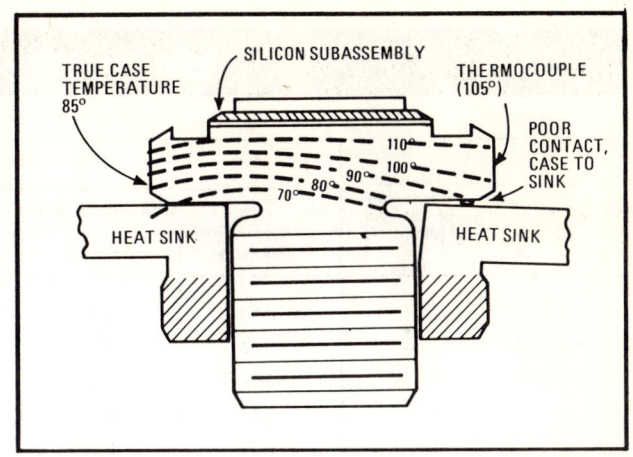

6. Poor contact. If a device's contact heat sink isn't uniform, eccentric temperature profile develops, which cause erroneous temperature measurements. Mating surfaces must be smooth so that the thermal interface will be uniform about the device's circumference.

tached by epoxy to a common convection-cooled heat sink. This sink, in reality the rear portion of the equipment enclosure, also provides a mounting surface for the fuse blocks and circuit-breaker shown. The blocks enlarge the heat-transfer area and thus offset the high thermal resistance of the epoxy.

Knowledge of the thermal-resistance limit and the constraints imposed by electrical isolation equips the designer to select the heat sink and coolant requirements. However, he must also consider carefully the thermal resistance at the interface between the semiconductor device and the heat-exchanger hardware.

The interface

The interface between the semiconductor device and the heat sink is a crucial thermal path. If the interface's thermal resistance is too high, it can render worthless the heat-transfer capability of a good heat sink.

As an example, the thermal-resistance values for the 10-32-thread, stud-mounted device listed in Table 2 range from 0.09°C/W at best to 1.2°C/W at worst—a spread of more than a full order of magnitude. In fact, 1.2°C/W is a far greater thermal resistance than that provided by a good heat-exchanging system.

Figure 5 illustrates the profile of the mating surfaces making up a typical interface. The roughness of the surface has been exaggerated to dramatize the mechanics of the interface. Applying force to the surfaces brings them into contact, but the net contact area is highly dependent on the ductility of the contacting metals, the surface finish, and the flatness, as well as the amount of force applied. A large void (Fig. 5) can be caused by a non-parallel fit between the device case and the sink. Thermal grease can help fill that void. And, as indicated in Table 2, an insulator would raise thermal resistance.

Although semiconductor manufacturers' guidebooks provide specific instructions, these guidelines can serve for all but the largest pressure-mounted disk packages:
- Mating surfaces should be flat to within 0.001 inch, and surfaces should be finished to a tolerance of 63 microinches or less.
- All paint and other impurities should be removed completely by a treatment with #000 fine steel wool and silicone oil.
- The surfaces should be cleaned and wiped free of foreign matter immediately before assembly.

Measuring temperature

Initial calculations should be confirmed by actual temperature measurements. However, if results are to be meaningful, a number of pitfalls must be avoided. It is crucial to determine the actual power dissipation of the semiconductor because temperature data that overlooks this factor will lead to false calculations of thermal resistance. More is involved than merely inserting a thermocouple and waiting for the device to reach thermal equilibrium before taking a reading.

During the design and prototype stages, the designer should verify his calculations by measurements in the operating environment, according to these guidelines:
- The semiconductor device should be mounted on the heat sink in the actual enclosure intended to contain the manufactured version of the equipment. Merely rotating an air-cooled heat sink 90° can make a vast difference in its thermal properties.
- Follow manufacturer's mounting instructions carefully.
- Limit thermocouple wire diameter to 12 mils to avoid localized cooling.
- Avoid placing high-dissipation elements, such as ballast resistors and transformers, near a semiconductor. If this juxtaposition is unavoidable, make sure that both devices are dissipating their full design-power levels during temperature runs.
- Block off local air currents.
- Prevent direct sunlight from contributing to the heat measurement.
- Make sure that electromagnetic fields do not couple to thermocouple leads and destroy measurement accuracy.

Consider the problem involved in measuring the case temperature of a stud-mounted rectifier or SCR. Assume that the thermal resistance—junction-to-case—is specified at 0.3°C/W maximum and that the ambient temperature is 45°C. Assume further that the device-case

235

TABLE 2 SOME THERMAL-INTERFACE RESISTANCES

Stud-mounted devices

Stud size	Hex size across flats or flat base (diameter)	Thermal resistance-case-to-sink θc-s °C/watt					
		With thermal grease			Dry		
		Minimum	Nominal	Maximum	Minimum	Nominal	Maximum
10–32	7/16"	0.09	0.3	0.8	0.2	0.5	1.2
1/4"–28	9/16"	0.07	0.25	0.6	0.15	0.4	0.9
1/4"–28	11/16"	0.05	0.15	0.4	0.10	0.25	0.6
3/8"–24	1-1/16"	0.02	0.06	0.15	0.05	0.1	0.25
1/2"–20	1-1/16"	0.02	0.065	0.2	0.05	0.12	0.3
3/4"–16	1-1/4"	0.025	0.08	0.2	0.06	0.15	0.35
3/4"–16	1-5/8"	0.015	0.04	0.10	0.03	0.07	0.15
Flat Based	1-7/8"	0.01	0.025	0.07			

Stud-mounted devices insulated with 5-mil mica washer

Stud size	Hex size	Minimum	Nominal	Maximum			
10–32	7/16"	1.2	2.5	4.5	—	—	—
1/4"–28	9/16"	0.9	2.0	3.5	—	—	—
1/4"–28	11/16"	0.7	1.5	2.5	—	—	—

Pressure-mounted disk packages — lubricated

Interface diameter (inches)	Nominal clamp force (lb)	Thermal resistance-case-to-sink θc-s °C/watt		
		Minimum	Nominal	Maximum
3/4	800	0.04	0.06	0.20
1	2300	0.02	0.03	0.10
1-1/4	2300	0.015	0.022	0.08
1-1/3	4000	0.014	0.02	0.07

temperature must not exceed 85°C, in accordance with the manufacturer's data sheet. This amounts to a power dissipation of:

$$P = (85°C - 45°C)/0.3°C/W = 133.4 W$$

Assume that the case temperature is actually measured to be 80°C. At first glance, the device appears to be below its maximum temperature, and all appears to be well. However, if the true dissipated power during the temperature measurement is less than the 133.4-w value, the actual case-to-ambient temperature is, in fact, too high. To illustrate why, assume that the power actually dissipated during the temperature measurement is 110 w. Then:

$$\theta = (80°C - 45°C)/110 W = 0.318°C/W$$

Therefore, at full rated power of 133.4 w, the true case temperature would be:

$$T_{case} = (P \times \theta) + T_{ambient}$$
$$= (133.4 W \times 0.318°C/W) + 45°C$$
$$= 88°C$$

or 3°C above the maximum rated case temperature at full power dissipation.

To avoid this pitfall, the designer should make sure that the device is dissipating the maximum power during the temperature measurement. To do this, a direct current is applied to the device, and the junction dissipation is adjusted by controlling the volt-ampere product.

An alternative method is to obtain a limit cell from the device manufacturer. This device, selected for worst-case power dissipation, is suited for thermal evaluation because, under actual circuit-operating conditions, its power dissipation provides a worst-case power-dissipation value.

Another factor can inhibit accurate temperature evaluation. As shown in Fig. 6, a local interface distortion can create a path of high thermal resistance directly under the thermocouple, thereby generating a false reading. If the maximum rated power dissipation and the ambient are the same as in the previous example, but the measured case temperature is 105°C, it does not necessarily mean that the heat sink is defective. It may be the local interface distortion.

To avoid this possible error, it is a prudent technique to use two or three thermocouples and to mate the semiconductor device carefully to the heat-exchanging surface to assure a uniform interface having low thermal resistance. □

THERMAL DESIGN

Forced air cooling in high-density systems

The trend toward miniaturization may seem to call for natural convection; but reliability analysis shows forced air cooling can boost system survival by 75%

by Gordon M. Taylor, *Rotron Inc., Woodstock, N.Y.*

☐ Heat sinks alone cannot dissipate excessive heat in a system when the air around them does not move rapidly. The problem is becoming more pervasive as system designers crowd ever larger numbers of circuit boards into ever smaller regions, reducing the number and width of possible air passages. However, air forced through the narrow passages by a fan will remove the heat and thereby raise the life expectancy of a high-density electronic system.

Two other factors that contribute to a system's tendency to overheat and so detract from its long-term reliability are: the increasing density of the circuitry on the chips inside the IC packages, and the increasing speeds at which this circuitry operates. These trends, too, are helping to spread the use of forced air cooling, which is also highly effective in smoothing temperature fluctuations at critical semiconductor junctions in densely packed, high-speed logic systems.

In the past, however, the pressure to optimize reliability has made packaging engineers hesitant to add an electromechanical fan to an electronic system that contains no moving parts. But the reliability of air-moving devices has recently risen an order of magnitude, improvements having been made in the insulating materials in stator windings and in the application of precision bearing design. The mean time between failures of a fan moving air at 100 cubic feet a minute at 158°F can by now reach over 50,000 hours.

The limitations of natural convection

Buoyancy is the driving force moving air in a natural convective air stream. But buoyancy can't deliver velocity much over 0.5 foot per second. The reason is that the specific weight of warmed air doesn't differ appreciably from that of the cooler air surrounding it. And when this small buoyant force must also overcome the counteracting viscous phenomena that develop along stationary air masses, the air flow rate is limited to a fraction of a foot per second. This is serious because the thermal path between a stationary wall and an air stream moving at velocities below 0.5 ft per second is relatively poor.

Figure 1 illustrates how the velocities and the thermal profiles of air moving past a stationary wall affect heat transfer. The velocity plot indicates that the speed of air at the boundary is zero because at the boundary air particles adhere to the wall. As the distance from the wall along the y axis increases, the velocity of the air also increases until it reaches the mainstream velocity. As for the temperature profile, notice that the air temperature at the wall is virtually the same as the wall temperature, and diminishes along the y axis to the value of the mainstream air temperature.

The shape of the velocity profile is crucial because the coefficient of heat transfer at the wall is a function of the rate of change of the temperature along an axis perpendicular to the wall. Increasing the flow rate enlarges this differential and thus the effective heat transfer from

Life saver. Typical modern fan can add 75% to the expectation of system survival. Device delivers 70 ft^3/min when driving a static load that's the equivalent to 0.1 inch of water.

the wall to the air stream. Since the natural-convection flow rate is limited, the value of this differential is also limited. However, forced air can develop velocities far in excess of those attainable with natural convection, enhancing the transfer of heat across the boundary.

Faster air flow yields a second benefit because speed increases the temperature differential between the moving air stream and the stationary wall being cooled. This is important because heat transfer is also a function of temperature differential. The larger temperature differential results because a molecule of air at higher speeds has little time to absorb heat, so it will not reach as high a temperature as a slower-moving air particle.

Thus faster air flow increases both the coefficient of conduction and temperature differential.

The goal behind improving heat transfer in a system is of course greater reliability. Proof that adding an air-moving device to a system that formerly relied on natural convection does extend a system's life is given by the following case history.

Bathtub curves

The curve labeled (a) in Fig. 2 is the survival expectation of a minicomputer packaged in a 36-by-12-by-12 in. envelope which employed natural convection cooling. The "early and chance failures" portion of the curve—the infant mortality region—describes the time interval immediately following manufacture, when marginal and defective components are weeded out. The second portion of the curve—"random and chance failures"—describes the useful life of the system. The final portion of the curve, "wearout and chance failures," signifies the wearout period of the equipment's life span where the failure rate climbs rapidly.

Originally the manufacturer had relied upon the natural convection of air to cool the ICs and other components and maintain temperatures below safe values. But the high packaging density of the equipment prevented the air flow from cooling all heat sources adequately, and average life expectation, as indicated in Fig. 2, was about 20,000 hours. This life span was too short, so the manufacturer turned to forced air cooling.

The forced-convection heat-transfer equation is:

$$Q = C_p W \Delta T$$

where Q = amount of heat dissipated, C_p = specific heat of air, W = air mass flow rate, and ΔT = temperature rise through the system. Incorporating conversion factors and specific heat for air at sea level yields an equation for the flow rate required to dissipate a given amount of power:

$$CFM = (3160 \times kW)/\Delta T°F$$

where CFM = flow rate measured in cubic feet per minute at an air density of 0.075 lb/ft^3; kW = power dissipated within the system enclosure, in kilowatts; and ΔT = average temperature rise of air passing through the system, in degrees fahrenheit.

For the minicomputer, the maximum allowable temperature within the cabinet and the maximum ambient were determined to be 113°F and 68°F, respectively. Secondly, the total power dissipated within the cabinet was computed as 1 kilowatt. These numbers, when substituted in the above equation, work out at (3160 × 1 kw)/(113 − 68)°F, or 70.2 ft^3/min.

The system was then subjected to an aerodynamic

1. Convective interface. Plots depict the velocity profile and the temperature profile of the boundary between a cooling air flow and a stationary surface. Optimized convective cooling requires the rate of change of temperature at the boundary and the temperature differential between the wall and the mainstream to be maximized along an axis perpendicular to the wall.

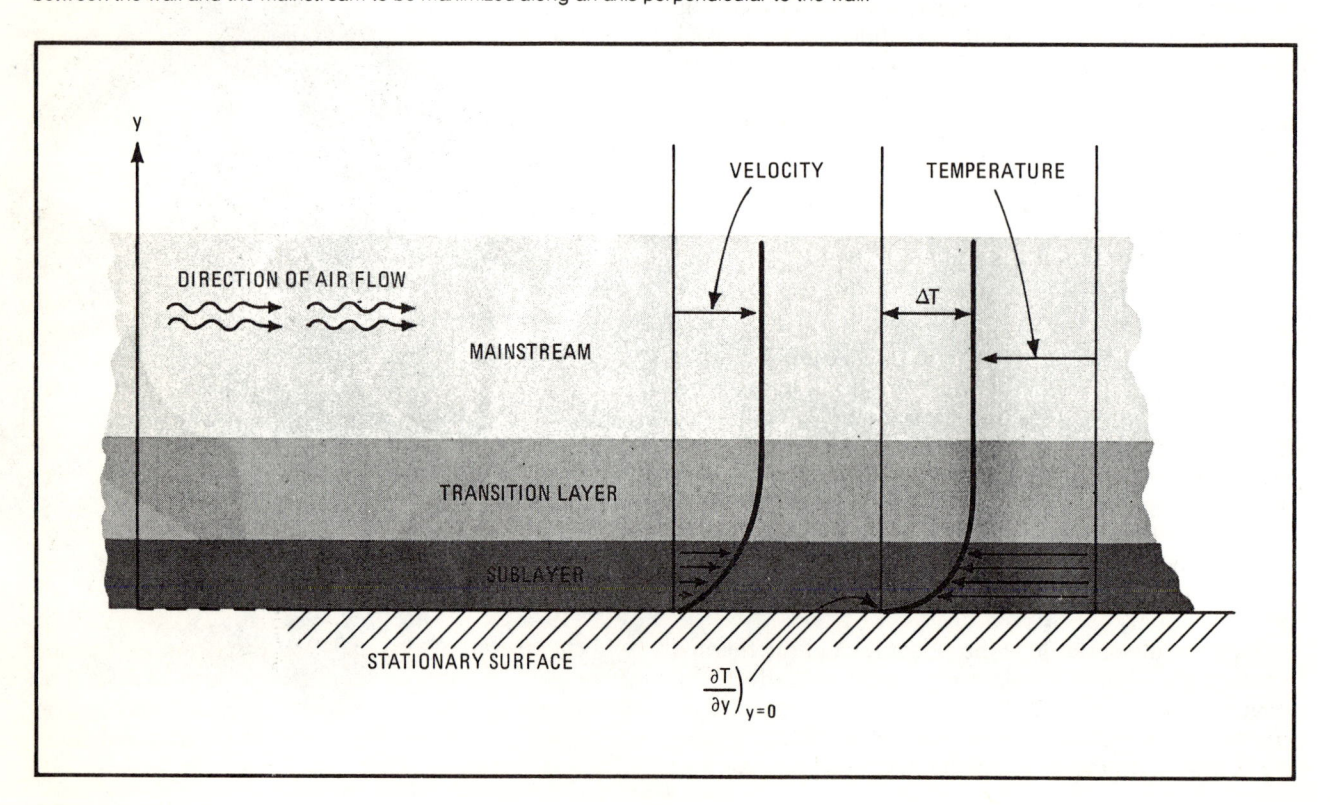

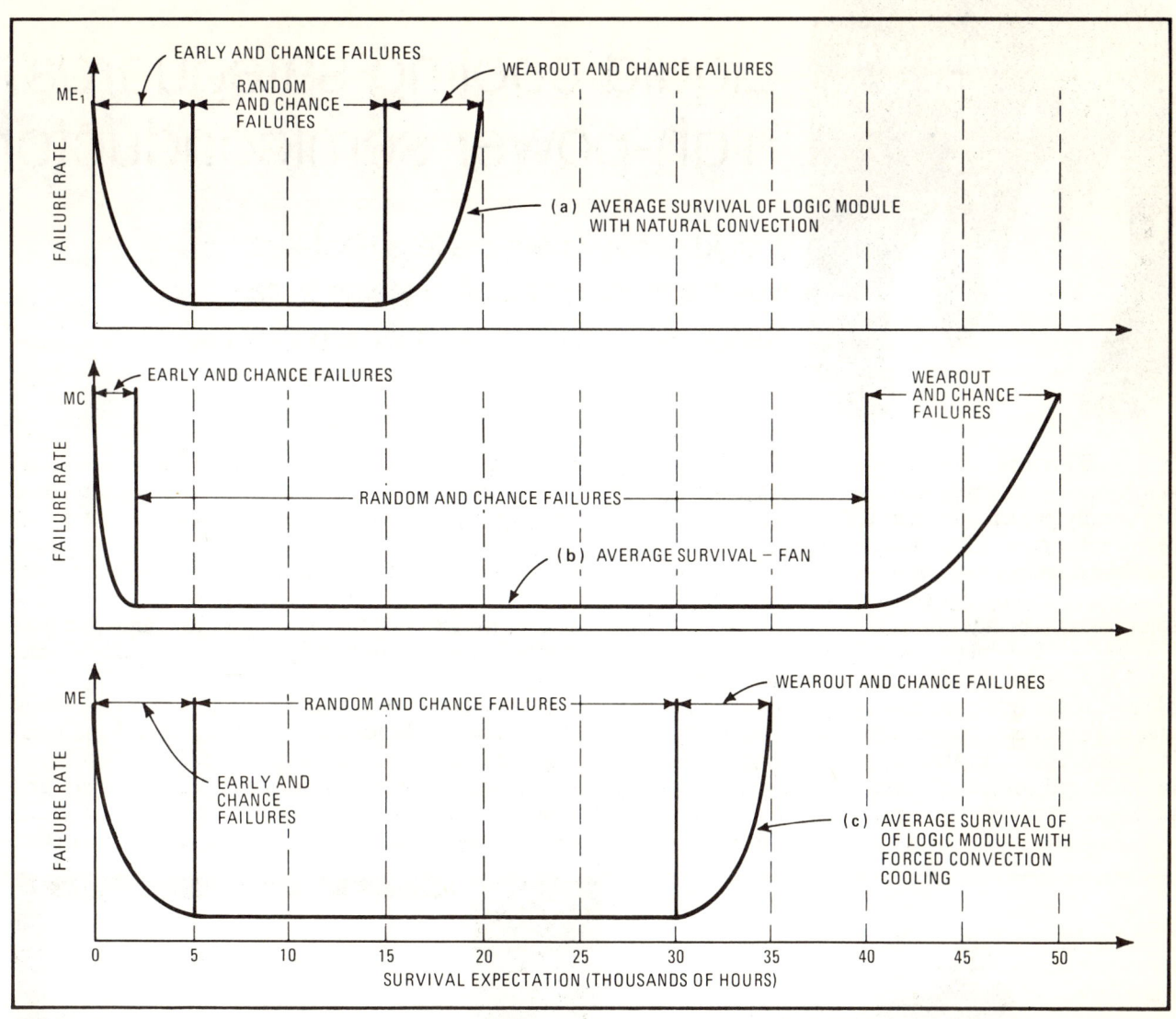

2. Stretchout. Survival expectation of a minicomputer (20,000 hours) was extended to some 35,000 hours by adding a fan. Such life extension runs counter to the common belief that a forced-air-moving device degrades reliability.

study to determine its resistance to air flow. This turned out to be 0.1 in. water-gauge static pressure at 70 ft₃/min.

On the basis of this data the fan shown on page 15 was selected. It measures 4-11/16 in. square by 1½ in. deep. Packaging engineers were able to accommodate it in the original equipment enclosure by rerouting some wire harnesses and moving several fasteners. The cost was less than one cent per watt dissipated. The fan occupied less than 0.5% of the enclosure volume.

Life tests of the fan indicate an average survival of 50,000 hours at 158°F, plotted as curve (b) in Fig. 2. The "early and chance failures" as plotted are really quite conservative. The reason is that electrical failures, which used to account for many of the early failures in air-moving devices, have been drastically reduced as a result of improved magnet-wire insulation and rigorous inspection procedures. Bearing failure is the principal wearout failure mode.

Increased survival

Addition of the fan raised the minicomputer's survival expectation from 20,000 to 35,000 hours—an improvement of 75%. This survival expectation is plotted as curve (c) in Fig. 2.

This curve is based on the formula:

$$ME = MC - [ME_1 + (MC - ME_1)]K \text{ hours}$$

where ME_1 is the average survival for the over-all system with natural convection cooling, MC is the average survival rate for air-moving device, and K is an empirically derived derating factor. Therefore, when ME_1 = 20,000 hr, MC = 50,000 hr, and K = 0.3, ME works out at 35,000 hr.

The result is conservative because the derating factor, K, generally used by system and air-moving equipment manufacturers, is in the neighborhood of 0.12 rather than 0.3 as shown. Thus survival values determined by the formula given with a derating value of 0.3 can be interpreted to mean "at least as good as."

The formula can be applied generally to ascertain the increase in survival attainable by adding an air-moving device. Should empirical data for the over-all system be lacking, the designer can combine the survival rates for individual components by employing traditional reliability analysis techniques. ☐

THERMAL DESIGN

Liquid cooling safeguards high-power semiconductors

Liquid coolants require less heat-sink volume than forced-air systems to carry away heat from kilowatt-level circuits

by John A. Gardner Jr., *Wakefield Engineering Inc., Wakefield, Mass.*

☐ Liquid systems have seldom been used to cool semiconductors in the past because convective-air cooling has usually proven adequate at low power levels. But as semiconductor rectifiers and thyristors grew in power capability, design engineers were confronted with the problem of cooling devices that dissipate hundreds and sometimes thousands of watts.

When semiconductors were first used, it seemed that large heat sinks, together with high-capacity fans and blowers could fulfill the most extreme cooling requirements. However, as power levels and device sizes were increased, space requirements for cooling grew exponentially and demanded precious space in the electronic packages; designers, consequently, turned to liquids.

Liquid provides a larger margin of reserve cooling power than other cooling techniques to cope safely with peak loads and transient conditions because thermal inertia enables fluid to absorb momentary heat pulses with only a slight temperature rise. Liquid cooling also minimizes acoustic interference, a persistent problem when cabinets are air-cooled. Noise can be readily abated by locating the heat-exchanger and pumping equipment at a distance from the electronic components being cooled.

If a designer had his choice, he would select natural convection cooling to reap the benefits of cost and reliability that must be sacrificed by adding the electromechanical components required for forced-air and fluid systems. However, costs and complexity of liquid-cooling systems cannot be denied. And many engineers

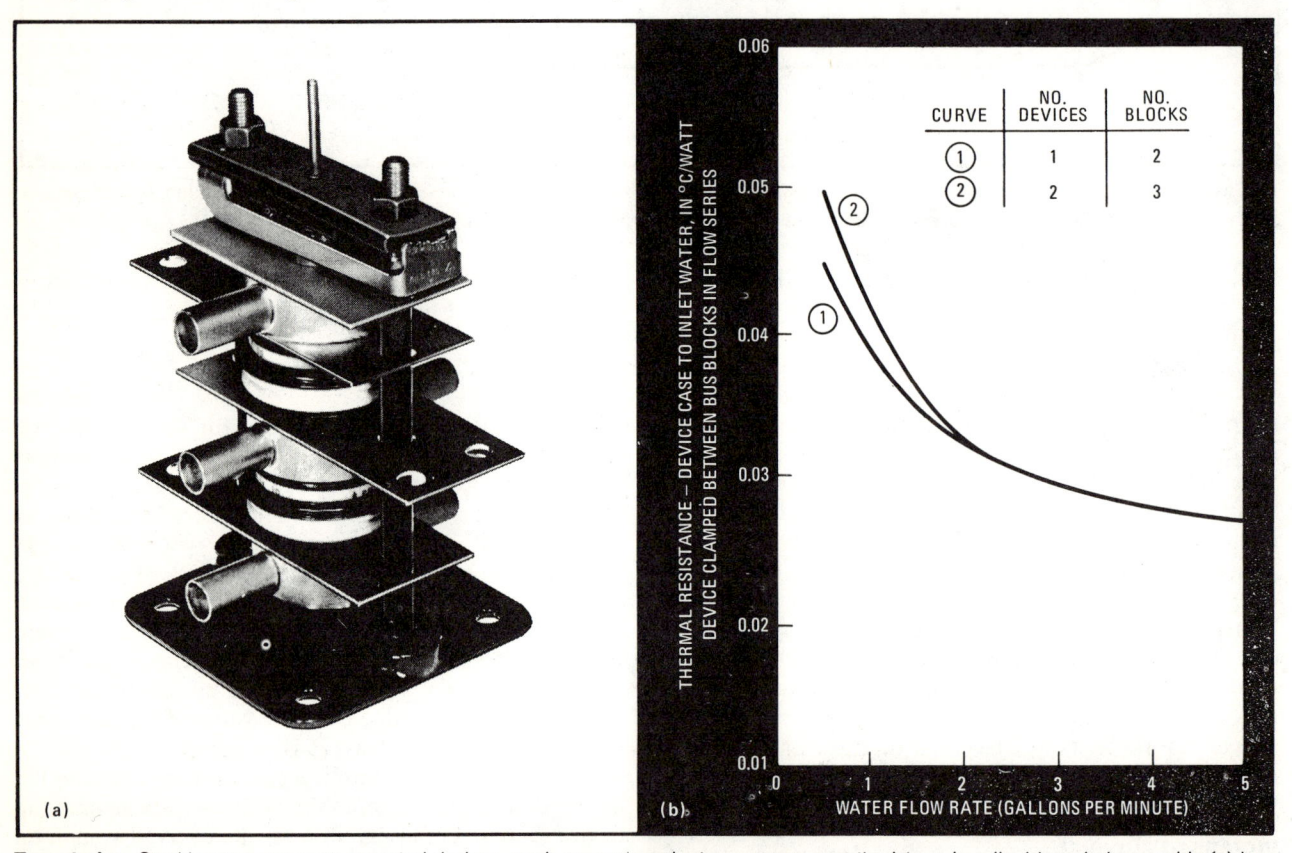

Two stories. Stacking two pressure-mounted devices, such as semiconductor rectifiers and thyristors, in a liquid-cooled assembly (a) is a compact technique for cooling kilowatt-level devices. Thermal resistance from the case to the coolant is low at low flow rates, and it diminishes little at flow rates above two gallons per minute (b). Adding the second device (curve 2) raises the thermal resistance a trifle (curve 1).

240

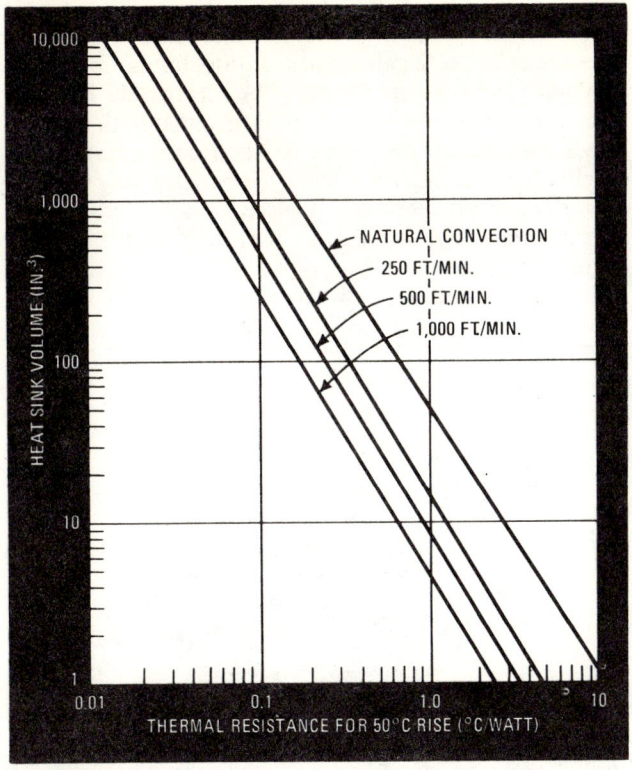

1. Gobbling space. The volume of an air-cooled sink grows enormously as a designer lowers the thermal resistance. These curves indicate the various sink volumes for cooling by natural convection and forced air. However, liquid-cooling requires much smaller sink volumes—desirable for compact electronic packaging.

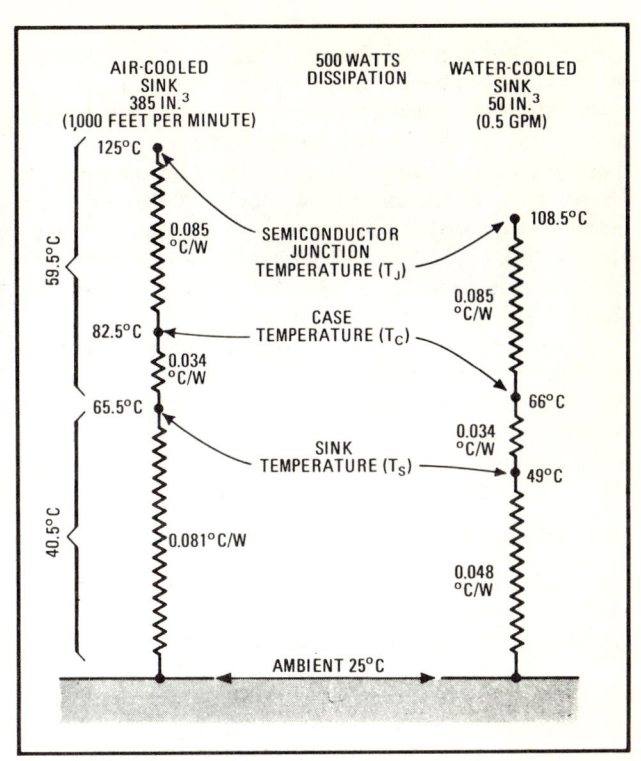

2. Cool the junction. Holding the semiconductor junction below a limiting temperature is crucial to thermal cooling—a key factor in long-term reliability. Here, circuit equivalents have been drawn for both an air- and a water-cooled sink dissipating 500 watts. The water-cooled system keeps the junction almost 17°C cooler and requires about one seventh the volume of the air-cooled sink.

are reluctant to plumb a liquid circuit into an electronic-equipment cabinet because of the likelihood of corrosion, leakage, and condensation.

Moreover, factors such as component reliability and maintenance demands all weigh in favor of air-cooled systems. But despite these drawbacks, liquid cooling is proving to be a highly satisfactory technique for compact, silent cooling of high-power semiconductors and electronic systems.

Thermal resistance

As power dissipation rises, the packaging engineer must whittle away at the case-to-ambient thermal resistance (Θ_{C-A})—the temperature rise in degrees for each watt of power transferred. As a rule of thumb, each time he halves the thermal resistance in a natural convective system, the designer must quadruple the heat-sink volume. Obviously, the demand for space becomes enormous when dissipation levels rise into the kilowatt range and thermal resistance falls below 0.1°C per watt.

Figure 1 illustrates the envelope volume required for several widely used shapes of heat sinks for a range of thermal resistances. The four curves show the relationship of volume to thermal resistance for natural convection cooling and forced-convection cooling at velocities of 250, 500, and 1,000 feet per minute, based on 50°C sink-temperature rise above ambient.

If natural convection is out of the question, then the designer must turn either to forced air or liquid cooling, and both are feasible in the range of 500 to 1,000 w. However, when volume is a crucial consideration, liquid has the edge. A forced-air cooling system requiring 500 cubic inches of heat-sink volume can't compete with liquid, which can deliver the same cooling capability in a sink volume from 60 to 120 cubic inches—an improvement of a full order of magnitude.

Whereas air-cooled systems require careful analysis of localized ambients within equipment cabinets to avoid interactive heating effects, analysis for liquid-cooling systems is relatively straightforward.

The arguments favoring cooling with liquids at higher-power levels emerge more clearly in an example.

A high-power example

Consider a 500-w pressure-mounted semiconductor rectifier with a maximum junction temperature of 125°C and junction-to-case thermal resistance, as shown on the manufacturer's data sheet, of 0.085°C/w. The case-to-sink thermal resistance is determined from experimental data to be an additional 0.034°C/w. Adding the two thermal resistances and multiplying by 500 w yields a rise of 59.5°C between the sink and junction.

These thermal resistances and the temperature rises across the resistances are shown schematically in Fig. 2. Such sketches help the designer to visualize how each portion of the thermal path contributes to the rise above the ambient temperature. If the ambient is assumed to be 25°C, the rise from ambient to sink is limited to 40.5°C. This means that the sink-to-ambient thermal resistance can be 40.5°C/500 w, or 0.081°C/w, at most. Figure 1 discloses that this thermal resistance requires a sink volume of about 385 cubic inches if air flow is to be

1,000 feet per minute. This requirement can be satisfied by two sinks of about 3 inches by 7 in. by 7.5 in.

By contrast, consider the requirements for liquid-cooling the same semiconductor rectifier. Figure 3(a) shows the device clamped between liquid-cooled blocks. Thermal resistance values for blocks of this type are shown in Fig. 3(b). At a flow rate of 0.5 gallon per minute, the sink-to-inlet water thermal resistance of the water system is 0.048°C/w—about half the value of the previously calculated forced-air system.

As for volume, the sinks occupy about 50 cubic inches, a mere 13% of the sink size required in the moving-air system. Liquid flows from the lower coolant block to the upper one in flow series, resulting in only a slight warmup of the water passing through the lower block. This heating is of little consequence.

Series flow

Stacking of devices, as shown on page 103, adds only slightly to the volume and causes little degradation of the case-to-inlet-water thermal resistance for the water system. At the same rate of 0.5 gallon per minute, cooling a second device in series degrades the thermal resistance only 0.005°C/w—the difference between plots 1 and 2 in the performance graph on page 103. Plot 1 is the case-to-inlet-water thermal resistance of a single pressure-mounted semiconductor. Plot 2 is the thermal resistance for each of two devices mounted in a stack so that both devices share a common pole piece.

Plots 1 and 2 are virtually coincident because the conductive path from the devices to the coolant-pole pieces is far more efficient than the thermal path between adjacent semiconductor devices. However, this is seldom the case in air-cooling systems.

Liquid cooling is also attractive for cooling groups of lower-power devices, such as semiconductor devices in TO-3 cases that have power levels in the range of 50 to 150 w, as shown in Fig. 4. The channel-plate cooler (Fig. 5) is designed as an inexpensive arrangement for cooling both stud-mounted and bolted semiconductor devices. Sink thermal resistances range from about 0.6 to 0.25°C/w, depending on the center-to-center spacing of the devices and coolant-flow rate.

On the other hand, the high-density cooler shown in Fig. 4 is virtually unaffected by thermal interaction because the devices are located directly over the coolant-flow lines, thereby optimizing the thermal path to the fluid. This arrangement is well suited for cooling large numbers of such smaller devices as those mounted in the popular TO-3s. This cooler, which measures 6 in. by 7 in. by 1 in. and occupies a volume of 85 cubic inches, is capable of dissipating 2 kw if the coolant-inlet temperature is 40°C or below.

Table 1, which lists the thermal resistivities of various liquid-cooler geometries, shows that the pressure-mounted assemblies offer thermal resistances an order of magnitude lower than the channel-plate mountings and the high-density cooler. However, the latter are more than adequate for clusters of lower-power, smaller devices. The pressure-mounted coolers offer an additional advantage. Since the bus plates and the bolts are cooled along with the device, the bus-current rating can be higher than in an air-cooled configuration. In general, thermal resistances diminish substantially as flow rates are raised to 2 gallons per minute, but not beyond.

When a semiconductor must be electrically isolated from a liquid-cooled sink, an interface material, such as beryllium oxide, offers high thermal conductivity and excellent electrical isolation. The penalty is a rise in over-all thermal resistance, caused by an increased interface thermal resistance, as indicated in the second column of Table 1.

Electrical isolation between pressure-mount cooling

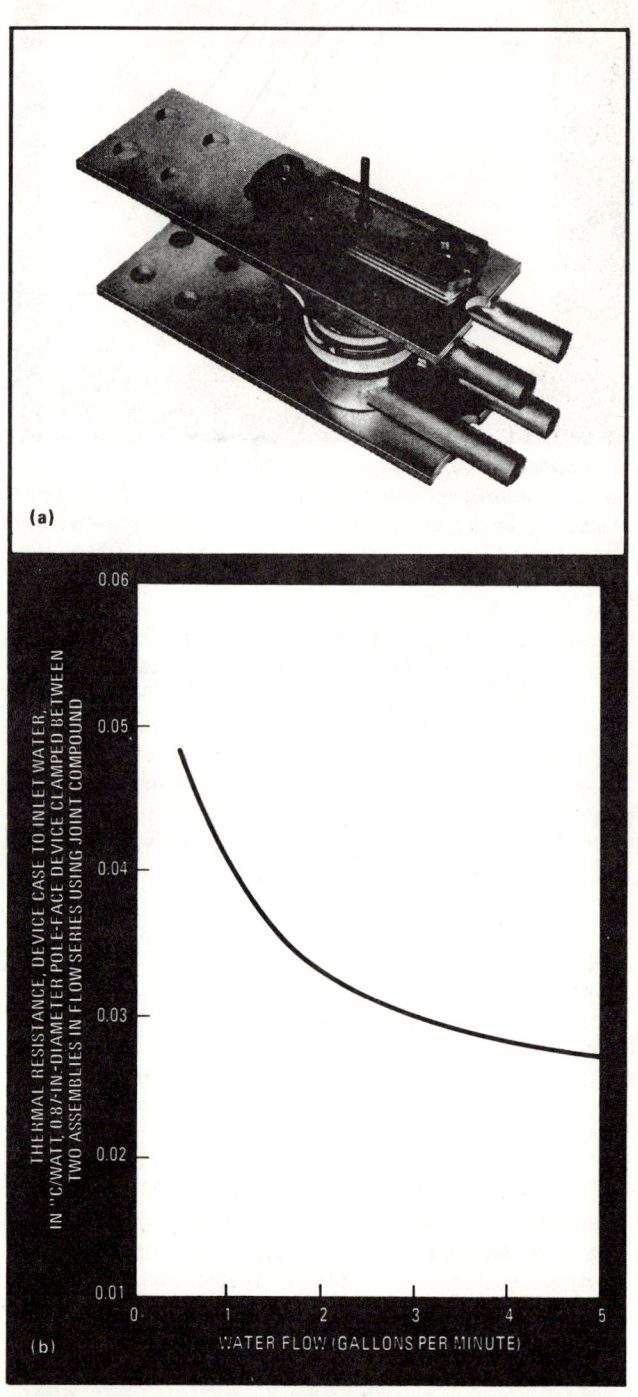

3. Cool cooler. The liquid-cooled solid-state copper blocks sandwiching the semiconductor device (a) provide two thermal paths (top and bottom) and dissipate 2 kilowatts. Clamp puts an 800- to 2,000-pound bite on the semiconductor device to assure a low thermal resistance—on the order of 0.034°C per watt at the interface (b).

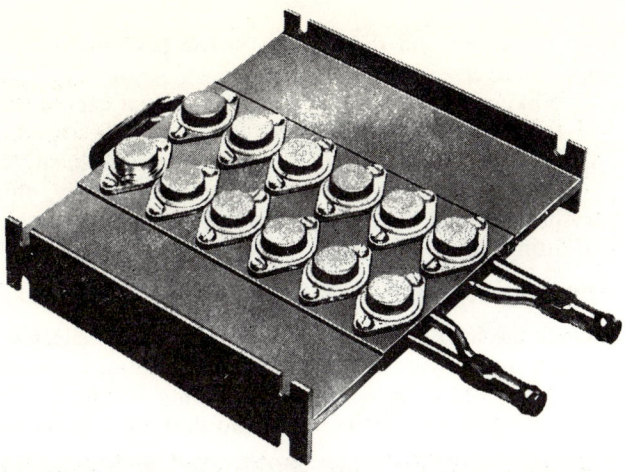

4. Dense package. This configuration is well suited for cooling TO-3 packages because the devices are placed directly over the coolant lines, ensuring short, efficient thermal paths to the coolant. Unit can dissipate 2 kilowatts, but occupies only 85 cubic inches.

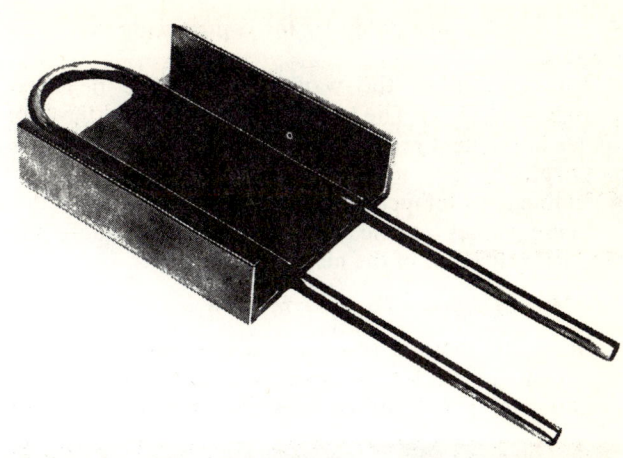

5. Channel chiller. This economical U-shaped aluminum plate, priced at $16, can cool a small number of relatively high-power devices. Thermal resistance varies from about 0.25°C/W to 0.6°C/W, depending on spacing between devices and coolant-flow rate.

blocks can be achieved with rubber liquid-transport tubing. A good rule of thumb is to employ one foot of tubing for each 1,000 volts of potential difference.

Open and closed loops

Liquid systems are commonly designed in an open-loop configuration in which tap water is fed to the cooler or cold plates through a pressure-reducing station, which ensures a constant flow rate. The heated water is then discharged into a drain, and no attempt is made to control water temperature. This is usually acceptable if the inlet water temperature never rises above 30°C. However, during humid summer months, condensation can form on cold plates and transport tubes, which may be troublesome.

By contrast, the closed-loop systems of Fig. 6 offer a number of advantages, including temperature control, water conservation, and reduced susceptibility to flow-rate variation. Moreover, by operating the coolant system so that the water temperature remains above the dew point, condensation on cold surfaces can't occur. Finally, a closed-loop system enables the user to add selected solutions to the water to attain desired coolant properites.

The hardware components of a closed-loop system include a cooler, a circulating pump to sustain the flow, an air-liquid heat exchanger to transfer the heat from the liquid to the surrounding air, and storage tank to allow for expansion.

The storage tank permits normal expansion and contractions that accompany temperature variations in fluids. It is also a deaeration point for the system and enables periodic sampling and replenishment of the coolant. The relative costs of forced-air, open-, and closed-loop systems are listed in Table 2. The entries, in all cases, apply to a dissipation requirement of 1 kw.

Heat exchangers

The air-liquid, high-efficiency heat exchanger with an attached fan, shown in Fig. 7, is typical of a compact series of exchangers suitable for closed-loop cooling of electronic components. Copper and brass lines are commonly used to carry the coolant and provide long-term high performance with most heat-transfer fluids.

This type of cooler is available with either single or double-pass flow on the water side. With double-pass flow—this means that the water makes a round trip through the exchanger region—thermal performance is enhanced. Also, inlet and outlet fittings mount on the same side of the exchanger, which is frequently a convenience. However, a larger-capacity pump is required to cope with the increased pressure drop that is characteristic of the double-pass system.

The fan of the illustrated exchanger draws the air through the exchanger core before the air passes through the fan itself, which produces even heat distribution across the core. Thus, the operating temperature of the fan assembly, including the bearings, will be elevated above the ambient temperature. And since the life of fan motors depends on their operating temperature, the temperature of the air leaving the core is a critical parameter in length of fan life.

Designing a liquid-cooled system

Generally, design requirements of a liquid-cooled system are less complex to compute than those of an air-cooled system because the string of thermal resistances from the device case to the coolant loop is less critical. That is because the thermal capacity of the liquid-cooling loop is large enough that the interacting secondary resistances among devices play a negligible role. This is not true when air-cooled sinks are employed.

Once the designer knows the power dissipation required in a cooler or cold plate and selects a flow rate, he can readily determine the rise in the cooler's water temperature by using the alignment chart in Fig. 8. As an example, for 1,000 w and eight gallons per minute, the rise is less than 1°C.

If the eight gallons per minute were to be split equally among four cold plates, each dissipating a kilowatt, the water temperature rise would only be 2°C. However, a careful analysis is mandatory because, in some cases, the temperature rise may be substantial.

Here are the parameters required to determine the thermal resistance (Θ) of a heat exchanger:

- Total power dissipated by the components that need cooling (P).
- Temperature of the water entering the heat exchanger ($T_{\text{water in}}$). The temperature drop from the cold plates to the heat exchanger should be subtracted if it is not negligible.
- Ambient air temperature ($T_{\text{air in}}$).

These values enable the designer to calculate the thermal resistance of the heat exchanger:

$$\Theta = (T_{\text{water in}} - T_{\text{air in}})/P \text{ in } °C/W$$

Once this thermal resistance is determined, the designer should check performance curves for various heat exchangers. These curves show that the flow rates of both the water and the air govern the performance of the exchanger.

It is likely that more than one type of heat exchanger will fulfill the cooling requirement. Selection can be narrowed by examining such factors as available space and position of inlet and outlet fittings. Finally, the pressure drop of the exchanger and all coolers, lines, and fittings must not exceed the pump capacity.

Selecting a pump

Once the heat exchanger and the coolers are selected, the drop in pressure through the cooler plates, the heat exchanger, and all interconnecting tubing and fittings is summed to determine the total head that must be deliv-

TABLE 1. LIQUID COOLERS
SOME REPRESENTATIVE THERMAL RESISTANCES

Type	Thermal resistance per device case-to-inlet water at 2 gallons per minute (°C/watt)
Liquid-cooled bus, 0.87-in. device interface, cooled on both ends	0.033
Same as above, but electrically isolated from the cooler	0.078
Channel plate 1 in.-diameter-devices, 2 in. center-to-center spacing, 0.038°C/watt case-to-sink impedance	0.38
High density cooler, TO-3 case style, thermal grease on device interface	0.30

TABLE 2. COMPARISON OF COOLING SYSTEM COSTS

System	Devices	Approximate cost
Forced-air convection	Two, pressure-mounted	$75
Liquid-cooled bus blocks, open-loop system	Two, pressure-mounted	$58
Liquid-cooled channel plate, open-loop system	Four, stud-mounted	$16
Liquid-cooled, high-density cooler, open-loop system	12, bolted	$21
Closed-loop system, cost to be added to liquid systems above	—	$138

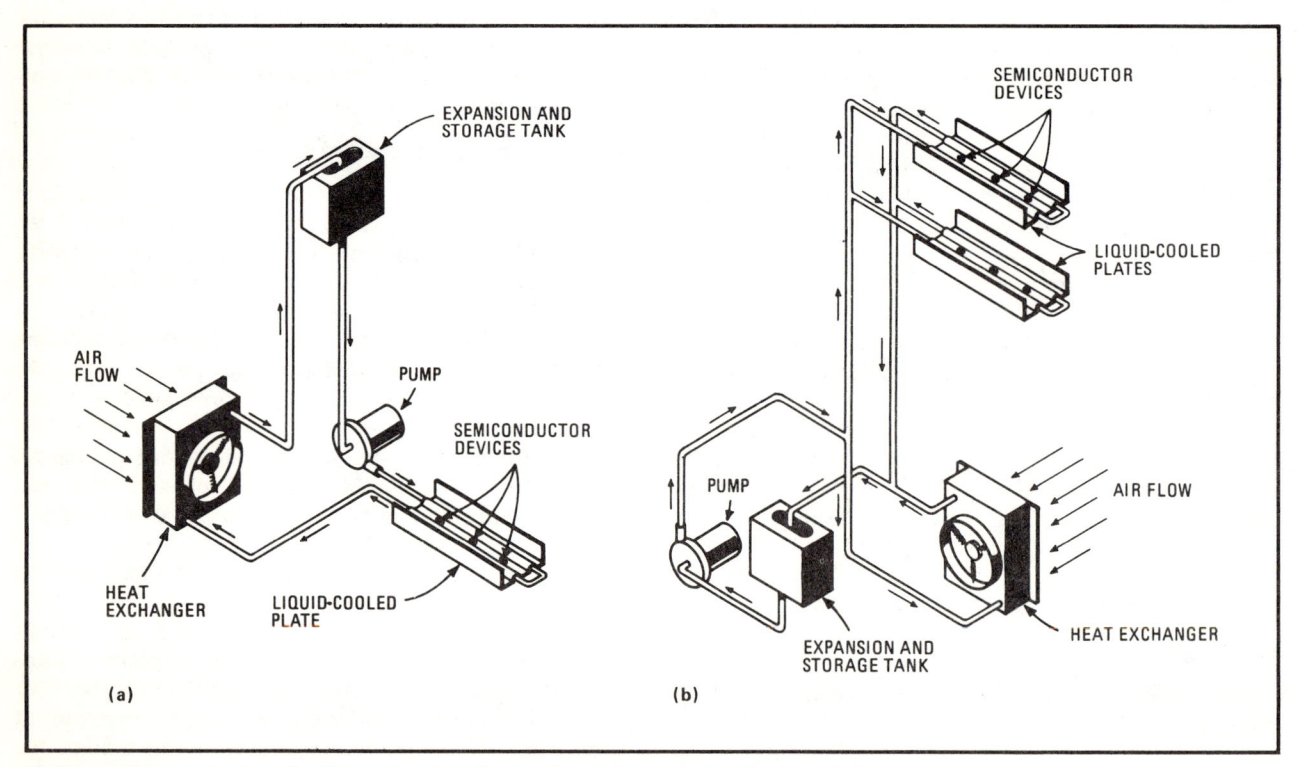

6. Flow system. A simple series-flow system is well suited for cooling a single cold plate (a). Connecting two or more liquid-cooled plates in a parallel-flow system (b) reduces the pressure drop so that a large-capacity pump isn't needed.

7. Cool exchanger. This double-pass heat exchanger transfers heat from the entering liquid to an air stream that is driven through the exchanger by a fan. The liquid enters and exits through the fittings above the fan. Raising the air and liquid-flow rates improves thermal performance. However, to minimize erosion and corrosion, liquid-flow rates through coolant lines should not exceed 10 feet per second.

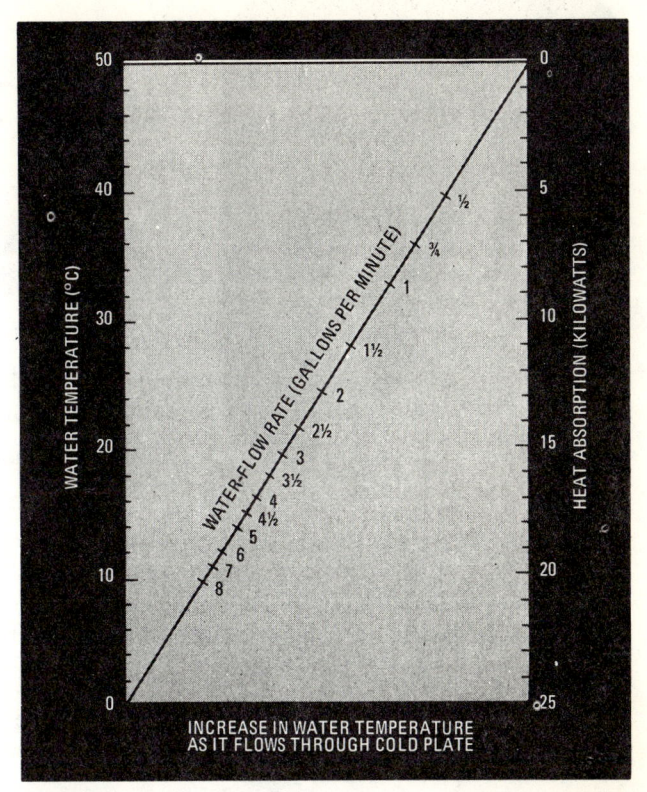

8. Warmup. By aligning a straight edge with values of the heat absorption and the water-flow rate in a cold plate, the temperature rise can be read from the scale at the left. A similar chart, found in heat-exchanger manuals, enables the designer to determine the temperature rise through a heat exchanger.

ered by the circulating pump. Head is the pressure, in pounds per square inch, delivered by a pump at a specified flow rate. Heat-exchanger manufacturers usually supply this necessary data. During the past several years, a number of centrifugal pumps have been developed to operate without rotating shaft seals so that long-term leakproof operation is assured. Flow capacities are sufficient for most electronic-package cooling.

If the pressure drop through a closed-loop system appears to be excessive, the parallel-flow system of Fig. 6(b) may be suitable. The advantage is that, unlike a series-flow system, the flow rate through each cold plate is not necessarily the same as the flow rate through the heat exchanger. And since drop in pressure through cold plates is usually much higher than it is through a heat exchanger, parallel connection limits pressure drop without significantly degrading cooling performance. Moreover, the flow rate of the heat exchanger can remain at a relatively high value, assuring high performance as a result of low thermal resistance.

Reliable transport of a liquid demands careful attention to both flow velocities and the materials contacting the fluid. Although copper tubing is relatively expensive, it offers the best envelope because the smooth wall surface resists corrosion in most environments. Copper also conducts heat well and resists the mechanical erosion and the chemical corrosion which are most severe at such points of high turbulence as sharp bends. Copper tubing, which is also easy to install, offers a good electrochemical and thermal match with other materials commonly employed in heat-exchanger and cold-plate construction.

Selecting the fluid

Water offers the best over-all coolant characteristics in terms of density, viscosity, thermal conductivity, and heat capacity. In closed-loop operation, where control of the content of the circulating fluid is possible, additions to compensate for losses of fluid can be made from time to time. Water that has been distilled, deionized, and demineralized provides the most efficient long-term performance. When both aluminum and copper-brass metals are present in a fluid circuit, specially—inhibited ethylene-glycol solutions can prevent deterioration of the fluid passages. However, because of their lower thermal conductivity, they do degrade thermal performance. Solutions of this type are mandatory where the ambient temperature can drop below the freezing point of water, or where surface temperatures exceed the boiling point of water.

Exotic dielectric oils are employed where severe electrical-insulation requirements team with freezing temperatures. Unfortunately, many of these oils, especially the chlorinated series, place severe demands on pump seals and plumbing joints in the fluid circuits. Again, even the best dielectric oils, as well as the series of silicone oils, require higher-performance heat exchangers than do water-cooled systems.

There are a number of variations in liquid-cooling systems, and one is the cold-sump system. This technique employs a refrigerant loop to cool a reservoir of refrigerated water, which is then circulated through a closed-loop cooling system. Cold sumps usually have a large cooling capability and may serve a number of heat loads simultaneously at remote locations. They are frequently selected for large complexes like computer installations. □

THERMAL DESIGN

Network analog maps heat flow

Equivalent lumped-parameter calculations by the finite-difference method resolves thermal factors early in the design cycle

by Carl J. Feldmanis, *Air Force Flight Dynamics Laboratory, Wright-Paterson AFB, Ohio*

□ It isn't easy to predict thermal conditions of an electronic system at the drawing-board stage. But knowledge of heat factors is crucial if a designer is to curtail the endless prototyping hours, which run up engineering costs. He must be confident that no temperature in the system will ever rise above some critical point. If it does, performance suffers, and devices fail. A well-engineered thermal system demands, then, that the designer have the analytical tools he needs to predict the thermal performance at every critical point.

A designer may approach thermal analysis from several directions. He may have designed a structure, has assumed a certain ambient temperature, and wants to check that no temperature in a device junction will exceed a critical value.

Or perhaps he may assume certain device temperatures and wants to compute the maximum thermal-resistance values that can be tolerated along certain thermal-resistance paths. Regardless of which variables are chosen as the dependent ones, the fundamental equations remain the same.

Fortunately, an effective method, called the finite-difference technique, is growing in popularity for solving such heat-transfer problems. The approach is to treat thermal relationships as analogs of electrical-network elements.

To proceed with this method of analysis, the designer first partitions the physical system into a number of subvolumes and places a node at the center of each volume element. Then he interconnects all nodes with the appropriate thermal impedances to simulate the flow of heat.

Thus, the thermal system takes on the appearance of an electrical network, and the designer can mobilize all the well-known methods for solving electrical networks, such as Cramer's rule, matrix inversion, and reduction by determinants.

The strength of the finite-difference approach presented here is that it enables a designer to form a circuit model of this thermal system and then vary resistances, power levels, and temperatures at will to achieve a satisfactory design. The network model need not be planar. It is crucial, however, for the designer to make sure that enough temperature nodes are selected so that the model becomes a workable approximation of the actual thermal system.

Computer simplifies calculations

A computer solution begins by converting the physical system into a network consisting of nodes connected by thermal impedances.[1] All the resistances and capacitances of the system are calculated and then presented in a form acceptable to the particular computer program.

Using this lumping process implies that the lump or node is at a uniform average temperature. By proper selection of the node size, any degree of accuracy can be obtained.

Program capacity should be considered, as well as expected temperature gradients and machine time. Unfortunately, there are no general rules for assigning the right volumetric size to each node. This consideration, then, becomes a matter of engineering judgment that improves with experience.

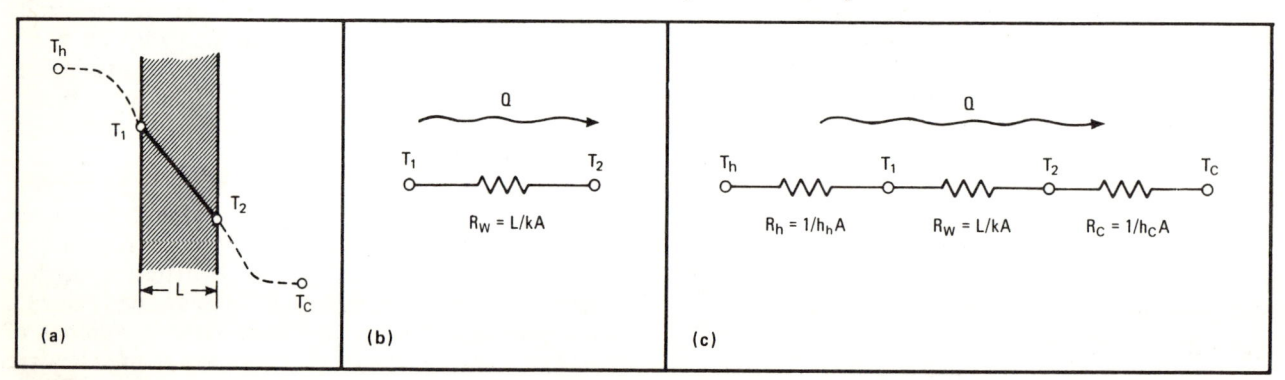

1. Willing wall. A planar surface of arbitrary area and thickness is the simplest geometry encountered in heat transfer (a). A resistor becomes the equivalent circuit element (b). Adding resistors for the convective interfaces (c) completes the circuit to the adjoining ambients.

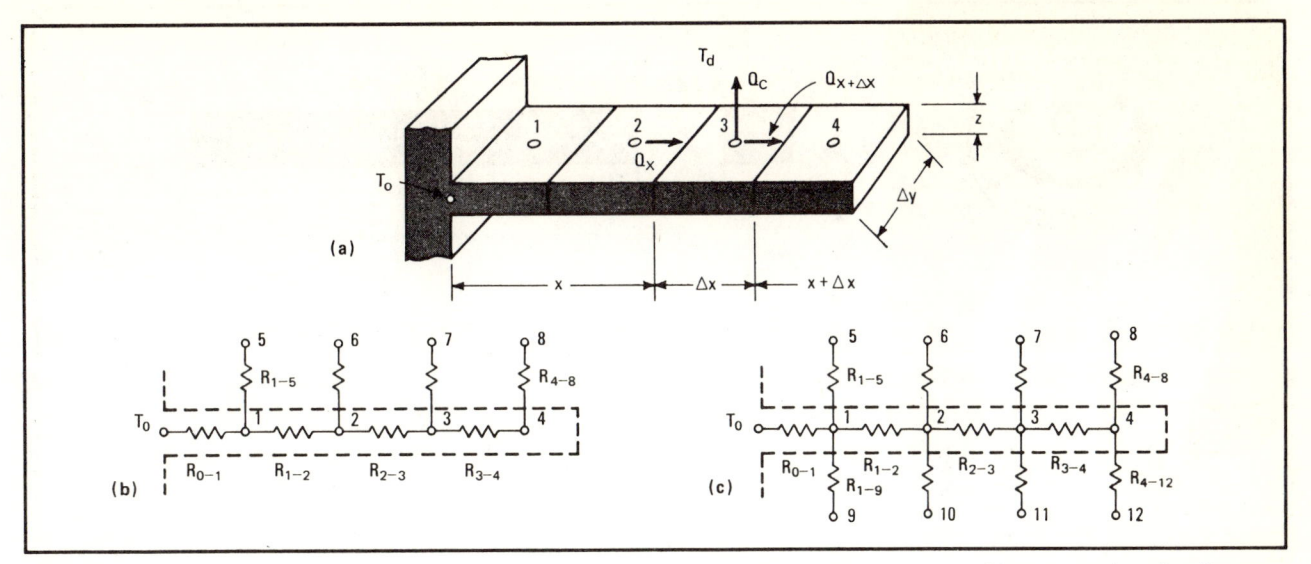

2. Fine fin. A thin wall projecting from a surface is an efficient convective interface (a). Network shown in (b) represents heat flow if convection occurs at top surface only. Adding resistors as in (c) simulates heat flow from top and bottom.

The heat-flow/current analog stems from a fundamental Kirchoff-law expression: the sum of the energy entering (or leaving) a node is zero:

$$\Sigma_i Q_i = 0$$

There are also terminal equations that relate the temperatures at the nodes to the heat flow between nodes. The flow is represented graphically by thermal impedances, or, in the steady-state case, thermal resistances are connected between the nodes.

The most time-consuming step is to convert the physical problem to a network equivalent, and a good starting place is to examine heat transfer in a simple geometric case—the plane wall.

Through the wall

Heat flow through a plane wall is analogous to current flow through a resistor. Figure 1 depicts flow from the warmer region at left to the cooler region at the right. The flow can be expressed as:

$$Q = (kA)(T_1 - T_2)/L$$

or

$$Q = (T_1 - T_2)/(L/kA) = (T_1 - T_2)/R$$

where Q is the heat-transfer rate in British thermal units per hour, k is the thermal conductivity in BTU/[(h)(ft)(°F)], T_1 and T_2 are the wall temperatures in °F, and L is the wall thickness in feet. L/kA is thermal resistance in (°F)(h/BTU).

Figure 1(b) is the circuit equivalent of the plane wall. In Fig. 1(c), two thermal resistors have been added to account for convection transfer at each wall. Note that the heat-transfer coefficients may be different (h_h, h_c), depending on the coolant flow and whether it is turbulent or laminar.

The convection-heat transfer to and from the wall can be expressed by the following equations:

$$Q = (T_h - T_1)/(1/h_h A) = (T_2 - T_c)/(1/h_c A)$$
$$= (T_h - T_1)/R_h = (T_2 - T_c)/R_c$$

Expression L/kA is called the conduction resistance, and expression 1/hA is called the convection resistance.

In any case involving heat transfer between two points at temperatures T_1, T_2, where $T_1 - T_2 = \Delta T$, the heat flow (analogous to Ohm's law) can be expressed as the equation:

$$Q = \Delta T/R$$

The relationship between the temperatures and the heat flow, therefore, for the plain wall with convection can be expressed as follows.

$$Q = (T_h - T_c)/[(1/h_h A) + (L/kA) + (1/h_c A)]$$
$$= (T_h - T_c)/(R_h + R_w + R_c)$$

A thin fin

The fin geometry is common in both natural and forced convection-cooling hardware because it offers a large surface area for convective-heat transfer. A fin attached to a hot wall or some heat-generating source is shown in Fig. 2. Heat from the hot wall flows along the fin, and from there, by convection, to the ambient air or other ultimate heat sink. Under steady-state conditions, heat entering the element must equal the heat leaving the element, or:

$$Q_{in} = Q_{out}$$

and

$$Q_x = Q_{x+\Delta x} + Q_c$$

where Q_c is heat lost by convection.

Heat entering the element at position x can be expressed in accordance with Fourier's law:

$$Q_x = -kA/\Delta x (T_b - T_a)$$

where $A = (Z)(\Delta y)$

Similarly, the heat leaving the element at position $x + \Delta x$ will be:

$$Q_{x+\Delta x} = -kA/\Delta x (T_c - T_b)$$

Heat leaving the element by convection will be

$$Q_c = (h\Delta A)(T_b - T_d)$$

where $\Delta A = \Delta x(\Delta y)$.

A designer may want to solve such a problem for any

247

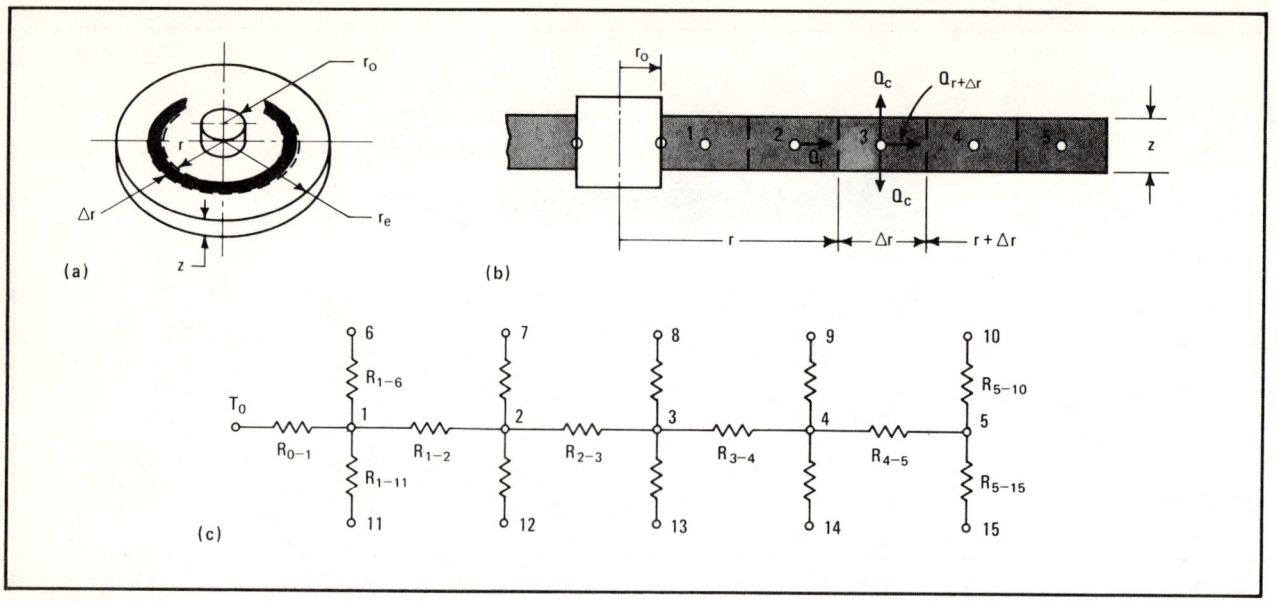

3. Cylindrical. Stacks of round, thin fins (a) dissipate heat efficiently. Analyzing designated concentric rings Δr wide (b) yields equivalent conduction resistors (horizontal) and convection resistors (vertical) arranged as shown in (c).

of several variables: heat flow in watts, temperature, or perhaps the heat-transfer coefficient (h). There are analytic techniques for the solution; however, solving a network analog is far simpler.

Figure 2(b) is the network representation of the fin when only one side of it takes part in transferring heat by convection; the other side and end are considered insulated.

Conduction resistances of the fin can be determined as follows:

$$R_{0\text{-}1} = \Delta x/2kA$$
$$R_{1\text{-}2} = R_{2\text{-}3} = R_{3\text{-}4} = \Delta x/kA$$

where $A = Z(\Delta Y)$.

Convection resistances are:

$$R_{1\text{-}5} = R_{2\text{-}6} = R_{3\text{-}7} = R_{4\text{-}8} = 1/hA$$

where $A = \Delta x(\Delta y)$.

If heat is dissipated by convection from both sides of the fin, then the circuit must be changed accordingly, as in Fig. 2(c). Thermal resistances are determined as in the previous example, except that half-thicknesses are assumed for determining the conduction resistances:

$$A = (Z/2)(\Delta y)$$

When heat is dissipated to the ambient air (assumed to be at constant uniform temperature) all the ambient nodes (5, 6, 7, 8, 9, 10, 11 and 12) can be combined into one node, which then represents the temperature of the ambient air.

A circular fin of rectangular cross section is another effective shape for transferring heat to a surrounding ambient. Its design ensures a large surface area for effective convection to the ambient.

Figure 3(a) shows such a fin with the heat source assumed to be located at its center. The easiest way to analyze the flow is to divide the fin into concentric cylindrical sections, or rings, with equal concentric thicknesses, Δr, as shown in Fig. 3$_b$. The network shown in Fig. 3(c) is a circuit equivalent when heat is convected from both the top and the bottom surfaces. Heat loss at the edge has been neglected. However, it could easily be included in the analysis.

A cold plate

The air-cooled plate has an excellent capability to sweep large quantities of heat away from board-mounted discrete devices. The plate shown in Fig. 4(a) consists of equipment-mounting surfaces with finned air-flow passages attached for forced-air cooling.

Figure 4(b) shows the network for both the plate section and the air stream. Note that resistors $R_{2\text{-}3}$, $R_{4\text{-}5}$, etc., have been drawn to account for the temperature rise of the moving air stream. The flow resistances are:

$$R_{2\text{-}3} = R_{4\text{-}5} = R_{6\text{-}7} \ldots = 1/c_p w$$

where c_p is the specific heat at constant pressure in BTU/lb/°F and w is the coolant flow rate in lb/h.

When the node regions are of equal size, then the convection resistances are:

$$R_{3\text{-}10} = R_{5\text{-}11} = R_{7\text{-}12} \ldots = 1/wc_p(e^\beta - 1)$$

where $\beta = hA_c/wc_p$. If the plate had no fins, then:

$$A_c = (\Delta x)(\Delta y)$$

However, since the air-flow passage has fins, the fin surface area must be added to the surface area of the plate. Figure 4(c) shows a section of finned plate. Note that heat from the baseplate flows along each fin that extends into the air stream, causing a temperature gradient to develop perpendicular to the flow. The gradient hinders effective heat transfer near the cover plate because convective transfer is dependent on temperature difference, and the difference between the cover plate and the coolant stream is quite small.

The term η is introduced to predict the convective heat-transfer capability of the structure. Then:

$$Q = hA\,\eta_0 \Delta T_m$$

where A is the total heat-transfer surface area.

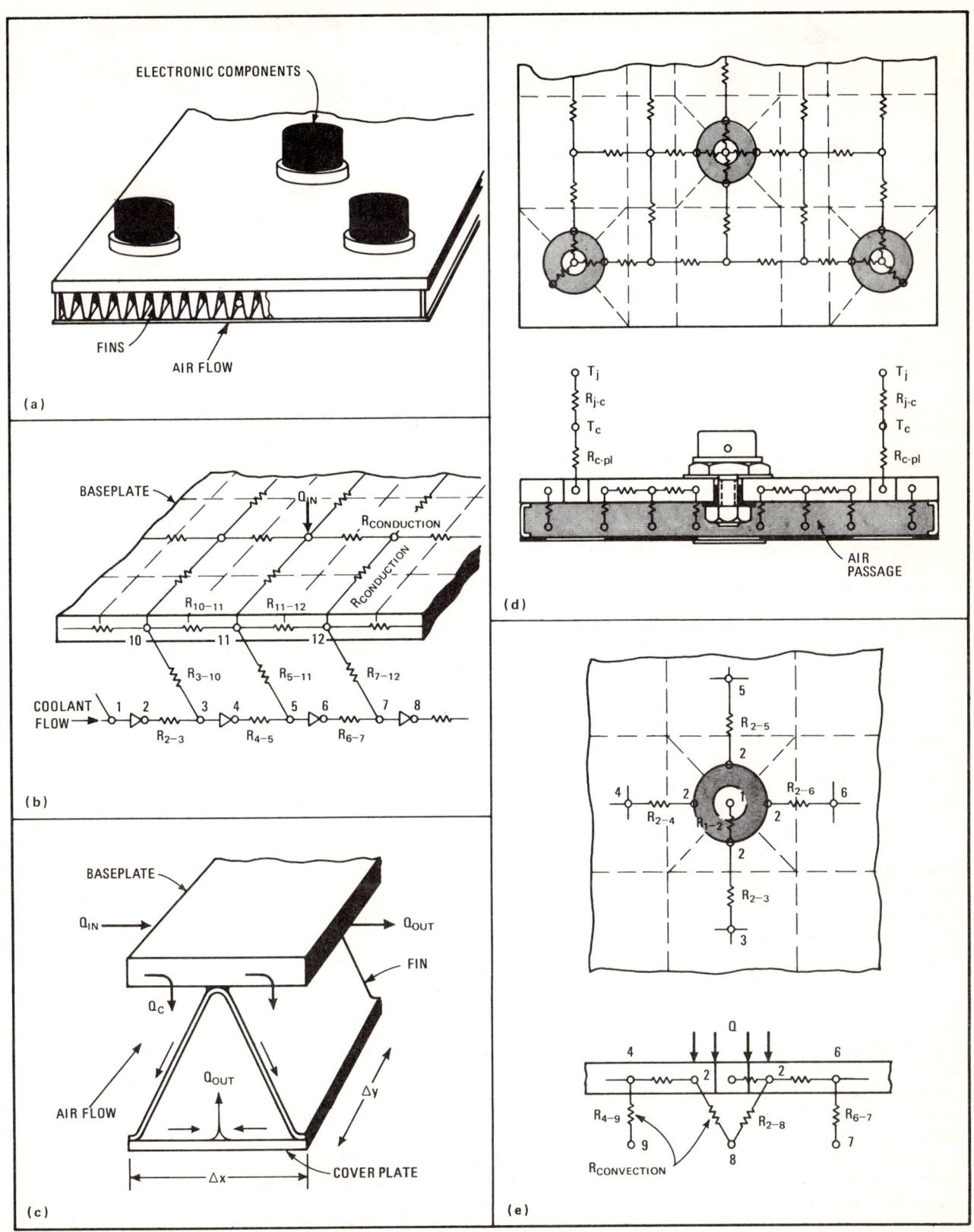

4. Cold plate. Fastening components to a plate, then forcing air through finned passages effectively dissipates the heat (a). Circuit diagram accounts for heat flow along the surface, as well as convective transfer to the coolant flow (b). Cross section of the passage denotes the various thermal paths (c). Top view of the plate (d) shows the thermal-circuit paths along the surface. Convective coupling resistors to the flow stream are drawn in the profile view. Cross-hatched region (e) is a critical path between the device and the coolant base plate.

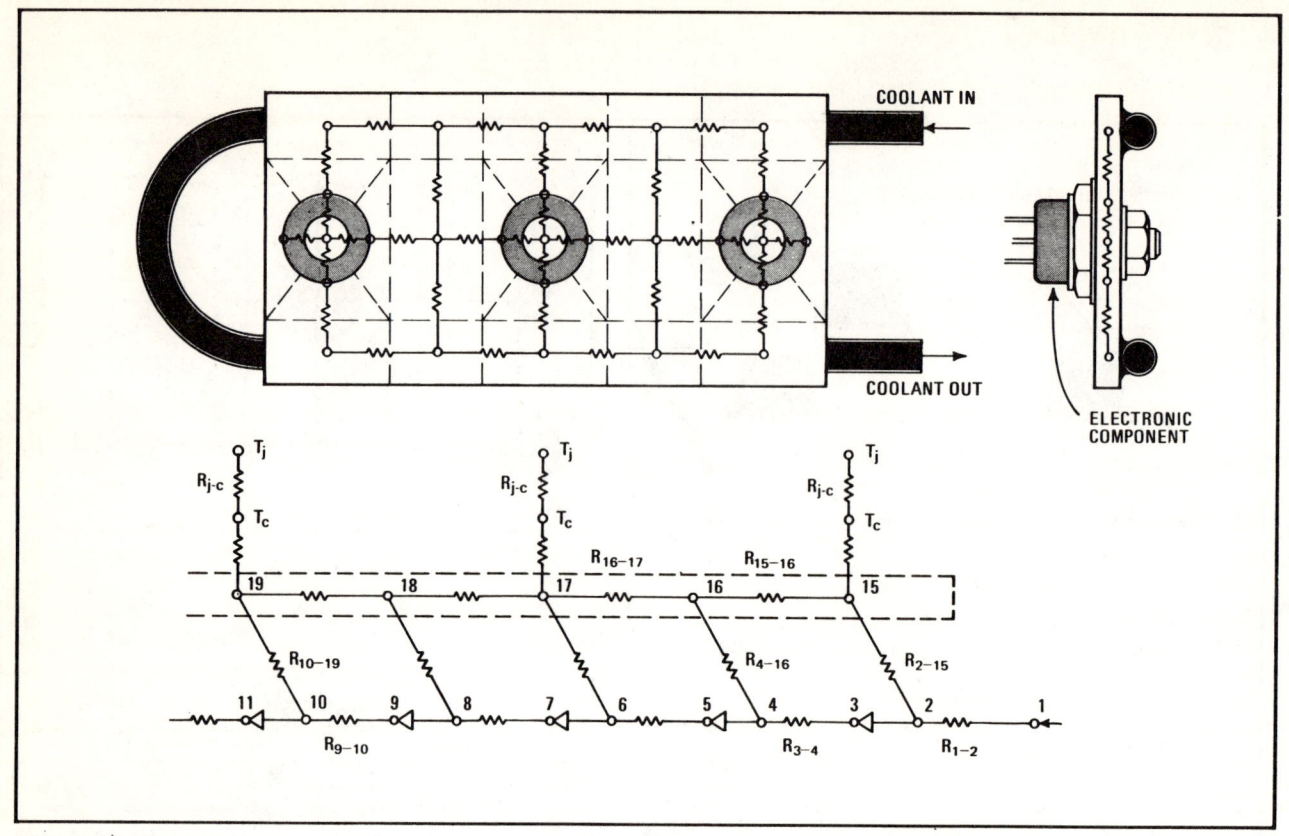

5. 'Round the bend. U-shaped coolant line attached to a plate forms a liquid-cooled cold plate. Circuit topography on surface is the same as that of air-cooled plate. Coupling resistors, drawn diagonally, represent heat flow from plate to coolant line.

For an asymetrically loaded cold plate, as shown in Fig. 4(a), the total heat-transfer surface area is expressed as:

$$A = A_b + A_f + A_c$$

where: A_b = baseplate surface, area, A_f = fin surface area, and A_c coverplate surface area

The over-all fin and cover-plate effectiveness can be determined from the following equation:

$$\eta_o = 1 - (A_f/A)(1 - \eta_f) - (A_c/A)(1 - \eta_c)$$

Where η_f and η_c are the effectiveness of the fin and base-plate, respectively.

Effectiveness of a flat, constant cross-section fin can be determined from an equation widely used in heat-exchanger design:

$$\eta_f = \tan h(ml)/(ml)$$

where $m = (2h/kz)^{1/2}$.

It is assumed here that the baseplate has an effectiveness of unity ($\eta_c = 1$) at each node.

Figure 4(d) is a network equivalent of a portion of the air-cooled plate. Once values are assigned to the known quantities, then network equations can be written. Techniques are covered in a number of volumes dealing with network analysis. The equations can then be solved by computer.

It may develop that such a network as is shown in Fig. 4(d) is not satisfactory for analyzing the configuration shown because it represents too crude an approximation. More nodes may be required.

Node count depends heavily on the thermal load placed upon the assembly. Concentrated high-heat loads may require more nodes, closely spaced around the component-mounting areas. If nodes are added to represent regions exposed to the cooling-air stream, they must be connected to the stream by convection resistances.

The boundary

A refinement that improves the accuracy of the network model is to model the thermal paths in the shaded region shown in Fig. 4(e). These paths are critical in the analysis because heat crosses the interface boundary along these surfaces and then develops a temperature gradient that lies in the plane of the baseplate. Four nodes have been assigned to the outer circle of the ring. However, these node points can be combined into a single node if the temperatures about the circle's circumference are assumed to be the same. When this is the case, this equation can be used to determine the thermal resistance of the annular region:

$$R_{1-2} = (r_2 - r_1)/(kA_m)$$

where A_m is a mean of the area defined as:

$$A_m = (A_2 - A_1)/[ln(A_2/A_1)]$$

The conduction resistances R_{2-3}, R_{2-4}, R_{2-5}, and R_{2-6} are the sums of the trapezoidal and rectangular node regions shown in Fig. 4(e). As far as the convection resistance is concerned, the part of the ring (node 1) covered by the washer and nut will not participate in the convective transfer. The active transfer area exposed to convection-heat transfer is the total rectangular area

> ## Why analyze?
>
> A designer embarks on the thermal-analysis course to make sure that no component in his system will exceed its maximum operating temperature. First thought usually goes to heat producers—semiconductors and resistors. But equally as important are nearby components, wires, and circuit boards.
>
> A useful analytic technique helps a designer answer the question: Is any critical temperature exceeded? If the answer is yes, he must alter parts placement and structural geometries; perhaps it may be necessary to substitute materials with lower thermal resistances so that heat transfer is improved.
>
> What complicates packaging architecture is that the thermal requirements are only part of the packaging designer's responsibility. He must also satisfy mechanical, electrical, manufacturing, and maintainability requirements—a huge order, indeed.

minus the covered position. This is represented by:

$$R_{2-8} = 1/wc_p(e^\beta - 1)$$

where $\beta = hA_2/wc_p$.

Natural convection and radiation also carry heat away from the external surfaces of the plate and the components, but they can be neglected if the forced-convection transfer coefficients are reasonably high.

Cooling with liquid

When powers levels soar into the kilowatt range, then designers often turn to liquid-cooling. This changes the analysis somewhat. Figure 5 shows a simple cold plate with three heat-dissipating components attached. Heat from the components enters the plate (shaded area) and travels to the liquid stream. Conduction resistances are determined in the same way as in the air-cooled systems. What differs is the convective-heat transfer. In an air-cooled plate, all nodes are exposed to forced-convection heat transfer, but in liquid-cooling, only the nodes adjacent to the coolant line connect to the coolant flow. In the network (Fig. 5b), the flow resistances are:

$$R_{1-2} = R_{3-4} = R_{5-6} = 1/wc_p$$

When temperature gradients through the coolant-line wall can be neglected, then the coolant-coupling resistances are:

$$R_{2-15} = 1/hA_c$$

where A_c is the convection-heat-transfer area.

A figure of merit, η, is introduced to weight the effectiveness of the area A_c:

$$R_{2-15} = 1/hA_c\eta$$

Heat capacity of the liquid must also be considered, and this capacity can be determined from the following expression:

$$C = \rho c_v A \Delta x.$$

where A is the cross section of the tube, c_v is the specific heat at standard volume, and ρ is the density of the fluid used for cooling.

Crossing the interface

Regardless of the mounting technique used, a thermal resistance always develops at the interface between a component and its mounting surface. There is a wealth of data available about idealized joints that have uniform pressure distribution, controlled flatness, and surface finish. However, since these ideal conditions are seldom realized, an analytical technique that considers these factors is required.[2,3,4]

Heat traveling across the boundary of two mating surfaces consists of the two components shown in Fig. 6. Heat flows through the contact points (k_c) and through the interstitial gaps (k_f). Therefore, the thermal conductivity is the sum of both terms:

$$K = k_c + k_f$$

The total heat transfer across the contact area is

$$Q = KA\Delta T = (T_1 - T_2)/(1/KA)$$

When only limited information about interface surface conditions is available, thermal conductance across the area of direct contact may be determined from the following expression:

$$K = 1.56k_A/(i_a + i_b) + 2n\bar{a}\,k_m$$

in which

$$k_m = (2k_1 k_2)/(k_1 + k_2)$$

where k_1 and k_2 are the thermal conductances of the mating materials, i_a and i_b are the root-mean-square values of surface irregularity (roughness plus waviness) for surfaces a and b, respectively, $\bar{a}$ is the average radius of the contact points, and n is the number of contact points per unit area.

If approximate values of i_a and i_b can be obtained, then $n\bar{a}$ values as a function of the stress distribution at the bolted interface may be determined from charts published several years ago.[3]

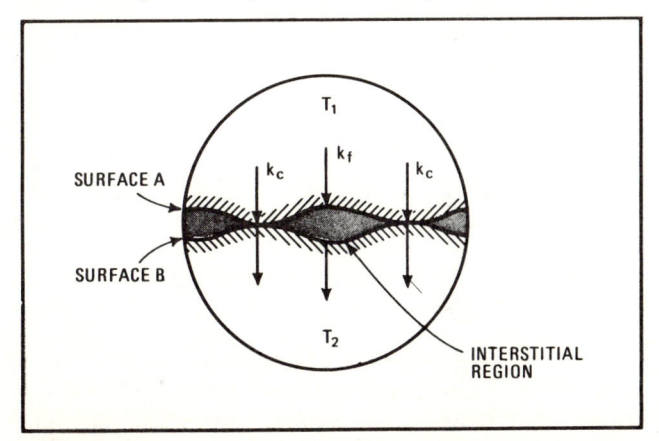

6. Bridgehead. Heat flow across mating interface of two materials travels both through contacting points and interstitial regions. Thermal coefficients k_c and k_f account for thermal coupling.

REFERENCES
1. NASA CR-65581, "Thermal Analyzer Computer Program for the Solution of General Heat Transfer Problems," NASA Manned Spacecraft Center, Houston, Texas, February 1967.
2. K.G. Lindh et al, "Studies on heat transfer in Aircraft Structure Joints," University of California, Report No. 57-50, May 1957.
3. J.E. Fontenot, "Thermal Conductance of Contacts and Joints," Boeing Document No. D5-12206.
4. C.A. Whitehurst, "The Thermal Conductance of Bolted Joints," NASA CR 94738, May 1968.

THERMAL DESIGN

Air through hollow cards cools high-power LSI

Providing parallel flow through hollow-core cards and wafer-mounted heat exchanger can cut temperatures at no cost in space

by Lou Laermer, *Singer Co., Kearfott Division, Wayne, N. J.*

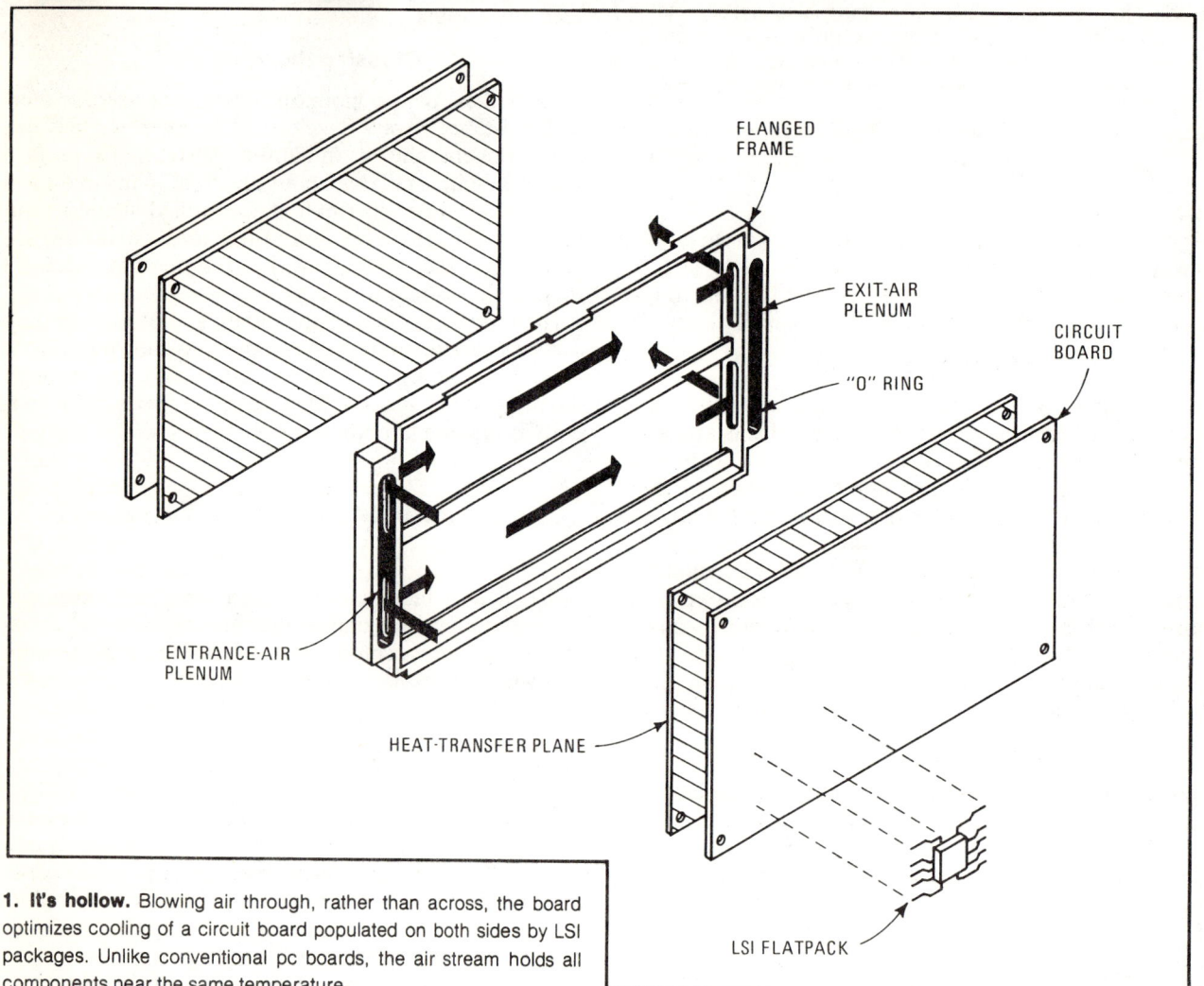

1. It's hollow. Blowing air through, rather than across, the board optimizes cooling of a circuit board populated on both sides by LSI packages. Unlike conventional pc boards, the air stream holds all components near the same temperature.

□ Circuit designers are excited about the tremendous functional capability of LSI, and they are constantly trying to increase the number they can pack on each printed-circuit board. However, dense packing of thousands of active, heat-producing devices on a square inch of circuit card places an awesome burden on the packaging engineer.

Traditional cooling methods are becoming increasingly inadequate, especially with densely packed high-power LSI devices, because power density, in watts per cubic inch, is much higher, and thermal paths from the heat-producing devices to the cooling medium are too long. It's not unusual for a logic card of 25 square inches, which once dissipated 2.5 watts, to dissipate 20 W when mounting LSI and MSI devices.

However, thermal paths can be shortened and temperatures of device junctions held well below safe values by using a patented hollow card so that the heat exchanger becomes an integral part of the circuit card. Air circulates through a channel between the two circuit

2. Potent package. Airborne computer packs 35 hollow cards containing over 2,500 flatpacks. Total power dissipated is over 400 watts, but device temperatures never climb above 75°C, thereby enhancing long-term reliability. Slots shown in the top and bottom of the cards form the entry and exit air plenums.

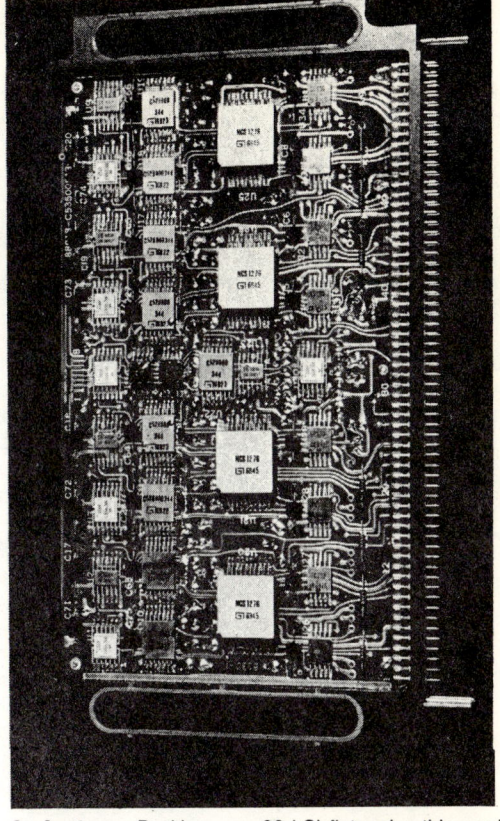

3. Anatomy. Packing over 60 LSI flatpacks, this card dissipates over 20 watts, almost 10 times more than its conventional printed-circuit-card ancestor. Cool air enters at left and exits from the plenum at right.

cards mounted back-to-back, and results are truly astounding. What's more, the hollow configuration weighs no more than a conventional card cooled in a conventional way.

Better still, thermally, is the basic building-block module (B³M), which Singer-Kearfott has designed for the U.S. Naval Air Systems Command. The module, which has eliminated the circuit card altogether, lowers the temperature of the IC junction from 141°C to a safe 68°C, while the ruggedized package is easily accessible and easy to interconnect.

Design objectives

Clearly, two factors discourage the conventional design approach: component temperatures are highly dependent upon card location in the chassis, and the temperature rise across the horizontal span of each card is too great because ICs near the center of the card build up intolerable temperatures.

True, a designer could use heat pipes, but they are expensive. A better and cheaper solution is to introduce cooling air at a common temperature to each card and then circulate the air directly through each card. This can be done most effectively by circulating air through a hollow-core card.

The configuration of the hollow-core card provides improved cooling by:
- Eliminating the thermal conducting path across the breadth of the circuit card.
- Replacing series air distribution with parallel air distribution.
- Increasing convection area per circuit card.
- Increasing convection effectiveness.

A design for a hollow-core card that satisfies design objectives for a high-power, high-density system is illustrated in Fig. 1. The assembly is actually a sandwich of two cards, mounted back-to-back on a flanged frame, which separates the cards to create a channel that allows cool air to flow across their rear surfaces. Bonded to the back of each card is a conductive heat-transfer plane, which serves as the convective interface.

The air enters a plenum on the left side of the frame and exits at the right. The reason that cooling air entering the inlet plenums of cards positioned at increasing distances from the air source does not get hotter is that the temperature gradient along the main air-cooling stream, perpendicular to the cards (the left side of Fig. 1), is virtually zero. The transverse flow rate at the entrance to each card is determined by careful selection of the cross-sectional area of the entrance and exit air plenums.

The hollow cards are clamped together by straps between their front and rear panels. The upper pair appears in Fig. 2. The lower straps (not shown) serve as a subchassis that supports the motherboard and the mating connectors for each card.

A close-up of a hollow-card assembly that mounts LSI flatpacks is shown in Fig. 3. Note that gaskets line the edges of the cooling-air entry and exit plenums to prevent air leakage. Besides assuring a uniform temperature at each card-inlet plenum, parallel cooling maintains a virtually constant air-pressure drop, regardless of the number of cards. In the traditional chassis-type heat exchanger, the card interface temperatures increase as cards are added so that the cooling effectiveness falls off

LSI turns up the heat

When it was populated by discrete components, the card shown below cooled the circuits on it admirably. But when large-scale integration multiplied the power density to as much as 500 milliwatts per square inch, this configuration could no longer fill the bill. Originally designed for an airborne computer, the chassis contained 35 cards that dissipated a total of 85 watts. A power supply raised the burden of thermal dissipation by another 65 W.

Circuit cards, some built on aluminum cores, conduct heat left and right to the air-cooled heat exchangers, which double as chassis walls. The card guides also serve a vital secondary role—carrying heat from the card to the exchangers. When each card dissipated 2.5 W, the cooling air could keep temperatures below a safe 75°C. However, when each card is packed with 60 LSI flatpacks, each measuring 0.25-inch square, the power on each card is boosted to 20 W, which drastically increases the amount of heat that must be dissipated.

If cooling air at 20°C is forced through the exchangers at three pounds per kilowatt, the temperatures at various points in the chassis will reach the temperatures assigned to the node points in the illustration. There is a colossal rise of 49°C laterally across the card. Although the edge temperature is only 39°C, the center of the near card is 98°C. The temperature of the last card at the rear, near the outgoing air, rises to 131°C—well above the tolerable levels for long-term IC reliability.

A designer could improve heat flow in the existing design by increasing the card-core thickness and by using a wedge-type card clamp, which would improve thermal conductivity. But this effort won't lower IC temperatures very much. Even a tripling of the card-core thickness fails to lower maximum IC temperatures below 92°C— too high for long-term reliable operation. Moreover, thickening the card is a costly tradeoff because it doubles the card weight and enlarges its volume by 30%.

Fortunately, the hollow-core card and the basic module with its integral exchanger are breakthroughs in thermal architecture. They both enable cooling air to circulate effectively and thereby provide the parallel air to hold densely packaged LSI devices at low operating temperatures.

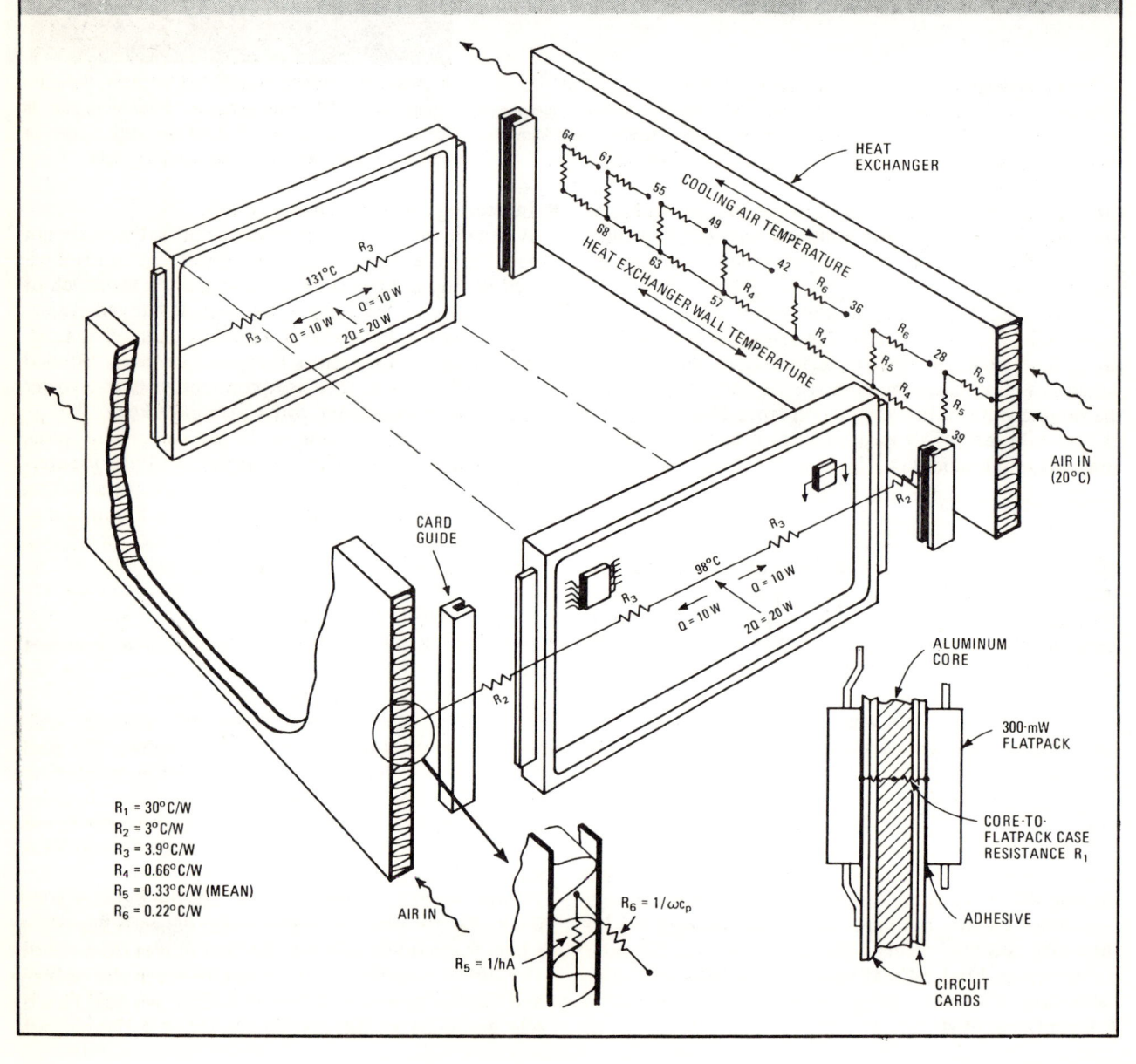

$R_1 = 30°C/W$
$R_2 = 3°C/W$
$R_3 = 3.9°C/W$
$R_4 = 0.66°C/W$
$R_5 = 0.33°C/W$ (MEAN)
$R_6 = 0.22°C/W$

$R_5 = 1/hA$
$R_6 = 1/\omega c_p$

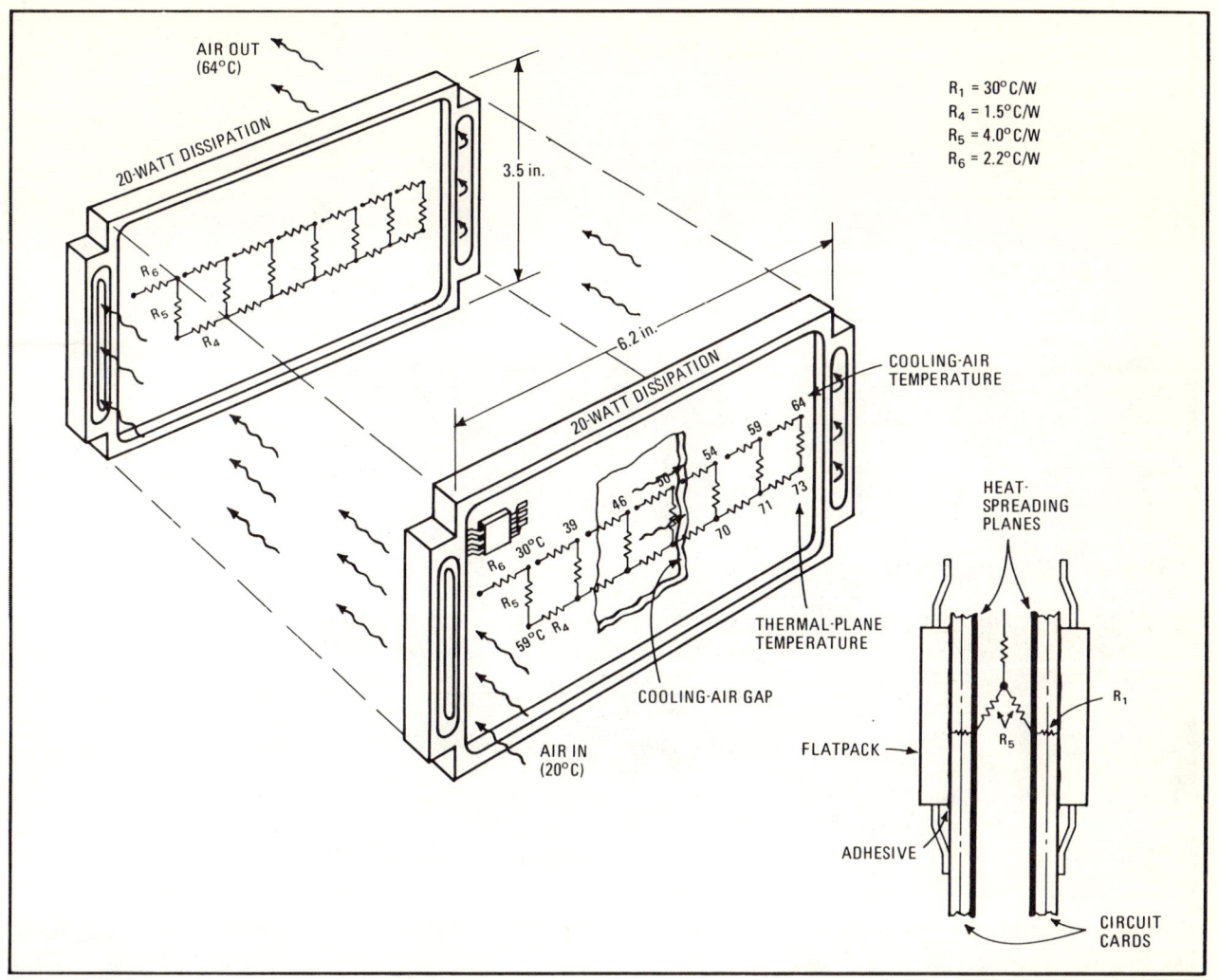

4. Flow path. Entrance air at 20° C distributes to each of the cards and exits at 64° C. Air temperature at entrance plenums of all cards is virtually the same. Circuits depict thermal paths. Resistor R_5 accounts for the thermal resistance of the convective interface.

as the distance from the air intake increases.

The hollow card successfully lowers component temperatures below what is obtainable in conventional designs. Using a design rule of three pounds per minute of cooling air per kilowatt, a card dissipating 20 W is allocated 0.06 lb of cooling air. As shown in Fig. 4, the air's exit temperature is 64°C if the inlet temperature is 20°C. The circuit card's thermal-plane temperature range is from 30°C to 64°C, and maximum and average component-case temperatures are 82°C and 74°C, respectively.

Table 1 compares case temperatures and indicates that the hollow card provides significantly lower component temperatures because the hollow region assures that cooling air is brought within close proximity of the heat-dissipating devices.

Convection efficiency increased

As an additional benefit, the geometry of air-core cards boosts convection efficiency. The convection coefficient, which is a function of air speed, rises because the shortened cooling path speeds air flow through the hollow card and also prevents buildup of a static boundary layer, which hinders heat transfer. The path in the air-core card is 6 inches long, compared to 15 in. for the conventional card.

Finally, the surface area of the hollow card presents 20% more convective area to the moving air stream than does conventional designs. If cooling is still inadequate, a designer can furthur enlarge the convective area by adding fins along the surface. A finned exchanger becomes practical when the power dissipation per card exceeds 25 w, not unusual in power supplies.

A system approach

Efficient as the hollow card is, one more improvement can be made. That's to reduce the resistance of the path from the chip to the thermal plane of the card.

In the LSI flatpack, junction-to-case thermal resistance ranges from 20°C/w to 75°C/w so that if a package dissipates 300 milliwatts, the junction temperature rises 6°C to 22°C above the case temperature. Junction-to-case thermal resistance is a major contributor to temperature rise, and if not lowered, can be a significant factor in loss of reliability.

Improving the thermal path within the flatpack is difficult because effective heat transfer depends heavily on a lateral spreading effect as the heat moves from the device junction toward the interface between the package and the circuit board. Attempts to improve heat flow by selecting a better thermal conductor or a thinner sub-

5. A cool wafer. Efficient thermal package houses a 3-inch LSI wafer (a). Component parts (b) include an alumina or beryllia heat exchanger that fastens directly to the base ceramic, optimizing cooling. Substituting more costly beryllia enhances thermal conductivity by a factor of 12.

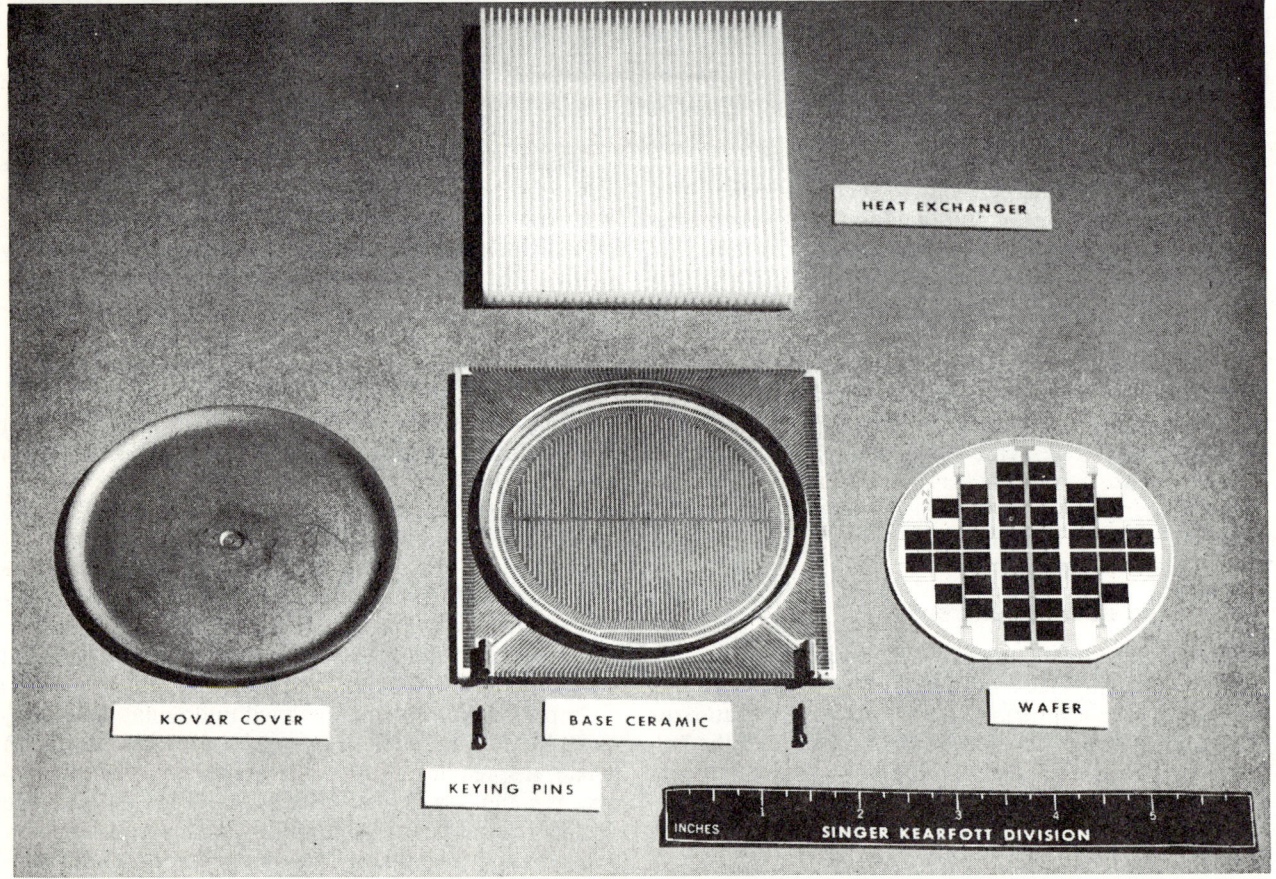

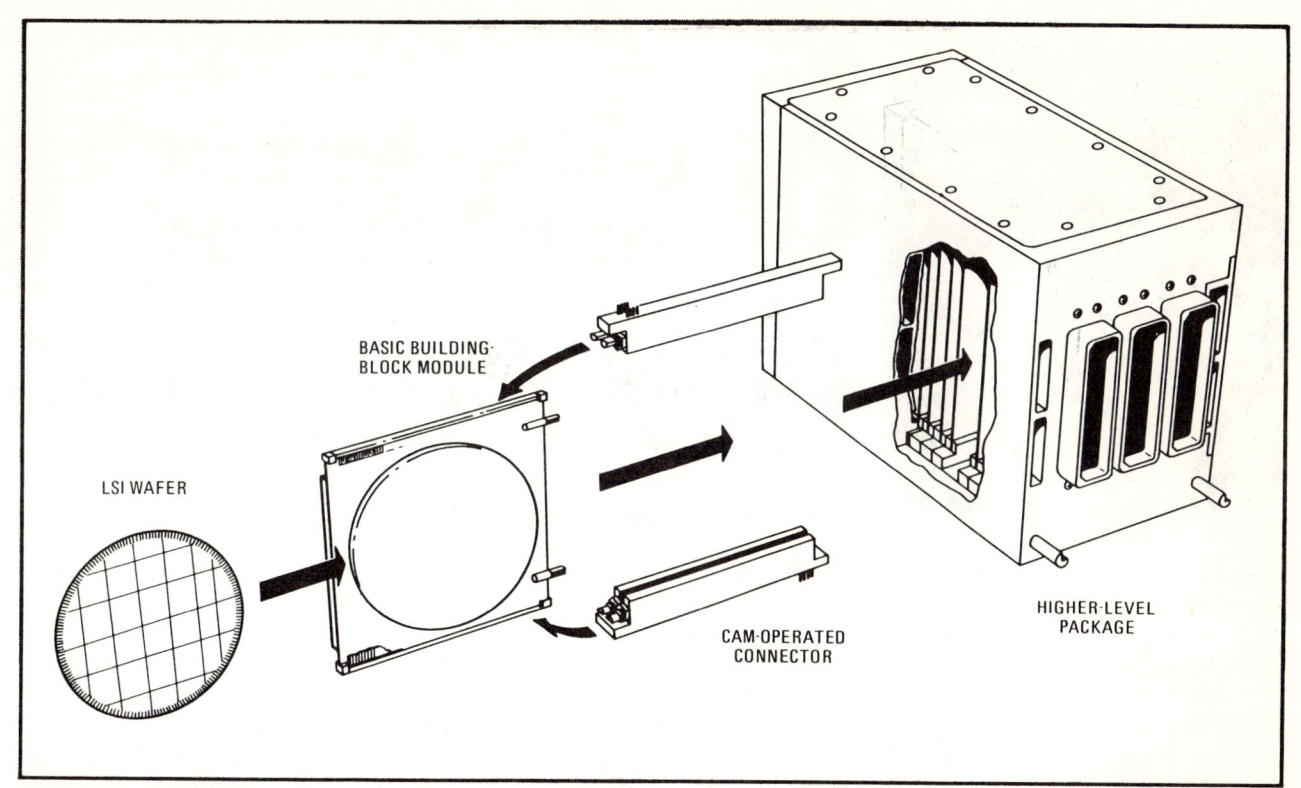

6. Full-wafer packaging. This airborne computer houses modules containing 3-inch wafers, doing away with the printed-circuit-card construction and holding device junctions below 68° C. Cam-operated connectors eliminate engagement force. At 400 W, higher-level package dissipates almost three times the power of an earlier computer—with no increase in package volume.

COMPONENT CASE TEMPERATURES — 20 WATT CARD		
	Hollow card (°C)	Conventional card (0.05-in. aluminum core) (°C)
Maximum IC case temperature	82	131
Average IC case temperature	73	115
Cooling-air temperature rise	44	44
Flow rate 3 lb/minute per kW — 20°C inlet air temperature		

strate material seldom lower thermal resistance very much. The problem requires a novel solution.

Extraordinarily potent in its ability to lower junction temperatures is the structure shown in Fig. 5(a). This package, the B³M, offers junction temperatures 23% lower than even the hollow card, and it can dissipate as much as 50 w. What is so unusual about this package is that it does away with the circuit board by marrying a heat exchanger directly to the active IC devices.

The (B³M) stems from a development program sponsored by the U.S. Naval Air Systems Command for the all-applications digital computer, designed to fulfill military and space requirements that are now anticipated for the latter part of this decade.

The module is designed to hold an LSI wafer 3 inches in diameter that has a complexity equivalent to more than 5,000 gates. Alternately, it can house a hybrid substrate 3 in. in diameter that contains a multiplicity of LSI chips and passive devices mounted on a multilayer thick-film substrate.

The key to the excellent thermal capability of this module is the ceramic heat exchanger shown in Fig. 5(b). The heat exchanger cements directly to the alumina-base ceramic, ensuring a very short thermal path from the chip to the cooling air stream. Interrupted fins can also be used to prevent static air boundaries from forming, and the reward is a high film-convection coefficient.

Substituting more-costly beryllia for alumina in the heat exchanger lowers thermal resistance still more—by a factor of 12—thereby lowering junction temperatures another 8°C. The combination alumina-beryllia heat exchanger lowers the lateral resistance so that hot spots are less likely to develop on the chip.

Singer-Kearfott's higher-level package, made up of basic building-block modules, is shown in Fig. 6. Airflow paths are much like those shown in Fig. 1. Air flows from left to right through the heat exchanger channels on each module. Again, flow rates and inlet-air temperature are independent of card placement, offering the designer great flexibility in arranging the configuration.

The basic module circulates cooling air where it belongs—in intimate contact with the IC. Doing away with the circuit card lowers IC-junction temperatures approximately 20°C. If one applies the rule of thumb that each 10°C of lower temperature doubles the mean time between failures, the life of each IC has been lengthened by a factor of 4. Such an enhancement clearly supports the role of sound packaging design in the development of high-power-density electronic systems. □

THERMAL DESIGN

Heat exchangers cool hot plug-in pc boards

When device power levels and packing densities rise, the thermal deficiencies of printed-circuit boards must be compensated by efficient heat exchangers

by Benjamin Shelpuk, *RCA Corp., Camden, N.J.*

□ On most counts, the plug-in printed-circuit board deserves its status as today's unofficial industry-wide standard. Mounting vertically in an equipment case, it can easily be withdrawn when replacement is necessary. Yet it is well protected from shock and vibration, being held rigidly in place by card slides.

Thermally, however, the plug-in printed-circuit board is much less impressive. Neither epoxy-glass nor paper-based boards are good heat conductors. Also, the thermal paths from hot devices on the boards to the outside world are often long and hinder cooling.

Proof of the board's inadequacy as a heat conductor is that a temperature gradient of 707°C is required to drive just 1 watt of heat through a piece of epoxy-glass board only 1 inch square and 20 mils thick. This determination was made from an equation that enables the designer to calculate thermal gradient whenever heat flow can be considered unidirectional. The equation is:

$$\Delta T_{max} = QL/8Ktw$$

where

ΔT_{max} = temperature gradient to the hottest spot in the board, in degrees centigrade

Q = heat transfer by conduction along the board, in watts

L = span of the board between card guides, in inches

K = thermal conductivity of the board, in watts per inch°C

t = board thickness, in inches

w = length of each interface between board and card guide, in inches.

The equation assumes uniform power dissipation over the surface of the board and is realistic if the designer has optimized both heat spreading and component location on the board's surface.

One way to improve heat flow through a board is to use the copper conductors on its surface to transfer heat. Being a fine thermal conductor, the copper lowers thermal resistance significantly—though precisely how much it is lowered is difficult to calculate because the pattern etched into the conductor markedly reduces heat transfer. For instance, if just 10% of the copper is removed from a fully-clad board, thermal resistance of the overall board can increase by a factor of 17.[1]

Materials other than epoxy-glass can be used for pc-board construction to upgrade their heat-transfer char-

1. Destined for the moon. Rarefied atmospheres deny package designers the advantage of convective cooling. This assembly, part of a radar system carried on the Apollo 17, employs a highly conductive frame to absorb heat from the printed-circuit board.

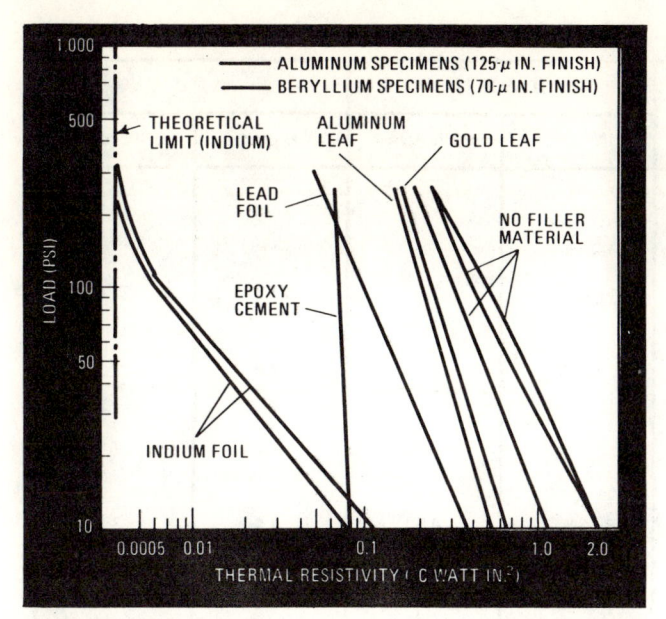

2. Beware of the boundary. Fillers between printed-circuit boards and the card slide, and high-compression forces aid heat flow across the interface. Data is based on research performed by MIT Instrumentation Laboratory.

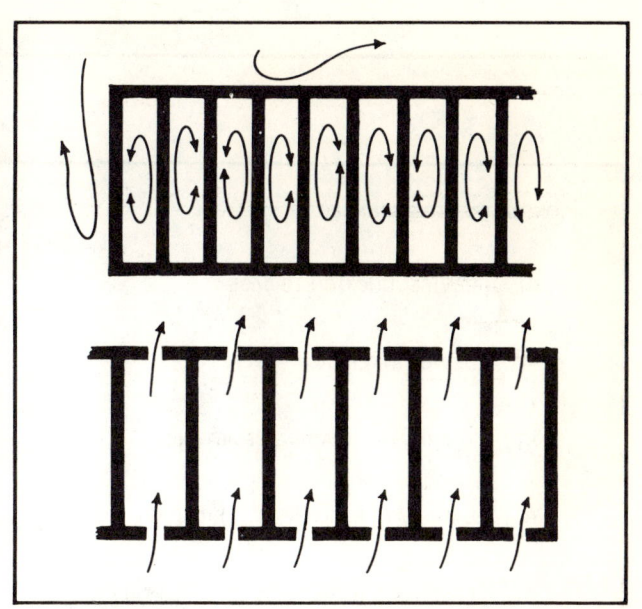

3. Alternatives. Closed convective system shown in (a) prevents contaminants from being swept in by a moving air stream. But switching to an open system (b) boosts cooling capability per unit volume tenfold—10 watts/in.³ versus 1 watt/in.³

acteristics. But generally they fail to improve heat transfer enough to compensate for the electrical constraints they impose. Instead, it is frequently better to switch to a heat-conducting frame to support the pc board.

The heat-conducting frame

This technique was used to good effect for the Apollo 17, in a pc-board assembly that was part of the coherent synthetic aperture radar (CSAR). Figure 1 shows details of that assembly. Effective conductive cooling is a must in space, where the lack of atmosphere robs the designer of convective cooling. In the CSAR assembly, heat flowed from the board to the housing through the threaded bosses on which the board was mounted. Maximum temperature rise was kept low because the thermal path to a boss from any heat-producing component was kept short.

The Standard Hardware Program (SHP) developed by the U.S. Navy also utilizes heat-conducting frames to guarantee adequate heat transfer from its modules.[2]

However, only a poor thermal path from board to frame is provided by the usual card slides. The problem is that ease of maintainability and accessibility demands boards that slip easily in and out of card slides—but the thermal interface between such boards and slides is not good. Fortunately, card slides can often be modified to provide a large positive area of contact that will optimize heat flow across the interface.

Figure 2 summarizes the results of some interface resistivity studies.[3] It shows that various filler materials can be used to lower the thermal resistivity of board/slide interfaces. Note also the negative slope of the plots, which denotes that high compressive forces along the interface also lower thermal resistivity.

Enter the ambient

Regardless of how effectively such conduction paths are enhanced, convective transfer to the ambient fluid (usually air) often emerges as the principal heat-transfer mode in electronic equipment.

Two geometries are common in convective transfer. Figure 3a illustrates a closed system in which transfer is in effect a two-step process. Heat is moved from the board surfaces to the surrounding air by natural or forced convection, and the air is then cooled by natural- or forced-convection transfer to the equipment case.

In the open system shown in Fig. 3b, the air is not entrapped, but enters the enclosure, sweeps across the pc boards, and then exits carrying the heat to the environment. There is no intermediate transfer to and from the equipment case. But such a system is often unacceptable because it can transport moisture and other harmful contaminants.

In either type of convective system, orientation and spacing of the boards play an important role in determining component temperatures. So do the flow rate and temperature of the cooling medium. Table 1 lists the range of typical heat-transfer rates for both open and closed plug-in pc-board designs. Note that the power density for a well-designed closed system where the exterior cooling is by natural convection ranges

TABLE 1: HOW COOLING MODE AFFECTS POWER DISSIPATION				
System type	Internal cooling mode	Exterior cooling mode	
^	^	Natural convection (W/in.³)	Forced convection (W/in.³)
Open		0.5 – 1	5 – 10
Closed	Natural convection	0.1 – 0.25	0.2 – 1.0
Closed	Forced convection	0.2 – 0.8	0.5 – 3.0
Closed	Conduction	0.4 – 1.5	2.0 – 4.0

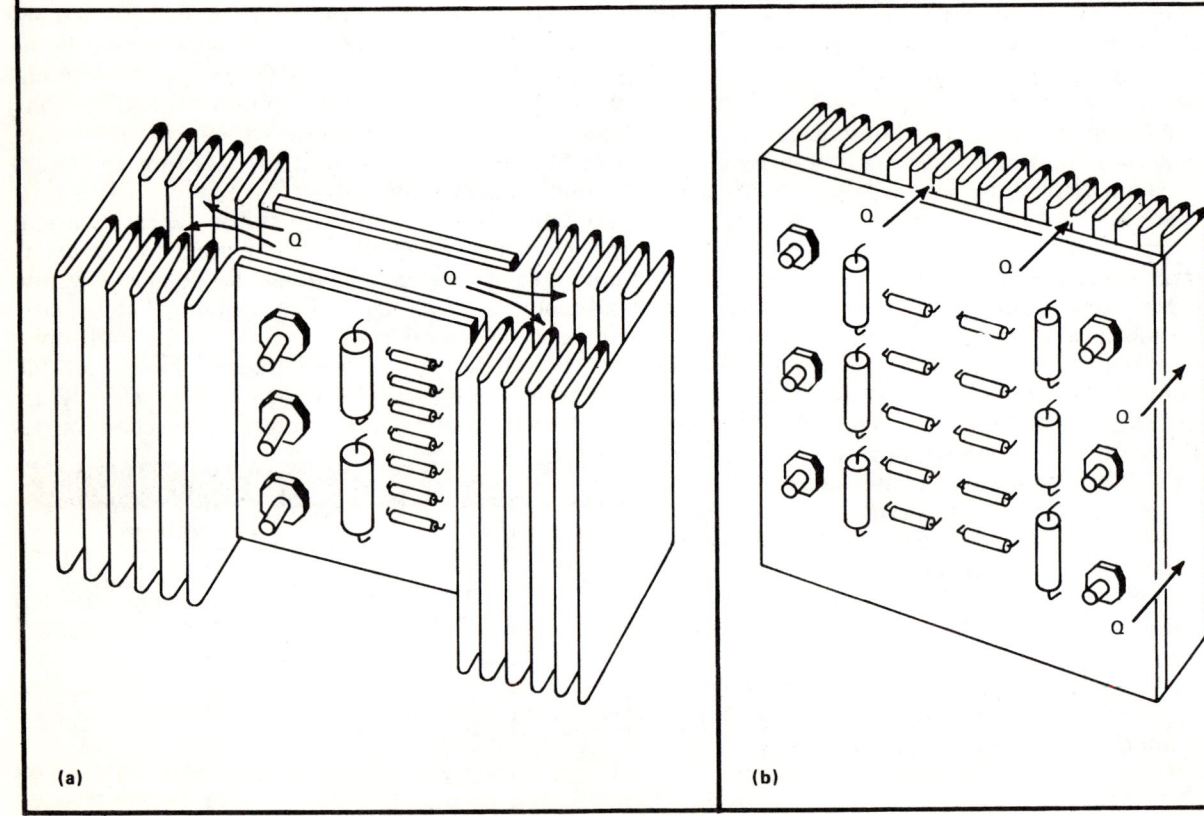

	RESISTANCE (°C/W)		ΔT°C	
	IN-LINE	COPLANAR	IN-LINE	COPLANAR
CONDUCTIVE RESISTANCES				
R_{TJ} DEVICE-JUNCTION TO CASE	2.5	2.5	27.5	27.5
R_{T1} DEVICE-INTERFACE	1.12	1.12	12.3	12.3
R_{T2} THERMAL SPREADING RESISTANCE	0.65	0.65	7.1	7.1
R_{T3} BASEPLATE RESISTANCE (VERTICAL)	4.83	4.83	3.4	3.1
R_{T4} BASEPLATE RESISTANCE (HORIZONTAL)	1.97	—	39.8	0
CONVECTIVE RESISTANCES				
$T_{SO} - T_{SI}$ COOLING AIR TEMPERATURE RISE				26.3
$1/hA$ CONVECTION TRANSFER RESISTANCE	—	0.53	—	29.0*
*INCLUDES $T_{SO} - T_{SI}$				
TOTAL TEMPERATURE RISE °C			90.1	79.0

4. Go coplanar. The in-line construction (a) accounts for the large temperature rise—90.1°C. By contrast, the heat-transfer path in the coplanar structure (b) is very short, and temperature rise is significantly less—66.1°C.

from 0.10 w/in.³ to 1.5 w/in.³. Also, for the internal forced-air cooling modes, total volume must not be so large that the space consumed by blowers and ducting becomes a significant fraction of the total volume. Otherwise, the listed values become invalid.

Looking at one design

Assume that a designer attempts to house a 100-watt uhf radio transmitter-receiver combination in a standard cabinet designed to mount printed-circuit boards. Detailed analysis of a particular design reveals that the maximum power dissipation that can be rejected in such a cabinet (4.87 in. wide by 7.62 in. high by 19.56 in. deep) is limited to 56 w at sea level and to 28 w if the equipment is operated at high altitude, where there is little convective cooling. Clearly, plug-in pc-board construction would not be appropriate for this equipment.

The power dissipation of the equipment, broken down by its component modules, is given in Table 2. Checking the power densities of each module against the criteria of Table 1 indicates that forced-air convection is necessary in two modules—the transmitter and the power supply. Since the equipment is intended for the military, however, an open system with forced-air convection directly over the circuit cards would be unacceptable because of possible contamination. So a closed-system, forced-air cold plate is a likely alternative.

Forced-air cooling differs from natural convection in that the driving force circulating the air is a mechanical pump rather than natural buoyancy induced by temperature gradients. This significantly increases the value of the parameter known as film coefficient (h), thereby upgrading the effectiveness of the surface area (A) of the heat-exchanging structure.

The basic relationship for convective transfer across a boundary is:

$$Q = hA\Delta T$$

where
Q = power, in watts
h = film coefficient, in w/ft²-°C
A = area, in square feet
ΔT = temperature gradient, in degrees centigrade.

It turns out that the film coefficient is about an order of magnitude higher in forced convective transfer than it is in natural convection—2.6 to 7.9 w/ft²-°C compared with 0.2 to 0.4 w/ft²-°C.

But this improvement has to be traded off against the energy that must be expended on forcing air past the surface that needs to be cooled. This usually translates as electric power driving a fan or blower and can be defined as:

$$P_f = KvH \qquad (1)$$

where
P_f = fan power required to deliver the necessary air velocity, in watts
K = a constant of 0.023 w-minute/ft-lb
v = air flow rate, in ft³/minute
H = frictional air pressure loss, in pounds/ft².

Thus design optimization comes down to the task of maximizing the heat transfer required in terms of Q and P_f.

As if this were not enough, the designer must usually restrict the physical size of the heat-exchanging structure to the smallest volume possible. In the case of the uhf transmitter-receiver, the space available for the rf power output stages, which dissipate 250 w, is 100 cubic inches, or roughly 8¾ by 5¾ by 2 in. The task requires that the junction temperatures of the rf power transistors be cooled to within safe limits.

Cold-plate considerations

Two cold-plate configurations were analyzed to determine the temperature fields which develop in each. Figure 4a is a straightforward variation of the plug-in printed-circuit board; the transistors are stud-mounted on an aluminum plate 90 mils thick that has integral heat sinks at both ends. Figure 4b shows how the board and the heat exchanger can be repackaged so that they become coplanar. The coplanar structure proved to be superior because it considerably shortened the conduction paths between each transistor and the heat exchanger.

The assumptions and design constraints used in the analysis of these two configurations are:
- Each transistor dissipates 11 w.
- Power is dissipated uniformly on the pc board at 2.3 w/in.².
- The equipment chassis is 90-mil-thick aluminum, with a thermal conductivity, K, of 4.4 w/in.-°C.
- Maximum transistor junction temperature is 150°C.
- Operating environmental temperature is 71°C.

The results of the analysis are listed in the table of Fig. 4. The critical ΔT, which is the temperature rise from the ambient to each device junction, can be expressed as:

$$T_J - T_A = Q(R_{TJ} + R_{T1} + R_{T2} + R_{T3} + R_{T4} + 1/hA)$$

The values and definitions of the thermal resistances are in Fig. 4. The subscripts represent thermal resistances which are conductive paths. The quantity $1/hA$ is the thermal resistance across the convective interface of the heat-exchanger surface.

If the conductive resistances are assumed to be known, then the design goal is to assure that the value of $1/hA$ will be small enough to hold T_J below 150°C. The film coefficient h is determined by the fluid dynam-

TABLE 2
THERMAL BUDGET FOR A 100-WATT TRANSMITTER-RECEIVER

Module	Peak power Transmit (W)	Peak power Receive (W)	Average power 50% duty cycle (W)	Power density (W/in.³)
Guard receiver	1.8	1.8	1.8	0.09
Frequency/control	18.9	18.3	18.6	0.19
Main receiver	4.8	4.8	4.8	0.05
Transmitter	272.0	25.9	149	2.72
Power supply	98.5	35.2	66.5	0.76
Totals	396	86	241.7	0.96

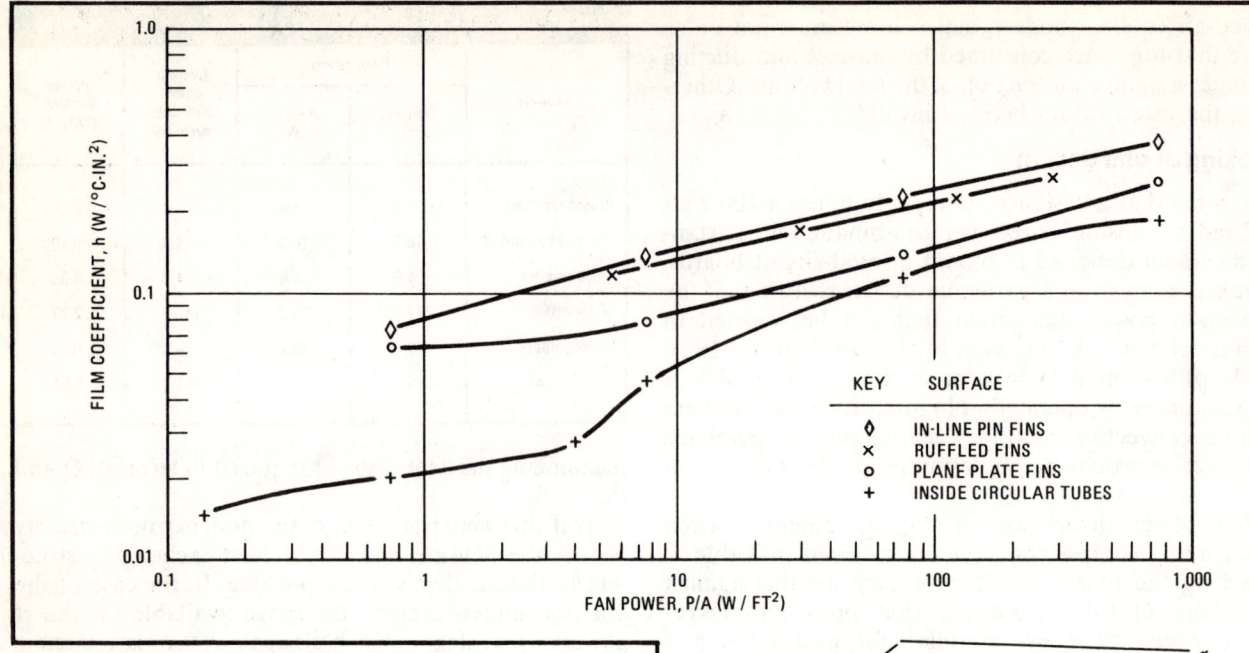

5. Different geometries, equal areas. In-line pin fin is the most effective convective surface, according to these plots of film coefficient, h, versus fan power.

ics of the system and is largely a function of fan input power. The heat-transfer surface area (A) is a function of heat exchanger type and volume. Thus the required value of 1/hA can be achieved by proper selection of heat exchanger type and size, and adequate fan power.

The analysis demonstrates that the in-line configurations of Fig. 4a won't do the job. If the in-line construction were selected, the 90.1°C rise would boost the junction temperature to 161.1°C, above the design limit of 150°C. Just how big this rise would be in actuality would depend on the value of 1/hA, because 1/hA has been assumed to be zero in the in-line case. But it really doesn't matter because the allowable gradient budget has been consumed in conduction drops. It is therefore impossible to maintain the desired temperature, regardless of the heat exchanger selected.

The horizontal baseplate resistance (R_{t4}) with a resistance of 1.97°C/w is the major contributor to the temperature rise. If a designer wants to stick with the in-line design he might reduce this resistance by using a thick chassis.

However, the coplanar design will certainly do the job. Here a value of 0.53°C/w is required for 1/hA, to maintain the hottest transistor below the maximum allowable temperature of 150°C. There is obviously a tradeoff between supplying more air to the heat exchanger and providing more heat exchange surface so that the exchange can make a closer approach to the exit air temperature.

Once the basic packaging structure has been selected, the next step in the design is to select a forced-air heat exchanger.

Exchanging heat

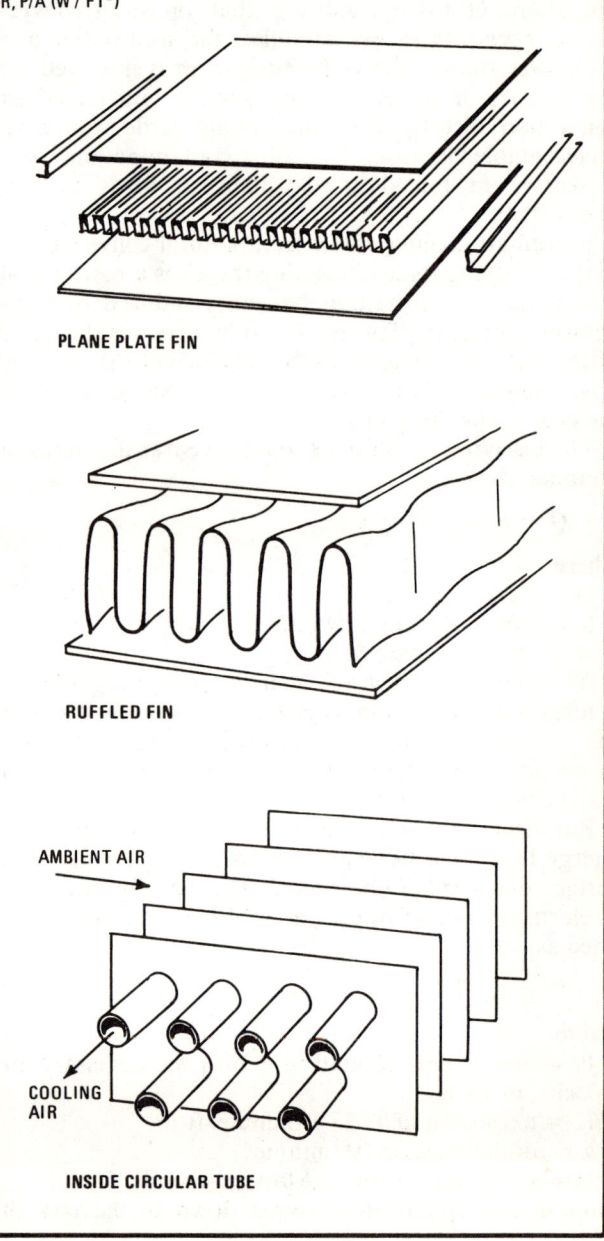

The heat exchanger enables heat to cross the boundary from a conductive region to a moving fluid such as air. Since its design is a major engineering challenge, it is worth summarizing the factors that go into a design analysis and to establish a design selection sequence.[4,5]

The prime considerations are the size and geometry of the heat exchanger structure. Heat transfer through the exchanger is expressed as:

$$Q = hA(T_H - T_S)$$

where
h = film coefficient of heat transfer, in w/in.²-°C
A = area, in square inches
T_H = heat exchanger temperature
T_S = cooling air temperature.

As has been shown in the example, the designer wishes to maximize both h and A so as to minimize the temperature gradient ($T_H - T_A$).

The relationship which determines the air temperature rise in the heat exchanger is expressed in the equation:

$$Q = mc_p(T_{SO} - T_{SI})$$

where
m = mass flow rate, in pounds per second
c_p = specific heat of the fluid at constant pressure, in w-sec/lb-°C
T_{SO} = cooling air temperature at the exchanger outlet, in °C
T_{SI} = cooling air temperature at the exchanger inlet, in °C.

The design goal here is to maximize the air flow rate (m) so as to minimize the temperature drop to be provided by the exchanger. However, a price is paid in electrical power required to energize the fan as can be seen from Eq. 1. In this case, air flow rate, v, as well as the air pressure drop, h, must be minimized for minimum fan power consumption.

The key variables in this group of equations—h, H and v—are interrelated for a given type of heat-exchanger surface. By carefully evaluating these variables, it is possible to tailor a heat-exchange system to a given application.[4,5]

Surface considerations

There is a considerable variation in the performance of various heat-exchanging surfaces. The value of h versus air power per unit cross-sectional area is plotted for a number of surfaces in Fig. 5. Note that the pin-fin exchanger delivers a value of h that is three and a half to four and a half times higher than the value of competing structures.

A useful figure of merit for evaluating a heat-exchanging surface is defined as the amount of heat exchanger surface contained in a unit volume or A/V. It is assigned the symbol β. From the standpoint of maximum βh, the ruffled fin provides the most heat transfer per unit of volume and thus offers the designer a very compact exchanger.

Figure 6 compares several surfaces on the basis of heat transfer per unit volume versus air friction power per unit volume. In effect, both the ordinate and abscissa in Fig. 5 have been multiplied by β. Thus the ordinate h becomes hA/V, expressed in w/in.³-°C. The abscissa is the frictional air power per unit volume dissipated in the heat exchanger, expressed in w/in.³. The values do not include other frictional losses or fan efficiency—typically 15% to 30% in small air-moving devices.

If the designer wants to include these losses, he can multiply the abscissa values by a number ranging from 7 to 13 to determine the approximate fan power. In the usual design operating range, this type of exchanger can reject 1.50 to 3.00 w/°C in.³ with a fan power requirement of 300 to 750 w/in.³.

The form factor, which is the width-to-height ratio of a forced-air heat exchanger, depends heavily on the quantity of air passing through a given cross section. A

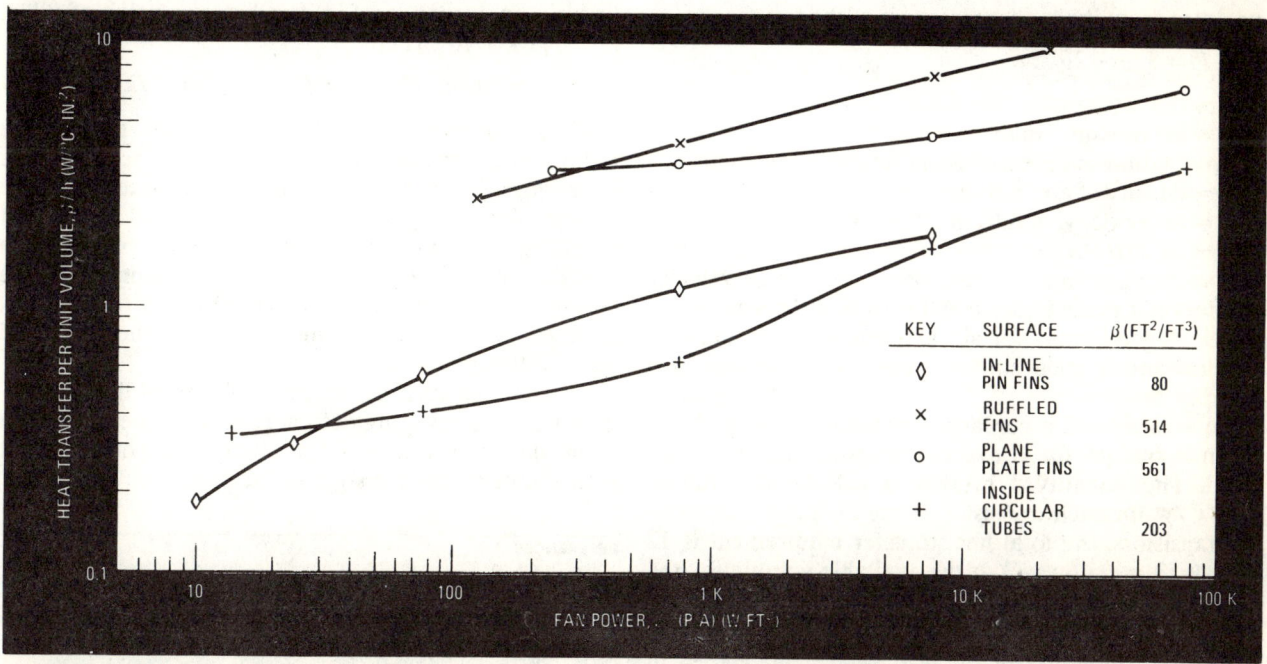

6. Equal volumes. By multiplying β (heat-exchanger surface area per unit volume) by the film coefficient and fan power, heat-exchanger surfaces can be compared on an equal volume basis. The ruffled-fin exchanger comes out on top.

7. Pin-fin exchanger. Die-cast heat exchanger safely dissipates 100 watts of power and fits into a volume of just 50 cubic inches. Stud-mounted transistors are in valleys between pins.

high-performance heat exchanger will generally require a large cross section to minimize air temperature rise and acoustic noise.

Pressure drops can build up quickly if there are long narrow ducts and many turns in the path or if there are expansions and contractions in the cross section. The pressure drop due to these effects is of the form:

$$P = k_1 \rho v^2/2g$$

where

P = air pressure, in lb
k_1 = a dimensionless constant related to geometry
ρ = density of air, in lb/ft^3
v = air velocity, in ft/min
g = 32.2 ft/s^2.

It is wise to keep air flow rate low so that the air velocity (v) does not exceed 500 to 800 ft to limit pressure-drop losses. A good value for air flow often used in military systems and a good starting point in any design is 4 lb/min/kw.

To return to the coplanar exchanger of Fig. 4b, a thermal budget for convective transfer can be calculated. The quantity 1/hA had a calculated value of 0.53°C/w for each transistor. If the exchanger contains 12 transistors, the total heat transfer requirement is 12 × 1/0.53 = 22.6w/°C. If the available volume for the exchanger is 50 in.,^{3} the required heat transfer per unit volume (βh) is 0.45 w/°C-in.^{3}. This value of h is well within the capability of the heat exchangers shown in Fig. 6. Pin fins are selected because they can easily be integrated into the module enclosure. Pins spaced at 5.35 per lineal inch facilitate die-casting the exchanger.

To determine the fan power required, the following relationship can be used:

$$P_{\text{fan}} = (\beta P/A)(V r_p)$$

where (βP/A) is the power required per cubic foot, plotted as the abscissa in Fig. 6; V is the volume of the heat exchanger in cubic feet; and r_p is the ratio of fan power to core friction, assumed in this case to be 13. Since the required βh is 0.45 w/°C-in.^{3}, then for the in-line pin fin exchanger, Fig. 6 indicates a value of βP/A of 0.045 kw/ft^3. Then:

$$P_{\text{fan}} = (0.045)(50/1728)(13)kW = 0.017kW = 17W$$

Thus a fan with 17 w of fan power will provide the required heat transfer. Figure 7 shows the actual design of the exchanger. Note that the fins are integral to the chassis, thus doing away with the thermal losses that would accompany an attempt to fasten pins on the chassis. The semiconductors are stud-mounted in the two rows in the spaces between the pins. A thermal test program has confirmed the validity of the predicted temperature profile.

By applying such design principles from the very beginning of a packaging design, equipment designers can avoid the compromises in reliability and power output which have plagued designs in the past. □

REFERENCES
1. Paul Dickenson, "Convenient Thermal Analysis Techniques for Printed Circuit Board Assemblies," Proc. National Electronics Packaging and Production Conference, 1967.
2. Standard Hardware Program, "Heat Transfer Considerations," Navord 37625, May 15, 1970.
3. R. M. Jansson, "The Heat Transfer Properties of Structural Elements for Space Instruments," Report E-1173, MIT Instrumentation Laboratory.
4. Kays and London. "Compact Heat Exchangers." (2nd ed.), McGraw-Hill Book Co., 1958.
5. W. H. McAdams, "Heat Transmission," (3rd ed.), McGraw-Hill Book Co., 1954.

THERMAL DESIGN

Thermal characteristics of ICs gain in importance

Consideration of several factors offers ways to enhance reliability; one testing method is preferred when high accuracy is uppermost concern

by Robert Bolvin, *Signetics Corp., Sunnyvale, Calif.* °

☐ All-out thermal analysis of an electronic design is usually reserved for the most sophisticated, expensive systems, and for systems that must operate in unusual environments. These days, however, as electronic packaging becomes more dense, as greater numbers of heat-producing active elements are placed on a single chip, as more ICs are mounted on a single board, designers must be increasingly concerned with the thermal aspects of systems.

Thermal resistance, therefore, is one parameter becoming more important in specifying packaging products. It is a measure of how effectively heat is dissipated from sensitive areas; the lower the figure the better. In emitter-coupled logic devices, for example, reference bias supply voltages are set internally by a diode-resistor network that is affected to a great degree by heating. If device power dissipation and package thermal resistance at operating conditions are known, the final junction temperature on the chip can be determined, making possible the correct design of the bias network.

A half-dozen factors can be considered in arriving at the thermal resistance of an IC:

■ Die size. The larger the die size the more heat sinking

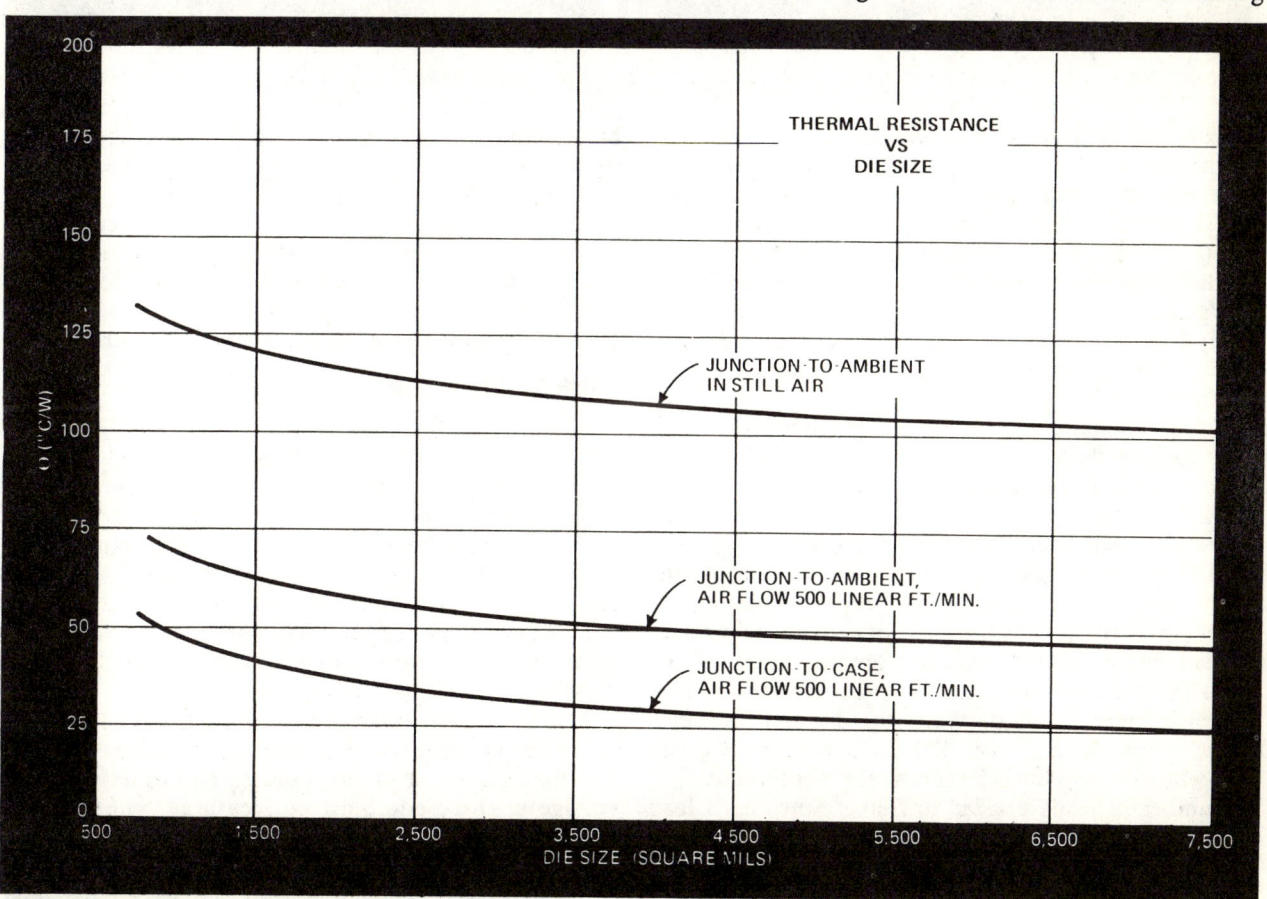

*Mr. Bolvin is presently with Monolithic Memories Inc., Sunnyvale, Calif.

1. Chip size counts. The thermal resistance of an integrated circuit drops as die size increases. Data plotted here is for a 16-pin ceramic dual in-line package with gold die attach and Alloy 42 frame, at 25°C ambient or bath temperature.

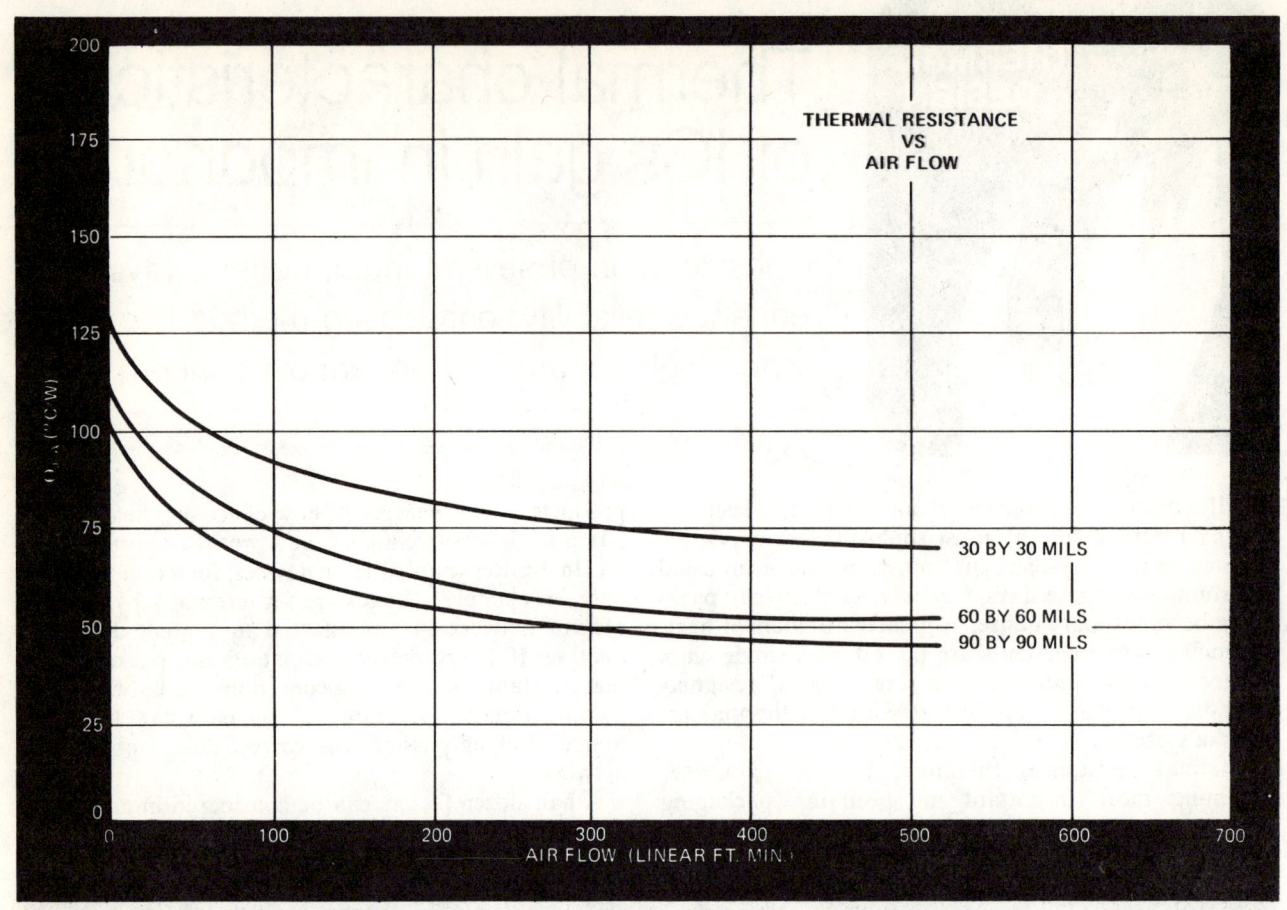

2. Air flow dependence. Thermal resistance also drops off with increasing air flow. Most noticeable when airflow is small, the effect is less pronounced when air flow becomes larger. Θ_{J-A} is plotted for three die sizes with power dissipation of 215 mW and an ambient of 25°C.

is provided for the junctions and the larger the area of contact for heat transfer. A number of thermal resistance measurements may be made on a package; in Fig. 1, these are junction-to-ambient without air flow, junction-to-ambient with air flow, and junction-to-case. As the graph shows, the increase in thermal resistance becomes more pronounced as the die gets smaller.

- Air flow. Even slight air flow can cause thermal resistance to fall off sharply (Fig. 2). Falloff continues as air flow increases, but not as sharply.
- Die attach method. Gold is a better conductor of heat than glass, so the gold eutectic die attach method yields a lower thermal resistance than does the glass attach method.
- Package material. In order of thermal resistance, from lowest to highest, are ceramic, epoxy and plastic. The superiority of ceramic over epoxy, moreover, is much greater than the superiority of epoxy over plastic.
- Lead frame material. As an example, nickel affords lower thermal resistance than alloy 42. As shown in Fig. 3, a nickel lead frame decreases the thermal resistance of the epoxy package from 120°C per watt to 88°C per watt when the air flow is 500 linear feet per minute.
- Number of leads bonded to frame. Since each lead acts as a heat sinking conduit, the more leads, the lower the thermal resistance.

Considerations such as these are practical because a device designer must see to it that maximum junction temperatures will not be exceeded at anticipated ambient temperatures. Testing becomes a problem here because limits on junction temperature are specified under "soak" conditions; in an emitter-coupled logic circuit, for example, thermal equilibrium has been established and transverse airflow of greater than 500 linear feet per minute is maintained. Under these conditions, production testing is not practical because of the long time it takes the device to reach thermal equilibrium.

More practical method

A more practical method on the production line is called "rapid" testing. Devices are checked quickly, without bringing them up to thermal equilibrium. But to prevent failures at untested higher temperatures, a correlation between soak and rapid testing must be established. Thermal time constants become important because the manufacturer can provide the proper guardbands if he knows, first, the changes to be expected in parameter values per degree of junction temperature, and, second, the equipment test time.

The device manufacturer can monitor junction temperatures by using actual on-chip functional devices like isolation diodes or clamp diodes. The initial forward voltage of the diode must be measured, *without operating power applied*, by forcing a fixed test current from ground to V_{CC} or V_{EE}. Operating power is then applied to establish thermal equilibrium. To record the final value of the diode voltage, operating power must then be removed and the fixed current applied once more.

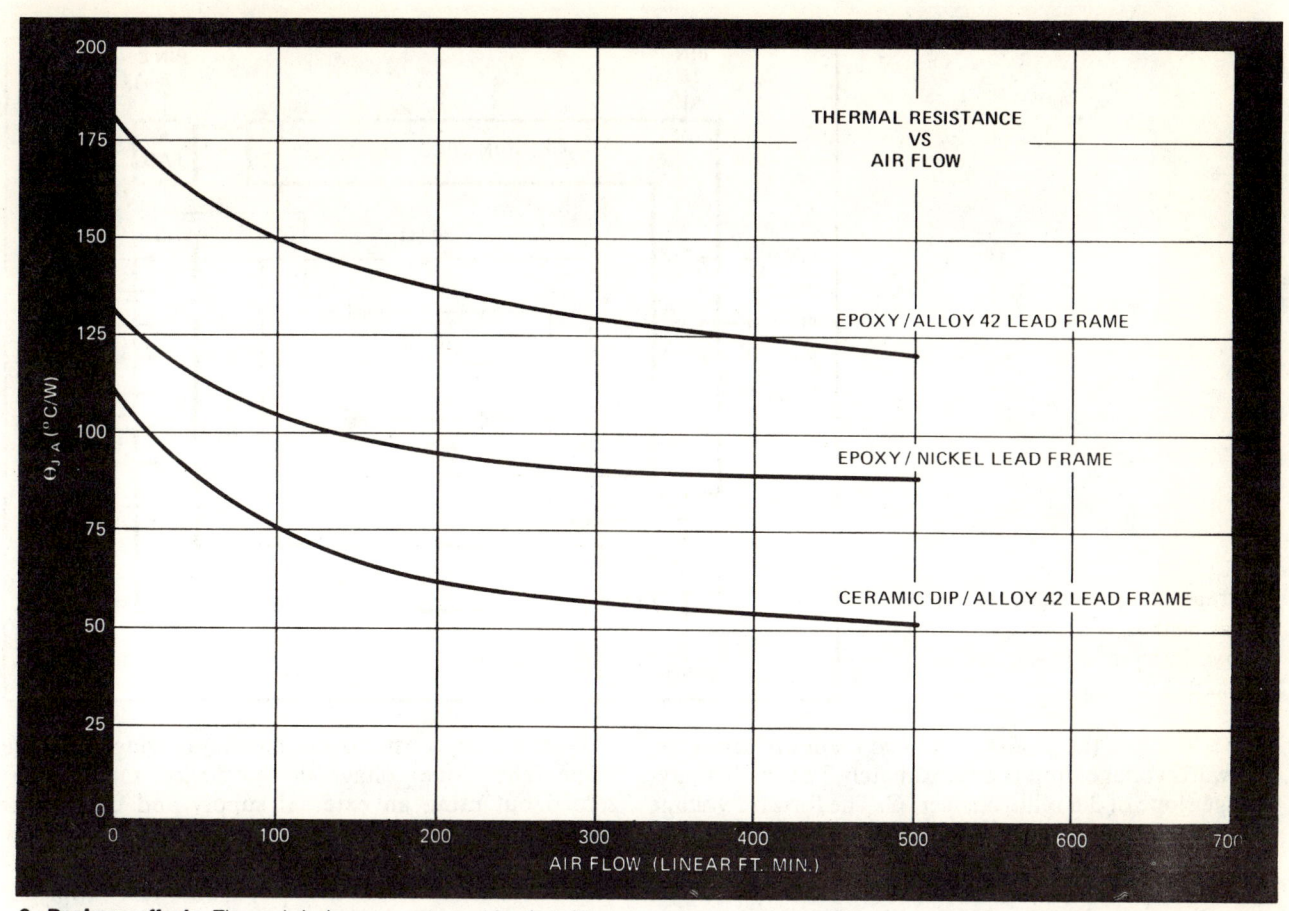

3. Package effects. Thermal designers must consider that different packages and lead frames have different thermal resistances even for the same die. Data here applies to a 60-by-60-mil chip dissipating 215 milliwatts. Ambient temperature is 25°C.

Calculation of thermal resistance (Θ), with this method is as follows:

$$T_J \text{ rise} = V_{FF} - V_{FI} / \text{slope}$$
$$\Theta = T_J \text{ rise} / P_D$$

Where V_{FI} is the initial forward diode voltage at zero power, V_{FF} is the final forward diode voltage after thermal equilibrium (in volts), T_J is the junction temperature, P_D is the power dissipation (in watts), and slope is the change in voltage due to change in temperature (in millivolts per °C).

Time span critical

The time span between the point when the power is turned off and the point when a final diode voltage is measured is critical. Switching from power on to power off and making a final measurement cannot be done fast enough to measure the true thermal equilibrium voltage. By measuring a lower value, the thermal resistance always appears smaller than it actually is.

The same problem is encountered with the clamp diode method and with techniques involving measurement of output voltage. In addition, clamp diodes and outputs may be so situated that they won't respond to the actual rise in junction temperature because of temperature gradients across the chip.

The most accurate method for measuring thermal resistance involves a set of test dies, an example of which is shown schematically in Fig. 4. Here, four series-parallel diodes are placed strategically around three 300-ohm resistors on a monolithic chip. The diodes and resistors are isolated from each other by individual back-biased diodes not shown in the schematic. The design permits a fixed diode current to flow continuously even though power is being applied to the resistors. This method eliminates the need to switch off power and switch on diode current, allowing an accurate measurment of V_{FF} at thermal equilibrium.

Fixed die sizes are used, starting with 30 by 30 mils and increasing by 30-mil increments up to the largest die size, 180 by 180 mils. Reverse breakdown voltages of the back-biased diodes allow up to 10 volts to be applied, which gives a maximum power of one watt.

The characterization of a package normally uses three die sizes, yielding three points on graph showing thermal resistance versus die size (Fig. 1). From this graph, the thermal resistance of any functional device may be found by knowing that device's die size.

Slope measured first

Forced diode current is normally between 3 and 5 milliamperes. A large diode current is used so that diode voltage measurements may be made on the essentially linear portion of the diode forward characteristic. This reduces the effect of increasing forward leakage of the diode due to rising junction temperature, which may affect the calibrated slope.

The slope characteristic is measured first to determine

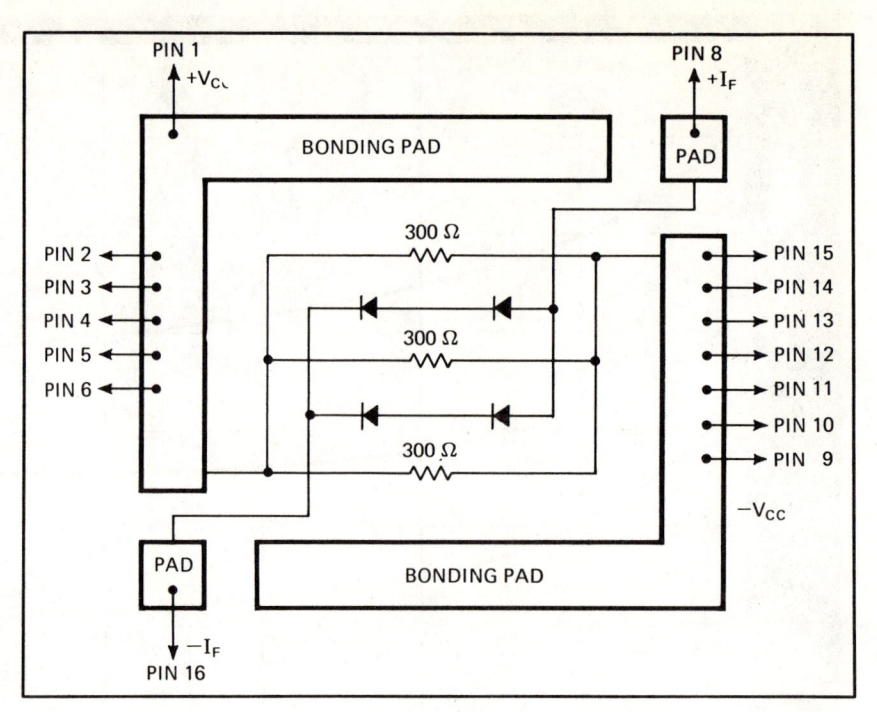

4. Thermal test die. When making thermal measurements, a special die, like the one shown schematically here, can yield more useful data than a functioning chip.

the $\Delta V/\Delta T$ of the diodes. With two diodes in series the forward voltage drop is approximately 1.5V with an average slope of 3.6 millivolts per °C. The forward voltage is measured at a minimum of three temperatures, usually +25°C, +75°C, and +125°C. Voltage and temperature measurements must be as accurate as possible. A 5-digit voltmeter should be used and the temperature held to ±1°C. A 5% error in slope measurement will be reflected as a 5% error in thermal resistance.

In preparation for making a set of thermal resistance measurements, a device is mounted on a pc board via socket pins which accept the device leads. These pins, and a rectangular hole in the pc board directly beneath the device, allow air or oil to flow freely around the device, a requirement for some of the tests that may be performed later.

To measure thermal resistance from junction to case at zero airflow, the pc board and device under test are mounted in an enclosure with holes drilled along the bottom edges. The enclosure prevents any air from flowing around the device.

Two fans are used

To make measurements with airflow, the device is placed in another enclosure where air can be forced across the device. Junction-to-ambient is measured first with 500 linear ft./min. air flow, then at 250 linear ft./min., then at 100 linear ft./min. Two fans separated by a partition are used, one to provide airflow across the device and the other to blow air at a downward angle, producing a wall of air in front of the device under test that prevents air flow variations across the device due to external disturbances.

To measure thermal resistance from junction-to-case, the device is immersed in an oil bath. Air is bubbled through the oil to maintain a flow of oil around the device under test. This flow holds the case temperature at the oil temperature.

Thermal time constants are measured using a storage scope. The diode voltage of the device under test is zeroed out using an external supply and differential scope plug-in. In this manner a vertical scale of as low as 1 millivolt per division may be utilized. By first setting the zero level on the scope and then applying power for a specified amount of time, the thermal time constant curve is stored. Several V_F readings are recorded after each power application. Power application time is increased each time to record from a full scale reading of 10 milliseconds to 50 seconds. The device under test is allowed to cool to the initial V_F after each power application. A stop watch is used to monitor time intervals over 50 seconds.

For all of the three measurements, a fixed method is used:
- Apply an I_F of 5mA and a V_{CC} of 1.00V; allow ample time for the device to come to thermal equilibrium. Thermal equilibrium is reached when a constant V_F can be measured.
- Record V_F, initial forward voltage.
- Record I_{CC1} at $V_{CC1} = 1.00V$.
- Apply increased V_{CC2} specified for amount of power required.
- Allow 15 minutes for thermal equilibrium.
- Record V_{FF}, final forward voltage.
- Record I_{CC2} at V_{CC2} specified.
- Using the previously calibrated diode slope, calculate Θ as follows:

$V_{FF} - V_{FI} / slope = T_J$ rise above ambient
T_J rise above ambient $/ (V_{CC} \times I_{CC2}) - (V_{CC1} \times I_{CC1}) = \Theta$

For the thermal time constant measurement, V_{FF} is recorded as V_F since V_{FI} has been zeroed out.

For junction-to-ambient with air flow, V_{FI} is recorded at 500 linear ft./min. and this V_{FI} used at all lesser air flows. □

THERMAL DESIGN

Heat pipes cool gear in restricted spaces

Fluid that cycles in tubes through evaporation and condensation process reduces temperatures around even high-power electronic subassemblies

by Alan J. Streb, *Dynatherm Corp., Cockeysville, Md.*

☐ Convective cooling of electronic components—even when heat sinks are used to increase the heat-radiating area—is not always enough to keep operating temperatures within limits. A simple alternative is the heat pipe, which uses fluids cycling through liquid and vapor phases to conduct heat away from temperature-sensitive components.

Although standardized heat-pipe cooling systems for electronic components and systems do not yet exist, heat-pipe technology has matured to the point where a wide range of practical configurations can be produced economically for many applications.

As shown in Fig. 1, a typical heat pipe consists of a sealed tube lined with a capillary pumping structure, or wick, which is saturated with a working fluid. As heat is added at one end of the tube, the working fluid is vaporized and moves down the tube until it condenses at a cooler site and reverts to the liquid state. As liquid is returned to the heat source via capillary action in the wick, heat is continuously transferred from one region to another through the nearly isothermal process of evaporation and condensation.

The efficiency of transferring heat energy is largely determined by the heat pipe's capability to conduct the vapor from the heat source to the heat sink and to return liquid from the condenser to the evaporator. Capabilities depend on fluid properties, wick configuration, and heat-pipe geometry and orientation.

Heat pipes have been successfully operated over the range from cryogenic to liquid-metal temperatures to convey heat from equipment dissipating power ranging from a few watts to thousands of watts. For any desired temperature range, a number of appropriate working fluids, containment-vessel materials, and wick designs are available. For cooling of electronic components, fluids such as water or organic fluids, contained in pipes made of such materials as aluminum, copper, or stainless steel, can dissipate as much as a kilowatt.

Water has proved to be the most desirable heat-pipe fluid for the temperature range of electronic compo-

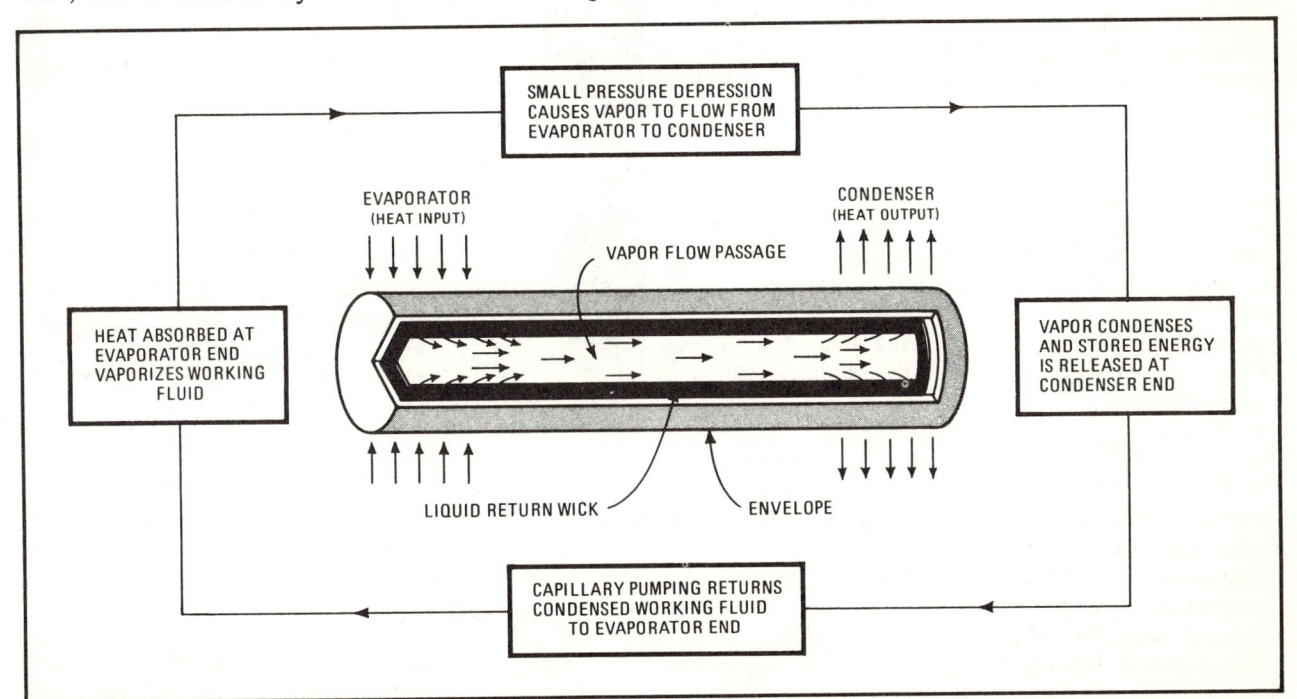

1. Heat Pipe. Using the properties of a fluid as it cycles between the liquid and vapor states, a heat pipe can cool the environment of temperature-sensitive components in enclosed spaces. Water in copper has proved the most desirable combination for electronic parts.

nents, 50°C to 200°C. Figure 2 shows the relative heat-transport capability of a number of fluids for this range. Since water is compatible with copper and copper alloys, but few other containment materials, copper is most widely used in the construction of heat pipes for cooling electronic components and packages.

The physical properties of the heat-pipe working fluid—especially heat of vaporization, surface tension, liquid density, and liquid viscosity—establish the heat-transport capability of the heat pipe. Since these properties are temperature-dependent, the performance of the heat pipe operating with a given fluid is temperature-dependent. One fluid will, therefore, function most effectively over any particular temperature range.

Although the temperature of the vapor within the heat pipe is nearly constant over the length of the device, the temperature varies at the evaporator and condenser regions, where heat is conducted through the walls and the liquid film that lines the walls. This results from the thermal resistance of the container and the working fluid. However, proper design of the containment vessel and the wick can minimize this resistance.

Designing a heat pipe

For a given heat-pipe geometry, the maximum heat-transport capability may be estimated by:

$$Q_{max} = ad^2/l$$

where Q_{max} is the maximum heat-pipe heat-transport capability in watts, d the inside diameter of the heat pipe in inches, l the heat-pipe length from midpoint of evaporator to midpoint of condenser in inches, and a is a constant.

The value of the constant is determined by the geometry of the heat-pipe wick, working fluid, and orientation. Values of this constant for a typical copper-water heat pipe configuration are presented in Fig. 3.

The maximum temperature gradient across a heat pipe can be estimated for a given heat pipe geometry from:

$$\Delta T = (Q/b)(1/A_{evap} + 1/A_{cond})$$

where ΔT is the over-all temperature drop in degrees fahrenheit, Q the heat transport in watts, A_{evap} the total area of the evaporator in square inches, A_{cond} the total area of the condenser in square inches, and b is a constant.

The constant in this equation has a value determined by the heat pipe's configuration. For a typical copper-water heat pipe,

$$b = 3 \, w/in.^2 °F$$

A first-cut selection of the size of heat pipe required for a particular application can be determined from these equations. But to optimize the heat pipe, the designer must also consider such other factors as the relative position of the evaporator and condenser, both with respect to each other and to the horizontal plane.

The pressure developed by capillary pumping within the heat pipe must balance the losses in viscous pressure within the system, as well as differences in elevation pressure that may be caused by locating the heat source

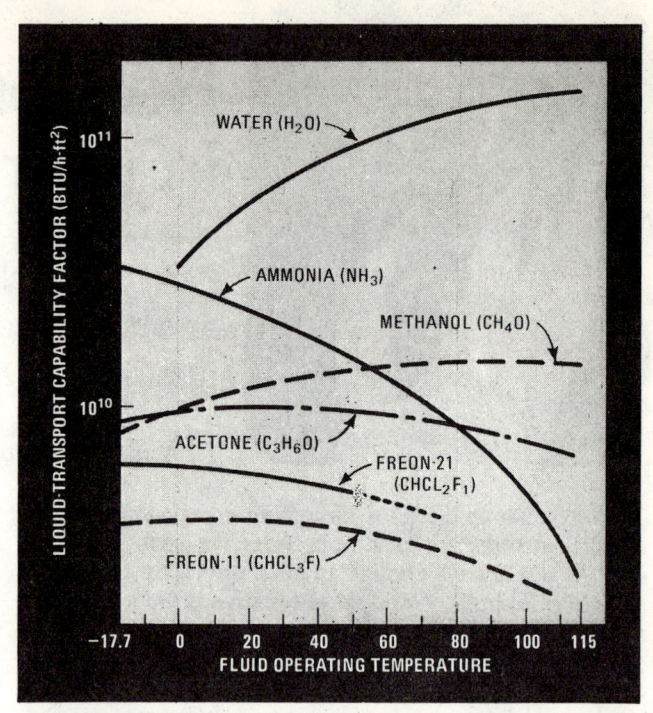

2. Working Fluids. Over a wide temperature range, water has a higher heat-transport capability than other fluids for heat pipes.

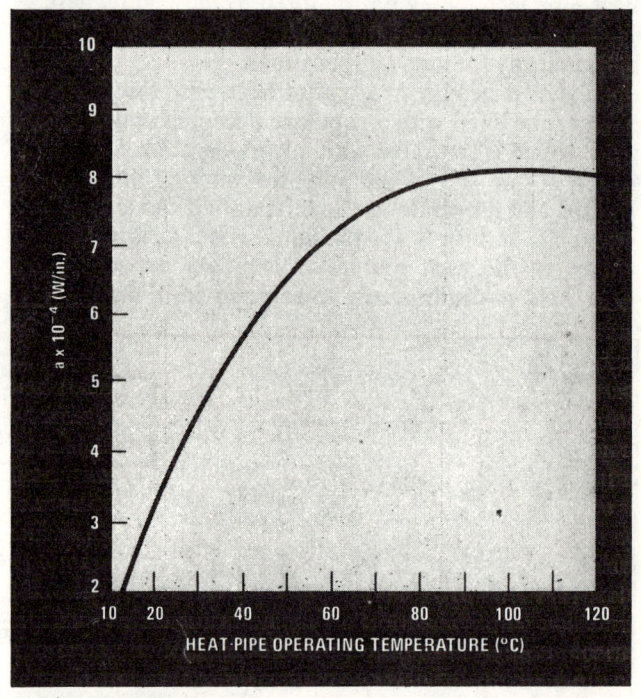

3. Heat transport. For a copper heat pipe using water as the working fluid, the heat-transport constant varies with temperature.

above the heat sink. Because the capillary-wick structure is only capable of a limited rise in liquid-pumping pressure, adverse orientation of the heat pipe can degrade its performance.

Calculating heat flow

Lowering the condenser end will degrade the steady-state heat-transporting capability of the heat pipe, and that capability will go to zero if the condenser end is lower than the evaporator by an amount in excess of a

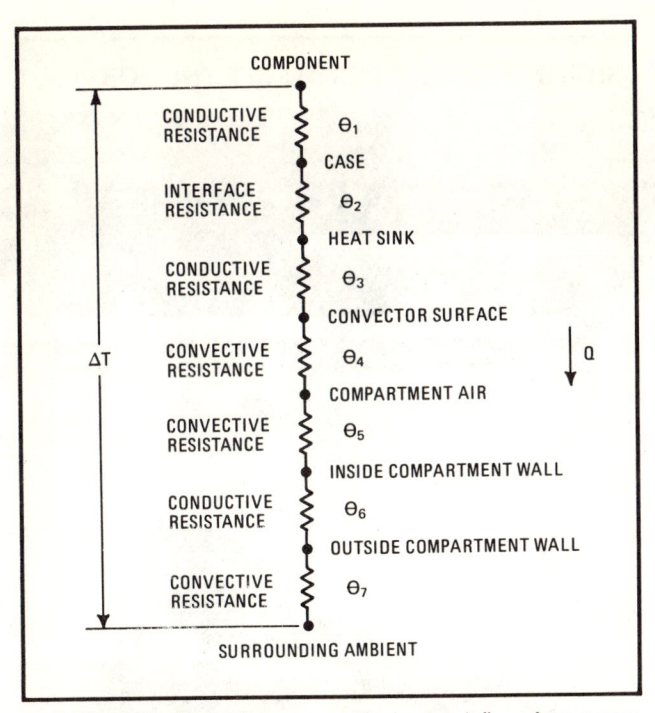

4. Analogy. Thermal resistances met by heat as it flows from a component to its surroundings are analogous to electrical resistances.

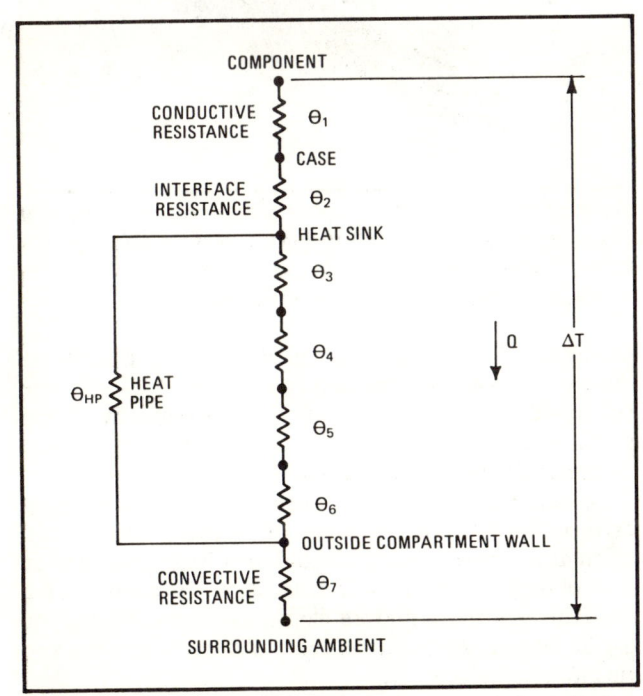

5. Short-circuit. A heat pipe can act as a thermal short-circuit to help heat flow to the surrounding ambient.

voltage difference is analogous to temperature difference, current to heat flow, and electrical resistance to thermal resistance.

In Fig. 4, the "circuit" for heat flow from a heat-producing component, through thermal resistances, to the surrounding ambient is shown. Heat (Q) is driven through thermal resistances (Θ) by a temperature difference (ΔT) so that

$$Q = \Delta T / \Sigma \Theta$$

The series-thermal-resistance path shown in Fig. 4 is idealized. In most practical situations, parallel paths are also present, but these resistances may be combined to an equivalent thermal resistance through an analog of Kirchoff's Law. The major contributors to the total effective thermal resistance in a system like that of Fig. 4 are usually those resistances external to the component itself, i.e., Θ_2 through Θ_7; therefore, techniques to improve heat transfer are often sought in these areas.

Because of its inherently low thermal resistance and the relative insensitivity of resistance to length, a heat pipe is an efficient conductor of heat. All of the resistances Θ_3 through Θ_6 can be bypassed by interposing a heat-pipe shunt between the component case and the compartment wall, as shown in Fig. 5.

If all the convective resistances (Θ_4, Θ_5, and Θ_7) are approximately equal, all the conductive resistances (Θ_1, Θ_3, and Θ_6) are approximately equal, and all of these thermal resistances are much larger that the thermal resistance of the heat pipe, Θ_{HP}, the total thermal resistance of the circuit with the heat pipe is approximately one third of the total circuit resistance without the heat pipe. The heat pipe acts as a thermal short-circuit.

Beating the heat

Heat pipes can substantially reduce the thermal resistance between the heat source and the heat sink in a wide variety of applications. They are especially valuable in cooling systems that have confined spaces, as well as in enclosed products. In one computer system, conduction had to be used to remove heat generated by large numbers of densely mounted components on a printed-circuit board. The glass-epoxy board itself did not have sufficient thermal conductivity, even when augmented with heavy layers of copper.

To solve the problem, a flat-plate heat-pipe panel (Fig. 6a) was designed to serve as a structure/thermal base for two board assemblies—one on each side. Heat absorbed by the panel surface from the card-mounted components is transferred to the two edges of the heat-pipe panel, where the panel interfaces with its mounting rails. The heat absorbed at the rails is conducted through the enclosure and dissipated to an external air or liquid cooling system. The effective over-all thermal conductivity of the panel is eight times that of copper.

Another difficulty arose when a high-density solid-state microwave amplifier for military applications had to be cooled by natural convection only, yet the circuitry had to be enclosed and sealed within an existing package designed for outdoor use. The ambient temperature extremes were therefore substantial, and the internal and external heat sinks were limited.

dimension called the static-wicking height. The static-wicking height, which is a function of the working fluid, the temperature, and the capillary size, ranges from eight to 16 inches for a typical water-filled heat pipe.

In an electronic system, heat flows from a source through a series of thermal resistances to the surroundings. The rate of heat flow depends on the total effective thermal resistance of the path and the total temperature difference between the source and the surroundings. This is similar, of course, to an electrical circuit, since

DESIGNING HEAT PIPES TO FIT THE JOB

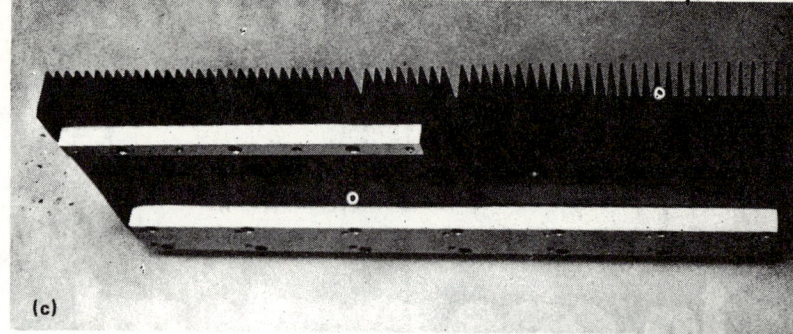

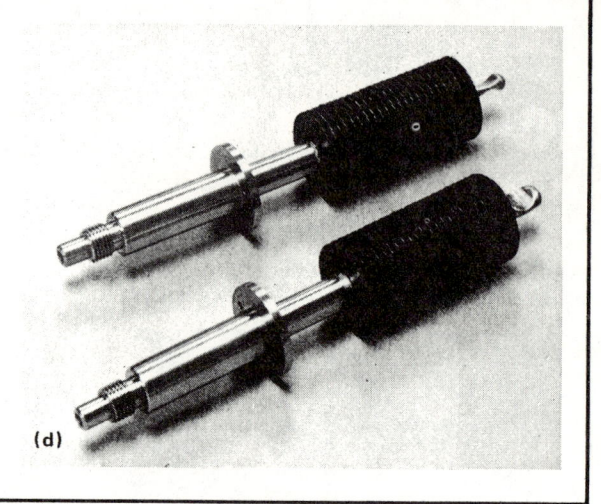

6. Configuring coolers. A flat-pole heat pipe (a) conducts heat away from printed-circuit boards in a computer system. Five heat pipes were brazed onto an aluminum baseplate to cool a microwave amplifier in an enclosed assembly (b). The entire rear wall of the enclosure in a sealed numerical-control system forms a heat pipe (c) to distribute the heat, and fins on the outer surface dissipate it. A heat pipe serves both as drive shaft and heat sink with integral fins for heat dissipation in a motor for a servo-control drive in an airborne system (d).

A heat pipe with integral fins (Fig. 6b) was devised to cool the amplifier assembly. The heat sink consists of an aluminum baseplate to which five heat pipes are brazed at equal intervals. Heat is transported by the heat pipes to copper fin plates outside of the sealed enclosure. Between a component mounted at any location along the 23-inch-long mounting plane and the ambient air near the fins, the thermal resistance totals 0.66°C/w. The unit is designed for a nominal heat throughput of 85 w.

Cooling a sealed enclosure

At another installation, a numerical-control system for a machine tool required a sealed enclosure to prevent contamination from dirt and oil vapor. As a result, heat removal became a problem. The conventional solutions were found to be either thermally inadequate or too complex and costly.

A large heat-pipe assembly was developed to serve as the entire rear wall of the enclosure (Fig. 6c). The outward-facing surface of the heat sink includes sufficient finned surface to dissipate the entire thermal load by natural convection. A single flat-cross-section heat pipe applied across the interior surface of the heat sink distributes the heat uniformly over the large panel area.

Heat-generation is predominantly associated with a power supply that is conductively coupled directly to the interior of the heat-sink panel. Over-all effective resistance of the heat sink from the subsystem mounting pad to the environment is 0.25°C/w. The unit is rated for a maximum heat input of 150 w.

To achieve the required performance in a compact drive motor for a servo control in an airborne system, efficient removal of a substantial heat load generated in the rotating armature was necessary. Circulation of a fluid through the armature would have necessitated the use of rotating seals and added undesirable volume and weight to the assembly.

A heat pipe was designed to serve both as a drive shaft and heat sink (Fig. 6d). The end external to the motor frame contains integral fins for heat dissipation. The armature winding is pressed onto the heat-pipe shaft. Heat generated within the armature winding is conducted through the armature to the evaporator end.

Heat is transported isothermally down the length of the shaft and delivered to the finned convector by the heat-pipe operating cycle. The heat is dissipated from the fins by convection and radiation. A thermal resistance of less than 5°C/w is achieved between the armature shaft interface and cooling air. The assembly is designed for a nominal rating of 15 w. □

Measuring thermal resistance is the key to a cool semiconductor

To improve performance, reliability, and yield of integrated circuits, knowledge of thermal design considerations and of measurement techniques is a must for the user, as well as the manufacturer, of semiconductor devices

by Bernard S. Siegel, *Sage Enterprises Inc., Mountain View, Calif.*

☐ Taking the heat off semiconductor devices is a sure way to improve their performance, reliability, and yield. For both user and manufacturer, a knowledge of their thermal properties and in particular their thermal resistance can spell all the difference between success and failure, whether in terms of equipment functioning or in relation to parameter yields. And new instruments are making this knowledge easier to come by.

Take performance. If an operational-amplifier chip capable of operation off an input bias current of 1 picoampere is not thermally packaged correctly, it may be unable to reach its bias specification. Yet in most attempts at improving electrical specifications, thermal design considerations come a poor second to changes in circuit design and fabrication methods.

As for reliability, numerous test programs since the transistor's invention have verified that every 10°C or so rise in junction temperature halves a device's operating life expectancy.

Finally, fluctuating yields at the manufacturing level and actual failures in users' equipment can often be traced to differences in parameter test results that are the direct consequence of variations in thermal resistance. In the case of most low-input bias op amps, for instance, a 10°C increase in junction temperature will just about double the input bias current required.

The ability to monitor both the absolute magnitude and manufacturing variations in thermal resistance in particular is clearly crucial. But first, it is as well to be clear as to the exact meaning of "thermal resistance."

Defining thermal resistance

Whenever electrical power is applied to a semiconductor device, some of the power is converted into heat, resulting in a rise in junction temperature. This occurs because the heat-generating portion of the device is separated from the outside world by the path shown in Fig. 1. The heat-flow path is made much more complex by the heat-removal paths associated with conduction and convection within the binding wires and by radiation from the heat source to the package. Such heat flow occurs with most semiconductor devices.

Each of the different portions of the path from the heat source (the active area or junction) to the outside world has two components. Of primary concern for most steady-state conditions is thermal resistance, shown as

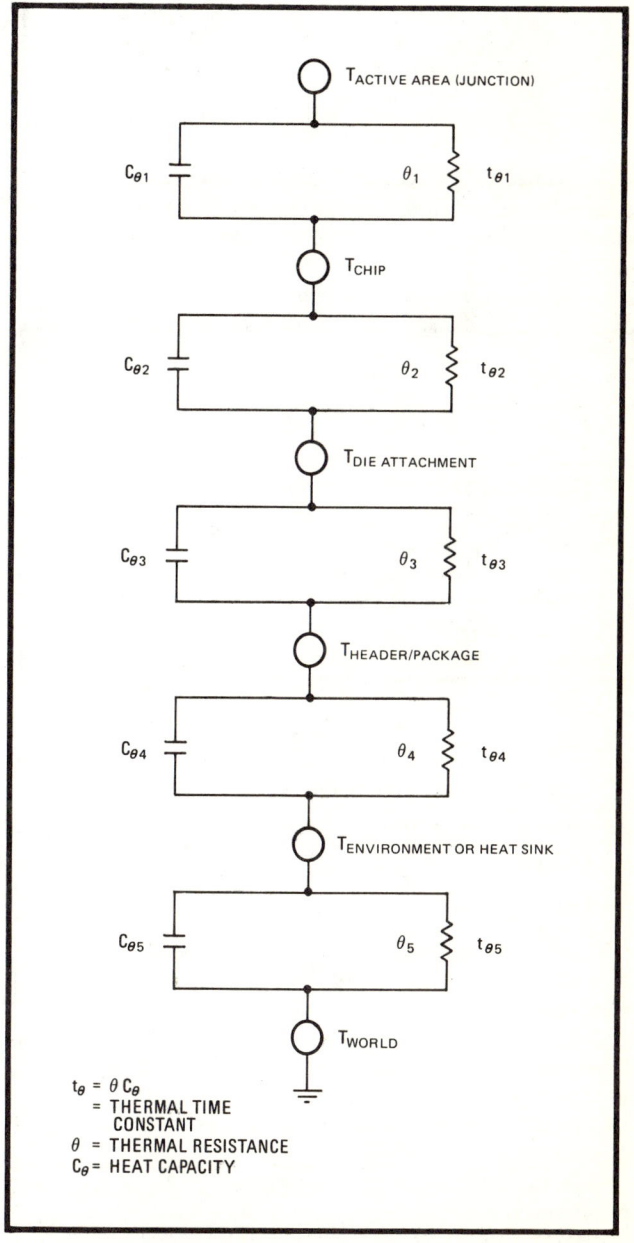

1. Thermal-equivalent circuit. The simplified electronic analog of the flow of heat from a semiconductor device to the outside world is a series of equivalent resistors and capacitors in parallel. For steady-state conditions, the thermal resistance is of primary importance.

$t_\theta = \theta C_\theta$ = THERMAL TIME CONSTANT
θ = THERMAL RESISTANCE
C_θ = HEAT CAPACITY

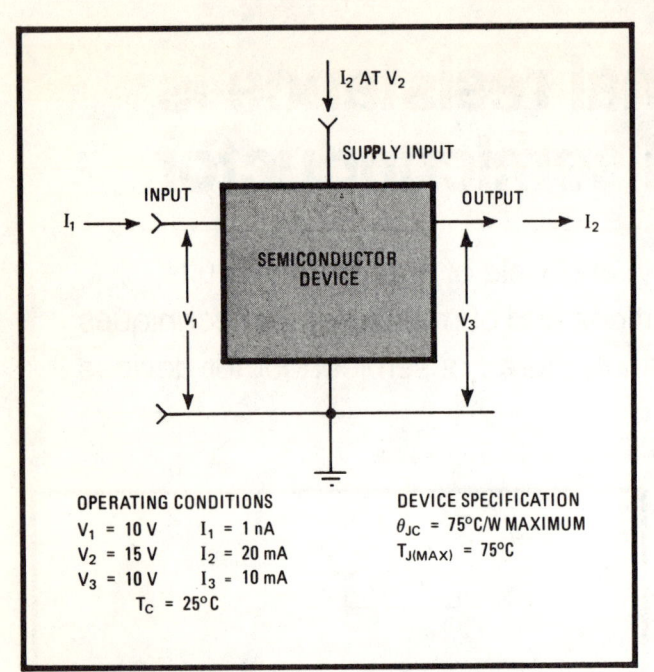

OPERATING CONDITIONS
$V_1 = 10$ V $I_1 = 1$ nA
$V_2 = 15$ V $I_2 = 20$ mA
$V_3 = 10$ V $I_3 = 10$ mA
$T_C = 25°C$

DEVICE SPECIFICATION
$\theta_{JC} = 75°C/W$ MAXIMUM
$T_{J(MAX)} = 75°C$

2. Junction temperature. With the operating conditions and device specifications shown, it is possible to calculate the device junction temperature, T_j. For the case shown, T_j equals 40°C, which is well below the maximum value of 75°C specified.

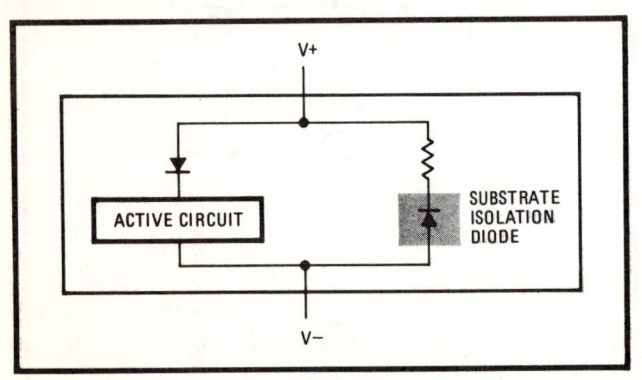

3. Substrate isolation. The forward voltage drop of the substrate isolation diode of most ICs can be used to make thermal measurements. An increase of power to the active portion of this circuit heats the substrate, decreasing the forward voltage drop of the diode.

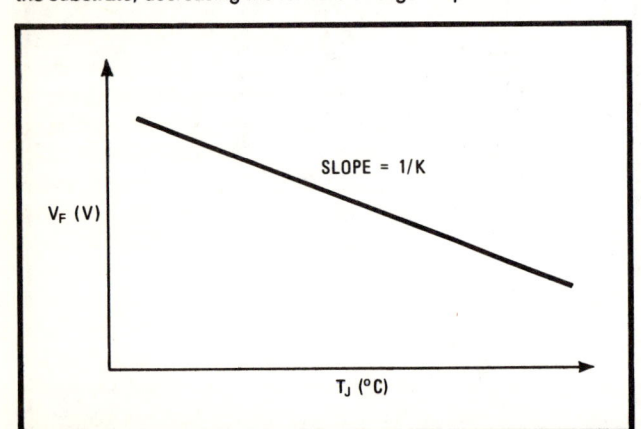

4. Negative slope. As junction temperature increases, there is a linear decrease in the voltage drop across a semiconductor junction driven by a constant current. The reciprocal of the curve's slope is directly proportional to thermal resistance.

θ in Fig. 1 but sometimes referred to as θ_{TH} or R_{TH}. The other component, heat capacity C_θ, becomes important when the device is subjected to pulse operation with short duty cycles, in which electrical power is applied for a period of time equal to or less than the longest thermal time constant.

Thermal resistance for any semiconductor is defined as the increase in junction temperature (T_J) due to the application of dissipative power (P_D) in the device:

$$\theta_{JX} = \Delta T_J/P_D$$

The two θ subscripts describe the points between which the thermal resistance is measured, with J always indicating junction and X being dependent on the reference point being used. For example, X could be C for case (header or package) or A for ambient or L for liquid. The two most commonly specified thermal-resistance parameters are junction to case (θ_{JC}) and junction to ambient (θ_{JA}).

Using the thermal resistance parameter

Deriving the thermal-resistance parameter is very simple provided that careful consideration is given to the power actually dissipated in the device. For a semiconductor device in the operating conditions of Fig. 2, junction temperature can be determined as follows:

$$\begin{aligned}
T_J &= T_C + \Delta T_J \\
&= T_C + \theta_{JC} P_D \\
&= T_C + \theta_{JC}(V_1 I_1 + V_2 I_2 - V_3 I_3) \\
&= 25°C + 75°C/W[(10 v)(1 \text{ nA}) + \\
&\qquad (15 v)(20 \text{ mA}) - (10 v)(10 \text{ mA})] \\
&= 25°C + 15°C \\
&= 40°C
\end{aligned}$$

In this instance, the case temperature could increase to 60°C before the $T_{J(max)}$ device specification was exceeded. If the semiconductor device is assumed to be an op amp and 1 nanoampere is the maximum current that can be supplied for I_1, then for a T_J of 40°C, the op amp should be chosen for input bias current of less than 350 picoamperes at 25°C.

Should θ_{JC} be specified as a typical value instead of a maximum, the thermal-resistance range must be estimated before it is possible to select a device properly. A range of 50° to 100°C/w causes a T_J variation of 10° to 20°C, dictating a maximum input bias current at 25°C of 250 pA. Also, I_1 will vary between 500 pA and 1 nA.

This simple example can be modified to cover a wide range of semiconductor devices, including discrete transistors and ICs, under varying operating conditions.

Comparison of measurement techniques

The advantages and disadvantages of the four basic techniques for the measurement of thermal resistances—optical, chemical, physical, and electrical—are summarized in the table. The power dissipation within the device is always measured electrically in each case, but junction temperature is measured differently.

Only optical and electrical techniques are in widespread use today. Infrared microscopes capable of resolving down to 0.5-micrometer spot sizes are particularly useful as an aid in the thermal design of high-power

	FOUR TECHNIQUES OF THERMAL-RESISTANCE MEASUREMENT		
Measurement technique	Procedure	Advantages	Disadvantages
Optical	infrared scanning of semiconductor chip surface	• provides temperature profile across chip surface, indicating particularly troublesome hot spot areas • is useful in device thermal-design efforts	• requires measurements be made on un-encapsulated devices • requires careful setup and data analysis if meaningful results are to be obtained • needs considerable expertise • very time-consuming and expensive
Chemical	coating of semiconductor chip surface with thin layer of temperature-indicating chemical such as liquid-crystal material	• same as above • is relatively inexpensive	• requires measurements be made on un-encapsulated devices • chemical material requires careful application and may contaminate chip • has poorer resolution and accuracy than optical technique • needs considerable expertise • very time-consuming
Physical	mounting of extremely small thermocouple(s) (or other temperature-measuring device) directly on chip surface	• is relatively inexpensive	• is not practical because of difficulty of mounting the thermocouple(s) • presence of thermocouple affects heat source temperature • can theoretically make temperature measurement as accurately as optical technique, but resolution is very poor • requires measurements be made on un-encapsulated devices • very time-consuming
Electrical	using precalibrated temperature-sensitive parameter of the device under test	• fastest measurement technique — using correlation techniques, very suitable for high-volume testing applications • entry-level operator can make accurate and repeatable measurements • meets the intent of MIL-STD 883A, method 1012 • measurements can be made on completely encapsulated devices	• provides weighted-average measurement, but lacks resolution of optical or chemical techniques • requires some circuit and device expertise to properly interface to measurement circuitry • may require individual device precalibration prior to actual thermal-resistance measurement • test equipment is moderately expensive

devices to prevent the occurrence of hot spots within the chip active area. However, careful attention to the emissivity of the chip surface materials is required to obtain accurate readings.

The electrical technique is usually implemented with standard laboratory instruments or individually designed and built in-house test equipment. It has suffered from the lack of definitive measurement standards, but the advent of commercially available thermal-resistance testers will go a long way towards providing *de facto* standardization within the semiconductor industry, for their ease of operation and relatively low cost should appeal strongly to both manufacturers and users of semiconductors.

The electrical technique determines thermal resistance in an integrated circuit by using a sequence of voltage and current pulses to both heat the device and measure the change in junction temperature. As an indication of this change, a temperature-sensitive parameter, such as the forward-biased voltage drop of a diode under low-value constant-current conditions, is used.

There are several different temperature-sensitive parameters associated with ICs that can be used to make thermal-resistance measurements. The most commonly used is the substrate-isolation junction diode that occurs in all semiconductors (bipolar, metal-oxide-semiconductor, and complementary-MOS) except for silicon on sapphire and dielectrically isolated units. Figure 3 shows a schematic representative of an IC for thermal-resistance testing purposes. If voltages of the proper polarity are applied to the voltage terminals, the active circuit, whether it be analog or digital, will be turned on and power dissipation will occur, causing the junction temperature to rise. If the substrate isolation diode is forward-biased by a small constant current insufficient to cause self-heating before power is applied, the diode voltage will be higher than a similar measurement made after heating, because of the V_F/T_J diode characteristic illustrated in Fig. 4.

Electrical measurement technique

Voltage waveforms across the device under test for a typical measurement are shown in Fig. 5. During period t_1, a small reverse constant current is applied through the supply terminals of the device under test, causing the substrate-isolation diode to become forward-biased and produce voltage V_1. Power is then applied to the device being tested during t_2 when the supply voltage goes positive to the heating voltage, V_H (shown as V_2).

At the completion of t_2, V_H is quickly removed from

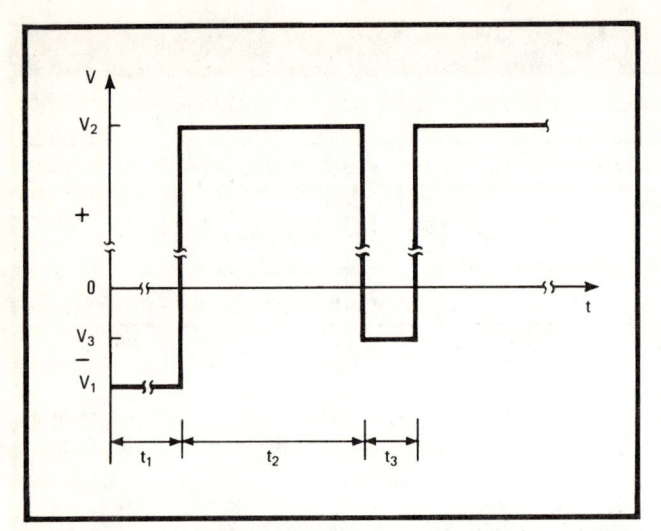

5. Timing. In a thermal-resistance measurement, the substrate diode is forward-biased during t_1. Power is supplied to the active device during t_2. Then it is shut off during t_3 when the diode is conducting. V_1-V_3 is proportional to junction temperature rise.

the device under test, and reverse constant current is again applied during period t_3. The diode voltage, V_3, will have a new value because of the higher junction temperature. The difference between V_1 and V_3 is proportional to junction temperature rise through the characteristic shown in Fig. 4, that is:

$$\Delta T_J = K|V_1 - V_3| \qquad (1)$$

where K is the reciprocal of the slope of Fig. 4. The power required to produce ΔT_J is the product of V_H and heating current I_H. The thermal resistance of the device under test from junction to some reference point (X) is:

$$\Theta_{JX} = \Delta T_J/P_D = K|V_1 - V_3|/V_H I_H \qquad (2)$$

The X subscript mentioned above can vary, as described previously, depending on what temperature reference is used for the device.

If the reference point is the device case, then the temperature of the case must remain constant during the entire test cycle. Similarly, if an ambient like air is the reference condition, then the air conditions surrounding the device under test must remain constant during the entire test cycle.

The timing of the various current and voltage levels used in this measurement has a critical effect on the accuracy of the Θ_{JX} measurement. Heating period t_2 must be long enough to assure junction temperature stabilization for the heating voltage applied. The voltage measurement, V_3, made during period t_3 must be made in a time interval as small and as close to the end of period t_2 as possible to avoid junction cooling effects. The transition between periods t_2 and t_3 must be very fast for the same reason.

Finally, t_2 must be very large compared to t_3 to ensure that the average and peak heating power are essentially the same, thus allowing direct computation of Θ_{JX} without a heating-power correction factor. Periods t_2 and t_3 should be repeated several thousand times to provide maximum system accuracy: period t_1 occurs only at the start of each test cycle.

A block diagram of a commercially available thermal-resistance tester, Sage's Theta 400, that uses the method previously discussed is shown in Fig. 6. The tester is built up out of seven basic blocks—heating source, measurement source, power supplies, switching, control logic, analog sampling and computation, and a display section.

Kelvin (four-terminal) contacts on the fixture holding the device under test eliminate the measurement errors caused by contact voltage drops under high current conditions. The heating and measurement sources feed the necessary test cycle voltage and currents required through one set of the contacts of the switch to the device being tested.

Voltage measurements are made across a second set of contacts and fed directly to the analog sampling and computation circuits. The display and indicators are driven by the control logic and analog sections. Test-cycle duration, voltage and current pulse timing, and display and indicator control circuits are all contained in the control logic section. The entire instrument takes up about 1 cubic foot of benchtop space.

Calibration

In order to measure thermal resistance, Θ_{JX}, it is necessary to insert a value of K, the reciprocal of the slope of the V_F versus T_J curve of a forward-biased semiconductor junction, into Eq. 2. The K parameter must be measured for a particular type of semiconductor and this value set into the thermal-resistance tester's K factor control. K is essentially the same for all semiconductors of the same type and number from the same manufacturer, even though the units may have been produced in several different manufacturing runs or batches. Typical variations in K are within 3% for devices within a given batch or from various batches.

The relatively constant value of K for a given device type and number allows it to be measured on a sample basis. Typically, once the average K value is determined for groups of 10 junction diodes taken from two or three different batches of semiconductor devices, then no further K-factor measurements are required except for periodic quality control.

The measurement system for the temperature-sensitive-parameter coefficient (K factor) is shown in Fig. 7. The current source should provide the necessary forward current, as specified by the thermal-resistance tester requirements, to an accuracy of ±5% or better. Forward voltage across the diode must be measured with millivolt resolution: a 3½-digit voltmeter with a 1-volt scale is recommended. The temperature monitor, which may be a thermocouple type or any other suitable instrument, should have a range of at least 150°C.

The linear nature of the temperature coefficient enables sufficient accuracy to be obtained from forward-voltage measurements at only two different temperatures. Room temperature may be one of them. The other can be some elevated temperature in the 75°-to-125°C range, depending on the type of parts to be measured.

The K factor for each diode is:

$$K = |(T_2-T_1)/(V_{F2}-V_{F1})|\, I_F = \text{constant}$$

where K = the temperature coefficient in °C/mv,
T_1 = the lower test temperature (usually room temperature),
T_2 = the higher test temperature (usually near the device maximum operating temperature specification),
V_{F1} = the forward voltage at I_F and T_1,
V_{F2} = the forward voltage at I_F and T_2,
I_F = the fixed forward measurement current, required by the thermal-resistance tester.

Once the average K value is determined, it is set on a thermal-resistance tester's K factor control whenever devices of the same type and number are to be tested for thermal resistance.

Measurement problems

There are two major problems in the previous thermal-resistance test method associated with the choice of the temperature-sensitive parameter. First, the use of the substrate-isolation diode for such a parameter produces a weighted-average indication of T_J because not all portions of the IC dissipate the same amount of power. Junction temperature gradients occur across the chip, and a lower-than-actual value of thermal resistance results.

Secondly, some IC fabrication technologies, such as dielectric isolation and silicon on sapphire, require some other diode for their temperature-sensitive parameter, because they lack the substrate diode.

The first problem can be partly overcome by sensing a temperature-sensitive parameter in the vicinity of the hottest portion of the IC chip, typically the output region. Any junction-isolated region that has a transistor in it will have its substrate diode, either directly or through a diffused resistor, available for use as the temperature-sensing element.

Diode connection

The other problem can be overcome if some form of diode connection is available through the pins of the device under test and can be coupled and decoupled during various portions of the test cycle.

A third potential problem occurs when attempts are made to test static devices, such as random-access or read-only memories, that dissipate significant power only when making the transition from one binary state to the other. External circuitry may be required to cycle these devices during the t_2 heating period.

Fortunately, for the purposes of thermal-resistance measurements, the thermal time constants for the various portions of the circuit differ by one or more orders of magnitude. To refer back to Fig. 1, for instance, $t_{\theta 1}$ is typically in the range of high tens to low hundreds of microseconds, while $t_{\theta 2}$ is in the hundred-millisecond range. Each of the other time constants is correspondingly greater.

The interface between die attachment and head-

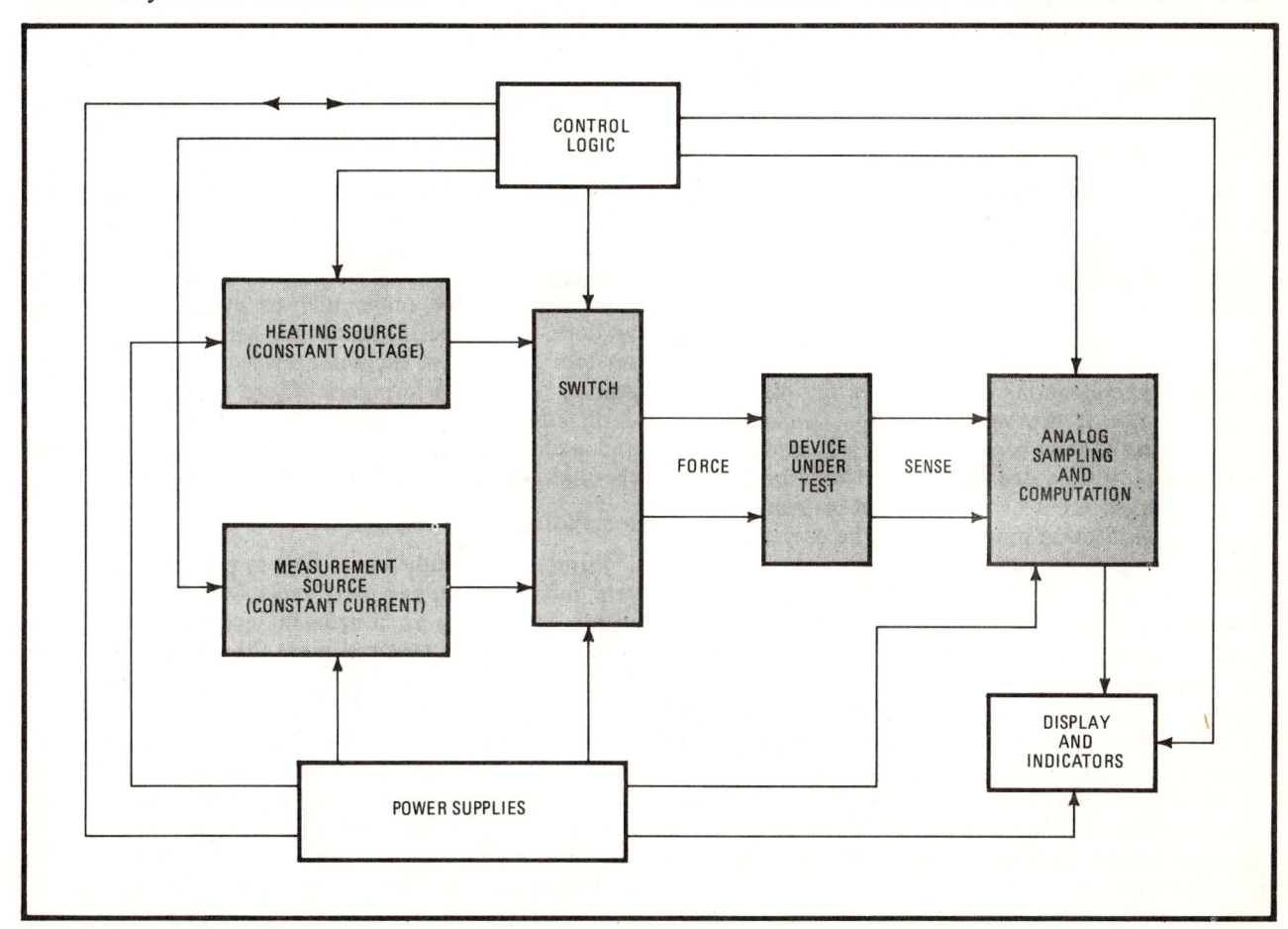

6. Thermal tester. Sage Enterprises' Theta 400 is an instrument designed for thermal-resistance measurement. All heating measurement sources, analog computation, and power sources are self-contained. Results are displayed digitally.

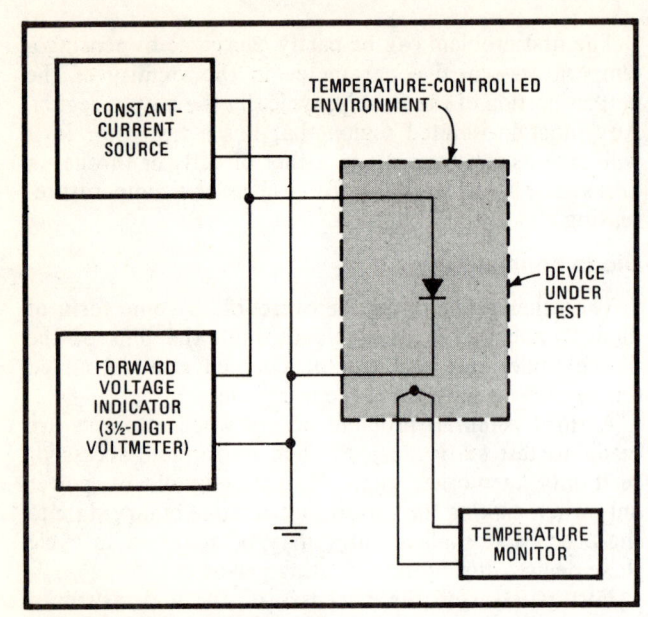

7. Temperature coefficient. The K factor, reciprocal slope of the V_F/T_J curve, is necessary to compute thermal resistance. Two measurements of diode forward drop—under constant current at room and at elevated temperature—are sufficient to calculate K.

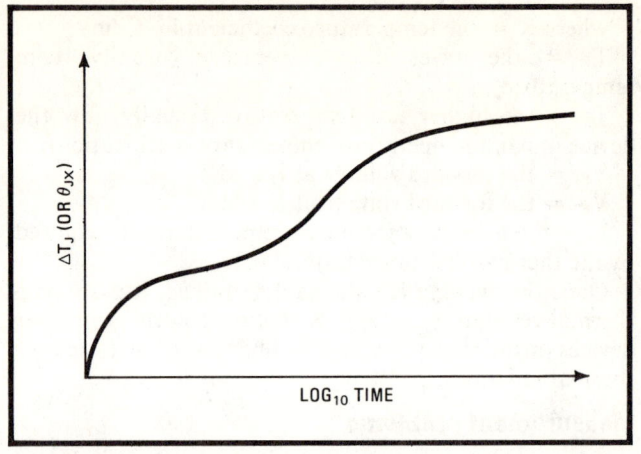

8. Temperature vs time. This curve displays junction-temperature change of a packaged IC as a function of time, assuming constant power dissipation. Differing time constants in the thermal circuitry cause junction temperature to vary as a function of time.

er/package starts at the bottom of the chip and ends at the top of the header/package. For mounting configurations in which the header and package are not the same, as in hybrid circuits where the chip might be mounted on some form of substrate, Fig. 1 would be modified to show additional thermal interfaces and their corresponding thermal circuitry.

Junction temperature variances

The existence of differing time constants in the IC's thermal circuitry causes the junction temperature (or thermal resistance) of the IC to vary as a function of time from the initial application of power to the device as shown in Fig. 8. The relatively short time constants from the IC active area to header/package cause the initial plateau to occur within several seconds. Then, depending on the package-mounting configuration, this plateau will either continue if the header/package temperature is fixed (because of good heat-sinking) or will lead to a new plateau as $t_{\theta 4}$ is exceeded and thermal equilibrium with the environment is reached. The second plateau typically occurs within 5 to 20 minutes when the environment is still air at room temperature.

The thermal circuitry from the header/package to the environment, whether the latter is still or moving air, liquid, or a heat sink, is relatively independent of the chip contained within the header/package. Thus, the thermal characterization of an IC chip/package configuration can be broken into two separate measurements: one for the IC active area-to-header/package, which only takes seconds for each measurement, and one for the header/package-to-environment configuration, which may take tens of minutes but does not have to be done for each device.

The semiconductor manufacturers should find thermal-resistance testers useful in many areas, including product characterization, manufacturing, and quality assurance. In the first area, thermal-resistance testers assist the device designer by providing thermal characterization of prototype devices as well as final versions. The effects of different packages (for instance, plastic versus ceramic dual in-line packages) can quickly be ascertained on meaningful sample sizes. The manufacturing area can use the thermal-resistance tester, or its derivatives, to quickly determine the consistency of die-attachment procedures and/or improvements. Quality-assurance operations would use the thermal-resistance tester to maintain an independent check on manufacturing and to insure conformity with high-reliability requirements.

The thermal-resistance tester can be equally well applied by semiconductor users. Once a satisfactory value of thermal resistance for an IC package in an actual module or subsystem has been established, the tester can be used to select suitable devices to meet the desired parameter range and to evaluate alternative vendors. Once specifications are set and a vendor or vendors are chosen, incoming inspection is required to monitor vendor performance. There is insufficient standardization among vendors, so users must select vendors and monitor them carefully to be sure of getting the thermal-resistance specification they want.

Instrument improvements

Efforts are already under way to provide more accurate, more flexible, and more versatile laboratory instruments. Ultimately an instrument accuracy of ±5% of reading should be possible, more than enough except in the most demanding device applications. For some semiconductor devices, a single heating voltage supply may not be adequate, necessitating an instrument capable of providing more supplies. Also, the problem of testing static memory devices can be overcome by clocking the inputs of the device under test. Versatility can be improved by building a single instrument with operator-controlled heating time to allow for package characterization, die-attachment evaluation, and paper-tape printout of sequential data for time plots of increasing thermal resistance and/or junction temperature. □

Countering the effects of vibration on board-and-chassis systems

A simple analysis tells the electronics designer whether the circuit boards in his system will need strengthening against vibration stress

by David S. Steinberg, *Singer Corp., Kearfott Division, Wayne, N. J.*

□ At some time in its life cycle, all electronic equipment encounters some form of vibration, at the very least when transported from manufacturer to customer. This vibration can cause fatigue failure in consumer systems, as well as in military or industrial systems, unless the electronics engineer has analyzed the design for its vulnerability to vibration stresses and built in an adequate safety margin against them.

The procedure is straightforward. It involves basic vibration theory, which is not normally part of the electronics curriculum. But "A guide to vibration analysis" explains the concepts involved, and the rest of this article tells how to combine these concepts into simple design rules for application to a board-and-chassis package.

The basic building block of today's electronic system is the easily serviced, plug-in printed-circuit board. These boards are mainly rectangular, epoxy-glass units with plated copper wiring. In general, a pc board acts like a flat plate in a vibration environment and is subject to fatigue failures, particularly at board resonance. Vibration fatigue failures in these boards usually take the form of broken leads, broken solder joints, and broken connectors. In extreme cases, parts have actually flown off a board under high vibration forces.

Applying the octave rule

Fatigue failures of these kinds can often be prevented by insuring that the boards and the chassis have different resonant frequencies. The goal is to avoid coincident resonances which can amplify acceleration (g) forces in adjacent structural elements very rapidly. For example, if the resonant frequency of a chassis is too close to that of a circuit board within the chassis, high acceleration forces may develop in the board. High accelerations produce large deflections resulting in stresses that can culminate in rapid fatigue failures.

To illustrate, the chassis and the multiple plug-in boards of Fig. 1a will behave like a system capable of vibrating along more than one axis simultaneously—in other words, like a system with multiple degrees of freedom. The chassis usually represents the first-degree-of-freedom system for externally induced

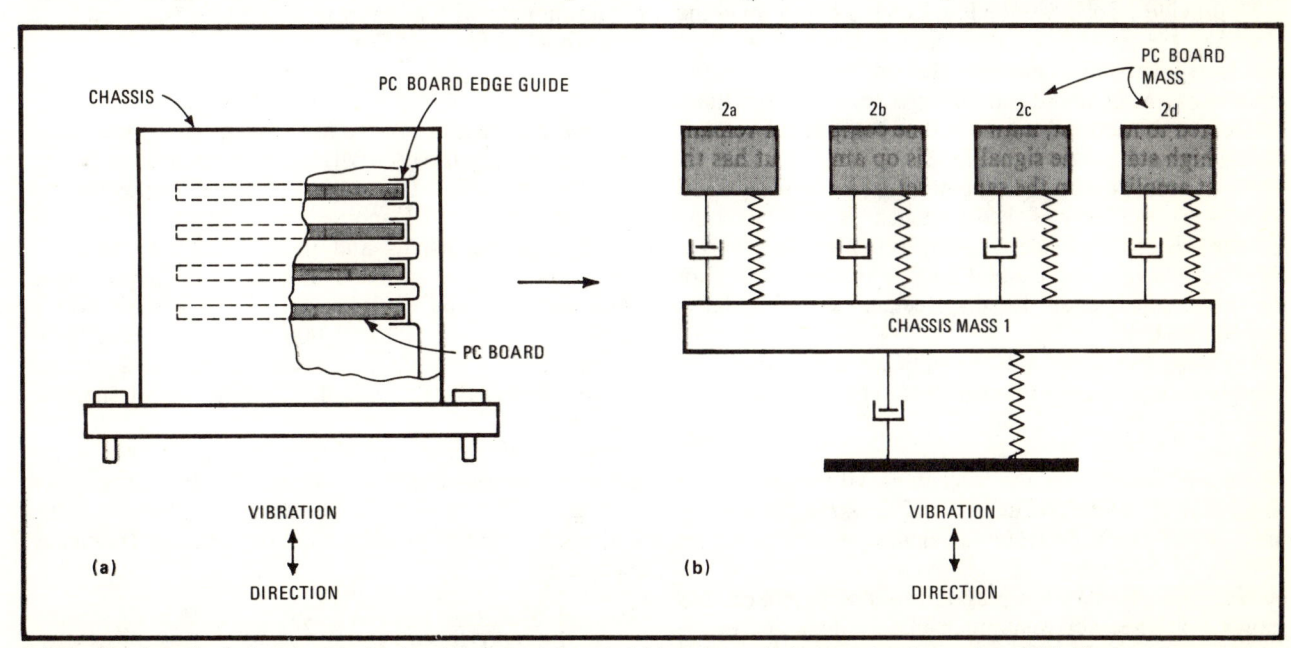

1. Vibrating boards. A chassis with plug-in printed-circuit boards (a) has the equivalent mass spring analog of (b). The boards represent a second-degree-of-freedom system with respect to the chassis, since dynamic forces must be transmitted through the chassis to the boards.

vibrations. The boards represent the second degree of freedom with respect to the chassis since the dynamic forces must pass through the chassis before they reach the pc boards.

Coincident resonances can generally be avoided in multiple-degree-of-freedom systems by following the octave rule. This rule states that in a series spring mass system (see p. 102), the natural frequency of an element should be doubled for each additional degree of freedom in order to avoid severe resonant amplifications caused by a coincident resonance. Thus in Fig. 1b if the natural frequency of the chassis (mass 1) is 100 hertz, the natural frequency of each pc board (masses 2a, 2b, 2c, and 2d) should be at least twice that, 200 Hz or higher if possible. This separation of resonances prevents the chassis resonance from amplifying the pc board resonance, which in turn reduces the dynamic stresses and increases the fatigue life of the system.

Where to begin

What is the best starting point for a vibration analysis of the packaging configuration of Fig. 1? Should the natural frequency of the chassis or the pc board be determined first?

Experience, along with extensive analysis and testing, favors starting with the pc board. Tests have shown that it is possible to provide a long fatigue life (greater than 10 million cycles) for plug-in pc boards by basing the requirement for the maximum dynamic single-amplitude displacement (Y) of a rectangular board on the length of the shorter side (b) of that board. The displacement relation is shown by the equation:

$$Y_{max} = 0.003 \, b \quad (1)$$

For example, the maximum dynamic single-amplitude displacement at the center of a rectangular pc board measuring 4 by 7 inches should be limited to a Y_{max} of 0.003×4.0 in. or 0.012 in.

To provide a good fatigue life, Eq. 1 is based upon the dynamic stresses that are developed in the lead wires of the electronic components mounted on the board. As the board vibrates up and down during a resonant condition, it forces the electrical wires on the components to bend back and forth, as shown in Fig. 2.

For a rectangular board, the most severe condition will occur when the body of the component is parallel to the shorter side of the pc board and at its center. The shorter side must have a more rapid change of curvature than the longer side because the displacement at the center is common to both.

The actual dynamic single-amplitude displacement of a single spring mass system can be determined from:

$$Y = 9.8 \, g \, Q/f^2 \quad (2)$$

where g is the acceleration force in gravity units, f is the sinusoidal vibration frequency, and Q is the transmissibility of the vibration from one element to the next (see p. 102).

Vibration test data on plug-in types of pc boards has shown that they act very much like a single-degree-of-freedom system when they are vibrating at their natural or fundamental resonant frequency. This test data shows

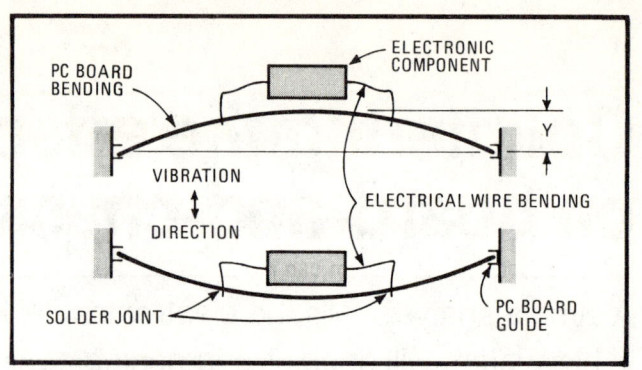

2. Over the waves. During a resonant vibration, a printed-circuit board moves up and down. These excursions can break leads and solder joints and even cause components to fly off.

that the approximate transmissibility for many different types of pc boards can be determined from this frequency, f_n:

$$Q = A \, (f_n)^{1/2} \quad (3)$$

where A is a dimensionless empirical constant dependent on the natural frequency, f_n, and acceleration inputs.

A large number of factors influence the transmissibility of a plug-in board. These factors will vary from one manufacturer to another because they each use different construction methods. Other variations include the component size, type of edge guides, type of electrical plug-in connector, conformal coating, and acceleration force.

Given acceleration inputs that vary from about 3 to 10 g, the value of A in Eq. 3 appears to be about 1.0 for pc boards with resonant frequencies between 100 and 400 Hz, drops to about 0.70 for boards with lower resonant frequencies between about 50 to 100 Hz, and rises to about 1.40 for those with resonant frequencies between about 400 to 700 Hz.

Equations 1, 2, and 3 can be combined to establish the minimum natural frequency required by a pc board for a long vibration fatigue life as:

$$f_n = (9.8 \, gA/0.003 \, b)^{2/3} \quad (4)$$

As an example of how to use this equation, suppose a typical plug-in printed-circuit board is required to have a long fatigue life—10 million fatigue cycles at its resonant frequency. Suppose also that the board has 5-by-8-inch dimensions and must operate in an electronic system that will be subjected to a prolonged sinusoidal vibration environment of 6.0 g peak over a frequency range of 100 to 1,000 Hz. What needs to be determined is the minimum required pc board natural frequency plus the maximum allowable chassis natural frequency.

Start by assuming the value of A is 1.0, to see where the pc board resonant frequency will be. From the above data, g = 6 and b = 5. Substituting these values in Eq. 4, the minimum required pc board resonant frequency becomes:

$$f_n = \left[\frac{9.8 \times 6.0 \times 1.0}{0.003 \times 5.0}\right]^{2/3} = 248 \text{ Hz}$$

Since the board's resonant frequency is between 100 and

A guide to vibration analysis

Any discussion of vibration will involve references to degrees of freedom, transmissibility, coincident resonances, and series spring mass systems.

The **degrees of freedom** of a vibrating system describe the coordinates necessary to locate the position of the vibrating element at any time. For example, a single-degree-of-freedom system can move along only one axis, in both directions. A two-degree-of-freedom system will require two coordinates to describe the position of the elements. A multiple-degree-of-freedom system generally has many elements that can move along many axes.

The **transmissibility** of a vibrating system is usually defined as the ratio of the maximum output force divided by the input force at a given frequency, or the maximum output displacement divided by the input displacement at a given frequency. In a lightly damped system, the maximum output force can easily be 100 times the input force. Such a system would have a transmissibility of 100. If a printed-circuit board has a maximum vibration input displacement of 0.001 inch at its edges at a frequency of 180 hertz, and the center of the board has a maximum vibration of displacement of 0.100 in. at its center at the same frequency, then the board has a transmissibility of 0.100/0.001 or 100.

A **coincident resonance** is a condition where two systems are joined together (or coupled) and both systems are vibrating near the resonant frequencies at the same time. It often happens that the amplified output from the first system turns out to be the input to the second system, which amplifies that input a second time. If the transmissibility of the first system by itself is about 50, and if the transmissibility of the second system by itself is about 50, the joint (or coupled) transmissibility of the second system can approach about 50 × 50 or 2,500. If this happens in a real system, and it often does, the fatigue life is very short.

A **series spring mass system** consists of several springs and masses attached to one another, in a string-like manner. Any force or motion in the outermost member must pass through each adjacent member until it reaches the support. The figure shows a series system with three springs and three masses.

Of course, an electronic box does not really have springs and masses bouncing around during vibration—the springs and masses are used only as mathematical models, to simulate a system that will have vibration characteristics similar to the box. The mathematics associated with a spring and mass system is relatively simple compared to the mathematics required to analyze a complete box. The simplified mathematics permits a quick evaluation to be made of the structure supporting the electronics, to see how well it will hold up under the pounding it receives in a vibrating environment.

---

400 Hz, the assumed value of 1.0 for A is valid.

The natural frequency of the chassis that supports the pc board must also be established in order to prevent higher transmissibilities from developing in the board. Following the octave rule, the maximum natural frequency of the chassis must not exceed one half the natural frequency of the pc board, in this case, 124 Hz.

Finding that the pc board's minimum resonant frequency should be 248 Hz is only half the solution. It is still necessary to determine its actual natural frequency. Knowing the weight and thickness of the board of Fig. 3, it is possible to determine board natural frequency:

$$f_n = \frac{\pi}{2}\left[\frac{D}{\rho}\right]^{1/2}\left[\frac{1}{a^2} + \frac{1}{b^2}\right] \tag{5}$$

where D is the plate stiffness factor in pound-inches, ρ is the mass per unit area of pc board, a is board length, and b is board width.

The equation for plate stiffness factor is:

$$D = Eh^3/12(1-\mu^2) \tag{6}$$

where E is the epoxy Fiberglass modulus of elasticity, h is board thickness, and μ is Poisson's ratio for the board. In this example, $E = 2.0 \times 10^6$ lb/in.^{2}, h = 0.090 in., and $\mu = 0.12$, so that:

$$D = (2.0 \times 10^6 \times 0.090^3)/12(1 - 0.12^2) \text{ lb in.}$$
$$= 123.3 \text{ lb in.}$$

The equation for mass per unit board area is:

$$\rho = W/gab$$

so that for the 0.5-lb, 5-by-8-inch board, W = 0.5 lb, a = 8 in., and b = 5 in., and acceleration due to gravity, g = 386 in./s^2:

$$\rho = (0.5)/(386 \times 8 \times 5) \text{ lb s}^2/\text{in.}^3$$
$$= 3.24 \times 10^{-5} \text{ lb s}^2/\text{in.}^3$$

Substituting the values of D and ρ in Eq. 5:

$$f_n = \frac{\pi}{2}\left[\frac{123.3}{3.24 \times 10^{-5}}\right]^{1/2}\left[\frac{1}{8^2} + \frac{1}{5^2}\right]$$

$$= 170.4 \text{ Hz}$$

Use of Eq. 5 results in a board natural frequency of only 170.4 Hz, which is below the minimum required natural frequency of 248 Hz. This design is not satisfactory, and its natural frequency must be increased. If pc board thickness is increased to 0.125 in., the natural frequency will be increased to 287.7 Hz, which is satisfactory. If for some reason the pc board thickness cannot be increased, then ribs can be added to stiffen the board. A single vertical rib 0.090 in. thick and 0.250 in. high can be made of epoxy Fiberglass and cemented across the center of the board parallel to the 5.0-in. dimension. This would raise the natural frequency to 275 Hz, creating a satisfactory solution for the vibration environment.

A design of this type might take perhaps one man-day, but it could eliminate many hours of time in the field spent servicing vibration-caused failures. □

Bibliography
David S. Steinberg, "Vibration Analysis for Electronics Equipment," John Wiley & Sons, New York, N.Y. 1973.

What happens to semiconductors in a nuclear environment?

For designers who must select components to survive high-energy radiation, it's important to know how each type reacts

by David K. Myers, *Fairchild Camera and Instrument Corp., Mountain View, Calif.*

☐ Of all the many ambient conditions to which semiconductor devices are exposed, from a computer's air-conditioned room to under an automobile's hood, none is as demanding as the nuclear radiation encountered in certain military and space environments or in the nuclear industrial field. Unhardened digital electronic equipment can fail when exposed to ionizing radiation doses of as little as 10^3 rads (Si)—out in space, for example, in the Van Allen Belt—or to a neutron fluence of as little as 10^{11} neutrons per square centimeter—near a nuclear reactor, say. (Rads (Si) stands for roentgens absorbed dose in silicon, while a fluence is defined as the time integral of neutron flux.)

Anyone engaged in designing circuitry for use in such environments must be knowledgeable about their differing effects on different semiconductor technologies. Exposure to high-energy radiation introduces primary structural defects into semiconductor materials (and hence changes their electrical characteristics) in ways that depend partly on the duration and type of incident radiation and partly on that particular semiconductor material's resistivity, impurity types and concentrations, temperature, and carrier-injection levels.

The nuclear environments to be considered here are:
■ Fast neutrons, which can permanently degrade gain in both bipolar and metal-oxide-semiconductor devices and increase the saturation voltage of bipolar transistors.
■ Steady-state ionizing radiation (the total dose), which can increase leakage current in bipolar devices and alter threshold voltages in MOS and particularly complementary-MOS devices.
■ The transient ionizing dose rate, which at a high enough level generates photocurrents in all reverse-biased pn junctions, causing changes of logic state in bipolar digital circuits and latch-up in C-MOS devices.

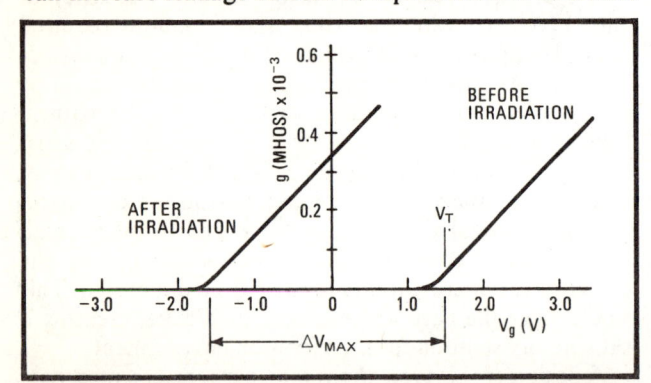

Neutron effects

In general, it is only when neutron levels rise to 10^{10} to 10^{12} n/cm² (E = 10 kiloelectronvolts) that silicon devices start exhibiting changes in their electrical characteristics. The base transit time and the base width of a bipolar transistor are the main physical parameters affected by exposure to fast neutrons, as can be inferred from the degradation in current gain (h_{FE}). Modern semiconductor manufacturing methods measure neither of these parameters directly, but do control h_{FE} and the gain bandwidth product (f_t), from which base width can be deduced.

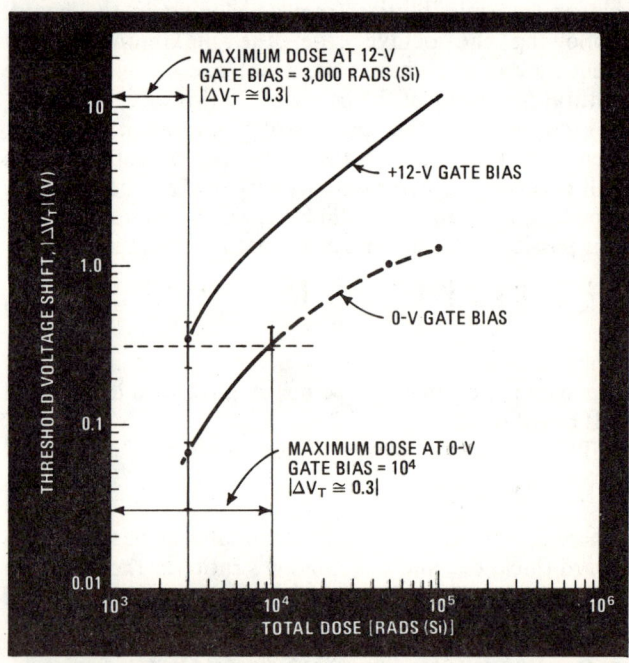

1. Radiation shift. In MOS and C-MOS devices, ionizing radiation alters gate turn-on voltage, changing the operating point radically. In the MOSFET curve shown, a threshold voltage change of almost 3 V is observed after exposure of the devices to ionizing irradiation.

2. Radiated RAM. Threshold-voltage shifts are a function of radiation dosage and bias conditions for 4,096-bit n-MOS random-access memories. Units with gate bias fail at lower radiation levels than zero-biased units. A 0.2-to-0.3-V change in V_T induces failure.

TABLE 1: COMPARING THE RADIATION SUSCEPTIBILITY OF VARIOUS SEMICONDUCTORS									
Radiation environment \ Semiconductor technology	Discrete bipolar transistors and J-FETs	Silicon controlled rectifiers	TTL	Low-power Schottky TTL	Analog integrated circuits	C-MOS	n-MOS	Light-emitting diodes	Isoplanar II ECL
Neutrons (c/nm²)	10^{10}–10^{12}	10^{10}–10^{12}	10^{14}	10^{14}	10^{13}	10^{15}	10^{15}	10^{13}	$>10^{15}$
Ionizing radiation — Total dose (rads (Si))	$>10^4$	10^4	10^6	10^6	5×10^4 – 10^5	10^3–10^4	10^3	$>10^5$	10^7
Ionizing radiation — Transient dose rate (rads (Si)/s) (upset or saturation)	–	10^3	10^7	5×10^7	10^6	10^7	10^5	–	$>10^8$
Ionizing radiation — Transient dose rate (rads (Si)/s) (survival)	10^{10}	10^{10}	$>10^{10}$	$>10^{10}$	$>10^{10}$	10^9	10^{10}	$>10^{10}$	10^{11}
Ionizing radiation — Dormant total dose (zero bias)	$>10^4$	10^4	10^6	10^6	10^5	10^6	10^4	$>10^5$	$>10^7$

Burnout by EMP

An electromagnetic pulse from a nuclear event can couple into a system's cables and antennas and create voltage/current spikes that may fuse the metalization on a semiconductor surface. Usually the interconnect system on a semiconductor device is a thin metal layer, only 10,000 angstroms or so thick, and will fuse at a current density of 10^6 amperes per square centimeter. Most integrated-circuit metalization stripe widths are designed to keep the current density below 10^5 A/cm² during normal operation, including worst-case testing.

EMP-induced burnout of semiconductor junctions is therefore a serious problem. But while it is a failure mode of semiconductor devices, it originates in the system design and not in semiconductor selection or reliability. The electromagnetic pulse must be shielded, filtered or shunted to ground.

Data available from Government laboratories and semiconductor manufacturers on semiconductors subjected to neutron irradiation indicates that double-diffused, epitaxially constructed integrated circuits, both digital and linear, will function within their original specification limits to neutron levels of 5×10^{13} n/cm² (E = 1 megaelectronvolt). This holds for diode- and transistor-transistor logic, as well as for low-power Schottky TTL. MOS circuits, whether n- or p-channel or C-MOS, are majority-carrier devices and not susceptible to neutron irradiation below 10^{15} n/cm².

The total ionizing dose

A steady state of ionization increases bipolar transistor leakage current most markedly in low-current, large-area devices. But even under worst-case conditions, these increased leakages are not enough to cause circuit failure at radiation levels below 10^5 rads (Si). Indeed, in many cases, bipolar integrated circuits have functioned well at levels in excess of 10^7 rads (Si). Tests run on DTL, TTL, and low-power Schottky TTL circuits reveal radiation-induced changes only above 10^6 rads (Si).

The effect of a total ionizing dose on MOS devices is more drastic. It permanently changes the crucial threshold voltage, V_T, which is applied to the gate of a MOS field-effect transistor to create the source-to-drain conduction path or channel. This change can be attributed to the buildup of a trapped positive charge in the gate-oxide insulator and to the creation of fast surface states at the interface of the silicon and silicon dioxide. The result is a marked shift in the operating point of a device (Fig. 1).

Recent radiation tests indicate that n-channel MOS dynamic random-access memories are very sensitive to ionizing radiation, having a nearly 100% failure rate at 3,500 rads (Si), regardless of manufacturer. For instance, the major failure mode of 4,096-bit dynamic n-MOS RAMs is the incidence of decoders stuck in the logic 1 state, which in turn is due to changes in threshold voltage that exceed the operating design tolerance.

Most current n-MOS test data is derived from these 4-k RAMs, but other large-scale integrated circuits like 16,384-bit RAMs, microprocessors, and similar complex n-MOS chips are also sensitive to continuous ionization, being manufactured according to similar design rules and processing. Failure threshold is 1,700 rads (Si) for 4,096-bit dynamic RAMs and 1,000 rads (Si) for n-MOS microprocessors.

Precise threshold voltage changes in n-MOS units are a function of dose and bias, as shown in Fig. 2. Under a normal +12-volt gate bias, shifts in V_T of 0.2 to 0.4 v have been observed at 3,000 rads (Si). Circuit analysis indicates that the n-MOS electrical designs will tolerate a change of 0.2 to 0.3 v in V_T without failing. Tests of MOS devices at zero gate bias (again, see Fig. 2) show they will fail only with the approximately 0.3-v shift in V_T caused by a dose of 10^4 rads (Si).

Radiation-induced shifts in threshold voltage similar to those described for n-MOS also occur with C-MOS semiconductors. The operating design tolerance $|\Delta V_T|$ is usually 1 v for commercial C-MOS products, indicating that a dose of 10^4 rads (Si) is required to cause failure. Actual cobalt-60 irradiations of Fairchild Isoplanar

TABLE 2· RADIATION SUSCEPTIBILITY OF EMITTER-COUPLED LOGIC			
Data source	Northrop	Sandia	Fairchild
Pulsed ionizing radiation			
Narrow-pulse transient failure level	3×10^8 rads (Si)/s	no tests performed	$5-7 \times 10^8$ rads (Si)/s
Wide-pulse transient failure level	$1.1-1.4 \times 10^8$ rads (Si)/s	"	no tests performed
Permanent-damage failure level	not determined — greater than 1.5×10^{11} rads (Si)/s	"	10^{11} rads (Si)/s maximum level (flash X-ray equipment*)
Neutron/gamma permanent damage			
Mean neutron failure level	2.2×10^{15} n/cm^2 (1-MeV equivalent)	1×10^{15} n/cm^2 (1 MeV equivalent)	1×10^{15} n/cm^2 * (1 MeV equivalent)
Total gamma dose	6.6×10^6 rads (Si)	2.5×10^7 rads (Si)	10^7 rads (Si)*
Observed neutron failure-level range	$1.9-2.6 \times 10^{15}$ n/cm^2 (1 MeV equivalent)	1×10^{15} n/cm^2 * (1 MeV equivalent)	1×10^{15} n/cm^2 * (1 MeV equivalent)
Device type tested	MC1678L 4-bit counter	custom IC fabricated by TRW	Isoplanar II, ECL F100101, F100117, F100102, F100141 and kit parts
Transistor gain-bandwidth product (f_T)	2.0–2.5 GHz	1.5 GHz	4.5–5.5 GHz

*Maximum radiation exposure level; no failures were observed at this level.

C-MOS and other available C-MOS devices cause device failures at above 5×10^4 rads (Si).

If the C-MOS design margin for V_T is reduced for electrical performance reasons, then the radiation tolerance would be sacrificed. This is the case for certain C-MOS circuits with a V_T of about 0.2 v, which all fail at approximately 2,000 rads (Si).

In general, improved tolerance of ionizing radiation could be realized by use of hardened oxide manufacturing techniques and circuit design modifications. However, though increasing the circuit V_T operating tolerance would raise the ionizing radiation failure threshold, it would also adversely affect the power, speed, component density, chip size, and yield.

Loss of memory and latch-up

A transient dose of ionizing radiation creates a photocurrent in any reverse-biased pn junction, such as the collector-base junctions of transistors and the pn junctions used for isolation in standard bipolar integrated circuits. These photocurrents can be large enough to cause digital circuits to change state, from a 1 to a 0. But though they may change the content of memories, they cause no permanent failure. For some programs, logic upset is acceptable, but survival at the specified transient radiation dose rate is required. Tests show that DTL, TTL, and low-power Schottky TTL devices will change logic state above a dose 5×10^6 rads (Si)/s and survive 10^{10} rads (Si)/s.

These transient photocurrents can induce another phenomenon, known as latch-up, in those types of IC that can be driven into silicon-controlled-rectifier action or second breakdown. In this situation, they force the device to latch into one state and remain there until the power is interrupted or the circuit destroys itself.

Such radiation-induced latch-up has not been observed in digital or linear bipolar ICs employing double-diffused epitaxial fabrication methods and operating within specifications. Fairchild has had over 140,000 low-power DTL devices tested by outside contractors to levels of approximately 10^{10} rads (Si) without a true latch-up failure. Nor has any been observed in TTL, Schottky TTL, or bipolar operational amplifier and comparator circuits.

Latch-up of triple-diffused ICs does occur. But this technology has not been used to manufacture commercially available circuits for over five years.

MOS resistance to latch-up, on the other hand, varies with the process used. The problem has not been found to afflict n-MOS devices, which lack the fourth junction necessary for SCR action. But junction-isolated C-MOS ICs, which have that fourth junction, have been observed to latch up at dose rates as low as 3×10^8 rads (Si)/s. When operated at 5 V in the latch-up state, C-MOS devices return to normal operation after the power has been interrupted. But when operated at 10 V, they fail catastrophically because of latch-up.

It is worth noting that the latch-up dose-rate level of C-MOS devices varies with the manufacturer, device function, and chip design: there is a very uniform susceptibility to latch-up for devices with the same function from the same manufacturer.

For bipolar devices, it is different. The radiation tolerance of a generic bipolar family can be determined by testing sample parts of the family, since circuit design rules and manufacturing processes are constant throughout. This theory has been tested and verified on low-power Schottky TTL ICs manufactured by Fairchild Semiconductor.

Low-power Schottky parts made by other IC firms have been tested using the same radiation criteria. These results, with the Fairchild data, add to the overall confidence that a specific bipolar part type has a specified radiation tolerance irrespective of its source.

Only limited testing has been reported on the radiation tolerance of high-speed emitter-coupled logic. A data summary of ECL failure threshold levels for various radiation environments is presented in Table 2. Nevertheless, a preliminary comparison of ECL with a number of other types of ICs and discrete semiconductor devices (Table 1) indicates that ECL easily excels them all in radiation tolerance. □

Part 7
Computer Aided Design

Computer-engineer partnerships produce precise layouts fast

With the evolution of design automation, a tool for the placement and routing of complex boards has been added to the designer's repertoire

by Richard Larson, Automated Systems Inc., El Segundo, Calif.

☐ Mating the ingenuity of man and the untiring rapidity of the machine is an old story, but only now is it coming into full flower in the design end of electronics. As digital circuitry steadily becomes more complex and more densely packaged, designers are finding computer-controlled techniques to be invaluable tools in laying out their circuit boards.

These tools have progressed beyond the point where the computer simply functions as an assistant to the designer, as in computer-aided design techniques. In many cases, the computers can take over the job of circuit routing, using design-automation techniques that have evolved to the point where software modules have given them a broad range of capabilities (Fig. 1).

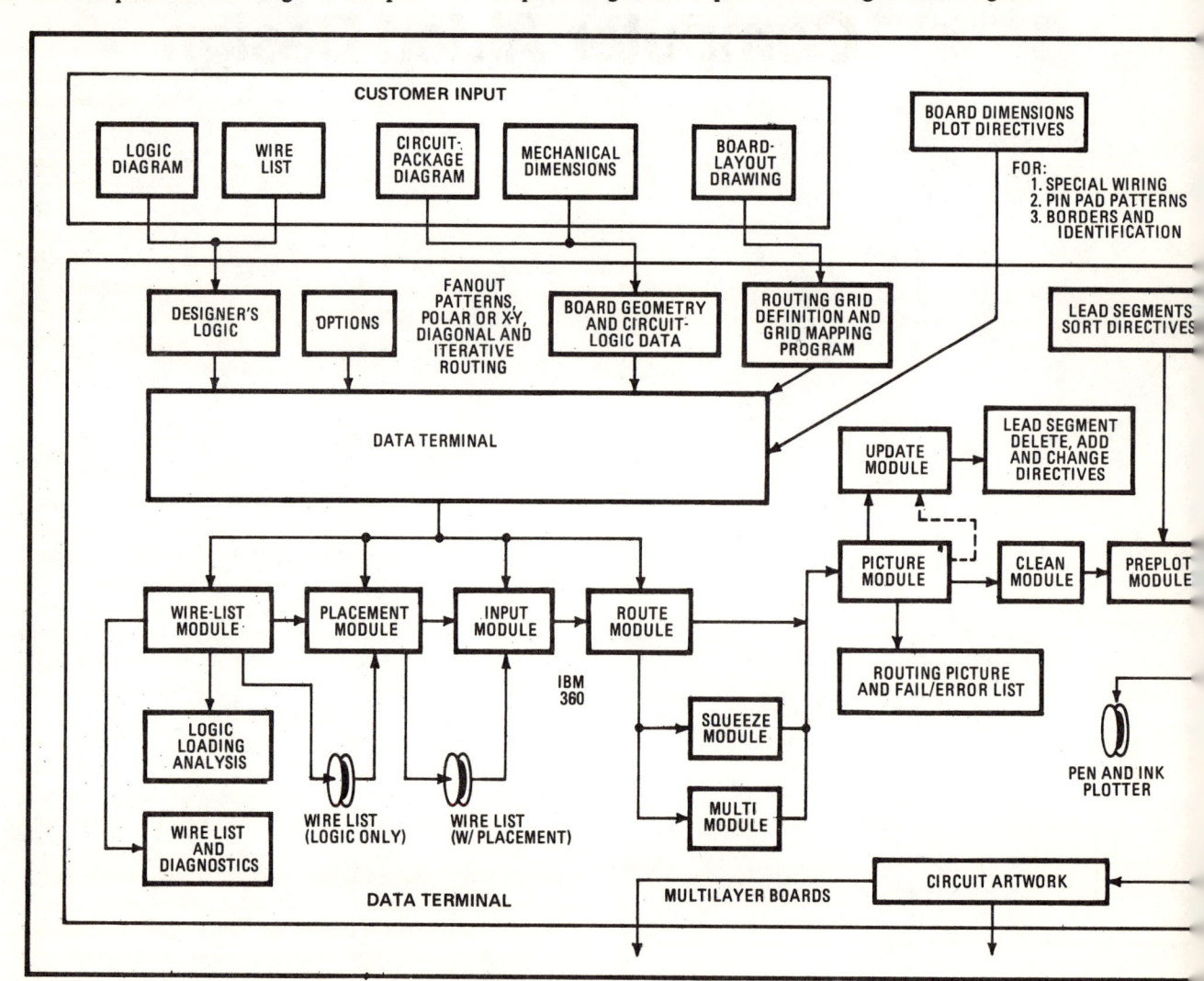

The lineage of design automation can be traced directly back to nonautomated techniques. To solve the problems inherent in drawing highly precise artwork, digitizing techniques appeared. Then CAD techniques came along to meet the constraints of time, cost, and output that grew in importance as large-scale integration developed. Each of these alternatives has its place in present-day design depending upon the factors involved (table). However, assessing these factors does require some comparisons of the four approaches.

Design without computers

Most printed-circuit boards are designed by engineers working without computer-controlled tools. The designers at the drawing boards place components and route interconnections by trial and error. They prepare the finished artwork either with pen and ink or by hand taping. This preparation is a tedious and time-consuming task, and finished art is frequently less than exact. However, the nonautomated approach will do the job—given a board featuring 50 or 60 mostly discrete components, densities in the area of 0.13 square inch per interconnect, and ample tolerance in the precision of the artwork that must be turned out.

The introduction of big multilayer boards carrying large numbers of digital integrated circuits in dual in-line packages or flat packs has relegated the nonautomated method to the design of simpler boards. The multilayer board's artwork with its requirement for registration of multiple layers, its increased component density, and the complex routing of its interconnections requires a degree of precision that sorely taxes the designer working, without computer-controlled tools.

To meet the need for precision artwork in the design of these densely packed boards, digitizing evolved. Sitting at a large back-lit drawing board, a designer works from the printed-circuit layout he has developed. He uses a digitizing pen, a cursor, connected to a minicomputer that encodes the X and Y coordinates of each circuit feature. The resulting punched paper tape or cards can be used with a simple line plotter or a photoplotter to generate artwork of the required quality.

Checking with digitizing

The latest generation of digitizers can make some design checks. In fact, it is now possible to encode both the wire list for a board and its design specifications. After the design has been plotted, it can be taken to a computer to be checked against this input to be sure that all interconnections have been made correctly. It is also possible to check such specifications as minimum clearances between pads and lines and so on.

However, these are historical checks. If errors have been made, it's back to the drawing board and to the digitizing board for redesign and replotting. Because of this factor and because digitizing does not help the initial design of the board, this design tool is probably best suited for the simplest of today's complex boards.

The next step up in the design hierarchy is the interactive CAD technique in which a cathode-ray-tube terminal, keyboard, and a small digitizer combine with a computer or minicomputer. This combination gives the designer what amounts to an automated drawing board from which to work during the design process.

In a CAD setup, a designer places and routes a board and sees his result displayed on the terminal. If a design specification programmed into the computer is violated, that fact is announced in real time on the CRT. Some systems have refined this process, with software programs that enable the computer to make suggestions about routing to the designer.

CAD helps out

Such systems can work with varying amounts of input, ranging from simple mechanical dimensions of the board and descriptions of devices to complete design specifications and wire lists. This flexibility is especially useful in the case of a design of a single board. The designer can bypass some of the initial setup costs of more sophisticated design-automation systems. CAD systems are also useful for analog boards with components of a wide variety of sizes and shapes, since topographic placement problems can be solved by the engineer as the design evolves towards completion.

Anyone with a CAD system in house should probably be looking for boards that can be designed interactively in about a third the time otherwise required. To achieve a two-year payback, the design volume must keep the

1. Design flow. A design-automation system accepts the five parameters shown on the customer-input block. The system's primary output is 1:1 printed-circuit artwork, but control tapes for component insertion, numerically controlled drill tapes, solder-mask and assembly drawings, and diagnostics also can be produced.

287

COST EFFECTIVENESS FOR VARIOUS PC DESIGN APPROACHES

Variables to be considered	Design by engineer and manual art preparation	Design by engineer and digitized art	Computer-aided design and computerized art	Design automation (computerized design and art)
Number of boards with mechanical commonality	1 – 3	1 – 3	1 – 10	3 and up
% discrete components	over 50%	over 50%	over 50%	under 50%
Number of ICs per board	up to 20	up to 50	up to 150	20 – 300
Discretionary cheating allowed on specs	yes	yes	yes	no
Production quantity per design	under 10	10 and up	10 and up	10 – 100
Need for computer-based tooling	no	yes	yes	yes
Need for computer-generated collateral documentation, engineering reports	no	no	limited	extensive
Need for computer-generated tape for interim wirewrap	no	no	no	yes
Available from service bureaus	yes	yes	no	yes
Capital investment for in-house system	n/a	$10,000 – $50,000	$100,000 – $250,000	n/a
Need for diagnostic checks such as loading analysis	no	no	yes	yes
Machine-readable input needed for in-house computer-aided design	no	no	yes	yes

system in use for the equivalent of 40 hours a week year round.

The next level of sophistication, design automation, embodies the concept of the computer as designer rather than as an aid to the designer. The goal for the developers of these systems is software programs enabling a mainframe computer to approximate the creative, problem-solving abilities of a skilled designer in assuming full responsibility for the placement and routing of complex boards. In addition, such systems must generate the precision artwork, error checks, manufacturing tools, and engineering reports provided by CAD systems.

Working with design automation

In early systems, a designer would first place components, while the computer was programmed to route as many of the interconnects as possible, within given design specifications. The technique used was a maze search. The computer would take one interconnect at a time, find the shortest route for its completion, and put it onto the board design.

Once in, the routing was there to stay. There was no way for the machine to go back and optimize the interconnections. Therefore it was common for first-generation systems to end their work with 10% to 20% of the interconnects not made. A designer would have to step in, undoing much of what the computer had done, to try to find paths for the unrouted connections.

Improving the routing

The routing problem made it readily apparent that programs with more sophisticated capabilities were needed. Iterative routing (Fig. 2) was one of the first breakthroughs. With this software, the computer can remove paths impeding later ones and find alternative routes for them, within localized areas of the board. This capability markedly decreased the number of nonroutings, especially on boards with discretionary vias: plated-through holes that connect plated wiring on different layers of a multilayer pc board. A discretionary via can be moved anywhere on the board to maintain minimum clearances of lines and spaces.

However, iterative routing falters on designs with fixed via sites or none at all, where it is not possible for the computer to unravel wiring crossovers simply by introducing a via. To cope with this kind of board, another software module was developed (Fig. 3). It enables the computer, temporarily and again in localized areas of the board, to disregard requirements for minimum clearances. Using this technique, almost all of the nonroutings could now be completed.

Ensuring manufacturability

With these additions to the capabilities of design automation, a computer can totally route most boards. But such designs share certain characteristics that make them more difficult to manufacture than equivalent boards routed by designers.

Because the computer works on a grid pattern, it will "staircase" the routing to get from one point to another on the grid, resulting in a longer path than needed. Because it looks only for the most accessible route that does not violate minimum spacing requirements, it may put paths close together, when they might have been more widely spaced for greater ease of fabrication. Moreover, paths are sometimes routed in an unnecessarily devious manner, and inessential vias are sometimes inserted by the computer.

To improve manufacturability, programmers have devised a series of software modules to go back over the routed boards, automatically correcting these complications. Staircased paths are turned into diagonals, spacing and line widths are expanded beyond the minimum wherever possible, unnecessary vias are eliminated, and paths are shortened and straightened out.

Placing parts precisely

Software engineers next turned to the problem of assignment of gates and placement of components, until then a procedure handled by circuit designers alone. They devised a module that could reconcile the logic diagram and the list of DIP types available. The software assigns similar gate types to a particular kind of package housing two to four of these gates. The criterion is the number of signals the gates have in common. Also, the program can place the packages on the board.

The computer then considers how to decrease the total wire length, by moving logic elements either from one slot to another within their assigned package or from one package to another, and by moving packages from one location to another on the board.

In a further enhancement of placement procedures, the computer analyzes crossing counts: the number of signals that have to cross given horizontal or vertical grid lines. Then it swaps pairs or groups of packages to minimize the crossings.

With the development of these sophisticated computerized placement techniques, design automation can take on the full layout of a board with little if any intervention from the designer. Given logic diagrams or wiring lists, plus board geometry and design specifications in machine-readable form, the computer can take over complete responsibility for figuring out a layout. The results are virtually indistinguishable from those designed by skilled engineers taking a far longer period of time. Typical turnaround time for the design of a board with 80 to 100 integrated circuits, for example, has dropped from 5 to 6 weeks down to 2 or 3 weeks.

The latest additions to the computerized design process enable it to tackle particular problems created by technological advances in the electronics industry. For example, special routers have been developed to search all layers of a board at one time. This searching eliminates very dense routing of the first few layers and far more sparse routing of the final layers.

Both computer-aided design and design automation can automatically produce a wide range of manufacturing tools from the same data used to produce the art. These tools include coded drill templates, assembly drawings, silk-screen artwork for outside cover layers, solder-mask artwork, numerically controlled drill tapes, parts lists, wire lists, control tapes for automatic-insertion machines, tapes for continuity test equipment, and network-description tapes for logic testers.

Fringe benefits abound

In addition, design-automation systems (but not CAD systems) can automatically provide computer-generated engineering reports detailing spare gates, unused connector pins, conductor length, parallelism, and thermal placement. They also can provide a capacitive-loading analysis for complementary-metal-oxide-semiconductor ICs and an emitter-coupled-logic routing analysis (including such information as stub lengths and maximum distances from source to source).

The advances in design automation do not render other techniques obsolete. For example, it is easy with CAD systems to make changes to a design in a highly controlled environment. If an interconnection must be changed, it is a very simple matter to call the design up, display it on the CRT, and make the change manually. With design automation, the process is equally straightforward. The system is asked to reproduce the reports. Then additions and deletions to the wire list are punched in and the revised wire list fed into the system, which will perform the rerouting. However, this revision will take more time and money than will the same change performed on CAD equipment.

On the other hand, CAD systems can be expensive to operate. In addition to the necessity for essentially nonautomated design of the board, they need a relatively even, predictable, high-volume flow of designs to ensure maximum utilization. If the flow is sporadic, a CAD user will find the payback period on the capital investment stretching out. Yet there may not be the capacity needed for peak periods.

Best of two worlds

A number of major equipment makers have established in-house CAD facilities that can handle their routine design load. In peak periods or for particularly complex groups of boards, they turn to a design-automation service bureau. Since design-automation software interfaces with most computer-aided-design systems, these firms can change the designs generated by the service bureaus on their own CAD equipment, at important savings in time and cost.

Some firms take maximum advantage of this interface by using their in-house CAD systems to prepare coded wire lists in card form from a digitized logic diagram;

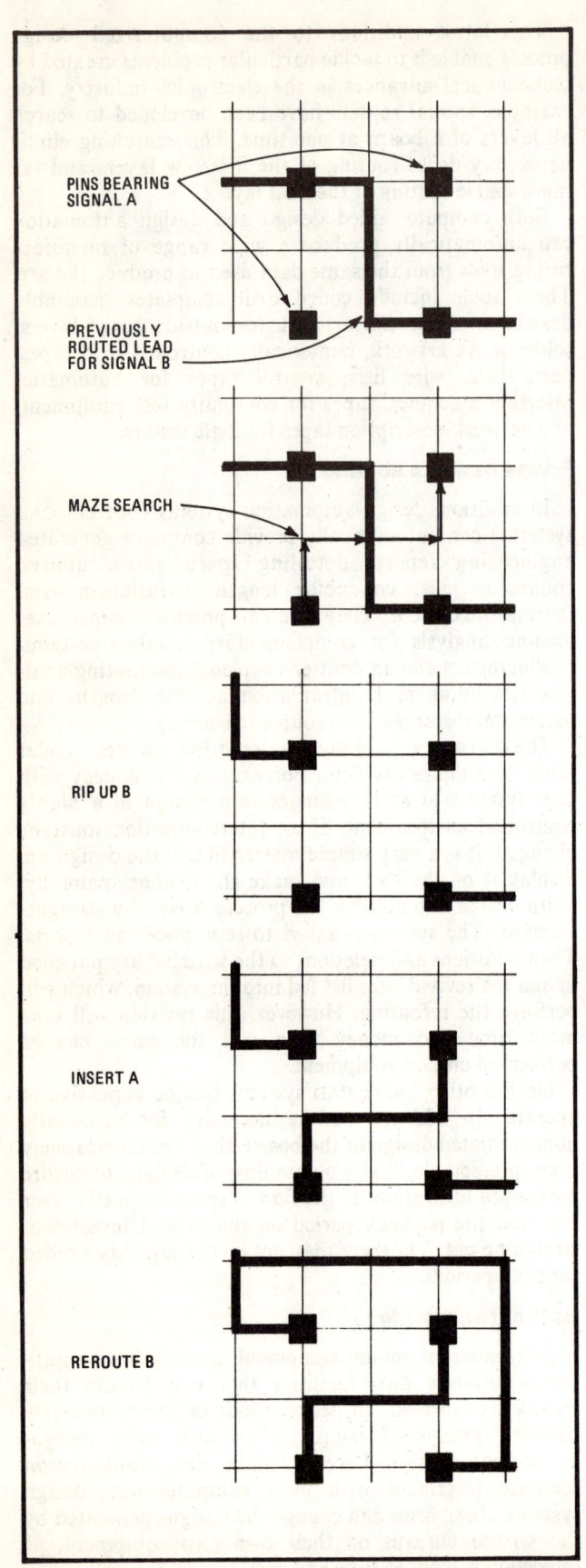

2. Software eraser. Iterative routing is a method for modifying computer-designed artwork. This scheme allows a design-automation system to rip up a previously routed lead to permit insertion of a new interconnection and the rerouting of the old path.

they send the decks of cards to the service bureau and get back decks that operate their photoplotters. Turnaround time is typically 10 days for a series of designs compared to 2 to 3 weeks for design automation alone, and cost savings are significant.

Assessing cost effectiveness

Because of the fixed overhead associated with computer-based systems, there is a threshold of design complexity before either design automation or CAD is cost-competitive with nonautomated design. Boards with fewer than 20 ICs usually should be laid out without the help even of digitizing. For boards with more than 150 ICs, design automation is imperative because of the difficulty of doing the layout by hand. The single exception to this rule is a design in which routing and use of via rules are so tight that it is impossible to place the components logically on a grid. Then the designer must do it himself or herself.

In the range between 20 and 150 ICs, several other factors influence the choice between design automation and CAD—for example, the number of boards of each type involved. In a one-board project, design automation can be ruled out as a cost-effective procedure. Every dimension and special ground rule of the single board must be described to the computer in detail as part of the start-up procedure. The cost of this process is typically about $500, and, amortized over three or more boards, it becomes increasingly nominal. For a single board, it can be a major part of the expense. However, with computer-aided design, board parameters are simply drawn in, and so initial setup costs are low.

Conversely, the greater the number of boards in a group, the more likely a CAD system is to become a bottleneck to the rapid work. Once the front-end programming is complete, design automation can rapidly turn out one layout after another. Any given CAD system has a maximum capacity, and the last board takes as long to design as the first.

Another point to keep in mind is the percentage of discrete components per board. Design automation can assign logic elements to multielement packages or place DIPs or flat packs of uniform package types to minimize the length of interconnections. If pc boards contain 50% or more discrete components, it is better to go with placement by the designer, perhaps aided by CAD.

Tightness of the design ground rules is another consideration. The computer cannot employ discretionary "cheating" as a manual designer can. Therefore, where ground rules are discretionary rather than absolute and crossing counts indicate the board may be difficult to route, higher densities may be achieved with nonautomated or computer-aided design.

Quantity is a key

The cost effectiveness of design automation vs CAD vs manual design and artwork may be directly dependent on the number of boards to be fabricated from each design. For less than 5 or 10 boards per design, it pays to tell the fabricator to take extra care in producing the boards. High-precision artwork supplied by the designer is not as vital as it would be for larger production runs,

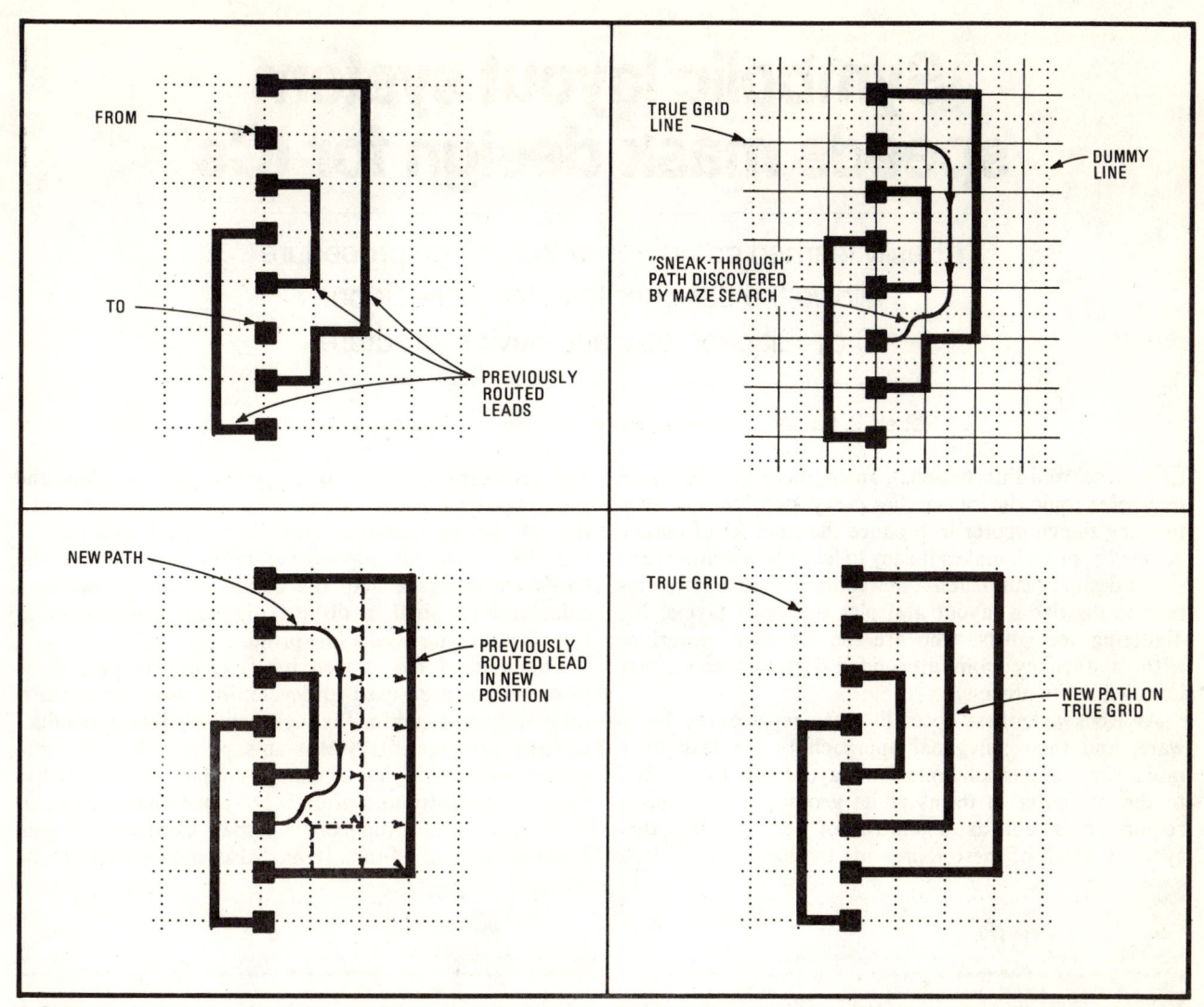

3. Squeeze. Another approach to inserting new paths is the "squeeze through" algorithm. This temporarily inserts an extra grid within an existing one (top right). Then previously routed lines are shoved aside (bottom left), and the new path is fitted to the new grid.

and design by hand may be adequate for the production of the boards.

With greater quantities, ease of manufacturability becomes a major cost-saving item. Computer input becomes the treatment of choice.

Using a designer

If there are more than 1,000 boards to be produced per design, manufacturing costs will be far lower if the board under consideration is limited to a two-sided, rather than a multilayer, design. Under these conditions, one can have a designer spend an additional month on a single design just to eliminate a few vias. (Given unlimited time and infinite patience, a skilled designer may be able to improve slightly on the densities achieved by design automation.) However, since very minor differences in the quality of the artwork can make big differences in yield, in all probability it will be valuable to digitize the design.

Firms that have tight delivery schedules cannot afford mistakes that demand extensive debugging and reworking further down the line. Therefore it is important to have diagnostic checks available along the way. Computer-aided-design systems are usually limited to physical design checks, such as adherence to line-width and spacing requirements. Design automation assures adherence to these requirements automatically, and it also can check conductor traces for conformance to the original wire list.

Obviously few boards and few operations are going to fit precisely the ideal requirements for any one of the four basic approaches to pc-board layout. A designer must consider all the variables, including delivery schedules and design loads, to discover which approach or combination of approaches will be most generally cost-effective for his operation.

The choice

In general, the more complex and highly dense the board and the larger the group of mechanically similar boards, the more likely is design automation the best choice. Moreover, unless the task involves very simple boards (20 ICs or less) to be produced in small quantities (5 or 10 per design), it probably will be profitable to investigate the various degrees of help obtainable from today's computer technology. □

Symbolic layout system speeds mask design for ICs

Easily learned computer-aided design procedure allows the designer to quickly check out IC circuit geometry and device structure

by Robert P. Larsen, *Rockwell International Corp., Electronic Devices Division, Anaheim, Calif.*

☐ At Rockwell International, an engineer who designs a microelectronic device handles every step himself, even to using the computer to produce the final set of masks. A novel approach makes it easy to learn how—an experienced designer can teach a newcomer in just 15 minutes how to digitize a layout and plot the mask layers. No digitizing specialists, who are the designer's interface with a turnkey computer-aided-design system, are involved in the process.

At Rockwell, these specialists are replaced by hardware, and their polygonal approach to the task by a much simpler symbolic one. The layout is not described to the computer in terms of its geometry and process requirements but as a matrix of letters and other symbols. Each of these represents the logical, electrical, and geometric functions of a specific piece of chip, and the computer is programmed to infer from the juxtaposition of the symbols the three-dimensional structure of the circuit and its topological properties. As for the hardware interface with the computer, it consists of a calculator terminal, a disk, a digitizer, plus cursor, a low-speed plotter, and a line printer.

The method has proven itself. Over the past three years, it has been used to develop hundreds of circuits, ranging from p- and n-channel metal-oxide-semiconductor and complementary-MOS-on-sapphire chips to integrated-injection-logic and charge-coupled devices. It has proven less costly and often more rapid than polygonal approaches like Applicon, Calma, Computer-Vision, Macrodata, and Mann. It has also shown itself to be

1. Symbolic-layout–topological schematic. Each layout symbol characterizes all mask geometries and represents a particular rectangular piece of the substrate. Associated electrical and logical functions provide the basis for a cost-effective node analysis.

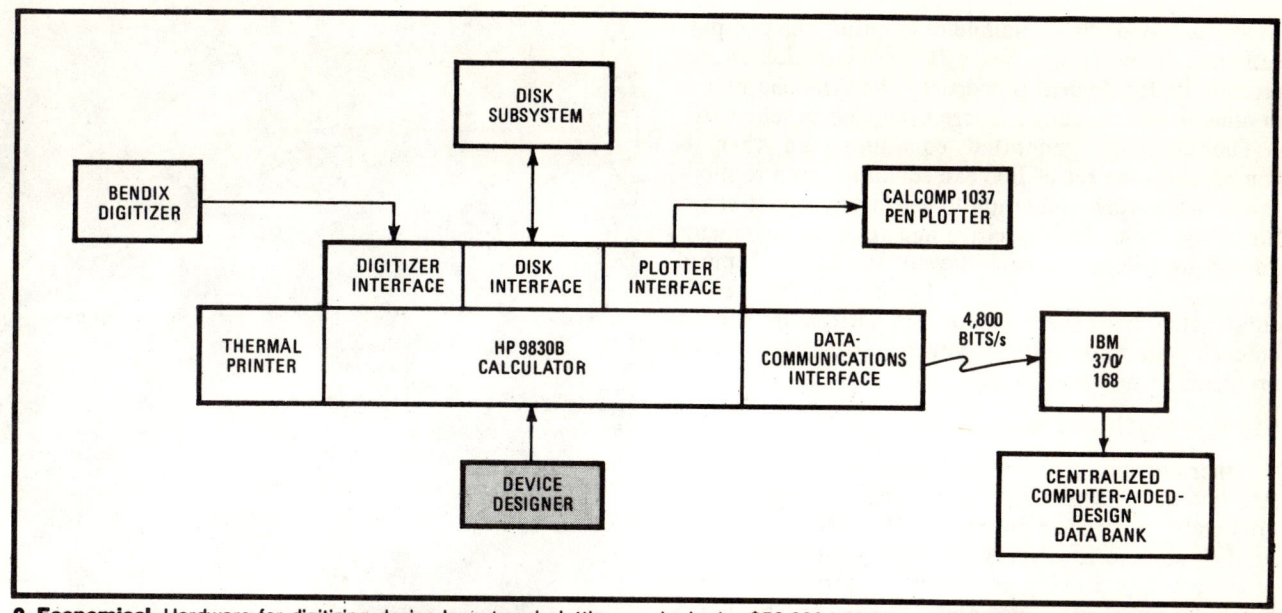

2. Economical. Hardware for digitizing device layout and plotting masks had a $50,000 initial cost, which was considerably less than any commercial turnkey CAD system. Batch and interactive processing services are provided by a corporate computer center.

readily adaptable to dramatic changes in the technologies of device production. But above all, it has proven satisfying to the device designer and efficient in management terms.

In contrast, the polygonal approach, which is standard in the industry and used by all commercial turnkey CAD systems, is frustrating and cumbrous. Conceptually, its starting point is the polygons in terms of which each mask layer can be characterized. Each polygon is represented by a set of vertexes defined by the X and Y coordinates of the layout grid. Grid scaling generally ranges from 0.1 to 2 micrometers.

Typically, the designer draws the polygonal geometries of each mask layer on a Mylar layout grid. After he has transferred the electrical and process requirements to his layout, he hands the design to a specialist for digitizing on the turnkey CAD system.

In the software, each mask is represented as a list of polygons and each polygon in turn is represented as a list of vertexes with implied polygonal edges. A memory pointer implying a direction, either clockwise or counterclockwise, links the data, so that a complete polygon can be quickly reconstructed from the X-Y coordinates of each vertex. All polygons making up the mask are similarly treated.

Geometry check

After digitizing the layout in this way, the specialist runs an inter- and intra-mask check on layer geometry to detect any geometric violations. Then he gives the device designer a complete set of pen plots of all the mask layers, drawn at 100, 500, or 1,000 times the true size. This may require several iterations, in part because of misinterpretations by the specialist, who frequently is unenthusiastic about his rather limited job. But eventually the plots do receive final verification from the designer, whereupon the specialist uses the turnkey CAD system to generate the numerical-control data needed for lithographic processing.

The CAD system developed at Rockwell International documents the layout of a device quite differently, by means of a schematic produced by a conventional computer-driven, high-speed line printer. For reasons that will become apparent later, this schematic may be described as topological. It is scaled to imply a grid structure, with a scaling range of 2 to 12 μm, that maps the silicon or sapphire real estate into small rectangular areas.

Symbolic utility

Each of these areas is denoted on the schematic by the presence (or absence) of a symbol or character taken from a set of symbols or characters representing such simple or complex entities as MOS field-effect transistors, contacts, metal or polysilicon conductors, diffusion areas, and so on (Fig. 1). Each character or symbol then is associated with a function having both logical and electrical properties, and from the juxtaposition of the symbols the structure of each node can be easily derived.

The layout appears as a matrix of symbols, which represents the complete three-dimensional structure of the specific microelectronic device. This form of notation allows the designer to synthesize his layout in accordance with the ground rules dictated by device and process physics. Thus he is freed from the intricate and boring details of mask geometries and able to concentrate on the logical and electrical aspects of device design.

The matrix of symbols can be quickly generated by a computer-driven, high-speed printer in segments that can be easily aligned and taped together. Furthermore, when analyzing design problems or exploring possible solutions, the designer can readily trace metalization paths, visualize diffusion structures, or derive intersymbol geometric relationships.

A major advantage of symbolic design is that the computer can economically perform an independent audit of the logical and electrical integrity of the layout.

Since each node has a unique description, the computer can trace it and reconstruct a Boolean equation characterizing its topological properties. The symbolic matrix, in other words, is really a discrete topological schematic.

The computer-generated equations can then be compared to the set of Boolean equations used to model the device during simulation to detect any layout errors. Similarly, from the capacitive and resistive parameters known to belong to each layout symbol, a lumped-parameter equivalent circuit for each node can be constructed. Now the computer can perform an analysis on each node to detect any potential speed and or noise problems in the layout design.

Storage requirements

Other advantages can be seen from the software standpoint. The discrete topological schematic has predictable storage requirements. Assuming 1 byte per symbol, storage requirements in bytes are equal to the product of the maximum X and Y coordinates. These requirements can be just about halved by the use of data-compression algorithms. The discrete topological schematic is also convenient for partitioning the data into pages for ease of access during computer processing and thus further reduces memory requirements. Rockwell has developed an extensive set of software to support the simulation and test, layout, analysis, verification, and mask-generation functions of device synthesis for p- and n-channel MOS, C-MOS-on-sapphire, I²L, and CCD devices.

The symbolic layout can be viewed as a shorthand, high-level method of documenting microelectronic device layouts, and one that stimulates design creativity. Because it standardizes the geometrical relationships between symbol and function and between symbol and mask, layout configurations can be documented quickly, allowing the device designer to evaluate candidate designs easily. Custom device design also becomes faster and more cost-effective.

What happens in practice

The nucleus of the simple Rockwell CAD system (Fig. 2) is a Hewlett-Packard 9830B calculator. Attached to the calculator are a Bendix 42-by-60-inch digitizer with free-moving cursor (Fig. 3) and a Calcomp 1037 pen plotter. A thermal line printer and disk storage of 4.8 million bytes are also used.

The calculator in turn is coupled to an IBM System/370 via a 4,800-bit-per second dedicated binary synchronous data link. The central CAD data bank is maintained and all primary computing done at the corporate computing center. The HP system does minimal pre- and post-processing computing.

From the standpoint of the IBM 370's operating system, the calculator is programmed to behave as a remote job-entry station. When transmitting data, it simulates a card reader and transmits the data to the IBM 370 job queue. When receiving data, it simulates a line printer and the data resides in its RJE station print queue as a print job. A command from the calculator activates the printer. As the lines to be printed are received, they are stored as a set of data on the HP disk.

3. Free-moving cursor. Layout digitizing by cursor facilitates symbol alignment and coordinate registration. The system permits digitizing from the manually drawn Mylar layout or from the computer-generated topological schematic shown here.

Since the data sets are relatively small and use a compressed data format, it normally takes only 3 to 5 minutes to transmit or receive data.

This hardware configuration has been extremely reliable and is set up in an informal environment (Fig. 4). No maintenance has been required on the Bendix digitizer and the Calcomp plotter in three years of service, and the HP calculator required only two minimal maintenance periods, totaling four hours, over and above its normal preventive maintenance.

How the layout is digitized

Since the device designer himself digitizes the layout, the procedures have been kept as simple and cost-effective as possible. Therefore there are only two steps: digitizing the Mylar layout (the data being stored temporarily on the HP disk), and then transmitting the data from the disk to the IBM 370 as a batch job.

In digitizing his layout, the device designer works with a menu, or list, of digitizing options. A prompting sequence tells him the order in which to enter the requisite data, such as device name, designer name, accounting charge number, and layout grid scaling. He then uses the cursor to digitize the reference points of the device layout, after which the software samples the cursor for normal digitizing of the Mylar layout.

In the normal digitizing mode, the appropriate command (point, line, area) and verb (layout symbol) are selected from the menu, and the corresponding

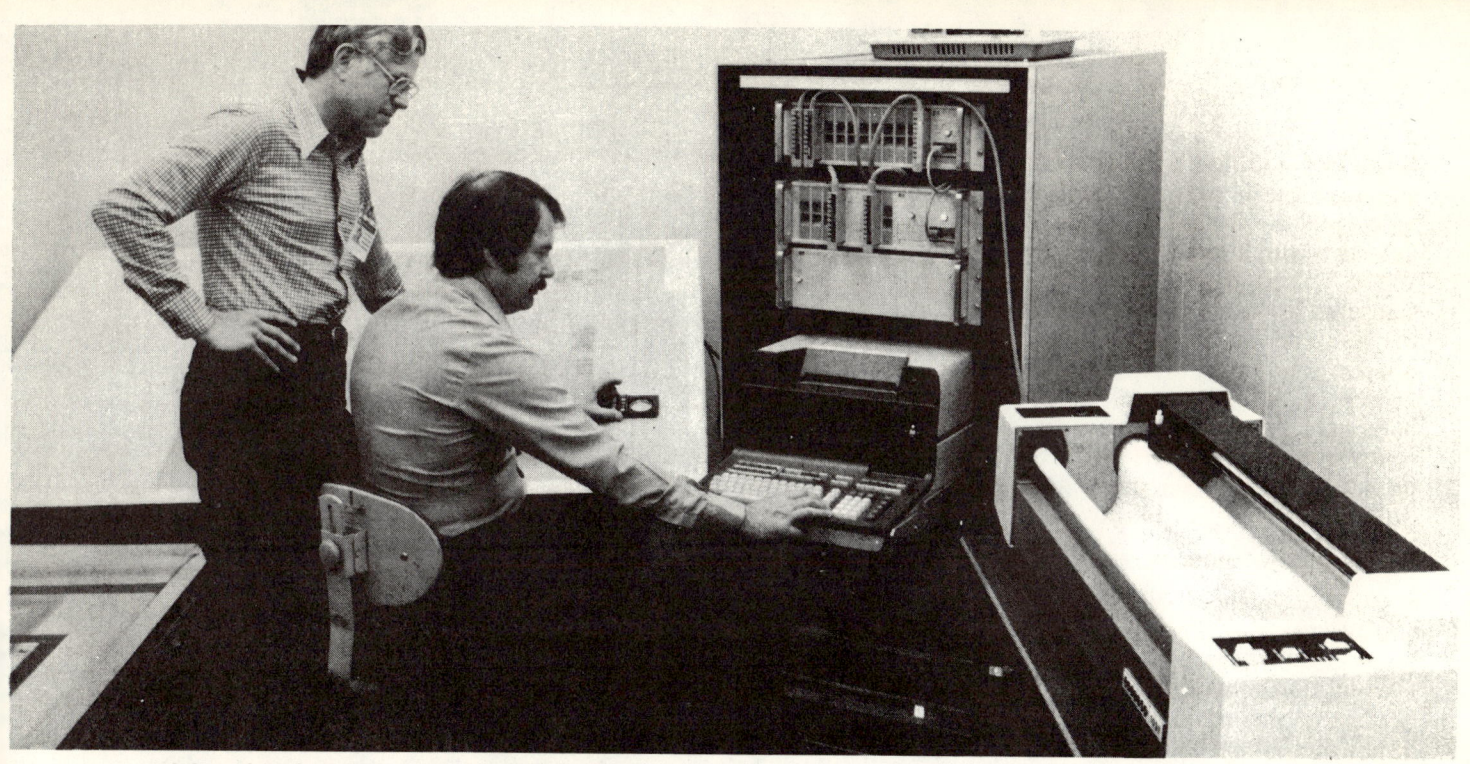

4. Informal environment. Author Larsen (left) shows device designer how to use the Bendix digitizer with free-moving cursor. The rack containing the Hewlett-Packard components is in the center; the Calcomp pen plotter is at the right.

points on the Mylar layout are then entered. These items could represent the vertexes of polysilicon or metal conductors, diffusion structures, or the points locating field-effect transistors or contacts. As the X-Y coordinates are received by the calculator, they are scaled and converted into relative coordinate data. The data is simultaneously formatted for subsequent processing on the IBM 370 and stored on the HP disk.

After completion of this stage, the device designer turns the calculator into a remote job-entry terminal by simply depressing a function key, so that the unit becomes active and is signed on. By next depressing a data-transmit function key, the designer enters the device name, and the data transmission begins. After the data set has been transmitted, he depresses the sign-off key to deactivate the terminal.

A symbolic schematic is then generated by the IBM 370 central computer and within 10 minutes is printed out on a high-speed line printer attached to an IBM RJE station. Since the Mylar layout grid is made to correspond to the grid of the schematic (0.125 inch by 0.100 inch), the designer can superimpose the schematic on the Mylar layout and note any differences. Subsequent updating is done from the corrected schematic.

The mask-plotting method

Once the schematic is complete, the IBM 370 turns it into a set of mask layers, storing numerical-control data for lithographic processing in a mask-plotting data bank that is a part of the centralized CAD data bank. The mask-plotting data bank is structured by device name and, under that, by mask layer name.

The mask plotting has three steps. First, the designer defines which parts of the plot he wants to examine and uses an interactive terminal to request the pertinent data from the IBM 370. Next, he stores the data received at high speed from the IBM 370 on the disk. Finally, he runs the plot off on the plotter.

The device designer is intimately involved in all these steps. His knowledge of the layout enables him to define the minimal plot windows needed to investigate mask geometry problems. In the plotting stage, his use of a low-speed plotter minimizes the capital expense: the Calcomp 1037 plotting speed is a mere 5 centimeters, or 2 inches, per second. Also, the entire mask is never plotted as in commercial turnkey CAD systems—only specific, needed windows are considered so that cost and time efficiency is maintained.

Suppose the device designer wishes to verify the interface between two building blocks having different geometric scales—say, a read-only or random-access memory array and its input/output logic. He identifies the plot window boundaries in relative layout coordinates, then logs on at an interactive terminal. He supplies the required device name and plot window boundaries, and he executes the program. This would be a high-priority job submitted for batch processing. The appropriate mask data is extracted from the mask-plotting data bank and sent to the HP calculator now set up for operation as a line printer.

As each line of print is received, it is stored as a data set on the disk, the device designer pressing the relevant function keys. Finally, he executes the pen plotting program by supplying device name, X-Y scaling factors, and plotting label and indicates whether or not the mask layers are to be plotted individually or overlaid. For plotting individual mask layers, the positioning is automatically to the right, layer by layer, to optimize paper use. For overlaid mask layers, the designer may choose a different color ink for each mask layer plotted.

Thus in Rockwell's CAD system, the design of a microelectronic device is the sole responsibility of the device designer. Motivation is high, as is productivity, in the informal design environment. The device designers cooperate among themselves in scheduling, with only minimal involvement by management. □

Part 8
Automatic Testing of Interconnections and Packages

Aerospace testing. This multi-bay DIT-MCO 772 circuit analyzer can test as many as 100,000 terminations for continuity, leakage, and high potential. A unit of this type is being used to check out the critical system wiring for a large space project.

Automated circuit testers lead the way out of continuity maze

Manufacturers are overcoming the bugaboo of astronomical numbers of circuits that can be shorted or open; they are now checking backplanes, printed-circuit boards, cables, harnesses, and even hybrid substrates

by Jerry Lyman, *Packaging & Production Editor*

☐ As designers push the state of the electronic arts to new heights, components are being shrunk to microscopic sizes, and their numbers are being multiplied to astronomic proportions. To make the simplest electronic product built of these microcomponents, they must be interconnected. The number and complexity of their interconnections can become an enigma within a mystery—especially when it comes to testing.

Interconnections come in many forms, chiefly the plated conductors on printed-circuit boards, backplanes, chassis wiring, cables, and harnesses. To illustrate the complexity of the wiring in modern equipment, the A-6 attack plane has 22,000 to 24,000 points interconnected by cables and harnesses. Backplanes in telephone exchanges have more than 80,000 points, and multilayer pc boards can have 25,000 connections. Even a "single" harness can have 700 wires. And every circuit must be checked for continuity and shorts.

Only automated equipment can cope with the number and complexity of these circuit configurations. Automated testing has brought huge savings to Grumman Aerospace Corp., where test personnel spend about 216 man hours testing about 16,700,000 wires a year. If these cables were manually tested, it would require an estimated 560,000 man hours, estimates Jerry Cronin, equipment-technology manager of the Bethpage, N. Y., aerospace company.

Automated continuity testers of various sizes and capabilities are being produced by several companies. It's not surprising that every manufacturer of ACT systems uses ACT equipment to test its own pc boards.

Examining the economics

To illustrate the need for finding defects as early as possible in a product's life cycle, the cost of repair increases tenfold at each step after the bare-board stage—after loading, at system test, and finally service in the field (Fig. 1).

One manufacturer estimates that it costs 30 cents for an ACT to find a fault on a bare pc board, in comparison

Testing by remote control

Two companies are using voice-grade telephone lines to link circuit analyzers to central controllers. For about a year, Sperry Rand Corp. has been operating a real-time control link between its headquarters in Great Neck, N.Y., and Waterbury, Conn. A DIT-MCO 810 in Great Neck has been controlling a DIT-MCO 610 in Waterbury.

Grumman Aerospace Corp., of Bethpage, N.Y., is planning to batch-transfer programing information between its central controller at Calverton, N.Y., and circuit analyzers at Bethpage, Houston, Savannah, Ga. and Stewart Field, Fla. The decision is based on a test recently conducted between the Calverton facility and the DIT-MCO plant in Kansas City, Mo.

At Calverton, Grumman has a DIT-MCO controller hard-wired to six automatic circuit analyzers on its flight-test line. A test program for cabling is stored on a magnetic disk. The data was transferred to Kansas City through modems by means of voice-grade telephone lines. DIT-MCO programmers edited the programs and sent them back to the 810's disk in Calverton. The revised data was then successfully used by the 810 to program a flight-line analyzer for testing cable.

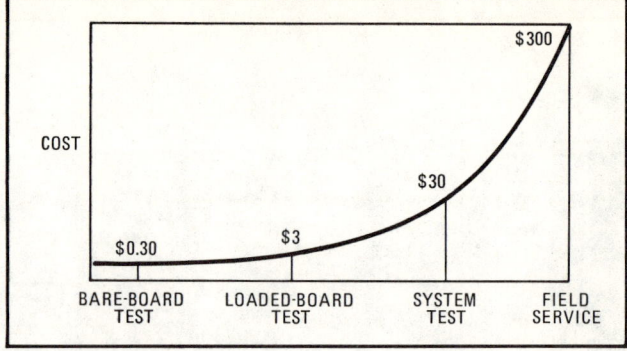

1. Testing costs. An error found in a componentless printed-circuit board can save a company 100 times as much money as it would to find it in a system test after the board is loaded, or 1,000 times as much as it would in the field. Obviously, faults must be found early.

to about $3 to find a faulty IC with a subassembly-logic tester after the board has been loaded. For testing backplane wiring, one company says the cost ranges from 2 cents a wire for less than 50,000 wires to 1 cent a wire for more than 100,000 wires.

When it is possible that automated continuity testing would not be economically feasible, costs should be weighed against the penalties for not testing. For example, a one-year payback on a $15,000 model N123 tester made by Teradyne Inc. of Boston that is used for testing 1,500-point backplanes works out to be 130,000 test points. It is assumed that wiring one board requires 15 minutes, the labor cost is $10 an hour, 90% of the wires are semiautomatically wrapped at an error rate of 0.3%, and 10% of them are applied manually at an error rate of 3%. The cost of finding a fault at the system-test stage is set at $30.

To find the number of test points (N) needed to pay the machine off in a year,

cost of not testing = machine + fixture + labor, or
$15,000 + $7,500 + N($0.25)($10)/1,500
= $30[0.9N(0.003) + 0.1 \times N(0.03)]$
N = 130,000 points.

Many manufacturers say their machines have paid for themselves in six months to two years. One did not have to wait long to discover that he had a bargain. In testing 12-layer pc boards, his Addison tester found 27 bad ones out of the first 52.

Recently, the engineers for a company in the Midwest were able to justify an expensive Teradyne N151 ACT to test five types of 14-layer boards, none of which had more than 900 units in a run. The projected savings were in the custom LSI chips, which account for 90% of the cost, that would have had to be scrapped along with the defective boards.

Although it is obvious that automated equipment is mandatory for testing cables and interconnections, the decision is not so clear-cut for testing backplanes and printed-circuit boards. If the manufacturer decides to automate, he should have an extensive knowledge of the capabilities of the various ACT systems so that he'll know which is desirable for his use.

Making the decision

Whether or not backplanes should be tested for continuity depends largely upon whether they are wire-wrapped manually, semiautomatically, or automatically. Testing is mandatory for manually wrapped boards because the error rate is 1% to 5%, which means 10 to 50 broken wires in a 1,000-pin panel. The number of pins and the number of boards in a run may determine whether or not ACT is necessary for semiautomatic or automatically wrapped boards.

But, even though the error rate is low enough to dispense with continuity tests of semiautomatically and automatically wrapped backplanes, usually a large number of connections must be wrapped manually, anyway. This situation dictates automated testing. In semiautomatic wrapping of panels, error rates are only 0.1% to 0.5% because a numerically controlled machine guides an operator with a manual wrapping gun to each pin in the wrapping sequence. In automatic wrapping, numerically controlled machines position the pins for wrapping in the desired sequence. The error rate for this type of operation is only 0.01% to 0.04%.

The necessity for manual wrapping on an automatically wrapped board has forced one large telecommunications company in the Midwest that builds exchanges to go to ACT. In an exchange that has 80,000 to 100,000 backplane points, half the points are wrapped manually and the other half automatically. Even at the lowest manual error rate of 1%, there would be a minimum error of 500 incorrect wires on a 100,000-point backplane.

However, the decision of whether or not to automate the testing of printed-circuit boards is not as simple as the decision on the larger, more-complex backplanes. Printed-circuit boards may be single-sided, double-sided with plated-through holes, of multilayer construc-

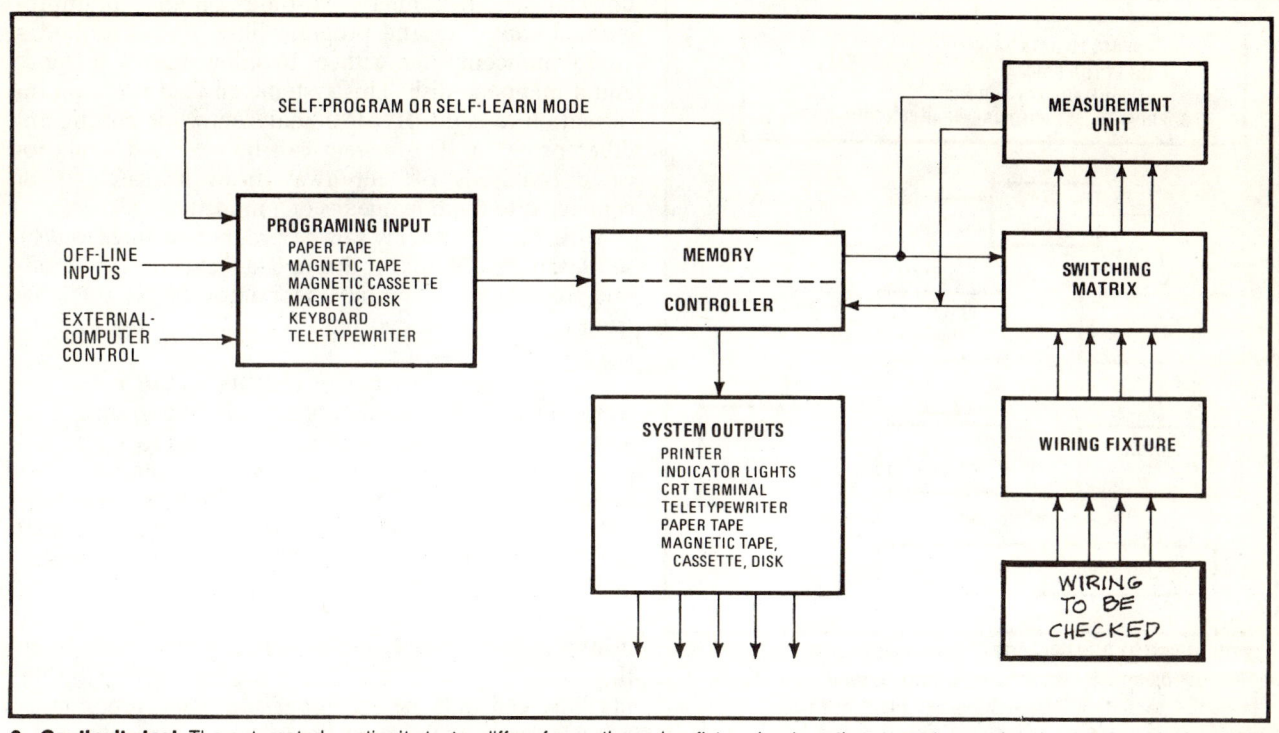

2. Continuity test. The automated continuity tester differs from other edge-fixtured automatic test equipment for circuit boards by its special point-to-point fixturing and a self-programing capability that enables the tester to "learn" the wiring of a known good unit.

tion, or even constructed on flexible substrates.

Continuity testing is a must for multilayer boards, which, because of their complex construction, are highly susceptible to internal shorts. And since 90% of the value of the multilayer board is in the components mounted on it, if a short or open in the inner layers of the board is not detected, 90% of the value must be tossed into the scrap-parts bin.

Continuity testing of bare printed-circuit boards can undeniably cut the cost of testing and repair at later stages of production. But even though the defect rate on single- and two-sided boards of equal complexity can vary from 1% to 7% from manufacturer to manufacturer, there is no general industry trend toward bare-board testing.

Many large aerospace companies like Sperry Rand Corp. in Great Neck, N.Y., Grumman, and commercial companies like Data General Corp. of Southboro, Mass., depend on a thorough optical inspection to weed out such board defects as plating slivers and cut lines in their one- and two-sided boards. All of these companies have extensive ACT installations.

On the other hand, most aerospace companies using multilayer boards test them 100% for opens and shorts. In manufacturing computers, Digital Equipment Corp. in Maynard, Mass., which formerly did not electrically test bare boards, has recently finished a pilot test run on a four-layer high-density logic board. Test results were positive enough for DEC to continue the program and consider testing other multilayer boards.

Considering circuit testers

The first automatic wiring analyzer was created for the military in 1953 by DIT-MCO, Kansas City, Mo. The electromechanical model 200, which is still being sold, can test 400 points or 200 wires for continuity, high-voltage breakdown, and leakage. Smaller, faster solid-state machines started to appear in 1971. They were designed for commercial, rather than military-aerospace applications.

ACTs have become much larger, and their capabilities have been expanded to meet the needs to today's complex circuitry. A typical ACT system (Fig. 2) has six major subsystems—programing inputs, controller, memory, measurement unit, switching matrix, and a wiring fixture.

Usually, the first step in testing is to connect to the wiring fixture the contact points of the unit to be tested. Then a program is loaded into the tester's memory. The stored program commands the controller to connect the measurement unit to the device under test by way of the switching matrix and the wiring fixture. Then the controller compares the measured continuity and leakage results with the stored results of the unit's wiring list. Results of the test are then fed out in one or all of the data formats shown.

Results can be presented either in machine language or product language. In machine language, the tester reads or prints out its results in an arbitrary numbering system that requires a look-up table to reconvert machine results to the nomenclature of the original wiring list. When equipped with a product-language translator, an ACT reads out its results directly in the customer's designations for his connectors and connector pins.

Automatic wiring testers or analyzers can be programed in one of three ways: by self-learning, off line, or by direct computer control. Self-learning is a method whereby an ACT system automatically learns the wiring pattern of a known-good unit connected to it. This test pattern, which is stored either in memory or on a pro-

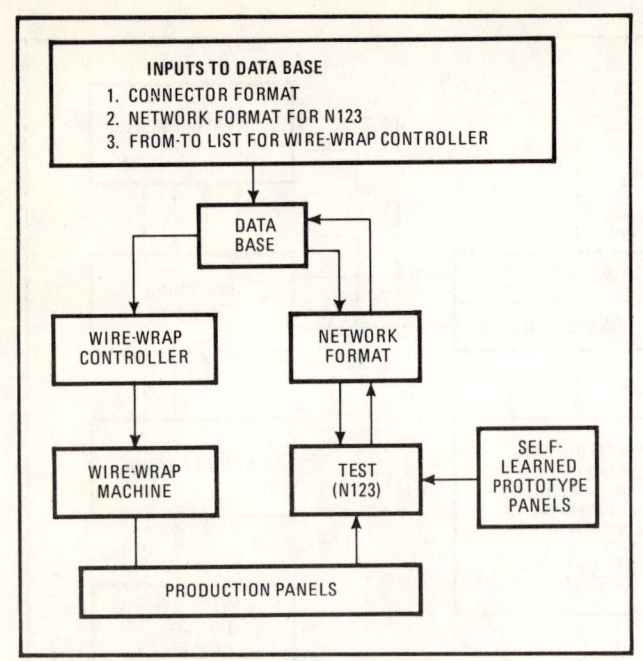

3. Programed by a tester. An automated continuity tester can store the correct pattern for a prototype wrapped panel and transfer this data back into a data base, where the information can be reconverted to a tape that controls wire-wrapping production.

gram tape, is then compared to the wiring of similar units.

In off-line programing or data-base translation, computerized wiring-run lists or digitized printed-circuit artwork are fed into an external computer that translates this data into a suitable programing format and stores it in a memory medium, such as paper tape or magnetic tape. These tapes and cassettes can then be inserted directly into the tester's input peripheral to program the ACT. In direct computer control, the translated data from the off-line computer is fed directly into the memory of the ACT's controller.

Manufacturers of ACT equipment produce four sizes. The largest systems are switched electromechanically by reeds and relays, the medium-size systems have solid-state switches, there are small benchtop cable and backplane testers, and the in-circuit testers can also be programed to test bare boards.

Controlling the ACTs

Automated continuity testers are controlled in various ways. Several of the low-cost units have hard-wired special-purpose controllers. Many of the medium-size and larger units are built around standard minicomputers. A few new machines are being controlled by integrated microprocessors. For instance, the Teradyne N123 is controlled by an 8080 from Intel Corp., Santa Clara, Calif., and the model 202 made by Faultfinders Inc. in Latham, N.Y., uses an Intel 4004. A whole new generation of automatic continuity testers using the 8080 and other microprocessors is already in the prototype stage at several companies.

Most DIT-MCO analyzers are controlled by hard-wired controllers, with inputs from such sources as paper tape and disks. However, DIT-MCO makes the model 810, an unusual eight-channel real-time on-line multiprogramed controller and program library built around a 16-bit minicomputer with a 16 kilowords of memory and a magnetic disk. This system can feed program information to eight DIT-MCO analyzers while editing another program. The system can be operated while located hundreds of feet away from its satellites or connected to them by means of a modem.

Most testing systems can be programed three ways—self-learning, manually through a keyboard, and off-line from either a wiring list or information available from a computer-aided data base. Most large systems are programed off line.

For instance, Sperry creates its automatic wire-wrapping information in an automated-program group, and this digitized information is converted to a DIT-MCO paper tape. A similar process is used at Grumman. The Bendix Corp. Test Systems division in Teterboro, N.J., has a computer-aided manufacturing data base that creates test tapes for wrapped wire, pc boards, cables, and chassis wiring.

However, many small to medium-size companies are using their testers only in the self-programing mode, either because of lack of software support or because the machine can only be self-programed. Self-programing presupposes a truly good board and a perfectly clean fixture. However, since most ACTs will print out a wiring list of the prototype in product language, a board's memorized wiring list can be checked against the known wiring list to ensure that there are no errors in the "learned" program. Realistically, self-programing is limited by the learning time of the internal computer. It would probably take thousands of iterations to get a correct self-program tape of a backplane that has as many as 25,000 points.

Despite its disadvantages, self-learning is now being used as a source of software for semiautomatic and fully automatic wire-wrapping machines. In the normal wire-wrapping process, a wiring run list is usually fed into a data base or computer-aided manufacturing flow to create both wire-wrap-control software and a test tape for checking the desired backplane or pc board.

In an alternate system now used by many companies and at least one wire-wrapping service, a prototype panel is hand-wrapped, checked out against a wiring list, and then the automatic continuity tester is programed to self-learn the wiring of the pattern board. This information is read out in paper or magnetic tape and transferred back to the data base, where it is converted into control software for the automated wire-wrapping machines. In effect, the automated test equipment is programing itself and a wire-wrapping machine. A flow diagram for a system of this type built around a Teradyne N123 is shown in Fig. 3. This system is successfully working at Teradyne's wire-wrapping facility.

Fixturing

Manufacturers of ACT and other products agree that the biggest problem in automated continuity testing is fixturing—connecting the tester to the cable, backplane, chassis, pc board, or even a hybrid-circuit substrate. Fixturing to a cable is messy but straightforward. A

4. Pin testing. If the wire-wrapped side of a backplane is unobstructed, it can be accessed by an air-driven bed-of-nails fixture. This unit, designed by DIT-MCO for an aerospace customer, can accommodate a 24-by-72-inch panel, and has 50,000 spring-loaded test probes.

tester is simply mated to the cable adapters. However, to work up all the cables necessary to test a plane may literally require miles of wire and plenty of time.

A unit that takes an hour to connect to a fixture usually takes only seconds to test. Because of this time-consuming operation, many companies have started to make their own fixtures, and several new fixture manufacturers have sprung up—Everett-Charles Inc., Pomona, Calif., Program Data Co., Irvine, Calif., Pylon Co. Inc., Attleboro, Mass., and Ostby & Barton in Warwick, R.I. Some ACT manufacturers, including Hughes and DIT-MCO also manufacture fixtures for their customers.

Backplanes and printed-circuit boards are most difficult to fixture. A basic backplane is a metal or epoxy-glass board with a matrix of square wire-wrap pins on one side and a collection of connectors on the other side. The test engineer has two choices—fixture to the pins or fixture to the connectors.

If a backplane has relatively few pins and relatively few input connectors, then the solution is simply to plug in mating connectors with harnesses that run back to the testers. If the backplane has many connectors and thousands of pins, fixturing is expensive—$300 to $25,000.

The best solution to this problem is to connect the wire-wrapped panel or pc board to a "bed of nails" wired back to the tester. This bed of nails, instead of being a fakir's home for the night, is a matrix of spring-loaded probes that is forced against the wire-wrap pins by pneumatic, hydraulic, or electrical pressure. The DIT-MCO fixture of Fig. 4 accommodates a 24-by-72-inch backplane with a maximum of 50,000 points. This is far from the largest bed of nails in use.

If the wire-wrapped side is inaccessible because of wire build-up, structural design, or other reason, the designer must build a fixture that can insert multiplex connectors simultaneously. Massive pressure is still needed to engage the connectors. One air-driven card inserter is shown in Fig. 5.

Normally, a bed of nails is the only way to fixture a pc board for continuity testing. A possible exception might be a motherboard, a layered glass-epoxy board with some printed wiring and many connectors. In general, using a bed of nails for any type of board requires consideration of the method of actuation, number of contact points, minimum center-line spacing, size of product, types of surface, desired load time, and cabling to the tester.

The extreme variations in pattern, size, shape, and number of terminations require that every fixture for a pc board must be custom-made. The number of points to be tested can range from 50 for a single-layer board to 25,000 for a large multilayer board. Dimensions can range from tiny boards that accommodate a single hybrid substrate to several feet.

No matter what the size or connection, all fixtures work along the lines of Figs. 6(a) and (b). Usually pressurized air or a vacuum either moves the test probes toward the board or moves the board toward the pins, and the unit to be tested is positioned by two locating pins on the test-probe head. A bare glass-epoxy board with holes for aligning the test pins is placed between the pins and the unit under test.

The designer must keep in mind that each test probe of a bed of nails requires somewhere around 4 ounces of pressure to be exerted. For a 40,000-pin bed of nails, this would correspond to a required force of 5 tons.

Choosing a universal fixture

To reduce the tremendous expense and amount of time required to fixture pc boards, a universal fixture is highly desirable. To meet that need, fixtures that approach that ideal have been designed by Pylon, Ostby & Barton, and Program Data.

Pylon recently introduced a $425 mechanically actuated U-1000 test fixture. This fixture can take a 9-by-7-inch pc board and access as many as 200 pins. To use the U-1000, the manufacturer drills test-pattern holes in a blank epoxy-glass board, mounts spring-loaded test contacts for the desired pattern, wires the probe and is ready to test. To test, the pressure plate is lifted and the board placed on guide pins. The plate is then closed and the lever rotated 180° to press the board against the test pins.

However there are several other variations. A quick-change universal fixture from Ostby & Barton requires less than 15 seconds to change to a different pc pattern. Still another Ostby & Barton fixture can test pc boards from 4 by 11 in. to 11 by 19 in. that have a maximum of 5,040 test points.

A Datamaster universal fixture from Program Data has removable probe heads that are simply slid into the fixture, which accepts one type of card with fixed dimensions. Some models can handle a card 24 by 26 inches with 10,000 test pins. However, a fixture that large can cost $25,000 plus $875 per probe head plus $1.40 per point.

Automation may even be instrumented into fixturing. The rotary test fixture shown in Fig. 7 was developed to test the circuitry of flexible pc boards. This fixture tests and sorts simultaneously. Passed boards are piled at one station and failed units at another.

5. Over the top. When wire-wrapping pins are inaccessible, the bed-of-nails fixture cannot be used. A special pneumatic fixture must be designed to automatically insert connectors or paddle boards into the connector side of a wrapped backplane.

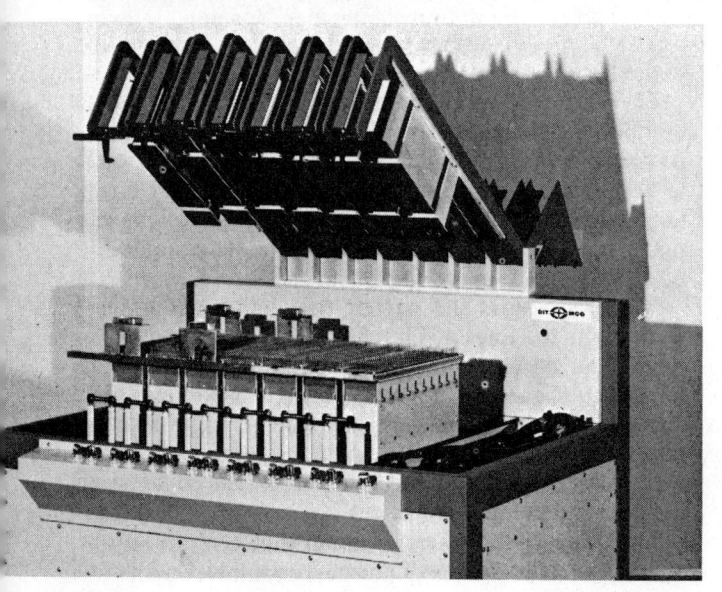

Checking the patterns on hybrid substrates calls for fixturing so small that 350 test points are accessed by probes that are only 0.10 inch center-to-center.

Three companies share the market for the high-priced, high-capacity relay-switched tester—DIT-MCO, Hughes Fact Systems, Los Angeles, and Automatic Production Systems of Pennsauken, N.J.

Evaluating the big ACTs

DIT-MCO's entries are the series 700, 730, and 770, which range in price from $50,000 to more than $100,000, depending on the installation. All systems are operated by special-purpose hard-wired controllers. These machines have no limit to their point capacity.

All models can be self-programed or programed off line from paper tape, magnetic disk, keyboard, ASR-33 Teletype or a master control computer. Their capabilities include testing continuity from 0.25 to 30 volts dc at 1 ampere and leakage as high as 1,500 v dc. Options include a built-in programable digital volt-ohmmeter, computer control, wire-resistance measurement, dielectric testing to 1,500 v dc, and impedance measurements. Switching speed is 1,000 tests per minute with proprietary 1,500-v relays and 6,000 tests per minute with 500-v dc reed relays.

The Hughes Industrial Products division produces the Fact II, another large automated system switched by reed relays and controlled by a dedicated minicomputer. This unit can be expanded to 100,000 points, test for continuity from 6 to 30 v dc at 1 milliampere to 3 A, and check leakage at voltages to 1,500 v dc or ac. Program input can be self-learning, paper tape, disk, magnetic tape, or teletypewriter.

Automatic Production Systems produces the Omnitester 2000, which is controlled by a DEC PDP-8. The 2000, which is switched by reeds or wire-spring relays for high-voltage systems, is expandable to 100,000 test points. It tests for continuity at 500 v dc, leakage and high dc potential. This machine can be optionally programed to measure impedances on loaded boards.

Reviewing the solid-staters

In general, ACTs switched by solid-state devices are stripped-down, higher-speed versions of their electromechanical cousins. Leading producers of these systems are DIT-MCO, Automatic Production Systems, and Teradyne. Because of the voltage limitations of solid-state switches, only continuity and leakage tests are implemented. The three large machines available are somewhat comparable to the larger electromechanical testers. They are the DIT-MCO 834, the Omnitester 3000 and the Teradyne N151.

The DIT-MCO 834, controlled by a built-in 16-bit minicomputer, tests wiring for leakage and continuity at 24 v dc. It can either be self-programed or programed off line. It performs 12,000 to 14,000 tests per second. The system, which can be expanded to the same point capacity as other large DIT-MCO testers, can be controlled by paper tape, disk, cathode-ray-tube terminal or magnetic tape.

Speed of 25,000 tests per minute is one of the leading capabilities of Omnitester 3000. Except that it tests only

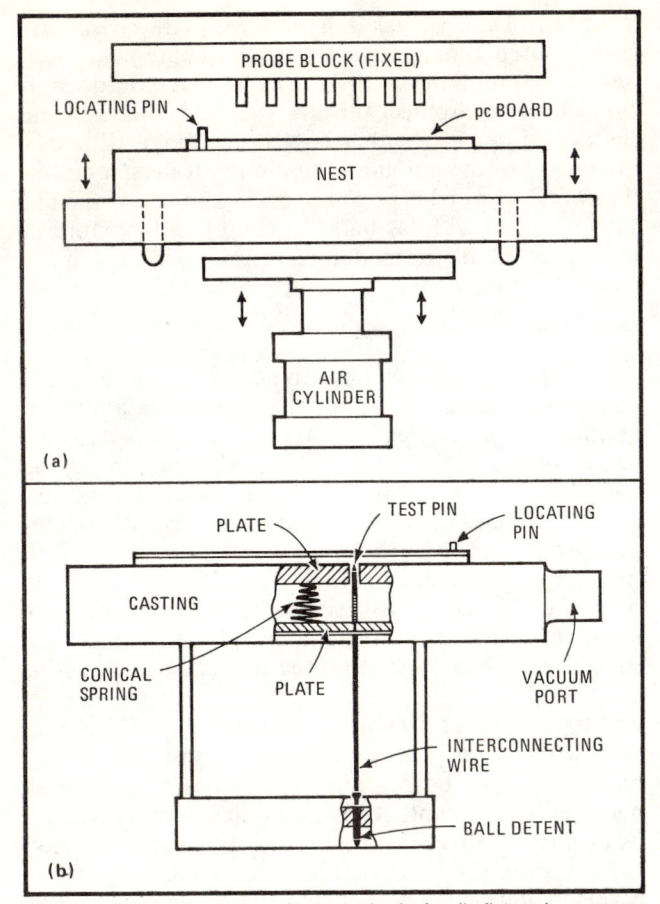

6. Up or down. No matter whether a bed-of-nails fixture is vacuum, pneumatic, or manually actuated, the printed-circuit card is either moved to the spring-loaded pins (a), or the pins are moved to the card (b). If the pins are above the card, they tend to keep cleaner.

for continuity and leakage, this all-solid-state machine is similar in system architecture, software, and peripherals to the Omnitester 2000.

Teradyne's N151 can test up to 200,000 points at a rate of 10,000 tests per minute. This machine can self-learn or be programed off line. Programs are usually stored in magnetic-tape cartridges. CRT terminals and printers are available as peripherals.

Teradyne's N151 and N123, as well as the DIT-MCO 834—solid-state units—have a system called daisy chaining (see Fig. 8), which reduces to two small cables the interfacing to the unit under test. This method, which is strictly limited to continuity testing, uses special active-fixture cards that are plugged into a backplane's connectors. In Teradyne's systems, one 14-lead cable runs from the tester to the first active-fixture card. Each card is then connected to the next card with a jumper of the same cable. The end card has a second 14-lead cable that returns to the tester.

Next in the ACT hierarchy are several smaller-capacity solid-state machines that can both self-learn and be programed off line. These machines are being offered by Plexus of Scottsdale, Ariz., the Muirhead Corp. Addison division, Mountainside, N.J., Data-Numerics Inc., Farmingdale, N.Y., and Teradyne. Plexus is offering three solid-state machines, the WA2K, 4K, and 65K. The largest, the WA65K, tests continuity for a maximum of 65,000 points at the rate of 20,000 tests a minute. These units accept magnetic cassettes or paper-tape input and print out results in product or computer format. The Teradyne N123, controlled by a microprocessor, finds opens and shorts in backplane wiring having as many as 5,000 points.

Two 10,000-point systems from Addison and Data-Numerics are the cheapest automatic continuity testers available. The $14,000 Addison BITs, which can only be programed by self-learning, stores results on magnetic cartridge. The $10,900 Data-Numerics ACT1 can either be programed off line, or it can self-learn and store the correct patterns on paper tape.

Testing cables

All of these machines could be used to test cables, but it would be wasteful to test only 100 to 1,200 wires with a high-capacity ACT. Smaller, lower-priced testers have been created to handle this task.

Glenair of Glendale, Calif., the Cablemaster division of Chatsworth Data in Chatsworth, Calif., and Automatic Production Systems are offering relay-switched machines that have capacities from 100 to 1,200 points. These ACTs can handle the continuity, leakage, and high-potential requirements of military, aerospace and telephone cables.

Solid-state cable testers from Addison and Data Numerics have capacities of 500 to 10,000 points. These are aimed at commercial cables that are only checked for wiring continuity and not cable leakage and hipot.

Is high-potential testing needed?

Makers of automatic circuit testers have opposing opinions on the necessity of high-potential and leakage testing, and they differ on the voltage levels required for continuity testing. Teradyne Inc. of Boston contends that neither high-potential nor leakage testing is necessary, and its machines don't offer those services.

The company has even sold its N151 to military suppliers who have managed to get the requirement for high voltage testing dropped. Some of Teradyne's customers claim that high-voltage testing damages equipment.

In the opposing camp, Al Summers, president of Automatic Production Systems Inc., Pennsauken, N.J., points out that his engineers have performed continuity tests on backplanes for their own equipment at 28 volts dc, 100 V dc, and 500 V dc. At 100 V dc, more faults were revealed than at 100 V dc, but above 300 V dc, few additional errors were found, he says. Automatic Production's equipment tests for continuity at voltages as high as 500 V dc.

Engineers of DIT-MCO, Kansas City, Mo., contend that 24 V dc is the minimum voltage required for accurate continuity testing. Bill Comer, assistant to the president, says, "we feel that at least 24 V is needed to burn away contact films on fixtures, so we even design our solid-state switch matrixes with 24-V excitation," Many solid-state testers based on transistor-transistor logic use +5-V dc excitation.

Before making a decision, it may be worthwhile to consider that one of the most pervasive industries based on electronics—the telephone company—operates on 48 V dc.

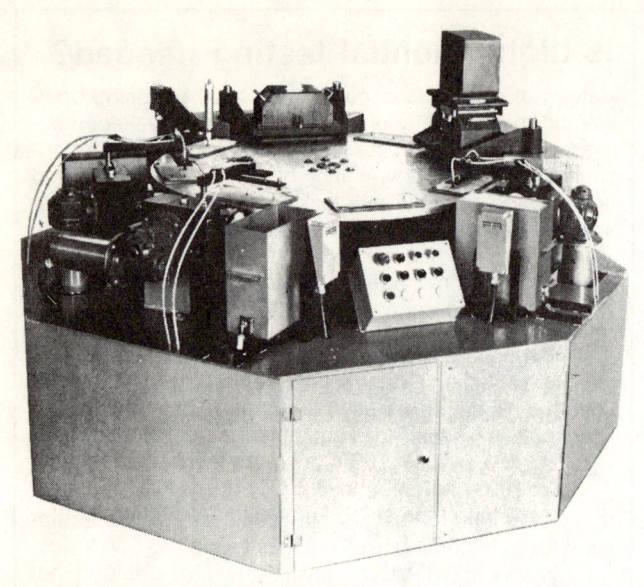

7. Merry-go-round. A 480-point bed-of-nails fixture is only one station of eight on this rotary test fixture designed by Ostby & Barton to test and sort flexible pc boards. By changing the test head of the built-in continuity tester, different pc boards can be accommodated.

A class of automatic testers is designed for guarded impedance measurements on loaded pc boards, as well as continuity testing of bare boards. All of the companies manufacturing this type of equipment—Systomation, Elnora, N. Y., Faultfinders, and Zehntel Inc., Concord, Calif.—admit that these machines could never be justified for bare-board testing alone, but should be used to test bare boards before testing loaded ones.

Machines like the Systomation Fixit and the Faultfinders FF101 can be programed to test as many as 600 points on a bare pc board, and a later Systomation machine, the Date II, has a hard-wired bare-board programing step. Still another approach to bare-board testing is Faultfinders FF202. This microprocessor-controlled, electromechanically switched unit can find shorts and opens in either loaded or unloaded boards. Like some of the automatic continuity testers, it can either self-learn or be programed off line. However this $19,500 unit with a built-in bed-of-nails fixture is mainly adaptable to medium-size pc boards and backplanes because of its 2,000-point capacity.

Manufacturers evaluate testers

Manufacturers of various types of equipment are happy with their ACTs. Among them is Cambridge Memories Inc., in Bedford, Mass. Two years ago, the company was compiling a 20% rejection rate in tests at the logic-test stage of seven-layer, high-density, emitter-coupled-logic boards on an extremely expensive computer-controlled logic tester.

Cambridge bought a Plexus WA65K and made its own bed of nails. Early screening by the continuity tester reduced the board-rejection rate at the logic-test stage to less than 5% and turned up such defects as cut edges, etch shorts, and even incorrectly layered boards. The tester paid for itself in six months.

Now the tester is being used on two shifts for testing bare multilayer boards and backplanes that have as many as 3,200 points. Cambridge has even developed its own inexpensive paddle boards to interface between the output of the Plexus cable connector and the backplanes. Cambridge lets its Plexus self-learn the unit under test, instead of programing off line. Ed Fassino, Cambridge test manager, says his company decided to buy the Plexus because "the price was right, and it was easy to fixture to."

Western Electric Co. is using an Omnitester 2000 in

8. Daisy chaining. With a solid-state continuity analyzer, active fixture cards plugged into the backplane under test allow the wires from a tester to the unit under test to be reduced from an extremely large bundle to a small multiwire cable.

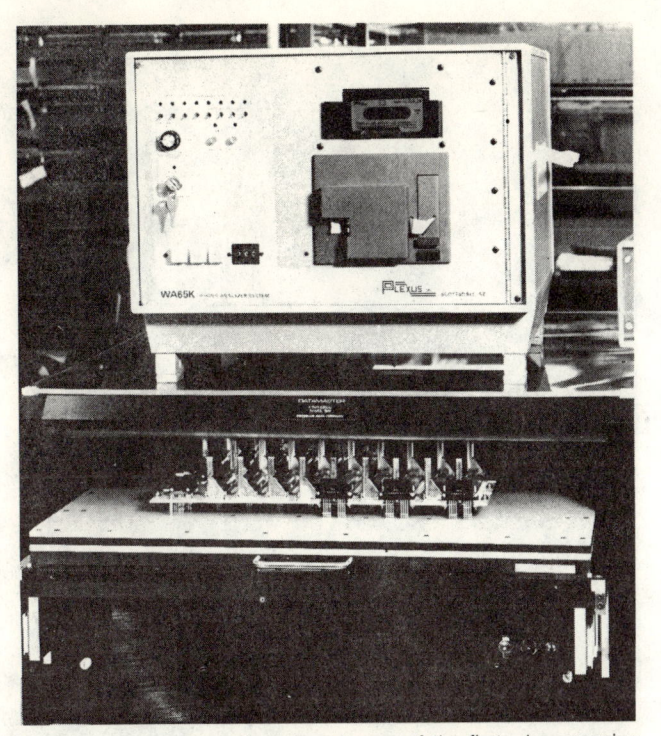

9. Tested backplanes. Interdyne is one of the first wire-wrapping services to supply both wrapped and tested panels. A Plexus WA65K analyzer and a 5,000-point Datamaster fixture are used in the test. Cost of testing runs about 2 cents per wire up to 50,000 wires.

Kearny, N.J. to functionally check a large relay rack. The system tests continuity in the de-energized system, pulls in the relays, applies a test tone, and then tests the rack for continuity in the energized system. This tester has proved to be twice as fast as equipment that had been designed in-house. Another DIT-MCO machine is being used at Grumman for the same general purpose—testing pc cards containing many relays.

ACTs are also being used by manufacturers of printed-circuit boards, cables, and harnesses, as well as wire-wrapping houses. Among the pc-board makers are Circuit-Wise, North Haven, Conn., General Circuits, Rochester, N.Y., TRW Inc.'s Cinch-Graphik, City of Industry, Calif., and the Litton Industries Inc. Advanced Circuitry operation in Springfield, Mo.

Circuit-Wise in 1974 purchased an Omnitester 3000 from Automatic Production Systems to check continuity of its single-sided and double-sided boards with plated-through holes. The continuity testing has reduced by about 90% the amount of optical inspection needed for each board.

Rowland Mettler, president of Circuit-Wise, expects eventually to completely replace optical testing by electrical testing. He says that, in the long run, it pays for his customers to spend the small premium to test the boards. In fact, two of his customers now specify testing of all boards purchased from Circuit-Wise.

The self-programed Omnitester is usually fixtured to the board under test by an Everett-Charles fixture, which is charged to the customer. Eventually, Circuit-Wise intends to make its own fixtures.

Although General Circuits extensively tests all types of its printed circuits, including flexible circuitry and multilayer boards, Paul Costello, quality-control manager of General Circuits, argues that optical inspection will always be needed because "no electrical tester can pick up cosmetic defects."

General Circuits began testing bare boards in 1964 with two DIT-MCO machines. Two years ago, it added two Teradyne N151s, which are used in the self-learning mode. Everett-Charles and Teradyne fixtures with as many as 2,400 pins are used to access boards as large as five by five feet.

Testing backplanes

Wire-wrapping service companies, like the pc-board suppliers, are beginning to supply tested panels. Four or five years ago, these companies were shipping panels that were either untested or manually "buzzed out" on a sample basis.

One of the first wire-wrapping companies to go to ACT is Interdyne Co. of Van Nuys, Calif., which has six automatic Gardner-Denver machines that wrap 50,000 to 75,000 wires a day. Marvin Mallon, vice president of operations, says that its Plexus WA65K has been such a powerful aid that it has increased the company's wire-wrapping business.

Interdyne says testing costs from 2 cents a wire for as many as 49,999 wires and drops to 1 cent for more than 100,000 wires. If a nonstandard fixture is required, the cost is $970 plus $2 per point, and a translation table is $75. Mallon says, "we test through a bed of nails, rather than through the connectors. With the large boards we test, we would need too many paddle boards, and our customers prefer to have their connectors untouched."

Interdyne fixtures panels to its tester with a Program Data bed of nails (Fig. 9) that will accommodate a panel 22 by 22 in. with 5,000 pins. The company is starting to test epoxy-glass panels that have press-fit pins. Extensive printed circuitry on the boards must be tested before pins are inserted. These panels are run through the Plexus before the pins are inserted. This service is equivalent to pc-board testing, which Interdyne is considering as an addition.

Checking harnesses

Harness-making is often thought of as a "garage" operation, but some companies have engineering and testing staffs that do nothing but turn out harnesses for business machines, computers, and peripherals. One of them, VIP Industries, of Paterson, N.J., found that these complex harnesses could not be tested adequately without automated equipment. In one instance, the 641 circuits in a braking cable could not be trusted to human error.

The testing service boosts the price of the cable to about 10% more than an untested one. However, pretested cables can save 30% by eliminating in-house testing by the purchaser, points out Bernard Lichtenstein, VIP president, Since installing an Addison BITs tester about a year ago, VIP is supplying defect-free cables that are 100% continuity tested for shorts and opens. Lichenstein says that after VIP supplied the first 1,000 cables for a Sperry-Univac peripheral, Univac waived the testing requirement. □

In-circuit testing pins down defects in pc boards early

A fixture that contacts conductors on printed-circuit boards operates with guarded impedance-measuring routine so that in-circuit tester screens out nearly all the faults before functional testing begins

by Frederick A. Schwedner and Stephen E. Grossman,
Faultfinders Inc., Latham, N.Y.

☐ Escalating costs of assembling printed-circuit boards have applied acute pressure to improve results by test engineers and quality-control managers. The challenge is to quickly and economically turn defective boards into fault-free deliverable products. However, isolating faults and trouble-shooting them can turn out to be an expensive proposition in itself unless efficient test systems and methods are employed.

A potent instrument to minimize costs is the automated in-circuit tester, which pinpoints the faults and prescribes the repairs for the mounted components, as well as the boards. What's more, the components can be tested to ensure that values fall within prescribed tolerances. Fortunately, inexpensive automatically generated functional test programs can be used by in-circuit test systems to test digital as well as analog pc boards.

In-circuit test systems follow simple routines, beginning with continuity tests, and are fast. Since testing a resistor takes only 6 milliseconds, 166 resistance tests can be performed each second. However, capacitors take longer to test because it takes a short time to charge them.

Presuming conformity

In-circuit testing presumes that if the unit under test is built to design specifications, it will function. The circuit should work if the conductor pattern on the circuit board and the component orientation agree with the schematic and if every component tests within design-tolerance limits. This premise has been fulfilled for more than 99% of all boards tested on a Faultfinders in-circuit component-test system.

In-circuit testing combines 100% component testing with a post-assembly test—virtually impossible with other test methods. And no amount of testing before assembly can ensure that a component will survive operations yet to come—insertion, crimping, and wave-soldering. To a manufacturer, it is important that tests be conducted at the post-assembly stage.

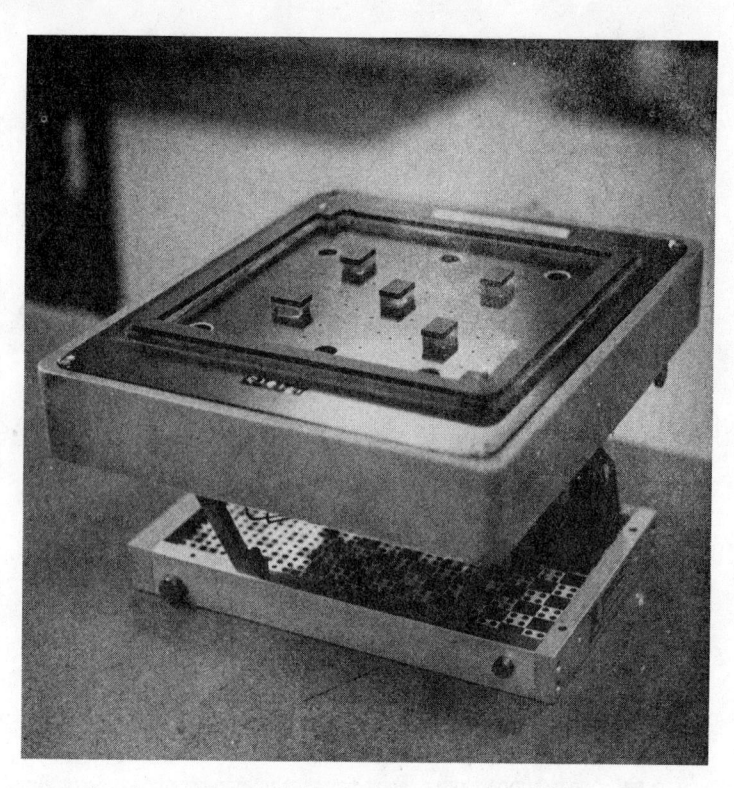

But, in addition to defective components, board failures can be caused by solder splashes, reversed components, shorts between conductors and open conductors on the pc board. Some components may even be missing. Whatever the defects, the challenge is to identify them—and fast.

Evaluating the in-circuit test system

The fine track record of in-circuit testing and fault diagnosis can be attributed to these capabilities of the equipment:
•Direct access to the pins of every component on the pc board.
•Guarding, which is the isolation that enables each component to be tested independently.
•Inherent fault-diagnosis.
•Nondestructive testing to verify the electrical integrity of the board and components before power is applied.

Because the test fixture connects to every device, the in-circuit test system is more effective than the functional type, which connects to the edge of the pc board. The major drawback of the card-edge tester is the complex software required to isolate faults.

The in-circuit tester tests the individual digital ICs on a board by applying 1s and 0s to the pins themselves and checking their outputs. With a card-edge tester, fault diagnosis is limited to points stuck at 1 (shorted to 5 volts), stuck at 0 (shorted to ground), and adjacent-pin shorts. Often a series of manually controlled measurements is necessary to isolate a common solder splash.

Because the card-edge tester "sees" logic transitions only at the edge outputs, fault diagnosis at intermediate logic levels depends on some type of a logic diagnostic,

1. Fixtured. Unlike the functional card-edge tester, the in-circuit tester uses a special vacuum-actuated fixture that contacts the lands and conductors of a pc board with spring-loaded pins.

which is usually implemented by a software package. If the functional test program is written in a high-level language, the test designer must isolate the error by means of a series of tests that almost invariably directs the tester through a series of manually probed measurements. In short, the capability of the card-edge tester to isolate circuit failures normally depends on the efficiency of the software the test engineer has the ingenuity to generate.

Gaining access

Direct access to components is what makes in-circuit testing possible. And this access is gained through the test fixture. The fixture, sometimes called a bed of nails, presses spring-loaded pins against the conductor pads, which become the test points, on the bottom of the printed-circuit board. In effect, the fixture provides visibility, which in card-edge test systems is limited by the relatively small number of points available in the card-edge connector. In-circuit testing enables measurements to be made on the board under test through switching between the test system and the fixture's test pins.

A typical fixture used in an in-circuit test system is shown in Fig 1. The board under test is held in place by a vacuum system that draws about 55 cubic feet per minute of air from the fixture. The fixture is wired and connected to a plug-board, which holds and interconnects the fixture to the test system. In operation, the plug-board is inserted into a mating connector called the receiver. The receiver, in turn, is cabled back to the tester's switching system. When vacuum is applied to the fixture, test pins rise and contact the solder pads on the bottom of the circuit board (Fig. 2).

Even though fixturing does indeed provide accessibility, it is not in itself sufficient to make in-circuit component testing possible. The reason is that the mounted components are nearly always connected to other components.

In general, measuring the impedance between points can identify a defect on a board. If the board contains a short or a missing or reversed component, the measured impedance will differ from the expected value. And when a measurement is out of tolerance, the fault still must be isolated and repaired. Measuring the impedance of the individual components makes this isolation possible. But in testing individual components, the load-

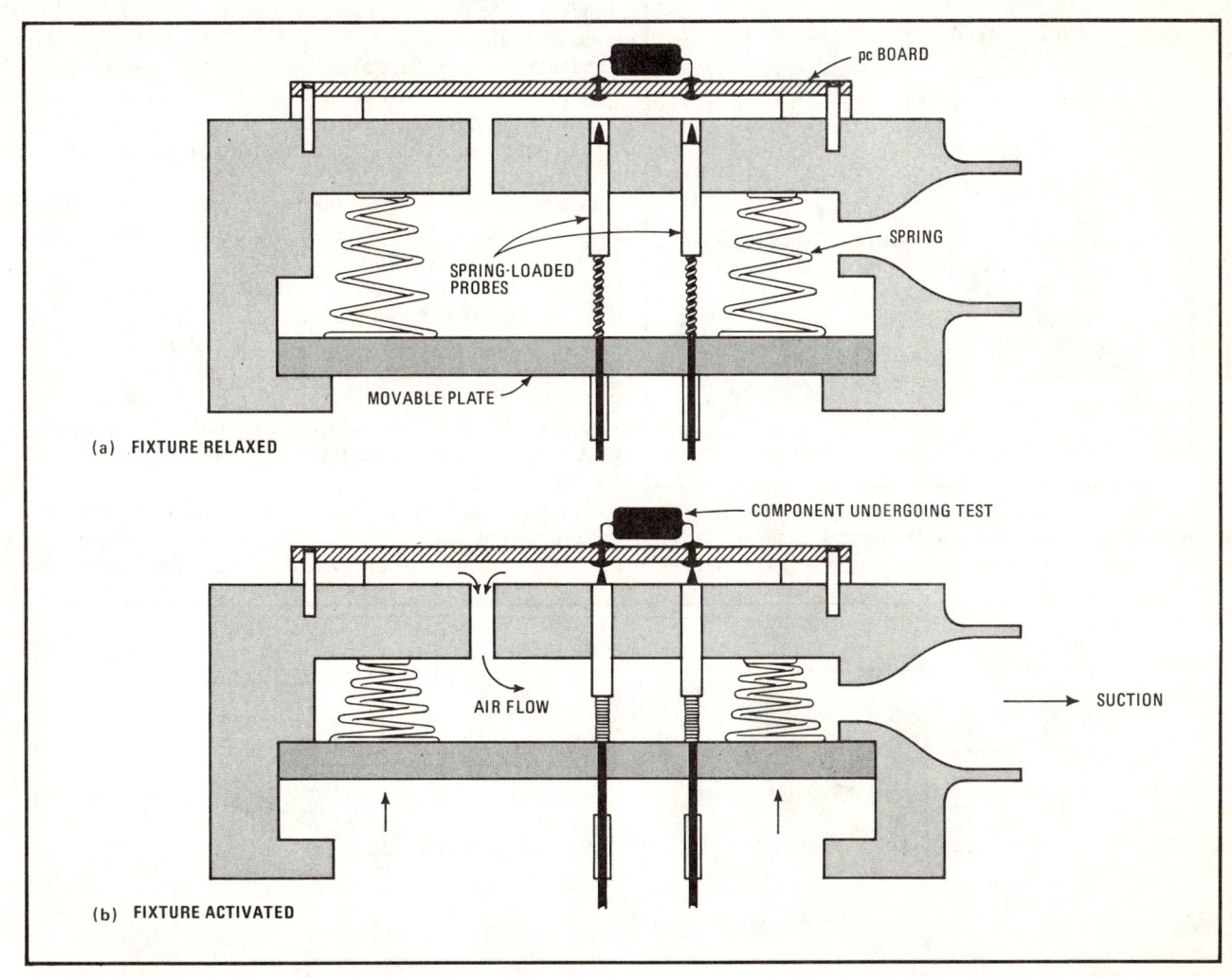

2. Vacuum-actuated. When the bed-of-nails fixture shown in the cross sectional view of (a) is evacuated (b) results and the bottom plate moves up to compress both springs and to place the spring loaded probes against the printed circuit board lands.

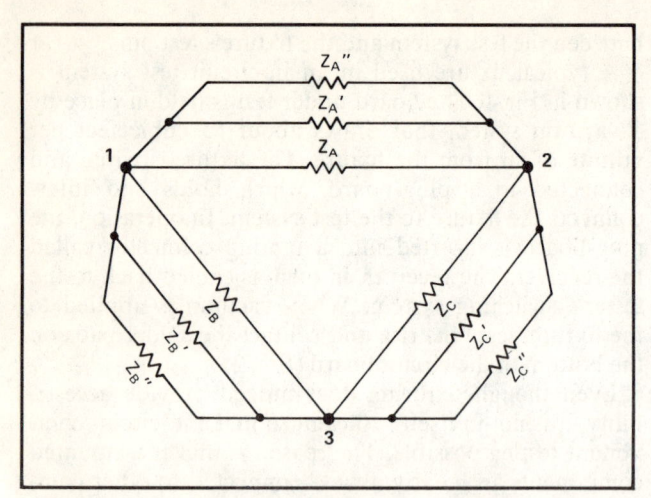

3. Isolated impedance. A measurement of impedance across points 1 and 2 by normal means will not yield the impedance of Z_A, which must be determined by a guarded measurement.

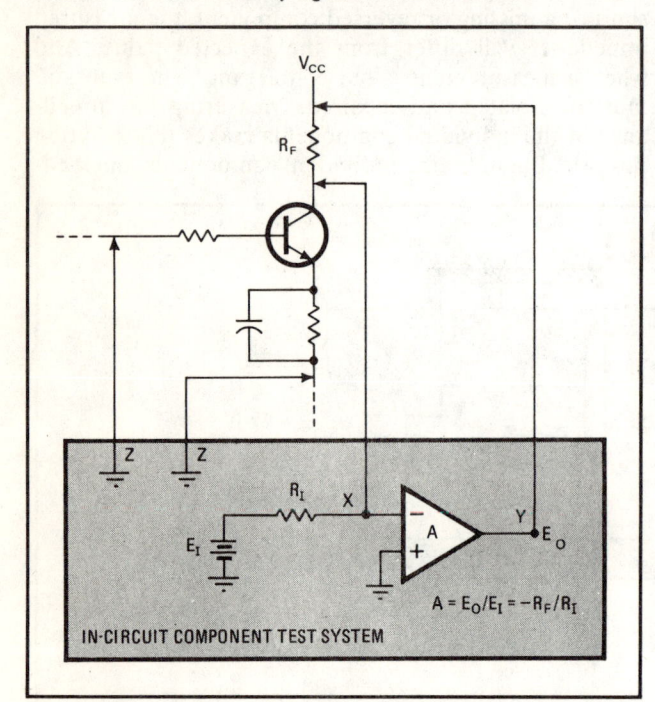

4. Guarding. By placing the impedance to be measured, R_F, inside the feedback loop of an operational amplifier, a guarded measurement of R_F can be derived from measured amplifier gain, $-R_F/R_I$.

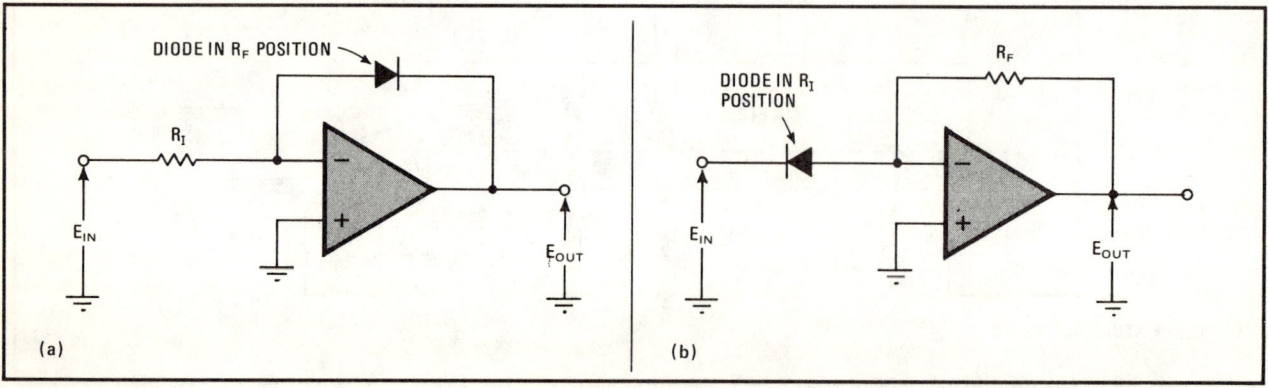

5. Diode testing. Diode forward voltage can be measured by substituting the diode for the operational amplifier's feedback resistor (a), and reverse leakage by substituting the diode for the op amp's input impedance and impressing a known reverse voltage across it (b).

ing effects of adjacent parts must be avoided.

This testing problem is illustrated in Fig. 3. Clearly, if the impedance is measured between points 1 and 2, the equivalent impedance is measured between points 1 and 2—not Z_A, the desired impedance. Although an isolated impedance can be measured and calculated through knowledge of equivalent impedances, the test technique becomes far more powerful if Z_A can be isolated from the remainder of the circuit.

There is a way this can be achieved, and it is the essence of in-circuit testing. This technique, known as guarding, ensures that adjacent components do not affect the accuracy of a component measurement.

Guarding

Guarding is the technique that enables a system to make in-circuit component tests. By simply testing the components, it is possible to check both component values and the workmanship of a pc board. Measurement of shorts, opens, and component impedance isolates faults so that it becomes easy to identify the conductors shorted by solder splashes, wrong components, missing components, reversed diodes, and polarized capacitors.

And, since a board can be tested independently of the way it functions, the remaining knowledge required reduces simply to a component's specifications and its location on the board. Consequently, the test engineer needs only to generate a test program that verifies part specifications, a function that can be performed quite well by manufacturing personnel. There is no need for design engineers to become involved in verifying the workmanship.

Guarding is made possible by an operational amplifier in the test system. Not only can the op amp effectively isolate a component, it also makes possible a number of different test configurations. Several properties of the op amp make it ideal for in-circuit testing.

As shown in Fig. 4, the component being tested in the unpowered printed-circuit board actually becomes part of the feedback loop of the op amp in the test system. The point labeled X in Fig. 3 is a virtual ground, which puts the collector side of the resistor R_F at ground potential.

The grounded test points labeled Z ensure that there are no current paths through either the base or the emitter of the transistor in the circuit being tested. Point Y is

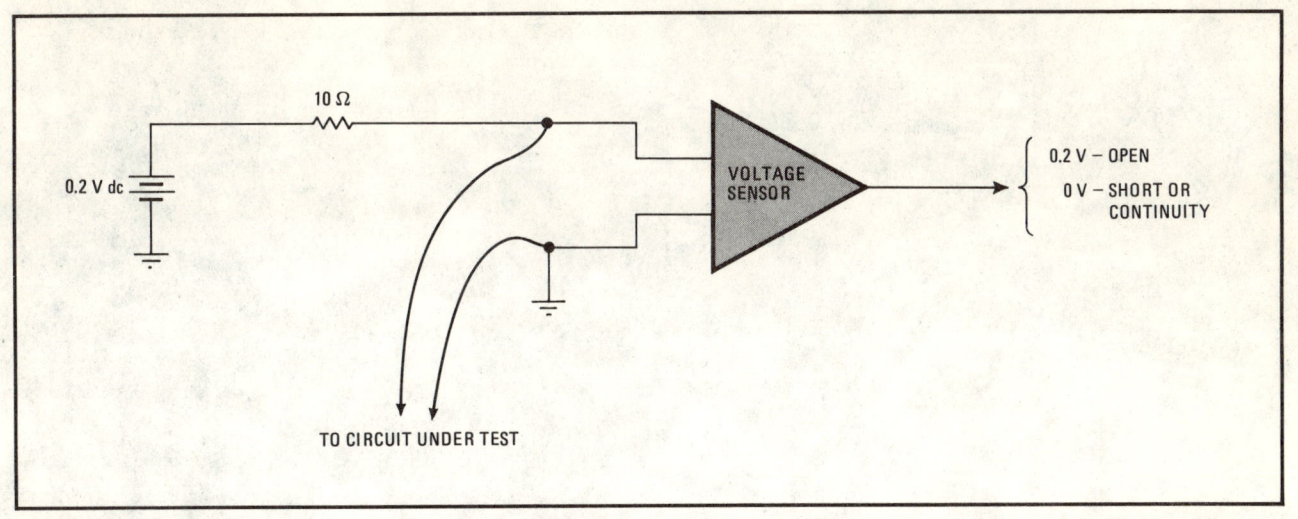

6. Opens and shorts. An in-circuit continuity tester consists of a current-limited low-voltage source and voltage sensor. The 0.2-V source is too low to forward-bias a silicon junction, and short-circuit current is only 20 milliamperes.

connected to the output of the operational amplifier, and since the operational amplifier has virtually a zero output impedance, the Y connection does not hinder measurements.

Finally, as shown in the illustration, the gain of the operational amplifier is $-R_F/R_I$, where R_I is a precision internal resistor and R_F is the impedance to be measured. In guarding, the component being measured becomes part of the measurement circuit, and all that's necessary is to measure the output voltage of the amplifier, E_O, to decide if resistor R_F is within tolerance.

A diode can be tested for several other characteristics. To check diode forward voltage, the diode to be tested is placed in the feedback loop of the operational amplifier and a known current is injected into the diode (Fig. 5a). Output voltage is measured, and if it is within the limits expected at the programed current, the diode passes this test.

To test a diode for reverse leakage current, the diode becomes the input impedance, R_I, of the amplifier circuit of Fig. 5(b), and a known reverse voltage is impressed across it. Leakage current produces a voltage drop across the feedback resistor, R_F, which, in turn, can be related to E_O. If the current is within acceptable limits, the diode passes this test.

Testing other semiconductors

By extending the techniques used to test diodes, other semiconductors may be tested. An in-circuit component-test system can also test such other components as transistors and field-effect transistors. By grounding appropriate test points, ICs of various logic families, including transistor-transistor logic and complementary-MOS, may be tested for truth-table conformity.

If the unknown impedance is excited by an ac voltage, the impedance of capacitors and inductors can be measured. Capacitors and inductors are tested in a circuit comparable to Fig. 5, except that the normal dc source voltage is replaced by an ac source.

Polarized capacitors are tested for proper orientation by applying 30% of rated dc voltage in what is expected to be the correct polarity and checking the leakage cur-

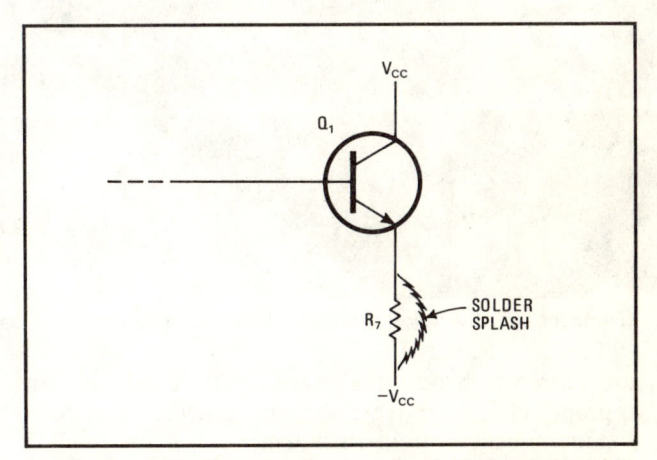

7. Solder splash. With a functional tester, it is likely that this solder splash would cause destruction of transistor Q_I. However, an in-circuit tester can detect the defect without risk of damage.

rent. If capacitor polarity is reversed, the current will be at least an order of magnitude higher than if the polarity of the charging current were correct. In addition to impedance measurements, the in-circuit test system can be configured for shorts and continuity testing.

Basic circuitry for finding shorts and opens is shown in Fig. 6. The test-circuit source delivers 0.2 v, and its short-circuit current is limited to 20 milliamperes. The 0.2-V potential is less than the cut-in voltage needed to drive a silicon-semiconductor junction into conduction so that the test may be performed without forward-biasing semiconductor junctions. However, the test voltage is ample to nondestructively find opens and shorts on a printed-circuit board.

Devising test strategy

In continuity testing, the test voltage is applied through two pins in the fixture to two pads or conductors on the board being checked. A voltage sensor is paralleled across these two points to identify both opens and shorts.

In-circuit component testing is facilitated by the test fixture, which provides visibility, and the operational

311

8. Into the board. A typical in-circuit tester is shown with its fixture. Defects are diagnosed, and results are delivered by a built-in printer.

amplifier, which does the guarding. By employing an appropriate test strategy, the manufacturer can reap maximum benefits from his system.

After the pc board is assembled, the test engineer should identify the gross defects, which are likely to propagate farther failures during functional tests if uncorrected. These defects should be repaired before powering up the board, which might destroy it. First come the shorts and opens, both in the printed-circuit conductors and the compnents.

Frequently he will encounter reversed polarized devices—capacitors and diodes—reversed ICs, missing components, and improperly oriented transistors. Historically, these gross defects account for 50% to 70% of all faults on an assembled printed-circuit board. When the gross faults have been cleared, the board may be tested for proper component values, followed by a check of the functional operation of active devices.

Since this logical testing sequence is so effective and easy to generate, test-equipment manufacturers, including Faultfinders, have been able to simplify the test-generation process so that the writing of a test program is reduced to entering a parts list and component-location information into an off-line computer.

Conducting the test

An in-circuit test program should begin with continuity tests to identify shorts and opens. But if a functional card-edge tester were used to test a conventional emitter-follower stage (Fig. 7), a short circuit of R_7 caused by a solder splash probably could not be detected until power was applied to the board which would drive the base into conduction. This condition would probably cause the transistor to exceed its power-dissipation rating and be destroyed.

To start an in-circuit test, (Fig. 8) the operator places a printed-circuit board on the fixture and presses the start button. A solenoid-operated valve applies a vacuum to the fixture and drives the pins against the conductor pads on the bottom of the printed-circuit board. The test system runs through the tests for continuity and shorts.

If defects are encountered, the printer identifies the defective conductor pair(s) in a fraction of a second. The operator tears off the printout and attaches it to the circuit board. Normally, a conditional end-of-test command is programed at the conclusion of these tests so that if there is an open or short, it can be repaired before testing continues. If so, the board is repaired and returned to the test system.

When the test system confirms that opens and shorts are cleared, it runs through the remainder of the tests, checking for component orientation and values. Finally, functional tests are performed on relays, logic circuits, and amplifier stages.

At the end of the test, the solenoid valve closes, shutting off the vacuum and releasing the board. If any failures have been detected in the later phases of testing they too will be identified by a diagnostic printout listing the failures. □

In-circuit tester using signature analysis adds digital LSI to its range

by Craig Pynn, *Plantronics/Zehntel Inc., Walnut Creek, Calif.*

☐ One of the most cost-effective ways of testing analog printed-circuit boards is in-circuit testing. But boards that include large-scale digital integrated circuits like memories and microprocessors are beyond the capabilities of present-day in-circuit testers, which can at best do limited isolated testing of medium-scale digital ICs.

One solution is to adapt signature analysis, a technique originally developed for functional testing of arrays of LSI devices, to use with individual LSI devices. In combination with some of the older analog techniques, this approach yields a new in-circuit tester capable of checking out everything on a board from LSI devices through discrete components to the printed wiring. Yet programming is much simpler than for the alternative, functional board testing. Moreover, program execution time is short, making the instrument suitable for high-volume testing.

Like all other in-circuit test systems, the new instrument tests components, not circuits, treating a loaded board as a set of discrete unrelated devices. It gains access to them through a bed of nails, a fixture that forces an array of spring-loaded pins directly against the plated conductors and lands of the pc board. Also, because most components are interconnected, it isolates each component from its neighbors during measurement.

For this purpose, the early digital in-circuit testers used a brief, low-impedance logic stimulus to the IC under test, to keep upstream logic states from affecting it. They then tested the IC by driving its inputs high or low and checking its output for compliance with the pertinent truth table. Such a state-by-state analysis is shown for a NAND gate in Fig. 1. The circuitry for measuring the output was designed to synchronize with the input stimuli, so that it sampled the output state during a short stimulus and measurement interval (typically less than 1 millisecond).

However, the number of states in a truth table soon climbs above 10 with LSI devices, and once that happens, test programs become so unwieldy that test coverage has to be compromised. Moreover, a short, single-state input stimulus cannot easily test sequential LSI circuits with their complex synchronous and asynchronous input state

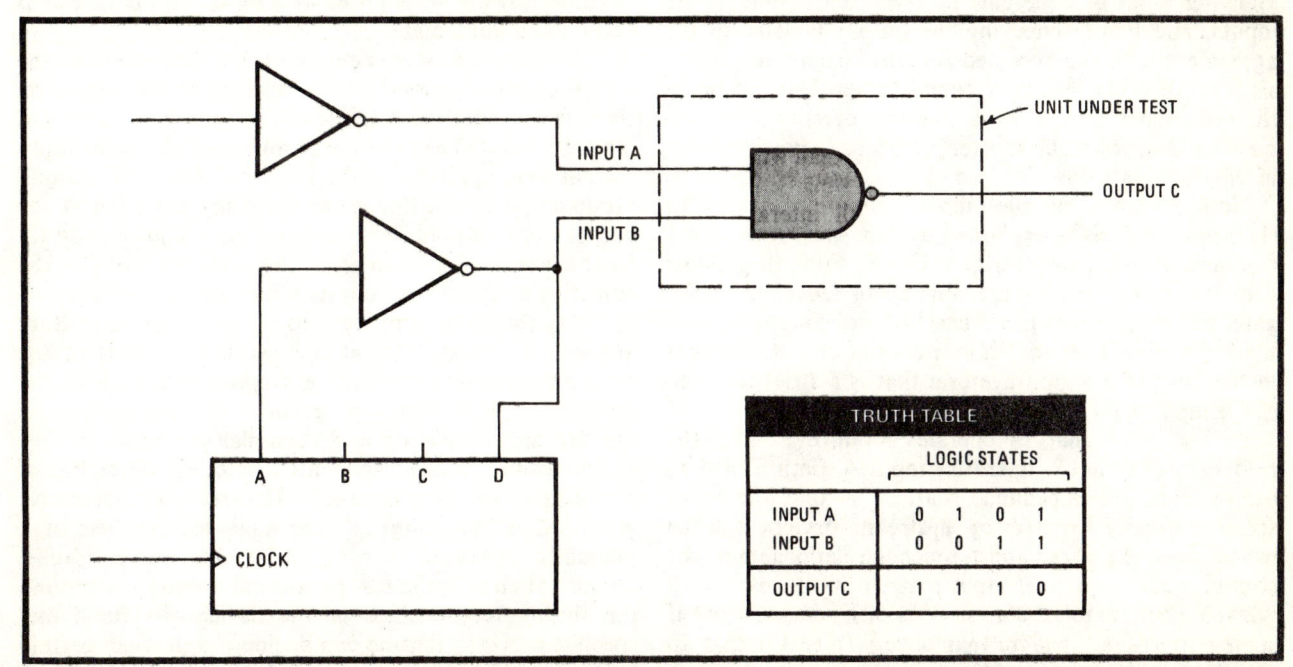

1. Digital isolation. In early digital in-circuit testing, each digital device was electrically isolated and then checked state by state. On this NAND gate, for example, inputs A and B would be driven to various logic states to check compliance of output with NAND truth table.

2. Coping with LSI. The Zehntel TS800 employs low-impedance drive for device isolation, coherent stimuli to excite device inputs, and signature analysis to detect faulty LSI devices. The system is also capable of conventional analog in-circuit testing.

requirements. Some more comprehensive but simpler-to-apply method of LSI inspection was called for, and time-domain stimulus and measurement—the compression of input and output information—provides the answer that is needed.

Comprehensive in-circuit testing

The instrument in Fig. 2 checks everything from shorts to chips. It begins by testing for faulty interconnections, proceeds to the analog components, and finally attacks the digital buses and chips (Fig. 3). For this last step it combines three techniques, first isolating each device by applying a low-impedance drive to it, then applying a set of coherent, time-varying stimuli to its inputs, and finally checking the output bit stream for agreement with the expected response (signature analysis). Truth table testing is complete yet, being done in the time domain, requires far fewer programming statements and hence much shorter programs than when done in the logic-state domain.

How coherent, parallel input stimuli interact with signature analysis is explained by Fig. 4a, which shows the same NAND gate as Fig. 1. Once again, all possible combinations of inputs are applied in sequence to the gate. But each output is not checked individually against the truth table. Instead, the entire set of outputs emerges in the form of a logic waveform that is a strict function of the input sequence.

In Fig. 4a, signal F_2 operates input B at half the frequency of signal F_1 operating input A. Both F_1 and F_2 derive from low-impedance sources, so that the input states normally created by upstream devices will be overridden. The phase and frequency relationship of the stimuli create a parallel input pattern that transmits all possible combinatorial states. (It is of no consequence if some patterns occur more than once.) To ensure that all the states in the truth table of the device are executed, the test interval must extend over at least one full cycle of the lowest-frequency stimulus signal. The resulting data bit stream is sampled at the output of the device under test and fed back to a signature-generating circuit in the test system for transformation into a four-digit signature unique to the device.

(The technique is sometimes called a cyclic redundancy check (CRC). It has been used extensively for detecting data errors in magnetic-tape and disk equipment and more recently in field service test instruments intended for digital systems.)

Since the measurement interval and stimulus pattern are exactly the same from board to board, a defective IC does not yield the same signature as a good device. Because just one IC is tested at a time, fault diagnosis is certain and immediate.

The program statement required to execute the complete test discussed above is shown in Fig. 4b. First the reference designation, the device type, and the output pin are named. Then the pin numbers at which the input stimuli were applied and the frequencies of these stimuli are listed. The last line is the signature expected at the output node and obtained earlier from a known good IC. In the event of a mismatch, the system identifies the defective IC by printing out its reference designation.

Thus, the programming effort involved in signature analysis is minimal. What is most significant is that a valid and rigorous test can be programmed without the need either to manually program each input state or to analyze each output state of a complex circuit.

The technique can also identify precisely which line of a multiple-line bus is stuck. However, to determine which IC is "hanging up" the node requires use of a manually operated current probe. In the case of a sequential circuit, the test programmer must understand the timing relationships of the device—its functional specification—so that he can decide if and when certain inputs should be defeated and which outputs should be monitored for proper response.

A vital issue is the thoroughness of a test, a particularly challenging consideration in relation to LSI testing. Here in-circuit digital inspection resembles the more traditional card-edge functional test techniques since in both cases, the correlation of test coverage with the detection of failure modes is generally performed empirically. Results to date indicate that a programmer using this technique of LSI inspection needs very few instructions to construct a test program that will yield a test confidence level exceeding 90%.

Applying in-circuit digital testing

The in-circuit digital technique just described can be applied to both combinatorial and sequential types of logic circuits, including decoders, counters, shift registers, universal asynchronous receiver/transmitters, random-access and read-only memories, and direct memory access chips, as well as microprocessors.

As an example, Fig. 5 pictures an inspection test for a 256-by-4-bit programmable read-only memory (PROM). Coherent stimuli, labeled F_1 through F_8, are applied as inputs to the address lines. Since each frequency is half the frequency of the previous one, all the PROM addresses will be exercised if the test duration equals or exceeds the periods (1/F) of the lowest frequency (in this case, F_8).

Static logic levels are applied to the chip-enable inputs CE_1 and CE_2, respectively. All other logic devices with outputs connected in common to the outputs of the device under test (the bus) have been previously disabled. Outputs O_1, O_2, O_3, and O_4 are monitored for consistent signatures.

In practice, the in-circuit test system would apply the stimulus bit stream to the PROM on a known good board, sample the output bit stream, generate a signature, and store it in memory as a characterization of a good PROM. The output bit pattern and consequently the signature, of a good PROM is determined by the program stored within the IC. If because of an engineering change or for some other reason the PROM program is modified, it is a simple task for the in-circuit test system to generate and store the new signature.

Dealing sequentially

Though the technique of signature analysis does away with the need for a state-by-state analysis of combinatorial devices like PROMs, the programmer does need an understanding of the functional characteristics of complex sequential circuits like microprocessors.

An LSI device such as a microprocessor requires a defined sequence of stimuli at appropriate inputs in order to be tested properly. An in-circuit test system cannot totally exercise all the states of such a device, but it should be able to interface with the inputs and outputs and exercise the IC through enough logical states to yield a high level of confidence that the device is operating in the way it should be.

Since tests must be performed that cause a change at the output pins of the device, altering internal combinatorial states such as occur in incrementing a microprocessor accumulator is not productive. It is better to execute two instructions in its instruction set that will

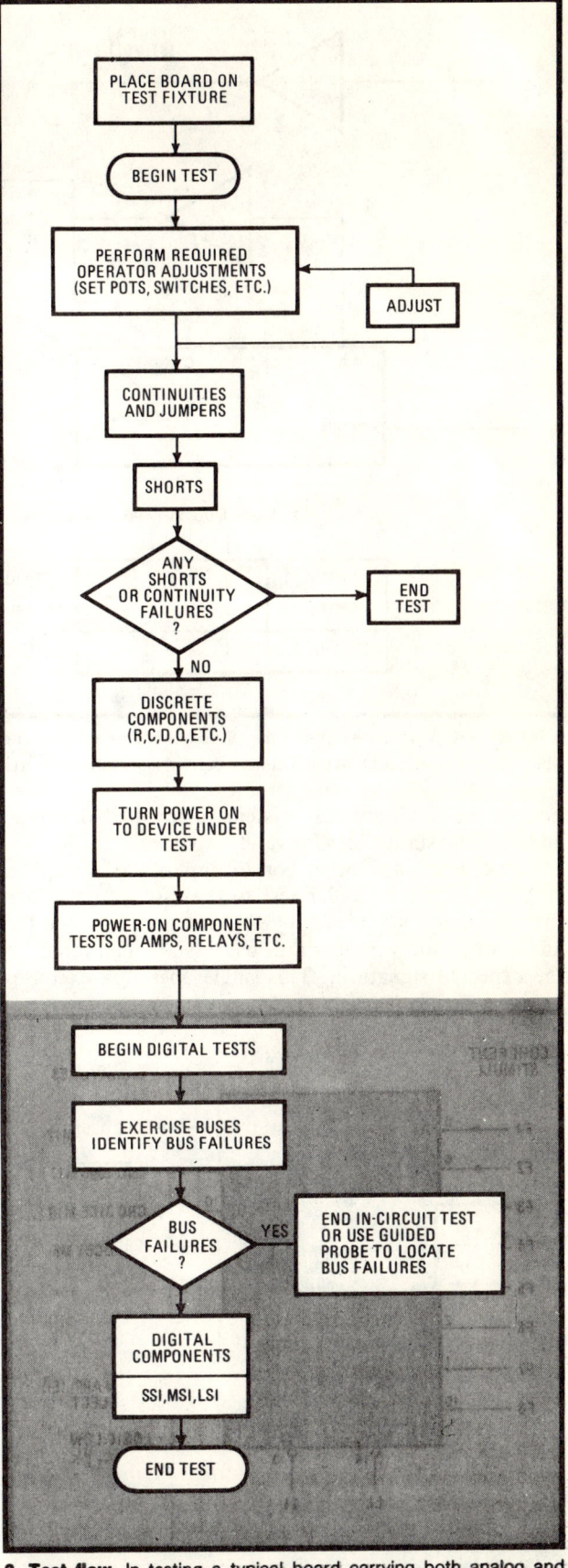

3. Test flow. In testing a typical board carrying both analog and digital circuitry, the in-circuit system first checks for shorts and opens in the wiring, then inspects nondigital components for proper placement and operation, and lastly begins exercising the digital elements.

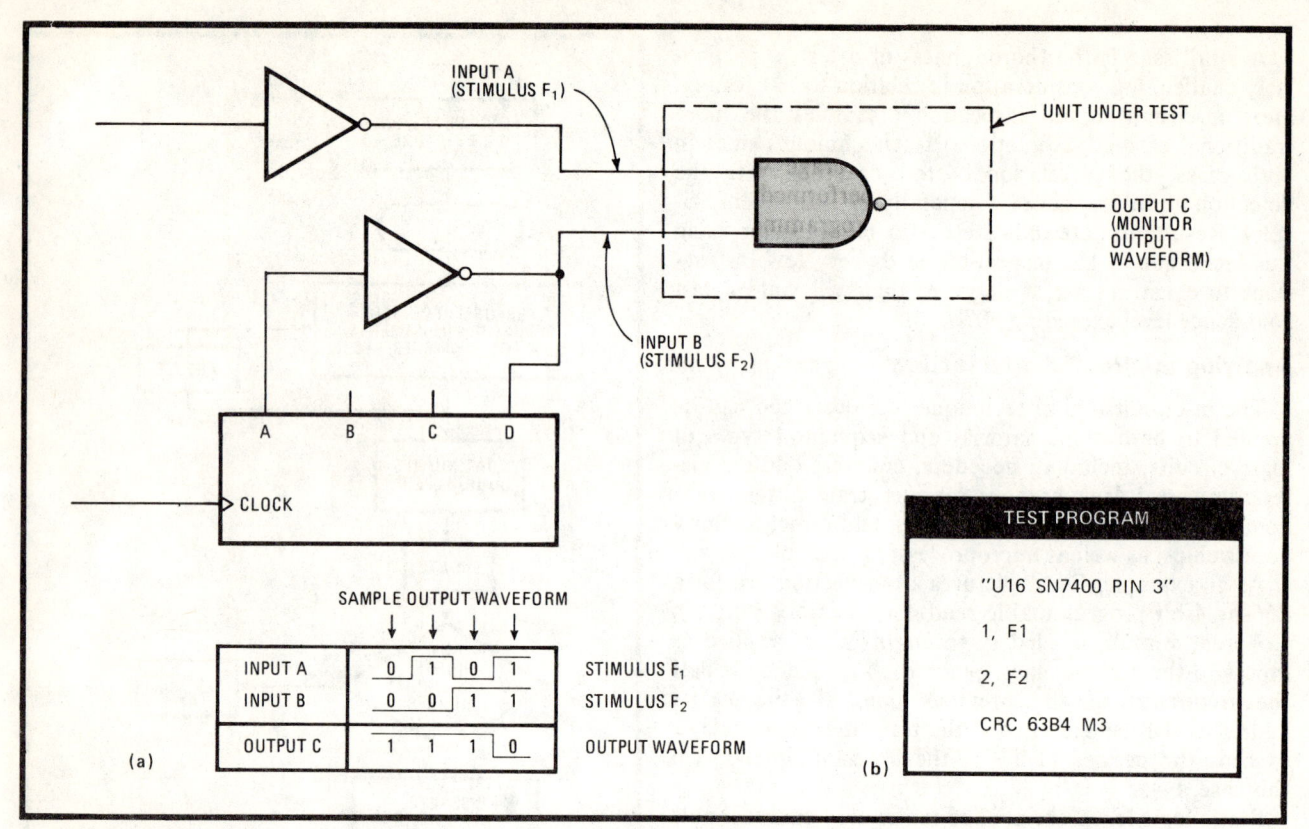

4. NAND test. Waveforms representing all combinations of logic states are fed into pins 1 and 2 of NAND gate by middle two lines of program, which then checks output at pin 3 against correct signature in fourth line. If gate is faulty, first line of program prints out, identifying culprit.

enable the in-circuit test system to monitor the data, address, and status lines for valid responses.

Clock, reset, and other control inputs are applied to synchronize the IC under test to the test system. Then the instruction NO-OP is applied to the data bus, and the address bus and various status outputs are monitored for the expected signatures. This single operation exercises the microprocessor's instruction decoder, program counter, address register, refresh counter, and various status control logic lines. Next, the HALT, or HALT-INTERRUPT, instruction is used to check the basic input/output registers and controls and also to exercise the jump and halt logic. In this test, the data bus first receives the instruction input and then provides the program counter contents (0001) so the bus can be read as data output.

The NO-OP test for a typical microprocessor is illustrated in Fig. 6a. Note the simplicity of the programming steps. F_1, the highest frequency, is the clock input. F_{13}, a lower frequency, initializes the microprocessor. The NO-OP instruction is produced by writing static logic lows (LL) onto the data bus. The microprocessor executes its entire NO-OP cycle once reset has duly been accomplished.

Figure 6b illustrates a typical HALT/INTERRUPT TEST. It resembles the NO-OP test except that a hexadecimal instruction is placed on the data bus to initiate the half cycle. The test system makes it possible to disconnect the stimuli at any time except during the read cycle. Here, multiplexing allows stimulation and measurement of bidirectional lines.

A DMA example

The Intel 8257 programmable direct memory access (DMA) controller is typical of the class of smart peripheral chips now appearing on printed-wiring boards. Designed to simplify and speed the transfer of data between memory and peripherals in microcomputer systems, it supervises the transfer of data between peripheral chips and memory, controls system buses

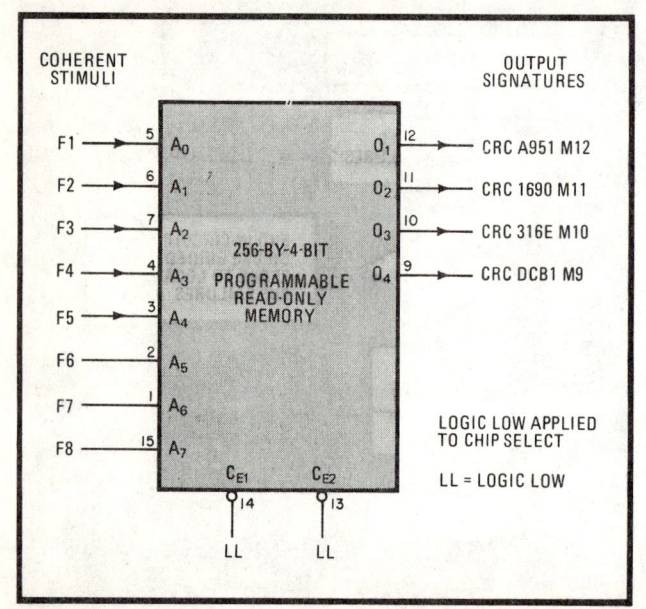

5. PROM test. Because the coherent stimuli, F_1 through F_8, of this 256-by-4-bit programmable read-only memory have powers-of-two relationship, all memory addresses are accessed so long as the test interval is at least equal to the period of the lowest stimulus, (F_8).

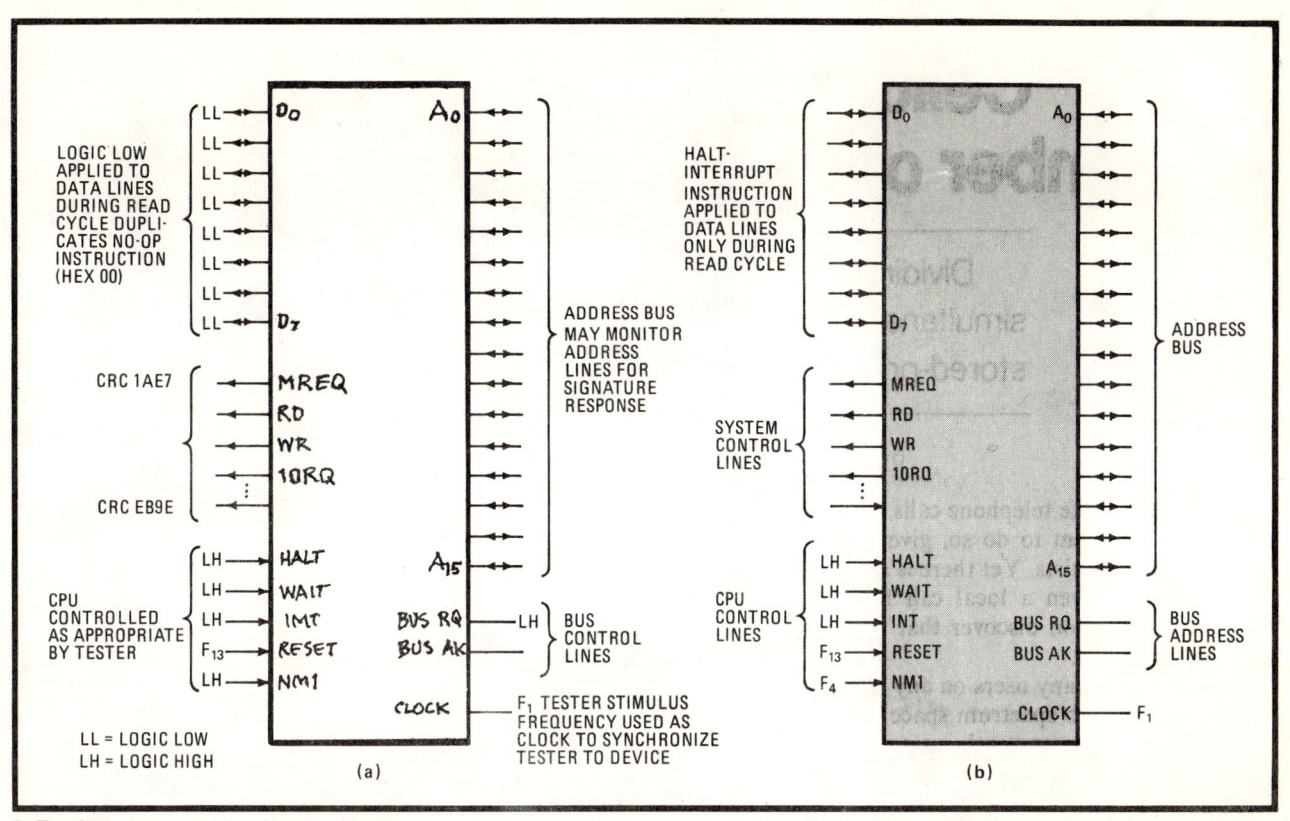

6. Testing one, two . . . Signature analysis can be applied to sequential circuits like microprocessors, on which it can carry out both a NO-OP test (a) and a HALT/INTERRUPT test (b). Output lines are monitored for conformity of test results with expected signatures.

when needed, and resolves the priority requests of external peripherals.

Before the operation of the DMA controller can be observed, it must be preconditioned. (The 8257 is a four-port device; each port undergoes identical tests.) To precondition port 0, it is necessary to:
- Reset the DMA chip.
- Ensure the chip to be in the slave (not selected) mode.
- Program the terminal count of port 0 (this takes 2 bytes, the memory block size).
- Program the beginning memory address of port 0 (also 2 bytes).
- Program the mode register to enable port 0 (1 byte).

When preconditioning is complete, the controller is given control of the address bus by being placed in master mode ($\overline{CS}$). It is then sent a DMA request (DRQ_0) so as to begin the transfer operation. During actual DMA operation its various outputs are monitored for a characteristic signature.

Test setup and execution

In the case of the 8257, terminal count, block size, and beginning memory address data are determined by the byte present on the data bus (D_0 to D_7). The port and corresponding functional information are specified by the current address present at A_0 to A_3. Data is latched by clock transitions at input/output write ($\overline{I/OW}$).

For the test of port 0, 2 bytes must be loaded for the terminal count, 2 bytes for the starting address, and 1 byte for the mode set. By choosing appropriate coherent stimuli from the tester and applying them to A_1, A_2, and A_3, all the programming addresses required for port 0 are produced. A higher-frequency stimulus is chosen for $\overline{I/OW}$ to produce strobes to latch the required number of bytes at each address. A data word is presented synchronously by the tester at D_0 to D_7 to load the desired information at each program state.

Once programming is complete, a low-frequency stimulus places the controller in the master state ($\overline{CS}$ high). Following this, the DMA request line (DMARQ) is taken high and DMA operation begins.

Any output may be monitored for transitions and a characteristic signature acquired for comparison with a signature obtained from a known good chip. The outputs of interest include: address bus (A_7); memory read ($\overline{MEMR}$); memory write ($\overline{MEMW}$); hold request (HRQ); and address strobe ($\overline{ADSTB}$).

In-circuit versus functional testing

An in-circuit test system with the digital capability described above is quite new, while digital card-edge functional test systems that interface at the card edge are well entrenched today. Is there room for both in the automated test equipment market?

The ability of in-circuit testing to test at the throw-away component level and the relative ease with which it can be programmed are undoubted advantages, particularly for high-voltage applications. Moreover, as LSI devices increase in complexity, modeling their functions will become a more formidable and hence more costly task. Simple programming, plus its simple one-device-at-a-time test strategy, would suggest in-circuit digital testing holds promise for years to come as a highly cost-effective production test technique. □

Index

ABS, 124
Acoustic interference, 240
Acoustic noise, 264
Aluminum, 245, 269
Analogs, thermal,
   of electronic systems, 246
   of Ohm's law, 224
Automatic continuity testing, 299
   of backplanes, 307
   of cables, 305
   cost of, 300
   DIT-MCO, 301
   programming, 302
   by remote control, 300
   solid-state, 304
Automatic-wiring techniques, 148–152
   impulse bonding, 160–166
   Multiwire, 148, 149, 150, 162
   planar stitch-wiring, 167–170
   Quick Connect, 148, 152
   Solder Wrap, 148, 152, 153–159
   Stitch Wiring, 148, 150, 151, 162
   Termi-Point, 148, 149
   Tiers, 148, 150, 152
   U-contact, 148, 152
   Wire-Wrap, 148, 149

Backplanes, 211
   testing of, 306
Basic building-block module (B^3M), 253, 257
Bathtub curves, 238
Batwing package, 228
Beam leads, 66, 67
Bed-of-nails fixture, 303
Beryllium oxide (beryllia), 234, 257
Bonding, 66
Breadboards, 126, 134
   with bus bars, 134–137
   types of
      circuit boards, 128, 129
      IC socket panel, 127
      insulation-piercing, 133
      low-noise, 134–137
      microprocessor, 139
      perf boards, 127, 130
      solderless, 130
Buoyancy, 237

Cable, flat, 103, 108
   crosstalk in, 108
   dual dielectric, 110
   planar transmission line type, 111
   shielded, 112
   woven, 112
Capacitors, 248
CEM-1 and -3, 123

Chip carriers
   ceramic, 68, 69, 173, 192, 201
      on hybrid substrates, 173
      IBM, 204
      in-house, 192
      leaded, 204
      vs other methods, 173
      on plastic pc boards, 205
      sockets for, 204
   film. *See* Film carriers
   Jedec family, 194–195, 204
   plastic, 192, 193
      AMP, 194, 196
Circuit models, 246
Closed-loop systems, 226
Cold dump systems, 245
Cold plates, 226, 248, 249, 251, 261
Computer-aided design (CAD)
   of hybrids, 74
   of IC masks, 292
   of pc boards, 287
Computer-aided manufacturing (CAM), 164, 165
Condensors, 269
Conductive elastomeric connectors, 217–220
   Zebra, 220
Conductive heat-transfer plane, 253
Convection cooling, 229, 237, 240, 241, 247, 258
Convection efficiency, 255
Coolants, 243
Coolers, 242
Cooling power, reserve, 240
Copper, 224, 243, 245, 269
   and brass, 245
   conductors, 258
   tubing, 245
Cramer's rule, 246
Current crowding, 224

Deep ultraviolet light lithography, 31–32
Dielectric oils, 245
Diffusion furnaces, 50, 51
   direct digital control (DDC) of, 52
Dual in-line package (DIP), 185
   ceramic, 188, 189
   CERDIP, 188, 189, 190
   cofired, 189
   on multilayer pc boards, 199
   plastic, 188, 190
   power dissipation of, 191

EBM II, 8, 9, 10, 48
Electrical isolation, 234, 242
Electrical path, 234
Electron-beam equipment, vacuum interlocking of, 17
Electron-beam exposure system (EBES), 9, 10, 11, 40
Electron-beam lithography, 3, 13, 23, 28, 35, 47

Electron beams
   deflection error of, 16
   formation and deflection of, 7
   pattern generation and control of, 7
   proximity effects of, 9
   registration of, 16
   square beam, 13
   and step-and-repeat process, 9
   vector-scanning, 9
   See also Electron-beam lithography
EL-1, 7, 10, 11, 13–18, 36
Epoxy encapsulant, 234
Epoxy-glass board, 258

Fans for forced-air cooling, 237–239
Film carriers, 68, 175
   bumped, 204
   chip attachment to, 176
   on copper, 178
   definition of, 175
   inner and outer lead bonding to, 178
   materials for, 177
   for plastic DIPs, 183
   processing of, 176
   used with hybrids, 181, 196, 202
Film coefficient, 261, 263
Finite-difference technique, 246
Flatpacks, 193, 200, 253
Flexible circuitry, 114, 116
   applications of, 123, 124
   connectors, 122
   fine lines with, 118
   and hardboard carriers, 121
   manufacturing of, 118, 120
   market for, 118
   materials for, 118
Flip chips, 67, 68
Fluids for liquid cooling, 245
Forced-air cooling, 225, 229, 237, 241, 261
Fourier's law, 247
FR-2 and -4, 123

Gasket, 253

Hardware, closed-loop, 243
Heat capacity of a liquid, 251
Heat-conducting frames, 259
Heat exchangers, 240, 243–245, 258–264
   aluminum-beryllia, 257
   ceramic, 257
   forced-air, 262
   hollow-core card, 252–257
   pin-fin, 264
   surface, 263
Heat flow, 247, 248
Heat pipes, 226, 269
   as drive shaft and heat sink, 272
   with fins, 272
   fluids, 270
   panel, 271
   temperature gradient of, 271
   thermal analog of, 271
Heat sinks, 227, 229, 237
   envelope volume of, 241

Heat transfer, 237, 238
    capability, convective, 248
    coefficient of, 237, 247, 248
    equation, 238, 247
    and surface area, 250
Hollow-core cards, 252-257
Hook in printed-circuit boards, 140-146
    definition and cause of, 140
    effects of, 141
    tests for, 141-143
Hot spots in electronic devices, 224
Hybrids
    computer-aided design of, 74
    multilayer, 70, 200
    substrates for
        ceramic, 65
        cofired, 71
        heat distribution of, 257
        nonceramic, 65, 66, 71
    testing of, 70
    thick-film, 62, 63
        processing of, 64
    thin-film, 62, 64, 75
        capacitors for, 81
        deposition of, 77
        inductors for, 81
        masks for, 78
        materials for, 76
        and monolithic networks, 75

In-circuit testing, 308
    digital, 315
    vs function testing, 317
    guarded measurements in, 310
    for signatures, 315
    text fixtures for, 309
In-line approach to wafer processing, 43, 44
Ion-beam milling, 5, 50

Kovar, 228

Large-scale integration (LSI), 252, 254
    packages for, comparison of, 188
    wafers for, 257
Leadless inverted device (LID), 67
Liquid-cooled systems, 240-245, 251
Lithography. See Electron-beam lithography; Optical lithography; X-ray lithography
Loops for cooling, 243, 245

Mating surfaces between heat sink and device, 235

Natural frequency of a chassis or a pc board, 280
Networks, 246, 248, 250
    equations for, 250
    equivalent, 250
    lumped, 246
Nuclear environment, semiconductors in, 282-284

Oils for liquid cooling, 245
Operating temperature, safe, of transistors, 227
Optical (UV) lithography, 3, 23, 28, 30, 45

Packages, LSI, comparison of, 188
Parallel cooling, 253
Parallel flow, 245
Plasma etching, 49-50, 54-58
    advantages and limitations of, 57
    anisotropic, 55, 57, 58
    cylindrical reaction of, 49, 56
    definition of, 49, 54
    isotropic, 55
    planar, 49, 56
    processing, 49, 54
    throughput of, 49
    variation of, 50
Plastic power ICs, 227
Polymer thick-film inks, 86, 87
Polysulfone, 124
Porcelain-on-steel substrates, 87, 94, 95, 199
    thermal characteristics of, 97
    thick-film inks for, 87, 94, 95, 96
    vs alumina, 97
Power density, 252
Power devices, geometries of, 232
Power dissipation of semiconductor devices, 231-232
Pressure drop of moving air, 245, 264
Printed-circuit (pc) boards
    computer-aided design of, 286
    design approaches to, 288
    design automation of, 287
    digitization of, 287
    etchless, 125
    fine-line, 88, 101, 199, 200
    hook in, 140-146
    mass-molded, 99
    materials for, 92, 123, 124
    multilayer, 99, 200
    processing of,
        additive, 90
        resistless, 89, 125
        semiadditive, 89
        substractive, 88, 89
    resists for, 90, 91
    vibration analysis of, 279
Pumps, 245

Quad in-line package (QUIP), 207
    vs ceramic chip carrier, 203

Radiation effects on semiconductors, 81, 283
    on ECL, 284
Reserve cooling power, 240
Resistance
    conductive, 260, 261, 271
    convection, 248, 260, 271
    coolant coupling, 251
    flow, 251
Resistors, 247, 248
Resists
    for electron-beam lithography, 4-5
    for printed circuits, 90, 91
    X-ray, 19, 20, 22, 27, 40, 41

Safe operating temperature of transistors, 227
SOT-23 package, 69
Stud-mounted semiconductors, 234
Survival expectation, calculating for minicomputers, 239

Temperature
    ambient, 228, 246
    case, 236
    gradient, 248, 253, 258, 263
    junction, 228, 246, 252, 261, 262, 265
    measurements of, 235
    profile, 237, 238
Thermal analogs
    of electronic systems, 246
    of Ohm's law, 224
Thermal analysis, 230, 246, 265
Thermal budget, 261, 264
Thermal circuit, 230, 231
Thermal conductivity, 251
Thermal design, 227
Thermal impedance, 246
Thermal inertia, 240
Thermal interface, 235
    resistance of, 236
Thermal management, 223
Thermal mismatch, 224
Thermal path, 225, 252
Thermal profile, 237
Thermal resistance, 228, 235, 241, 242, 246, 247, 248, 250, 251, 258, 261, 265, 273
    calibration of, 276
    electrical measurement of, 275
    equivalent circuit for, 273
    of a heat exchanger, 243, 244
    of an IC, 265, 266, 268
    of liquid-cooled geometries, 242
    measurement of, 274
        techniques for, 275
    paths, 246
Thermal runaway, 233
Thermal stability, 233
Thermal temperature rise, total, 260
Thermal time constant, 268
Thick-film inks, 64
    for porcelain-on-steel substrates, 87, 94, 95, 96
    polymer, 86, 87
Thin-film hybrids. See Hybrids, thin-film

Vacuum interlocking of electron-beam equipment, 17
Velocity profile of fans, 237, 238
Vector Scan One (VS1), 8, 9-10, 11
Vibration analysis of pc boards, 279

Water, 245, 269
Wick, 269

X-ray alignment, 27
X-ray distortion, 22, 23
X-ray lithography, 19-20, 21-28, 40-41, 48-49
XXXP, 123

Zero-insertion-force (ZIF) connectors, 211-216
    definition of, 211
    stacked, 215
    types of, 212, 213